Structures

Theory and Analysis

Other titles of related interest:

Structures

Theory and Analysis

M.S. WILLIAMS
J.D. TODD

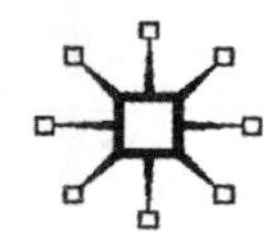

Published by
PALGRAVE MACMILLAN
Houndmills, Basingstoke, Hampshire RG21 6XS and
175 Fifth Avenue, New York, N. Y. 10010
Companies and representatives throughout the world

PALGRAVE MACMILLAN is the global academic imprint of the Palgrave Macmillan division of St. Martin's Press, LLC and of Palgrave Macmillan Ltd. Macmillan® is a registered trademark in the United States, United Kingdom and other countries. Palgrave is a registered trademark in the European Union and other countries.

ISBN 0–333–67760–9

This book is printed on paper suitable for recycling and made from fully managed and sustained forest sources.

A catalogue record for this book is available from the British Library.

10 9 8 7 6 5 4 3 2
10 09 08 07 06 05 04 03

Printed and bound in Great Britain by
Ashford Colour Press Ltd, Gosport

Contents

Preface

This book is the successor to Joseph Todd's *Structural Theory and Analysis*, first published in 1974 and reprinted frequently throughout the 1970s and 1980s. It is not a new edition of that text, but rather a completely new book, radically different in content, organisation and style. We have, however, tried to keep as close as possible to the principles that underpinned Todd's original text: that the material should be presented in as clear and concise a way as possible; that both qualitative understanding of structural behaviour and a firm grasp of the mathematical theory are important; and that a sound knowledge of basic structural theory is an essential prerequisite to the introduction of computer methods.

The book is deliberately broad, aiming to cover in reasonable detail most of the structures topics that an undergraduate is likely to meet in a civil or general engineering degree course. Given the broad scope it is not, of course, possible to go into enormous depth on all topics. Reading lists are therefore included for topics where more in-depth study may be desirable.

The opening chapter sets the scene by introducing the topic of structures in an entirely non-mathematical way. Using well-known structures as examples, it explains how the most common structural forms work and provides the motivation for the remainder of the book. The following five chapters lay the foundations of structural theory, covering statically determinate structures and the basics of the theory of elasticity. The book then moves on to cover analysis methods appropriate for indeterminate structures, where both statics and elasticity are required in order to obtain solutions. It is in this form of analysis that the use of computer methods is now near-universal. We have therefore kept the account of traditional methods such as virtual work and moment distribution as concise as possible and gone into rather more detail on the computer-oriented stiffness matrix and finite element approaches. The book concludes with three chapters on topics of fundamental importance to structural engineers – stability, plastic analysis and structural dynamics.

With the exception of Chapter 1, which is entirely qualitative, an attempt has made to integrate mathematical analysis techniques with explanations of structural behaviour throughout. It is our belief that these two must be taught in parallel – a qualitative appreciation of complex structural behaviour cannot be developed without knowledge of the relevant theory, and conversely the choice of an appropriate analytical approach must be guided by a physical understanding of what is going on.

The issue of sign conventions in structural engineering is a particularly thorny one – there is no convention that pleases everyone, or that entirely eliminates the occasional occurrence of unexpected and irritating minus signs. We have used a consistent sign convention as far as possible in this text and, to minimise confusion, have explained it in detail in Appendix A. Where numerical examples are given, the units are either kiloNewtons and metres, or Newtons and millimetres; we have tried to avoid mixing the two systems within a single example. A glossary of terms frequently used in structural engineering has been provided in Appendix B, though every effort has been made to keep jargon to a minimum.

M.S. Williams
J.D. Todd

Acknowledgements

We wish to thank Malcolm Stewart and Christopher Glennie of Macmillan Press, for their help and encouragement during the preparation of the text. We are grateful to colleagues in the Department of Engineering Science and at New College, Oxford, for their advice and support, particularly Guy Houlsby, David Clarke and Michael Burden. The finite element mesh in Figure 10.17 is reproduced by kind permission of Charles Augarde. We are grateful to Manti Mendi for his assiduous checking of the worked examples and problems, though the authors alone are responsible for any remaining errors. Lastly, we are indebted to E.J. Milner-Gulland, whose unfailing support was essential to the completion of this project.

Notation

The notation used accords with accepted conventions as far as possible. This has the unfortunate result that a few symbols have more than one meaning. However, the correct meaning should be obvious from the context. SI units are given in square brackets.

A	Cross-sectional area [m^2]
A_e	Area enclosed by mean perimeter of closed, thin-walled section [m^2]
a	Acceleration [m/s^2], or Robertson constant (empirical factor used in strut design)
$\mathbf{B}$	Element strain matrix
b	Section breadth [m]
c	Damping coefficient [Ns/m]
c_{ij}	Carry-over factor for bar ij – ratio of moment at end j to that at end i
$\mathbf{D}$	Element stress matrix
d	Section depth [m]
d_{ij}	Distribution factor for bar ij
E	Young's modulus (constant of proportionality between direct stress and strain) [Pa]
$\mathbf{E}$	Equilibrium matrix
e	Extension [m]
F	Concentrated load (or W, or P) [N]
$\mathbf{F}$	Load vector (global coordinates)
$\mathbf{f}$	Load vector (local coordinates)
f_n	Natural frequency [Hz]
G	Shear modulus (constant of proportionality between shear stress and strain) [Pa]
H	Horizontal reaction [N] or warping constant [m^6]
h	Height [m]
I	Second moment of area [m^4]
J	Polar second moment of area [m^4]
K	Bulk modulus (constant relating mean stress to volumetric strain) [Pa]
$\mathbf{K}$	Stiffness matrix (global coordinates)
k	Stiffness
$\mathbf{k}$	Stiffness matrix (local coordinates)
k_A	Axial stiffness [N/m]
k_B	Bending stiffness [Nm/rad]
l	Length [m]
l_e	Effective length [m]
M	Bending moment [Nm]
$\mathbf{M}$	Mass matrix
M_c	Critical moment at which lateral-torsional buckling occurs [Nm]
M_{ij}^D	Displacement-related moment at end i of bar ij [Nm]
M_{ij}^F	Fixed-end moment at end i of bar ij [Nm]
M_p	Plastic moment [Nm]

M_y	Yield moment [Nm]
m	Mass [kg], or modular ratio of two materials ($= E_1/E_2$)
P	Concentrated load (or W, or F) [N]
P_c	Critical buckling load for a strut, or collapse load for a structure [N]
P_E	Euler buckling load for a pin-ended strut [N]
P_R	Rayleigh buckling load for a strut [N]
p	Pressure [Pa]
q	Shear flow (shear stress × thickness) [N/m]
R	Radius of curvature [m]
r	Radius [m]
r_y	Radius of gyration about the minor axis [m]
S	Shear force [N]
$\mathbf{S}$	Shape function matrix
S_i	ith shape function for a finite element
s_{ij}	Rotational stiffness of bar ij when loaded by a moment at end i [Nm/rad]
T	Axial force [N], or natural period of vibration [s]
$\mathbf{T}$	Transformation matrix between local and global coordinate systems
t	Thickness [m]
$\mathbf{t}$	Bar force vector
t_{ij}	Tension coefficient for bar ij (axial force divided by length) [N/m]
U	Strain energy [J]
$\mathbf{U}$	Displacement vector (global coordinates)
$\mathbf{u}$	Displacement vector (local coordinates)
u, v, w	Displacement components in x, y and z directions (or u_x, u_y, u_z) [m]
$\dot{u}$	Velocity [m/s]
$\ddot{u}$	Acceleration [m/s^2]
u_g	Horizontal ground displacement [m]
u_x, u_y, u_z	Displacement components in x, y and z directions (or u, v, w) [m]
V	Vertical reaction [N], or volume [m^3], or potential energy [J]
W	Concentrated load (or P, or F) [N]
w	Uniformly distributed load [N/m], or displacement component in z direction [m]
$\mathbf{X}$	Nodal coordinate vector for a finite element
$\mathbf{x}$	Coordinate vector for an arbitrary point within a finite element
x, y, z	Axis directions
$\bar{y}, \bar{z}$	Coordinates of centroid of cross-section [m]
Z_e	Elastic modulus [m^3]
Z_p	Plastic modulus [m^3]
z_i	Generalised coordinate for the ith mode [m]
α	Coefficient of thermal expansion [/°C], or shape factor ($= Z_p/Z_e$)
γ	Shear strain, or safety factor (= yield load ÷ working load)
δ	Deflection of a point [m], or logarithmic decrement (a measure of damping)
ε	Direct strain
$\boldsymbol{\varepsilon}$	Element strain vector
$\varepsilon_1, \varepsilon_2$	Principal strains
ζ	Damping ratio (proportion of critical damping)
η	Perry factor for a strut (also known as the initial curvature parameter)
θ	Angle of rotation [rads], or temperature [°C]
λ	Slenderness ratio for a strut ($= l_e/r_y$), or load factor (= collapse load ÷ working load)
ν	Poisson's ratio

σ	Direct stress [Pa]
$\boldsymbol{\sigma}$	Element stress vector
σ_1, σ_2	Principal stresses [Pa]
σ_c	Critical axial stress at buckling [Pa]
σ_h	Hoop, or circumferential stress [Pa]
σ_l	Longitudinal stress [Pa]
σ_r	Radial stress [Pa]
σ_y	Yield stress [Pa]
τ	Shear stress [Pa]
τ_y	Yield stress in shear [Pa]
ϕ	Angle of twist, or angle between local and global axes [rads]
$\boldsymbol{\Psi}$	Modal matrix (comprising a set of mode shapes written as column vectors)
$\boldsymbol{\psi}$	Mode shape vector
Ω	Frequency ratio (= loading frequency ÷ natural frequency)
ω	Circular frequency [rad/s]
ω_d	Damped circular natural frequency [rad/s]
ω_n	Undamped circular natural frequency [rad/s]

CHAPTER 1

Introducing structures

CHAPTER OUTLINE

This book is concerned with methods of analysing structures – working out how they carry loads, what internal stresses are developed within them and how they deform. Such analysis is an essential part of the structural design process; we cannot choose appropriate dimensions or materials for the structural elements without being able to work out what forces they are required to carry.

Before getting into the detail of structural analysis techniques that form the bulk of this book, it is worth stepping back and taking an overview of the broader topic of structures. This chapter outlines the most important structural concepts in a non-mathematical way, and introduces some of the basic terminology and notation that will be used throughout the book. The chapter should enable the reader to gain an understanding of:

- what engineers mean by the word *structure*;
- the different types of loads that structures may have to carry;
- how structures are supported, and how supports are idealised for analysis purposes;
- how some of the most common types of structure carry loads;
- the role of structural analysis in the design process.

(a)

(b)

Figure 1.1
Two different types of structure: (a) The Renault Building, Swindon, UK; (b) a chair.

1.1 What is a structure?

A structure can be most broadly defined as an object whose purpose is to carry a set of loads, or *forces*, from one place to another. In most cases the aim is to transmit the applied loading from somewhere in space to the ground without collapsing and without deforming excessively. Thus, while the building in Figure 1.1(a) is immediately recognisable as a structure, the above definition would encompass a much wider variety of objects, such as the chair in Figure 1.1(b).

The chair exhibits all the principal characteristics of a structure. It is designed to transmit an external load (the weight of a person) down its legs to the ground. The load generates forces within the various members of the chair, and reaction forces at the contact points between the legs and the floor. To perform its intended function, the chair must not break under the person's weight, and neither should it deform excessively, since this would make it uncomfortable and rather alarming to use.

Now consider a rather different structure, an aeroplane. In level flight, the weight of the plane must be balanced by the uplift force (that is, the reaction) provided by the air pressure on the wings. The structure of the wings must be designed so as to transmit this force to the fuselage. The wings must not bend too much under this force as the aerodynamics of the plane would be affected.

We have thus identified some common characteristics of structures:

- they are designed to carry loads in space;
- they are usually supported either on the ground or on another structure (though exceptions such as the aeroplane are possible), with reaction forces generated at the support points;

- the applied loads and reactions cause forces to be generated within the members of the structure;
- the structural members must not collapse or (in most cases) deform excessively under these forces.

We shall consider some of these characteristics in more detail in the following sections.

1.2 Elements of the structural system

1.2.1 Loads on structures

Since the purpose of a structure is to carry loads, it is important to consider the nature of those loads. There are many ways of categorising loads, of which a few of the more important ones will be briefly discussed here.

Dead and live loads

One possible distinction, often used in design codes of practice, is between *dead* and *live* loads. The dead load is due solely to the weight of the structure itself, and can usually be estimated quite accurately. The live load (also sometimes called the *imposed* load) is the sum of all the other loads on the structure, which may arise from a wide variety of sources and can often only be estimated. Examples of live loads are the weight of traffic on a bridge, wind loading on the side of a building and the weight of snow on a roof. Obviously live loads may come and go, or at least fluctuate in magnitude. It may be necessary to examine many different load combinations in order to come up with the most critical design case.

To allow a margin of safety in design, loads are often multiplied by *safety factors*. Because of the greater uncertainty, safety factors for live loads tend to be larger than those for dead loads.

Static and dynamic loads

A dynamic load is one that varies significantly over a relatively short time span, so that the response of the structure to the loading is affected by *resonance*. Many readers will be familiar with the example of an army marching over a bridge; if the pace of marching is close to the natural frequency of vibration of the structure, then very large deck movements may occur. This is why soldiers are instructed to break step when crossing lighter bridges.

Strictly speaking, a static load is one that does not vary at all over time. Of course, very few loads stay constant indefinitely. For our purposes, therefore, it is more useful to define a static load as one that varies sufficiently slowly that it does not induce any resonance in the structure. With this definition, most loadings can be classified as static. In this book we shall deal mostly with the analysis of structures under static loads, though we will take a quick look at dynamic loadcases in the final chapter.

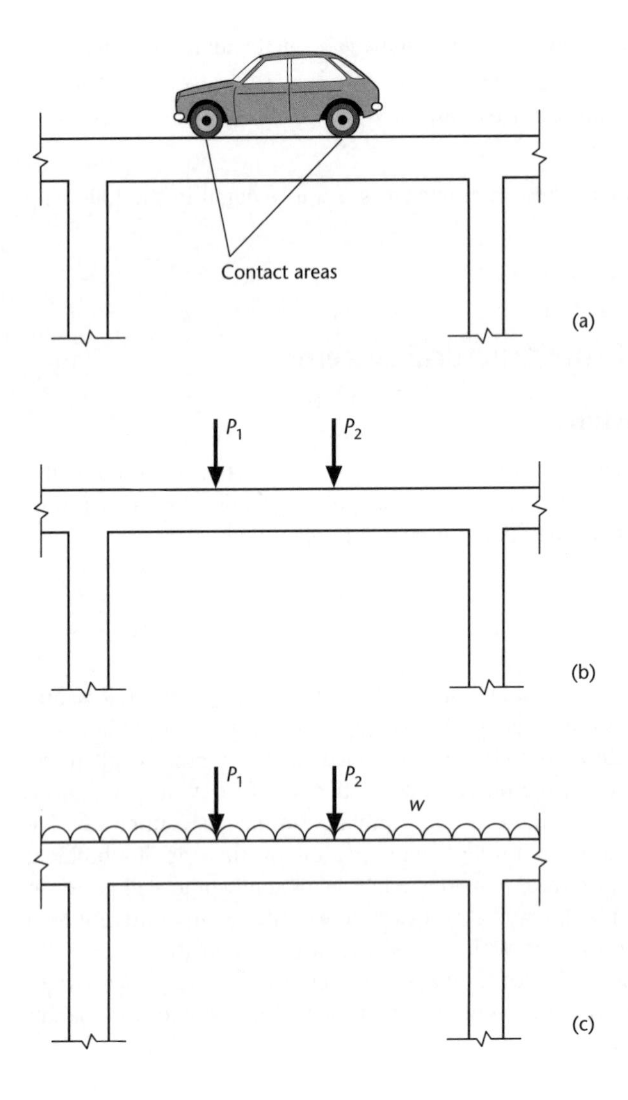

Figure 1.2
Idealisation of external loads.

Idealisation of loads

A load can be defined in terms of its magnitude, the direction in which it acts, and the area over which it acts. There are many different types and combinations of loads, but for analysis purposes it is helpful to simplify them into a few idealised cases. Some of the most common ones will be briefly discussed here.

A concentrated load, usually denoted by the symbol W or P, is assumed to act at a point; in reality this is an impossibility as there would be an infinite stress in the member at the point of load application. In practice, a load must always act over a finite area, and when very high stresses are induced it is quite likely that this area will be increased somewhat by local deformations of the structure. Nevertheless, the concentrated load assumption is a very useful approximation for analysis purposes. A simple example is shown in Figure 1.2. The car imposes vertical loads on the bridge through its wheels. In reality, both the tyres and the bridge deck deform slightly, so that the load is spread over a small contact area, as shown to exaggerated scale in Figure 1.2(a). However, when analysing this structure it would be sufficiently accurate to assume concentrated loads acting at the centres of the contact areas; the loads are denoted by arrows in Figure 1.2(b).

Another very common load type is the uniformly distributed load (UDL), usually given the symbol *w*. As the name suggests, the load is distributed over the surface of the member with a constant value per unit length, or per unit area. For a uniform structural element, the self-weight will generally be a UDL. So returning to the bridge example, the total loading is as shown in Figure 1.2(c); the self-weight of the deck creates a uniform load of intensity *w*, denoted by the series of short arcs, in addition to the concentrated loads due to the weight of the vehicle.

It is, of course, possible that other load distributions will arise in practice. For example, it is quite common for distributed loads to have linearly varying intensity over the length of a member, particularly in complex frame structures where floor loads are shared between adjacent members. These more complex load configurations will be discussed in more detail as they arise later in the book.

1.2.2 Supports

Except for very unusual cases such as spacecraft, structures generally rest on and transmit forces to either the ground or another structure. The contact points via which the forces are transmitted are called *supports*, and the nature of the supports plays a vital role in determining how the structure carries the loads. Like loadings, structural supports come in a wide variety of types but are generally simplified to a few idealised cases for ease of analysis.

Pinned support

A pinned support is one which prevents the structure from moving translationally in any direction at the support point, but provides no resistance to rotation. An example of a mechanical bearing that provides such support is shown in Figure 1.3(a). The horizontal beam is connected to the foundation by interlocking steel leaves pivoting around a steel pin. Assuming the assembly to be frictionless, there is no resistance to rotation about the pin. However, this bearing does prevent rotation about an axis perpendicular to the pin. Bearings are also available which allow rotation about any horizontal axis.

When analysing a structure, we generally use simple line diagrams. Figure 1.3 (b) shows the standard diagrammatic representation (sometimes called an *icon*) of a pinned support, with the solid line representing the beam and the pinned

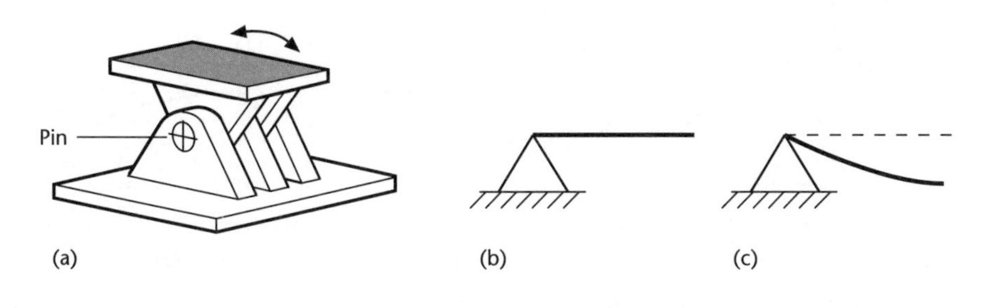

Figure 1.3
Pinned support.

Figure 1.4
Roller support.

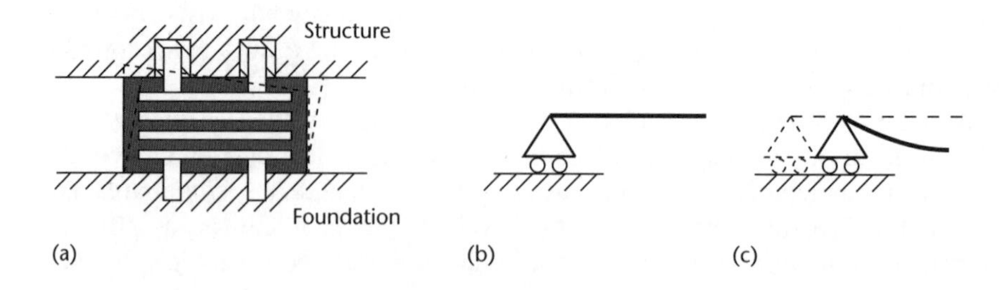

support denoted by the triangle. The diagonal shading underneath the triangle represents the ground, which is assumed completely rigid. Figure 1.3(c) shows the type of deformation that can occur at a pinned support.

Roller support

A roller support allows some translational as well as rotational movement. An example is the laminated elastomeric bearing shown in Figure 1.4(a). This consists of a block of very flexible elastomer bonded to and reinforced by several layers of horizontal steel plates. The plates make the elastomer very hard to deform in the vertical direction, but it remains easy to deform sideways and rotationally. Thus the bearing effectively prevents vertical movements, but allows rotations and horizontal translations. The symbol used for a roller support is a triangle resting on open circles, as shown in Figure 1.4(b), and the possible movements of a structure at a roller are shown in Figure 1.4(c). Note that, despite the form of the symbol used, the roller prevents *all* vertical movements, both upwards and downwards.

Fixed support

The last type of support we will consider is the fixed or *built-in* support. As its name suggests, this prevents all movements of the structure at the support point, both translations and rotations. Complete fixity can be provided, for example, by casting a member into a heavy concrete foundation (Figure 1.5(a)). The symbol for a fixed support and the nature of structural deformation at the support are shown in Figure 1.5(b) and (c) respectively.

Of course, all the above are idealisations of real support behaviour. For example, pins are never truly frictionless, while built-in supports can never be completely rigid. Nevertheless, it is possible in construction to get reasonably close to these

Figure 1.5
Fixed support.

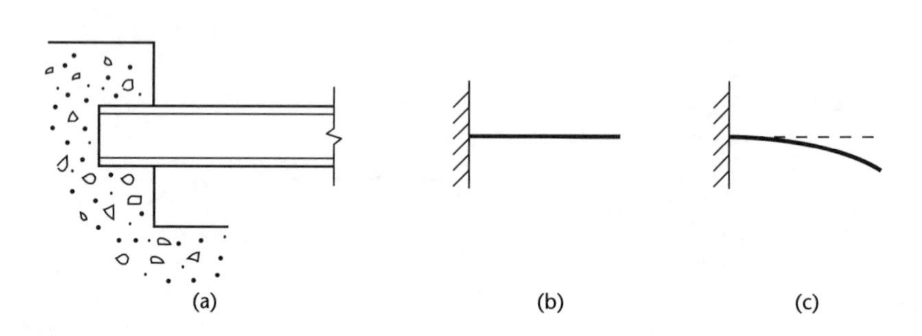

three cases, so that the errors introduced by assuming ideal support behaviour are usually tolerably small. In some instances it is necessary to use more complex models of support behaviour, for example when a foundation rests on very soft ground, whose deformations cannot be neglected. In this book, however, we shall use only the simple idealisations introduced above.

It might seem obvious that, since we want to prevent the structure from deforming excessively, we should provide as much restraint as possible at the supports. In practice this is not usually desirable. All structures move to some extent, for example due to changes in temperature, and if such movements are limited by the support restraints then significant additional forces can be set up within the structure. In addition, the provision of high levels of fixity at the supports can be difficult and expensive, and can create undesirably large stresses and deformations in the ground. For these reasons, pinned and roller type supports are very widely used.

1.2.3 Members, joints and structural actions

Having considered the loading on a structure and how it is supported, we are now in a position to look at the structure itself. A structure consists of one or more elements, usually known as *members*, connected together at *joints* and resting on supports. The form of both the members and the joints will vary depending on what type of forces they are intended to carry.

The basic principle of how a member carries a load is very similar to that of a simple spring. The force in the spring is related to its extension by the spring constant k. When a load is hung from the spring, it extends by an amount x such that the spring force kx exactly balances the load, so that the system is in equilibrium. So it is with structural elements – for each possible mode of deformation it is possible to determine a 'spring constant' or *stiffness coefficient* relating the amount of deformation to the force within the member. When a load is applied, deformation occurs so that the product of the deformation and the stiffness coefficient balances the loading.

In reality all members are three-dimensional bodies, but it is convenient in analysis to idealise them as one-dimensional *line elements* or two-dimensional *surface elements* whenever possible.

One-dimensional elements

Most of the structures we shall consider in this book are made of elements that are essentially one-dimensional, that is, their length is large compared to their cross-sectional dimensions. Some of the most common one-dimensional member types are illustrated and discussed below.

Elements that are intended only to carry forces acting along the member axis are usually referred to as *bars*. These axial forces are termed *tensile* if they tend to stretch the bar, and *compressive* if they tend to shorten it. A bar in tension, often referred to as a *tie*, behaves in exactly the same way as the simple spring mentioned above – it stretches by an amount proportional to the applied load. Tension elements are usually made of metals such as steel, and tend to have very simple, slender cross-sections. Common types include simple circular rods, angle sections and T-sections, see Figure 1.6. Sometimes *cables* are used. These are very

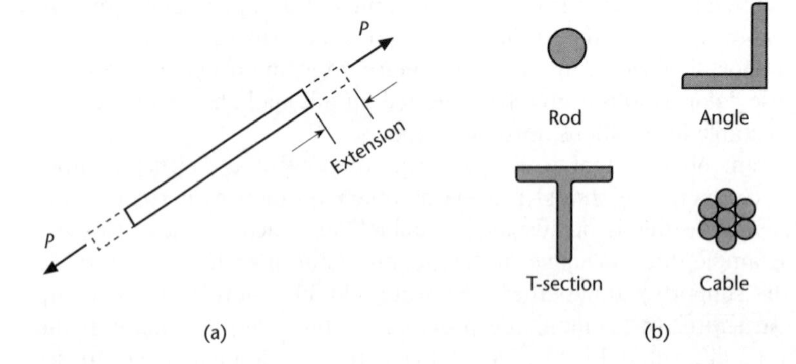

Figure 1.6
Tension members:
(a) structural action;
(b) typical cross-sections.

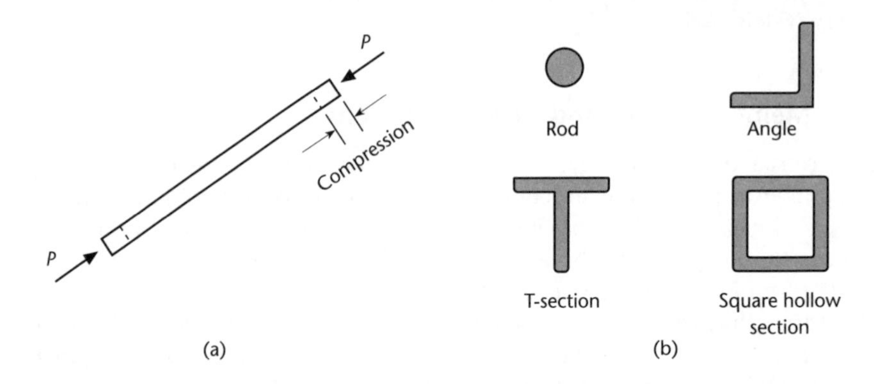

Figure 1.7
Compression members:
(a) structural action;
(b) typical cross-sections.

flexible elements, usually made up by winding together small-diameter wires. Because of their flexibility, they go slack when subjected to compressive loads, and therefore can *only* carry axial tensile forces.

Compression bars (Figure 1.7), commonly known as *struts*, are essentially similar to ties, but with the direction of loads and deformations reversed. These elements tend to be rather more stocky than tension members – the ratio of cross-sectional dimensions to length is generally higher and, where thin-walled sections are used, the wall thicknesses are greater. This is because struts are prone to a form of instability known as *buckling*. A vertical strut is often called a *column*.

Probably the most common form of structural element is the *beam*. A beam is designed to carry loads perpendicular to its axis, and does so primarily by *bending*. It may also carry loads along its axis (that is, tensile or compressive forces, as discussed above), but it is the transverse loads and bending action that dominate its behaviour.

Figure 1.8(a) shows a beam resting on pinned and roller supports (known as a *simply supported beam*). If a set of vertical lines is drawn on the side of the beam then, as the beam bends under the vertical load, the lines get closer together near the top of the beam and further apart near the bottom. In other words, the bending action causes the beam to shorten along its top edge and extend along the bottom edge. Now we have already seen that shortening of members is caused by compressive forces, and extension by tensile forces. It follows that there must be a

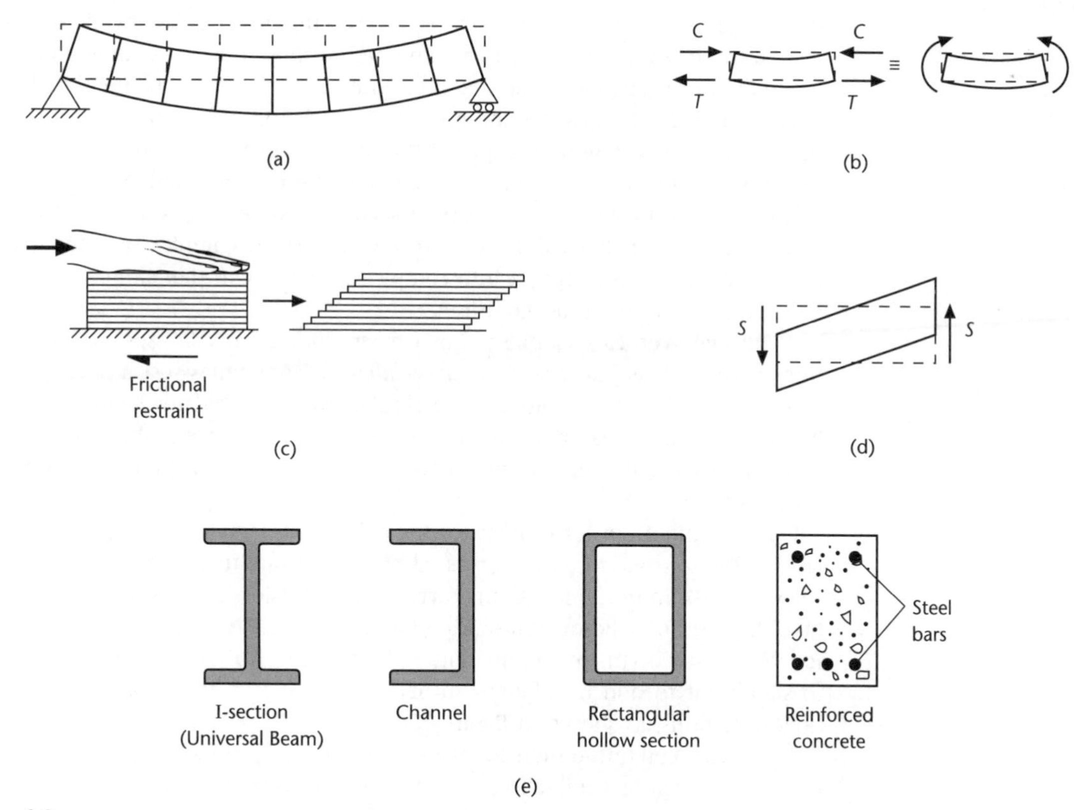

Figure 1.8
Beams: (a) bending of a simply supported beam; (b) bending forces; (c) shearing action in a deck of cards; (d) shear in a beam; (e) typical beam cross-sections.

compressive force acting along the top part of the beam and a tensile one along the bottom.

Figure 1.8(b) shows the forces acting on a short element of the beam. In the absence of any applied axial loads, the resultant compressive force C and the tensile force T must have equal magnitude but act in opposite directions. This means that the sum of the forces acting on any vertical cross-section is zero and the effect of the two opposing forces is to produce a couple, or *bending moment,* within the beam.

Bending is nearly always accompanied by another mode of structural behaviour known as *shearing.* Whereas tension pulls the particles within an element apart and compression pushes them together, shearing causes them to slide across each other. Imagine a deck of playing cards resting on a flat surface. We can impose a shearing action on the deck by pushing horizontally on the top surface (Figure 1.8 (c)). If the loading is applied completely uniformly then each card slides relative to the next one so that the deck, which initially looked rectangular in elevation, deforms into a parallelogram. If we look at a short length of beam, a similar phenomenon occurs. Vertical *shear forces S* are set up within the beam which cause sliding between adjacent planes, so that an initially rectangular beam element distorts into a parallelogram (Figure 1.8(d)). In most beams the deformations caused by shearing are very small compared to those due to bending and can be ignored, but the stresses generated are significant and cannot be neglected.

The bending action causes the greatest stresses and deformations in a beam to occur at the top and bottom edges of the section. For this reason it is common to use cross-sections in which most of the material is concentrated in these locations (Figure 1.8(e)). In steel structures the I-section, also known as the *Universal Beam* section, is very widely used. In reinforced concrete, the complexity of the material and the difficulty of casting it into complex shapes mean that simple rectangular cross-sections are the most common.

A (usually vertical) element that carries both axial compressive forces and transverse loads is known as a *beam-column*, or sometimes just a *column*. Cross-sections used for beam-columns are similar to those used for beams. However, because of the greater influence of compression forces, they tend to be less elongated, with the use of square or circular sections quite common – see Figure 1.9.

One last important form of structural action is *torsion*, or twisting, in which the cross-section at one location is rotated relative to one further along the member (Figure 1.10). In machines, shafts carrying pure torsion are common, and usually take the form of solid or hollow circular sections. In civil engineering structures, torsion usually occurs in conjunction with other types of load. Members subjected to significant torsion in addition to bending often take the form of rectangular hollow sections, as shown in Figure 1.8(e).

Joints between one-dimensional elements are generally idealised as one of two types. A *pin joint* is similar to a pinned support, in that it prevents translational movements between the elements being joined, but allows rotations. Simple bolted connections in steel frames, such as the one shown in Figure 1.11(a), can be thought of as pinned. The cleat joining the beam and column together is very flexible, and the bolts can slip under load, since they are positioned in holes

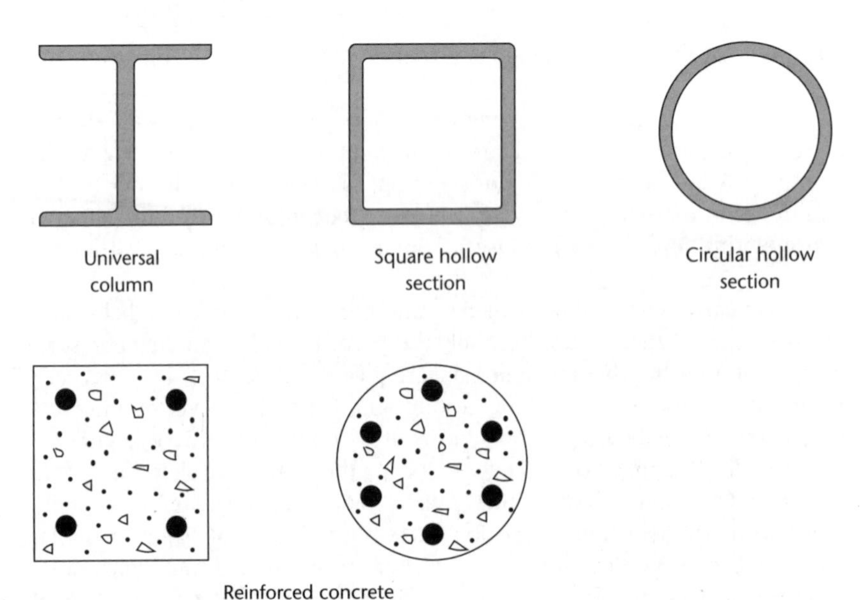

Figure 1.9
Beam-column cross-sections.

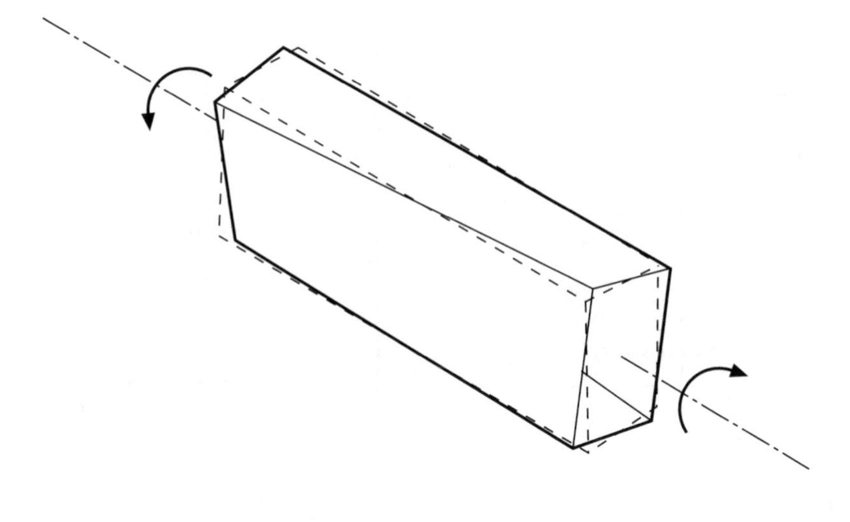

Figure 1.10
Torsion.

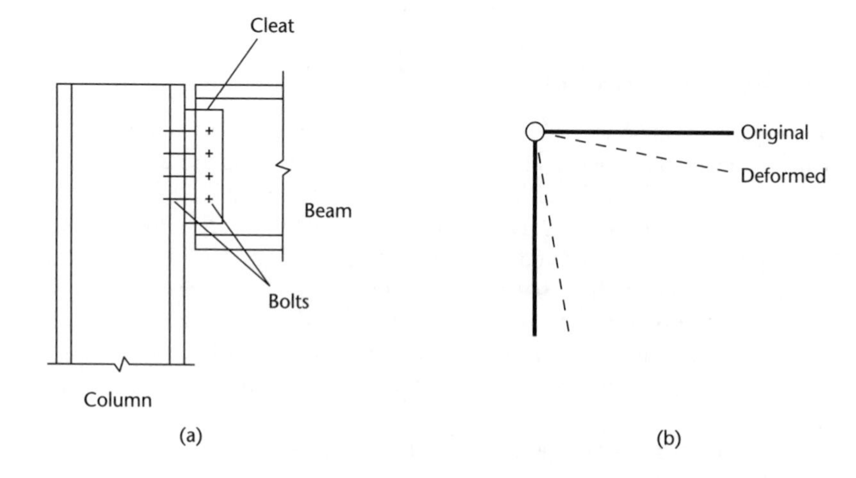

Figure 1.11
Pin joint.

slightly larger than the shank diameter. Significant rotation between the beam and column can therefore occur at quite low loads. The diagrammatic representation of a pin-joint is a large, solid dot see Figure 1.11(b), which also shows the possible deformation at such a joint, with the members no longer at right angles to each other.

Similarly, a *rigid* joint can be thought of as analogous to a rigid support, in that it prevents all relative movements between the members meeting at the joint. Two rigid joints are shown in Figure 1.12(a). In the first, two steel members are joined with extensive use of welded connections and stiffening elements; in the second, two concrete members are cast together integrally. A rigid joint between members is drawn as solid lines meeting at a point, as though they formed a single continuous member. Note that if one member meeting at the joint undergoes a translation or rotation, then all the other members at that joint must experience exactly the same movement. This means that, for instance, members initially joined at right angles must still be at right angles to each other after the structure has deformed – see Figure 1.12(b).

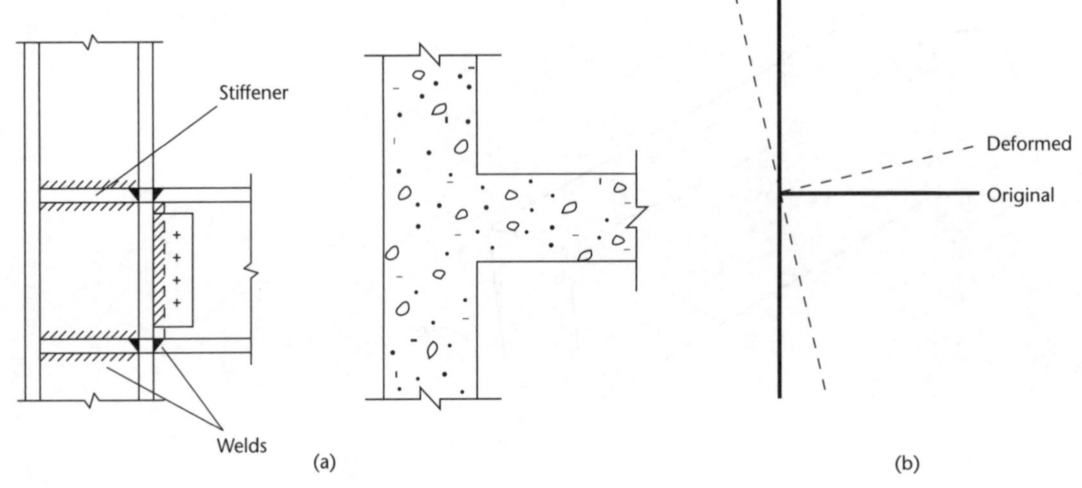

Figure 1.12
Rigid joint.

Two-dimensional elements

Elements with one dimension much smaller than the other two are referred to as two-dimensional or plane elements. They can be classified according to their shape and the orientation of the external loading. A *plate* is a flat element loaded perpendicular to its plane surface so that, like a beam, it carries loads primarily by bending. The two-dimensional nature of the element makes the structural action much more complex than that of a beam, but it can be visualised by resolving the deformation into perpendicular components, as shown in Figure 1.13(a). Many structures, such as floors, roofs and bridge decks, can be idealised as plates, though often their configuration is rather more complex than the one shown here.

If the surface is curved rather than flat, then the element is called a *shell*; an example is shown in Figure 1.13(b). The curvature has a dramatic effect on the structural action, as it causes a significant proportion of the load to be carried by direct tension or compression in the plane of the shell. This is called *membrane action*.

Sometimes a plane element is required to carry a load parallel to its surface. An example is the *shear wall* (Figure 1.14), which is widely used for resisting horizontal loads in buildings. As its name implies, this element carries the in-plane load primarily by shearing action between its top and bottom edges. This provides a very stiff element, which is excellent at minimising lateral deflections.

Three-dimensional elements

Comparatively few structures are made up of members that cannot reasonably be idealised as one- or two-dimensional. Very large solid structures such as dams behave in a genuinely three-dimensional way, but even in this case it may well be possible to reduce the analysis to two dimensions by making certain assumptions about the structural behaviour.

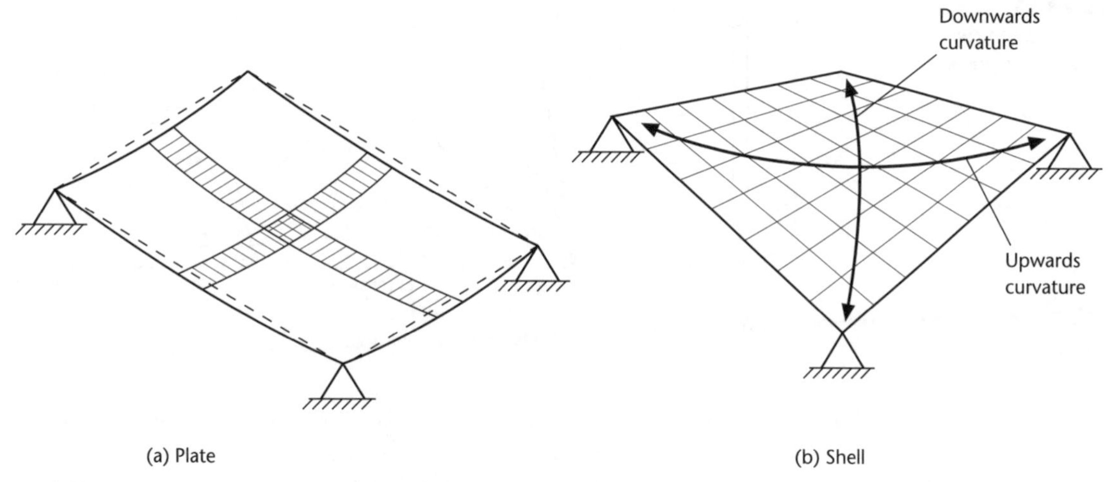

Figure 1.13
Plate and shell elements.

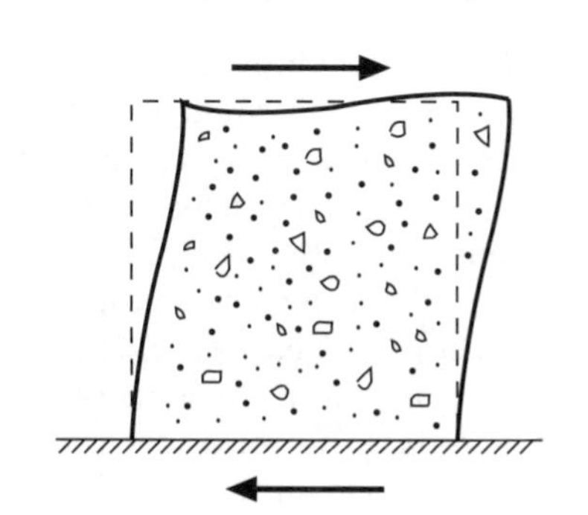

Figure 1.14
Shear wall.

1.2.4 Structural materials

The materials out of which a structure is made play a vital role in determining its ability to perform its desired function. We have already noted that a structure must satisfy two principal criteria: it must carry loads without collapsing and without deforming excessively. These two criteria are highly dependent on two different material properties:

- the *strength* is defined as the load required to cause failure;
- the *stiffness* is the load required to cause a unit displacement.

To illustrate exactly what is meant by these terms, consider the simple bar shown in Figure 1.15(a), which is rigidly fixed at one end and loaded by a tensile force at the other. As the load is increased, the bar stretches and eventually breaks. If we plot a graph of the applied load against the resulting extension, its form will depend on the dimensions of the member and on its material properties. Figure 1.15(b) shows possible curves for three different materials, all based on the same member dimensions. In each case the strength is simply the maximum attainable value of load and the stiffness is the gradient of the load–deflection line:

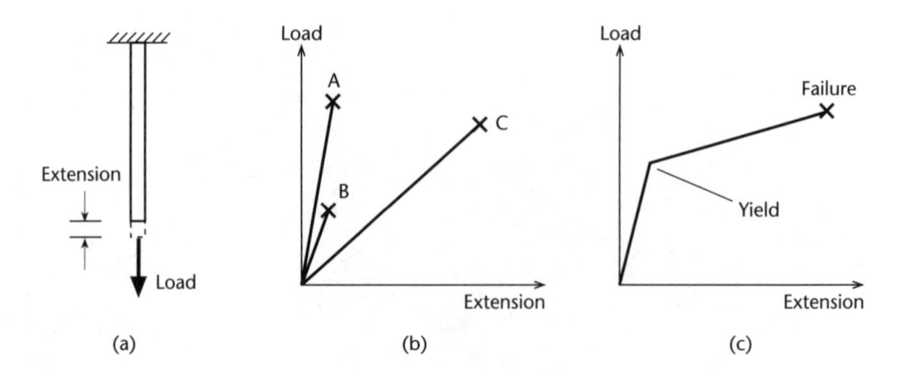

Figure 1.15
Load–extension curves for materials in tension.

- Material A has both high stiffness, that is, a steep gradient, meaning that a large force is required to produce a relatively small displacement, and high strength – a large load is required to cause failure, indicated by the cross.
- Material B also has a high stiffness, but in this case the strength is much lower.
- Material C has a much lower stiffness but quite a high strength.

It is important to distinguish very clearly between these two properties, and to realise that high strength and high stiffness do not necessarily go together. For example, many polymers are extremely strong (difficult to break) but have low stiffness (easy to deform). Heat treatment of metals can result in enormous improvements in strength, but usually has little or no effect on stiffness.

A further complexity must now be introduced. The behaviour in Figure 1.15(b), in which the load–extension curve is linear right up to the point where the specimen breaks, is typical of *brittle* materials such as glasses. A more normal load–extension curve for a metal is shown in Figure 1.15(c). Here, the behaviour is initially linear up to a *yield point* at which the stiffness suddenly reduces. The specimen then undergoes very large extensions with little increase in load, until it eventually breaks. This is referred to as *ductile* behaviour. For a ductile material it is possible to define two different strengths:

- the *yield strength* – the load at which large deformations start to occur; and
- the *ultimate strength* – the load at which the material fails completely.

Ductile materials are generally considered preferable to brittle ones, as they reduce the risk of catastrophic collapse if the structure is overloaded, and they give some warning of impending failure.

It might seem obvious that it would always be desirable to select materials that combine the virtues of high strength, stiffness and ductility. In reality, the choice is usually constrained by many other factors, such as durability, appearance, availability, weight and cost.

In civil engineering the principal construction materials are steel, an alloy of iron, and concrete, a surprisingly interesting material produced by mixing sand, gravel, cement and water. Since concrete is prone to cracking it is not used for tension members. For bending elements, it is normally reinforced by embedded steel bars in the areas where tensile forces are expected, producing a highly complex composite material. In mechanical and aeronautical structures, while the use of steel is widespread, many alternative materials are also used, including ceramics, polymers and a wide range of alloys.

Materials science is a huge subject in its own right, which cannot be dealt with adequately in the space available here. The reader is referred to specialist textbooks for further information (see the reading list in Section 1.6 at the end of this chapter).

1.3 Structural systems

This section briefly introduces some of the more common structural types and considers why they are used and how they work. Structures come in a huge variety of shapes and sizes – a full description of them all is not possible in the space available here, and the interested reader is referred to texts devoted to the subject.

1.3.1 Pin-jointed frames

One of the simplest structural types is the pin-jointed frame, or *truss*, comprising bar elements joined by pins. The bars can be positioned so as to form a flat, two-dimensional *plane truss*, as in Figure 1.16, or a more complex, three-dimensional *space truss*, such as in Figure 1.17. Pin-jointed frames are easy to analyse and since

Figure 1.16
The Granville Bridge, Vancouver, Canada.

Figure 1.17
Interior view of the Pyramid of the Louvre Museum, Paris (an exterior view of this structure is shown on the front cover).

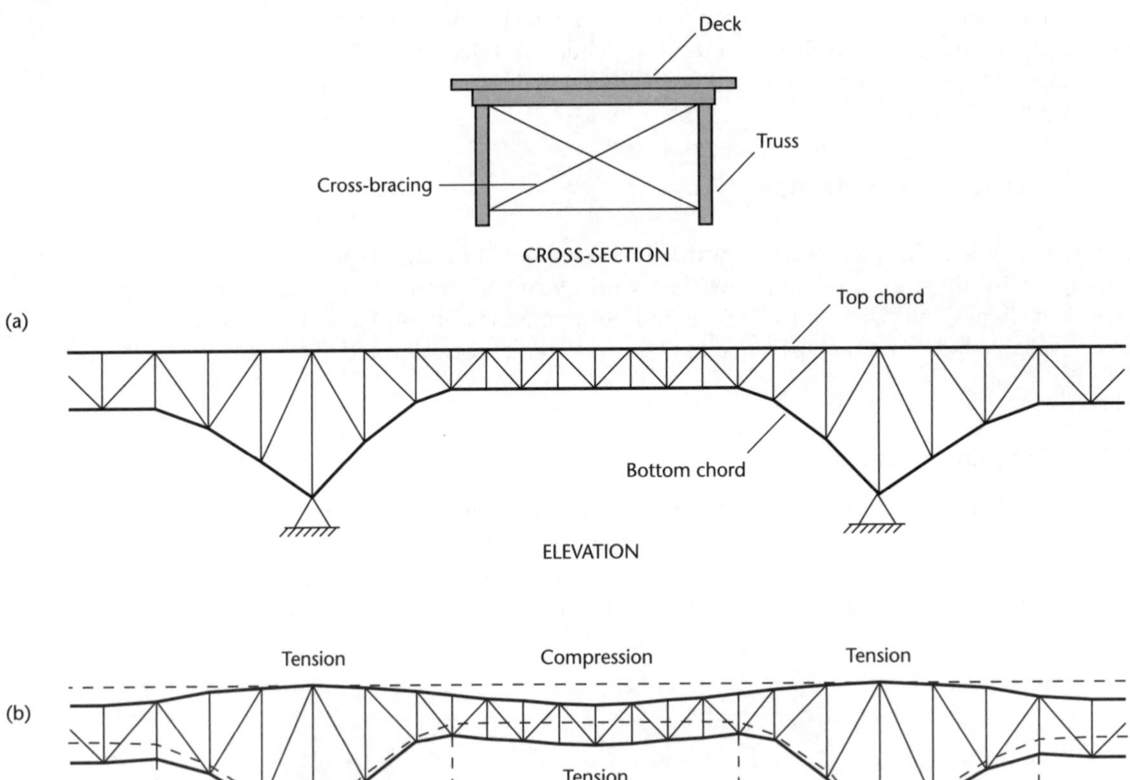

Figure 1.18
Structural behaviour of the Granville Bridge.

the bars carry only axial forces their cross-sections can be kept quite simple. Probably the most widespread use of pin-jointed trusses is in roof structures, though they are also used for some bridges.

Consider for example the bridge pictured in Figure 1.16. The structure of the bridge can be seen more clearly in the elevation and cross-section of Figure 1.18(a). The deck is supported along its edges by two parallel, plane trusses, connected by some cross-bracing members. When vehicles cross the bridge they impose vertical loads on the deck which are transferred by bending action into the two trusses – these are the main load-bearing elements of the structure, transmitting the forces they receive from the deck along the length of the bridge and into the supports.

Each truss is made up of a number of steel bars, joined together to give a triangulated framework. Assuming that the load is transmitted into the truss at the joints, then each bar develops only axial forces and remains straight as the structure deforms, with all rotation concentrated in the joints. If the load from the deck is distributed evenly over the span then the deflected shape is as shown at exaggerated scale in Figure 1.18(b). The structure deflects downwards at most points along the span but curves back upwards around the supports in order to satisfy the zero deflection requirement at the pins.

Consider first the middle section, between the dotted lines. In this region it is quite easy to see that members along the bottom chord have extended as the truss has deflected downwards, suggesting that they are subjected to tensile forces, while those along the top have gone into compression. It is a bit harder to see at a glance what is happening in the diagonals; in fact, some are in tension and some in compression. In the regions around the supports the behaviour is quite different. The structure is forced to reverse its curvature so that the top chord is in tension and the bottom in compression.

The overall behaviour of the truss is in fact very similar to that of the beam in Figure 1.18(c). Under the action of a uniform downward load, the beam deforms symmetrically. In the centre it curves downwards so that the bottom edge goes into tension and the top into compression; this is known as a *sagging* curvature. At the supports the vertical deflection must reduce to zero. This, together with the fact that the beam must take up a continuous, smooth shape, causes the curvature to reverse, giving tension at the top and compression at the bottom. In this region the beam is said to be *hogging*.

In both the truss and the beam, the amount of curvature will depend on the overall depth. If we tried to make the truss very shallow, as we might if we wanted to make the bridge less obtrusive, then this would reduce its overall stiffness, so that it would deflect further under a given vehicle load. This in turn would cause larger forces to be generated in the individual bars. So for the truss to work efficiently, it must be proportioned roughly as shown.

The magnitude of the compression forces in a truss is likely to be the limiting design parameter; if the forces are too large then extremely stocky bars will be required to prevent buckling problems, making the structure both unsightly and uneconomic.

1.3.2 Moment frames

A moment frame differs from a pin-jointed frame in two principal respects:

- loads may be applied to the structure at points other than the joints; and
- the joints are capable of resisting relative rotation between the members meeting there.

These differences cause the frame to resist loads primarily (though not exclusively) by bending action, rather than by the development of axial forces. Moment frames are particularly useful for multi-storey building structures, where large trusses can be inconvenient and wasteful of space. In addition, one-storey *portal frames* are very widely used for industrial buildings, where a single, very large enclosed space is required. Figure 1.19 shows some typical configurations of moment frames.

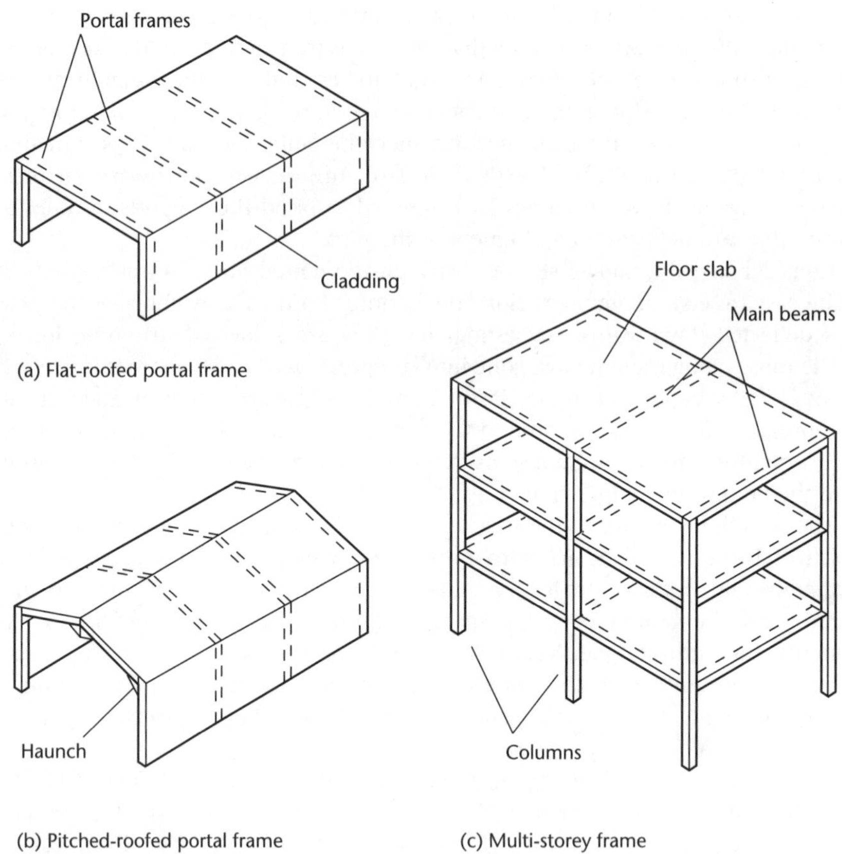

Figure 1.19
Moment frames.

We shall concentrate on the simple case of the flat-roofed portal frame. The importance of the rigid connections to the way the frame behaves can be illustrated by comparing it with a similar frame, but with pinned connections. Figure 1.20(a) shows the deformations of the two frames when a vertical load is applied to the roof. In the pin-jointed structure, the pins allow freedom of rotation between the roof beam and the supporting columns. This means that bending is confined to the beam, and the columns carry only an axial compressive load. With rigid joints, on the other hand, the rotations at the tops of the column and the ends of the beam must be identical, and the deformed shape therefore involves bending of both columns as well as the beam. This results in a substantial reduction in the vertical deflection of the beam. However, the loading on the columns is now more complex, involving both axial compression and bending, making the analysis and design of these elements more difficult.

Under a horizontal wind load (Figure 1.20(b)), the difference between the frames is far more striking. The pin-jointed frame has no resistance at all to this load and simply folds up. It could be strengthened by adding diagonal elements between the supports and the top corners, but these would interfere with the use of the internal space. The rigid-jointed frame, however, is able to resist the load by a combination of bending and axial forces. The beam acts in compression, transmitting half of the applied horizontal load into the right-hand column, and the load is transferred down to the supports by bending of the columns. Because of the rigid joints, there is also some rotation in the beam.

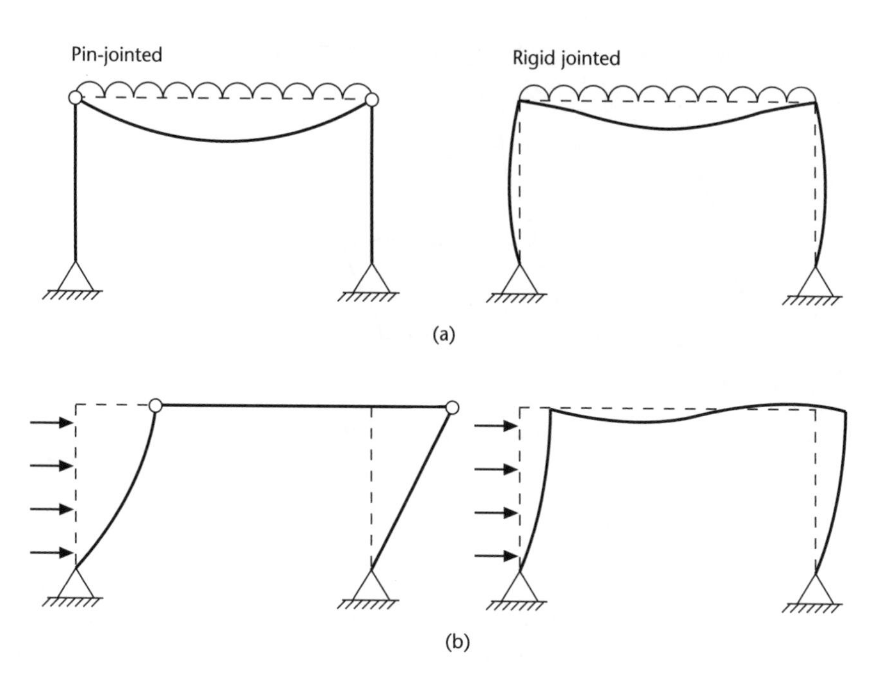

Figure 1.20
Comparison of portal frames with pinned and rigid joints.

Note that the frame with internal pins is inherently unsafe as it will collapse under an infinitesimally small horizontal load. A frame such as this one would never actually be used, even if only vertical loads were expected, as it would be impossible in practice to ensure that the loading remained purely vertical at all times. In fact it is not strictly a structure at all, but a *mechanism*, that is, an assemblage of members which can move without offering any resistance to applied loads. The distinction between structures and mechanisms will be discussed further in Chapters 2 and 3.

As well as preventing the structure from collapsing under the applied loads, the designer must ensure that it is adequately attached to the ground, so that the structure as a whole does not move. Under horizontal wind loading two types of movement are possible. Firstly, the applied force can push the building horizontally, causing it to slide along the ground – this must be resisted by generating horizontal reaction forces at the supports (Figure 1.21(a)). Secondly,

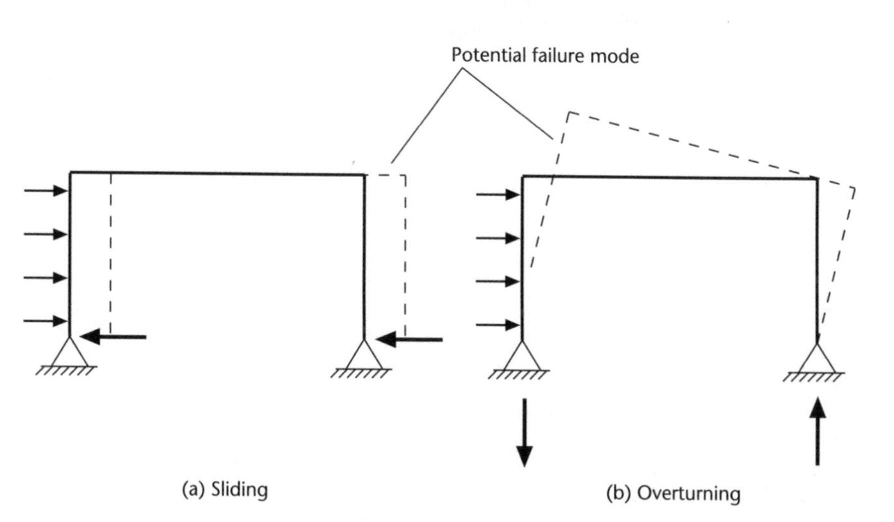

Figure 1.21
Resistance against sliding and overturning failure modes.

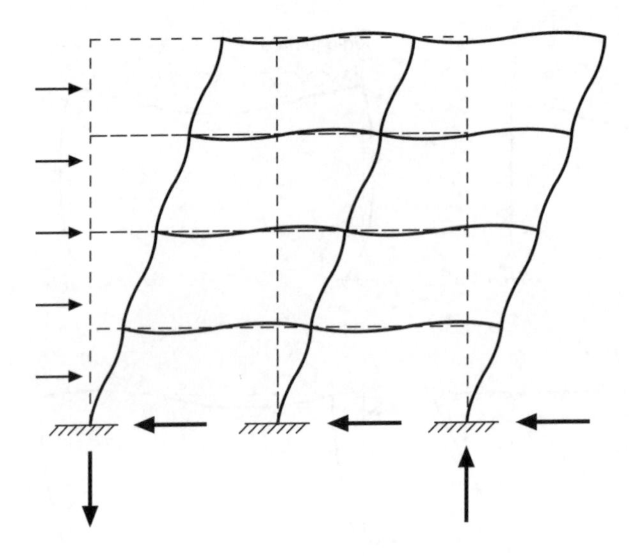

Figure 1.22
Multi-storey frame under wind loading.

the fact that the resultant wind force occurs at some distance above the ground means that it has an overturning effect on the structure, causing it to topple (Figure 1.21(b)). To prevent this the supports must provide an opposing set of overturning forces – this can be achieved by an upward reaction force at the leeward support and a downward one at the windward support. These reactions in turn impose axial forces on the vertical members – compression in the leeward column and tension in the windward column.

For the multi-storey moment frame in Figure 1.22 similar principles apply. Loads are carried by bending action of the entire frame. Horizontal wind loads

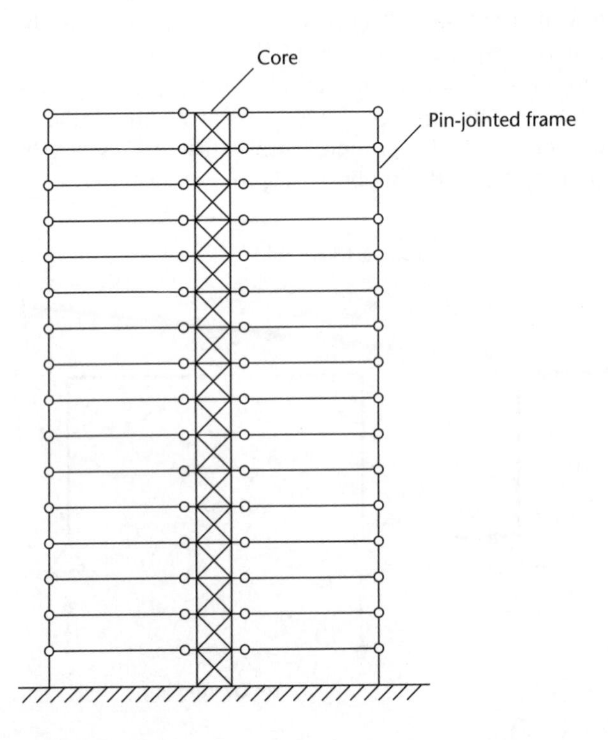

Figure 1.23
Cored frame building.

cause overall sway of the frame and, in addition to bending moments, generate compressive forces in the leeward columns and tensile forces in the windward columns.

Figure 1.24
Empire State Building, New York, USA.

1.3.3 High-rise frames

As building height increases, the effect of the horizontal wind loading rapidly becomes more severe. For heights above about twenty storeys simple moment frames are too flexible, suffering very large sway deformations in high winds. They also become very expensive to build because the rigid joints have to carry extremely large forces. A more efficient system must therefore be used.

The traditional approach to this problem has been to separate out the horizontal and vertical load-carrying systems as shown in Figure 1.23. A very stiff, central core provides a high resistance to horizontal loads – most tall buildings have such a core anyway, containing lift shafts, service ducts and so on. The stiffness is provided either by very heavy diagonal bracing or by using solid concrete *shear walls*. The remainder of the structure is then constructed as a simple pin-jointed frame. We have already seen in Figure 1.20 that a pin-jointed frame of beams and columns can carry vertical loads quite efficiently, with the beams acting in bending and the columns subjected to pure compression, but is unable to sustain horizontal loading. So when the pin-jointed frame and the core are connected, the core carries the horizontal loads, leaving the frame free to carry the vertical loads. This is the system that was used in most early skyscrapers, such as the Empire State Building (Figure 1.24).

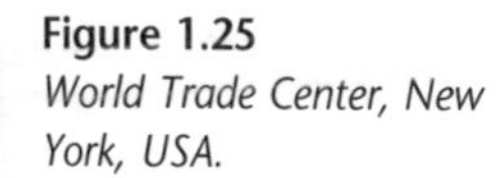

Figure 1.25
World Trade Center, New York, USA.

Another way of providing high horizontal stiffness is the use of the *tube* concept (Figure 1.25). Here, the perimeter of the building consists of numerous, closely spaced columns, connected together by beams at each floor level using rigid joints. In this way the outer surface of the building forms an extremely stiff tube, perforated only by small window openings between the columns. The whole building is like a giant beam with a rectangular hollow cross-section. Under wind load it behaves as a cantilever, that is, a beam rigidly fixed at one end and unrestrained over the rest of its length (Figure 1.26).

A further refinement is the use of the

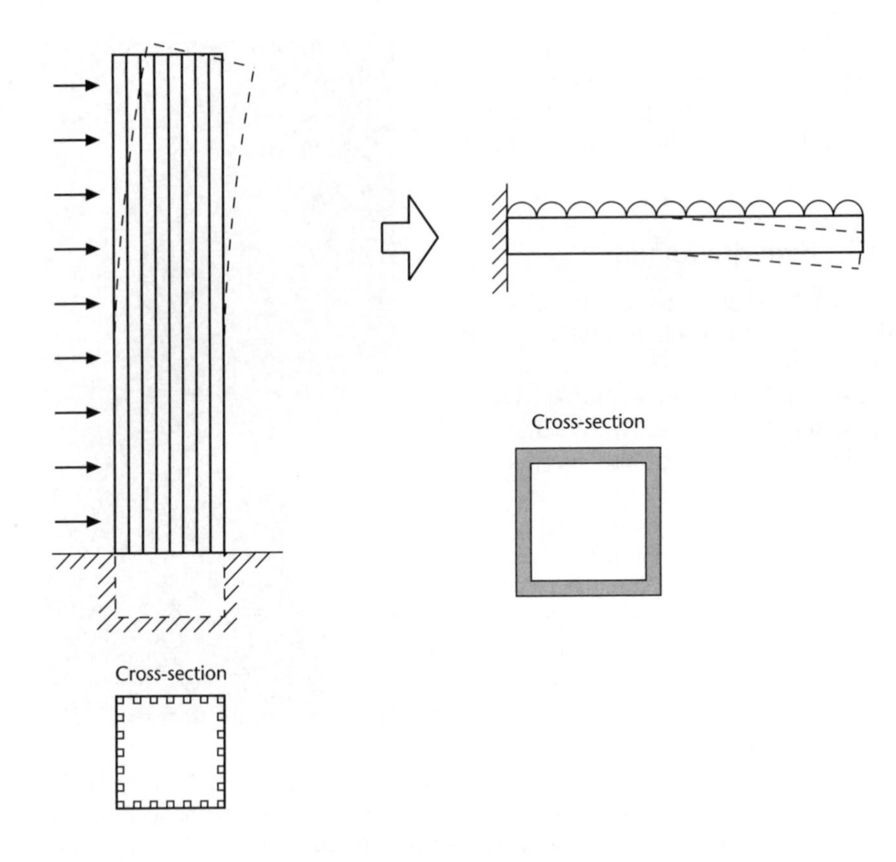

Figure 1.26
Representation of a skyscraper as a cantilever beam.

Figure 1.27
Bank of China, Hong Kong.

bundled tube, where the building comprises several tubes of different heights (Figure 1.27). At the top the structure has a relatively small cross-section, and so does not attract much wind loading. At the bottom it is much broader, providing stability against overturning and a high resistance to bending.

1.3.4 Cable and arch structures

In the last hundred years or so, the availability of high-strength steels has enabled the construction of some spectacular cable structures, of which probably the best known is the *suspension bridge*. Figure 1.28(a) shows the principal elements of a suspension bridge. Loads on the deck are carried up vertical hangers into the suspension cable, which hangs between two large towers.

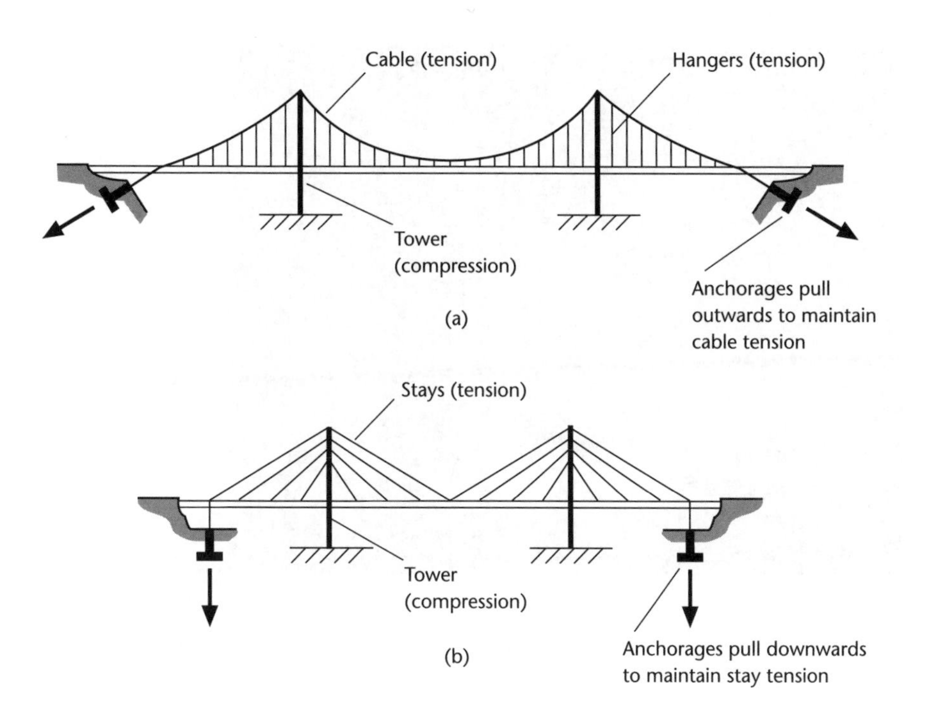

Figure 1.28
Suspension and cable-stayed bridges.

The cable is made by winding together a large number of small-diameter wires, giving very high strength and stiffness in axial tension, but virtually no resistance to compressive or bending loads. Therefore, when loaded, the cable takes up a shape that enables it to carry the applied loads in pure tension. If the loading applied to the cable is uniform over the span of the bridge, then it is quite easy to show that this optimum shape is a parabola.

Besides the cable, there are two other vital elements in the suspension bridge system. First, the towers that support the suspension cable must transmit the forces from the cable down to the ground in axial compression. Second, to prevent the cable from sagging down in the centre under the applied load, it is necessary to apply an outwards pull at either end. This requires very substantial anchorages, which can be formed either by tunnelling into the rock, or simply by providing enough dead weight to resist the cable force.

Because it acts in tension, the cable is extremely strong and does not suffer the stability problems associated with compression elements. Suspension bridges are therefore able to span huge distances – Japan's Akashi-Kaikyo Bridge, completed in 1998, has a span between the towers of 1991 metres!

One of the problems with suspension bridges is that the deck is very flexible. If it is not carefully designed it can experience large twisting motions due to wind gusts. On the Brooklyn Bridge in New York (Figure 1.29), the deck is stiffened by diagonal cables or *stays* running directly from the tops of the towers to the deck. The stays eliminate the twisting motion by preventing large vertical movements of the edges of the deck. More recently, many bridges have been constructed with the deck supported by the stays alone, without the suspension cable (Figure 1.28(b)). These *cable-stayed bridges* are rather stiffer structures than suspension bridges, and provide an economic alternative for intermediate spans, up to about

Figure 1.29
Brooklyn Bridge, New York, USA.

1000 metres.

All of the structures introduced so far involve tensile stresses, either due to direct tensile forces or due to bending. They are made possible by the ability of materials such as steel and reinforced concrete to resist tension. These are, however, comparatively new materials, only widely available since the late nineteenth century. Prior to this, the main construction materials were stone and timber. Timber can withstand a certain amount of tension but lacks the strength and stiffness necessary for large-scale construction. Therefore, most major structures built prior to the mid-nineteenth century were made of stone, a material with high compressive strength but little or no strength in tension. Engineers were therefore required to design structures to carry loads almost entirely by axial compression.

One very widely used compression structure is the arch. An arch can be understood very easily if one realises that it is really just an upside-down cable. Suppose we could fix the shape of the cable in Figure 1.28(a) and then flip it over, so that the supports are at the lowest rather than the highest points. Now, instead of a pure tension the arch acts in pure compression. This makes it prone to buckling, and so a rather more substantial cross-section must be used. Lastly, whereas a cable needs an outward pull at its ends to stop it from sagging, an arch needs a large inward thrust at the supports to prevent it from spreading, as shown in Figure 1.30(a).

The Romans built many superb arch structures, such as the three-tier aqueduct in Figure 1.31. Unfortunately, they did not realise that the optimum arch shape is an inverted cable, and instead used circular arches. This meant that significant bending was developed in addition to the axial compression (Figure 1.30(b)), and this limited the size of arch that could be built safely. Modern arches are more optimally shaped, and are made from materials with considerable tensile strength, so that they can sustain some bending loads in addition to the axial thrust. This enables them to bridge quite large spans, hundreds of metres in some cases.

The arch principle is used in many different kinds of structures. The dam in

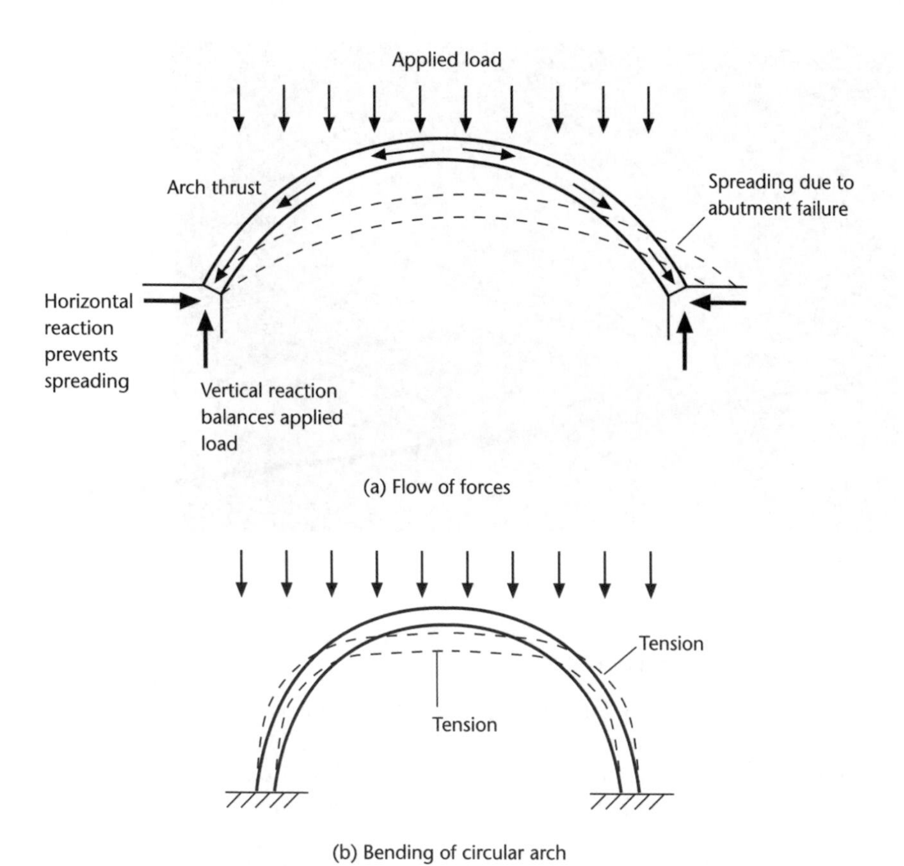

Figure 1.30
Arches.

Figure 1.32 is simply a very broad arch turned on its side. Load is applied by the pressure of the retained water on the convex face of the dam, and is resisted by a compressive thrust acting along the line of the arch. The sides of the valley act as rigid abutments, preventing the arch from spreading.

Figure 1.31
The Pont du Gard aqueduct, Nimes, France.

Figure 1.32
Ross Dam, Washington State, USA.

1.3.5 Continuum structures

Continuum structures may take the form of flat or curved surfaces, such as roofs, or three-dimensional solids, such as large dams. Probably the most interesting continuum structures are those made up of shells, that is thin, curved elements which carry loads primarily by developing tension and compression forces in the plane of the shell.

Of the various shells that can be generated, traditionally the most popular has been the dome. Like the arch, this is predominantly a compression structure. Indeed, it is tempting to think of a dome as simply a series of arches set on a circular base. However, this is not strictly accurate. While the radial lines, or *meridians,* do act like arches, they are restrained from bending outwards by the action of the horizontal *parallels,* which act like the hoops on a barrel – see Figure 1.33. This enables a hemispherical dome to be made much thinner than a set of unconnected circular arches. The disadvantage is that significant tensions are developed in the parallels, particularly near the base. In older structures such as St Paul's Cathedral (Figure 1.34), this was resisted by making the base very thick. Nowadays it would be normal to use a tension element such as a concrete ring beam.

With the recent development of high-strength materials such as fibre-reinforced plastics, there has been a rapid growth in the use of continuum structures that act in tension. Often these are made up of thin, flexible sheets of fabric, incapable of sustaining any compressive or bending loads. When loaded, the material takes up a shape which enables it to carry the applied loads purely in tension in the plane of the fabric, rather like a continuum version of a suspension cable. If the fabric is initially slack, then it will need to undergo a very large deformation in order to develop sufficient in-plane tension to carry the applied load. However, if we first of all pull the fabric taut then it becomes extremely stiff, and can carry quite substantial out-of-plane loads without much further deformation. This is the principle used in the grandstand roof in Figure 1.35, in which the fabric is held taut by a network of bars and cables.

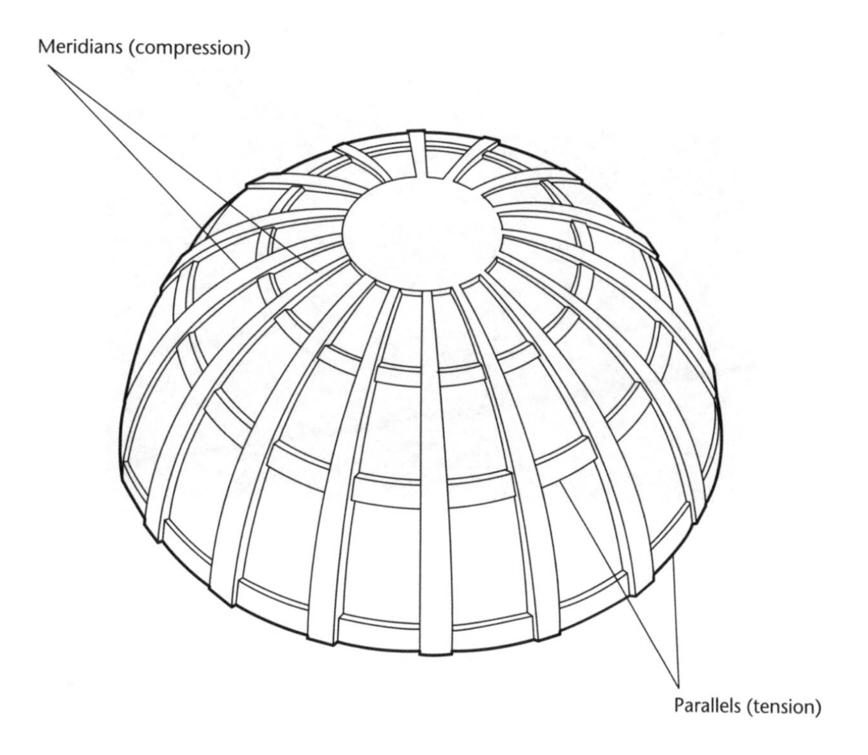

Figure 1.33
Components of a dome.

1.4 Structural analysis and design

A structure is normally designed to satisfy two principal requirements:

- It must be capable of carrying the biggest load it is ever likely to encounter without collapsing or sustaining irreversible damage. This is often referred to as the *ultimate* load condition. This requirement may occasionally be relaxed somewhat for very extreme loadcases. For instance, it is commonly accepted that structures in seismic regions may be designed to allow limited damage (but not complete collapse) under very large earthquakes.
- It must be able to sustain everyday loadings without deforming excessively or in any way becoming unfit for use. This is known as the *serviceability* loadcase.

Figure 1.34
St Paul's Cathedral, London, UK.

Figure 1.35
The Mound Stand, Lord's Cricket Ground, London, UK.

It is important to remember that structural analysis, while important, is just one element of structural design. The full design process can be summarised as follows:

(a) Define the problem – what loads are to be carried and where will the structure be located?
(b) Choose an overall structural system – this will depend on numerous factors, including the loading, where it is possible to place the supports, how easy or difficult the structure is to build, aesthetics and constraints related to the use of the structure.
(c) Analyse the structure in order to determine the member forces.
(d) Choose appropriate dimensions and materials for the members, to ensure they do not collapse or deform excessively under the forces calculated in (c).

It would be nice to be able to work through these steps in the order set out above, but unfortunately things are rarely so simple. The information available at the start of the design process is usually rather limited. In most cases, it will be necessary to make some educated guesses, work them through and then reiterate until an optimal design is reached. For instance, since the self-weight is often a significant load, some estimate may have to be made of the member dimensions and weights before performing the structural analysis. Also, there will often be many possible structural solutions to a design problem. It may be necessary to work several of them through to quite an advanced stage before the preferred solution becomes clear.

It should be clear from the above that structural design is a complex, iterative process, requiring considerable creativity and ingenuity in addition to analytical skills. While the ability to analyse structures is essential, this is not sufficient by itself to make a good designer. Equally important is the understanding of structural behaviour that a good engineer should be able to develop through application of analytical techniques.

1.5 The role of computers in structural analysis

It should come as no surprise to the reader that the vast bulk of structural analysis is nowadays done by computer, and the proportion is sure to increase in the future. A wide range of *stiffness matrix* computer programs is available for the analysis of framed structures, while *finite element* software is widely used for the analysis of continuum structures. The algorithms on which these programs are based are derived from, and closely related to, the more traditional approaches, but by writing them in matrix form they are rendered amenable to computational solution.

It is, of course, impossible to learn computer analysis techniques without first understanding the basic structural theory on which they are based. The early chapters of this book therefore concentrate on the fundamental principles of structures, while later sections place an increasingly strong emphasis on computer applications. In particular, Chapters 9 and 10 are devoted entirely to computer methods.

With so many computer programs for structural analysis already available, it is likely that most engineers will be involved in using existing software rather than writing their own. Making effective use of analysis software is not the trivial task that many imagine it to be; without a clear understanding of what the program is doing and what the user is trying to achieve, the results can be disastrous. The sections on computer methods therefore do not aim to give the reader all the technical detail required to write a program from scratch; this would take an entire textbook in itself. Rather, they attempt to give software users sufficient understanding of the underlying theory of analysis programs, their strengths and limitations, and what the user needs to do to ensure an accurate solution.

1.6 Further reading

The following give excellent introductions to the huge variety of structural forms in use, how they work and why they sometimes don't work:

- M. Salvadori, *Why Buildings Stand Up*, Norton, 1990.
- M. Levy and M. Salvadori, *Why Buildings Fall Down*, Norton, 1992.
- T. Y. Lin and S. D. Stotesbury, *Structural Concepts and Systems for Architects and Engineers*, Wiley, 1981.

For those with a real enthusiasm for structural analysis and design, the following are fascinating accounts of how these topics have developed, and how they relate to each other:

- W. Addis, *Structural Engineering: The Nature of Theory and Design*, Ellis Horwood, 1990.
- J. Heyman, *Structural Analysis: A Historical Approach*, Cambridge University Press, 1998.

The properties of engineering materials are dealt with in a very large number of texts. Among the most authoritative and enjoyable ones are:

- M. F. Ashby and D. R. H. Jones, *Engineering Materials 1: An Introduction to their Properties and Applications,* 2nd edn, Butterworth-Heinemann, 1996.
- M. F. Ashby and D. R. H. Jones, *Engineering Materials 2: An Introduction to Microstructures, Processing and Design,* 2nd edn, Butterworth-Heinemann, 1998.
- J. E. Gordon, *The New Science of Strong Materials, or Why You Don't Fall Through the Floor* 2nd edn, Penguin, 1991.

CHAPTER 2

Plane statics

CHAPTER OUTLINE

The subject of *statics* covers the analysis of bodies subjected to sets of forces that are in equilibrium. If the forces are not in equilibrium the system is termed *dynamic*, since its movement varies significantly with time. Of course, static systems also move, but not continually. When loads are applied to a structure, it deforms and internal forces are generated within it. Soon, the structure reaches an equilibrium position, where the internal forces balance the applied loads, and then it becomes stationary.

The simple concepts of statics form the basis of all structural analysis, and it is therefore vital that they are fully understood. In this chapter, brief reference will be made to three-dimensional systems, but we will concentrate mainly on applying the methods of statics to structures lying wholly within a two-dimensional plane. This is reasonable, as many structures can be idealised as planar for analysis purposes, and most of the important concepts can be more easily understood in two dimensions. First, it is necessary to introduce some basic principles and terminology. We will then go on to look at the external forces imposed on structures (applied loads and support reactions) and the resulting internal forces generated within structural members. On completion of this chapter the reader should be able to:

- understand the fundamental concept of force equilibrium;
- draw free body diagrams for simple structures;
- understand the concept of statical determinacy, and assess the determinacy of sets of support reactions;
- calculate reactions using equilibrium of forces and moments;
- distinguish clearly between external forces on structures (loads, reactions) and the internal forces generated in the members (axial forces, shear forces, bending moments, torques);
- calculate and sketch shear force and bending moment diagrams for statically determinate beams.

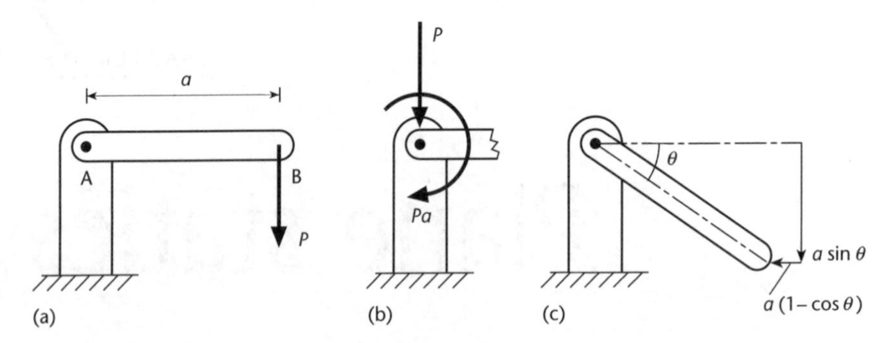

Figure 2.1
Forces and displacements for a simple mechanism.

2.1 Basic structural principles

2.1.1 Forces and displacements

Forces on a structure can arise from many sources, such as the structure's own weight, any objects placed on it, wind pressure and so forth. Force is a vector quantity, that is, it has both magnitude and direction. The SI unit of force is the Newton (N), which is defined as the force required to impart an acceleration of one metre per second per second to a mass of one kilogram (that is, $1\ \text{N} = 1\ \text{kg m/s}^2$). An object placed on a structure will thus impart a vertical force equal to its mass multiplied by the acceleration due to gravity ($g = 9.81\ \text{m/s}^2$).

The forces on a body can also give rise to *moments*, which tend to cause the body to rotate about an axis. The moment of a force about an axis is simply equal to the magnitude of the force multiplied by the perpendicular distance from the axis to the line of action of the force. Consider, for example, the lever AB shown in Figure 2.1(a). The effect of the force P acting at B is to impart both a direct force P and a moment $M = Pa$ on the hinge at A, as shown in Figure 2.1(b).

The loads acting on a structure cause internal stresses, and so cause it to deform. The deformations are usually expressed in terms of deflections (that is, straight-line movements) and rotations about a point or axis. These deflections and rotations are, of course, closely related to each other. Returning to the lever AB, suppose that the loads generate no bending or extension of AB, but cause only a rotation θ about A (Figure 2.1(c)). Then B has deflected vertically by $a \sin\theta$ and horizontally by $a(1 - \cos\theta)$.

Lastly, it is worth mentioning that the term *forces* is often used in practice to encompass both direct forces and moments, and likewise the term *displacements* is often taken to include both translational movements, or *deflections*, and rotations.

2.1.2 Sign convention

Systems of forces, moments, displacements and rotations must be analysed using a logical and consistent sign convention. It is extremely important that the convention used is clearly stated and adhered to at all times, otherwise confusion and errors are sure to arise. The choice of a sign convention is far from straightforward. No system is ideal for all circumstances; in particular, it proves impossible to prevent irritating minus signs from appearing in formulae and analyses. The system used in this book has been chosen so as to accord with most engineers' views on what quantities and directions should be regarded as positive.

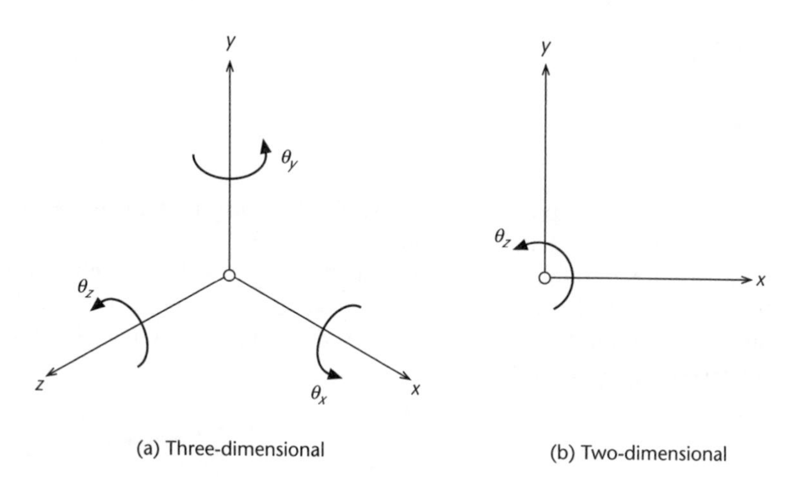

Figure 2.2
Positive axis directions and rotations.

The system will be briefly outlined here and will be added to as more concepts are introduced. For ease of reference, the full convention is summarised in Appendix A.

For three-dimensional problems, the right-handed (x, y, z) coordinate system shown in Figure 2.2(a) will be used. The y-axis is taken as positive upwards. Rotations obey the right-hand screw rule, which states that a positive rotation about an axis is in the direction of twist of a right-handed screw advancing in the positive axis direction. Thus to perform a positive rotation θ_x about the x-axis (that is, a rotation in the y–z plane) we turn from the positive y towards the positive z-axis. Similarly, a positive rotation about the y-axis implies turning from positive z towards positive x, and a positive rotation about the z-axis implies turning from positive x towards positive y.

When considering structures that can be idealised as two dimensional, we will normally assume that they lie within the x–y axis system shown in Figure 2.2(b). Note here that the positive z-axis is coming out of the page, so that θ_z is positive going *anti-clockwise*.

2.1.3 Equilibrium of forces

A system of forces is said to be in static equilibrium if the sums of all the forces and of all the moments at one point are equal to zero. In statics, all systems must satisfy the requirements of equilibrium. If equilibrium is not maintained, then the out-of-balance force will cause an acceleration, so that the system is no longer behaving statically. A brief introduction to dynamic systems is given in the last chapter of this book.

For a three-dimensional system with a set of mutually perpendicular axes x, y and z, as in Figure 2.2, six conditions must be satisfied; the force components in each direction must sum to zero, and the moments of the forces about each axis must sum to zero. These can be expressed in mathematical form as:

$$\sum P_x = 0 \tag{2.1}$$

$$\sum P_y = 0 \tag{2.2}$$

$$\sum P_z = 0 \tag{2.3}$$

$$\sum M_x = 0 \tag{2.4}$$

$$\sum M_y = 0 \tag{2.5}$$

$$\sum M_z = 0 \tag{2.6}$$

where P_x is the component of any force in the x direction, M_x is the moment of a force P about the x-axis, and so on.

A two-dimensional, or *coplanar*, system is taken to lie entirely within the x–y plane, so that it can be expressed in terms of force components in the x and y directions and moments about the z-axis. For a coplanar system there will be only three conditions of equilibrium:

$$\sum P_x = 0 \tag{2.1}$$

$$\sum P_y = 0 \tag{2.2}$$

$$\sum M_z = 0 \tag{2.6}$$

Some simple examples of equilibrium of coplanar forces are shown in Figure 2.3. For a system of two forces acting on a body (Figure 2.3(a)), equilibrium requires that the two forces share the same line of action, and are equal in magnitude but opposite in direction. If three forces act on a body, they must either be parallel, as shown in Figure 2.3(b), or *concurrent* (that is, their lines of action pass through the same point) – Figure 2.3(c). If the forces are neither concurrent not parallel, then they cannot be in equilibrium, since one of the forces must create a resultant moment about the point of intersection of the other two.

2.1.4 Free body diagrams

A free body diagram is simply a sketch of all or part of a structure, with all of the forces due to external loads, reactions etc. indicated by arrows. It is extremely important always to draw a free body diagram prior to commencing an analysis, and to ensure that all the forces acting on the structure are shown on the diagram.

As an example, Figure 2.4(a) shows a man standing on a plank, which rests on walls at either end. Assuming that the self-weight of the plank is negligible, and that there is no friction between the plank and the walls, the free body diagram for the plank is as shown in Figure 2.4(b), with the vertical load due to the weight of the man balanced by vertical reactions from the walls at each end.

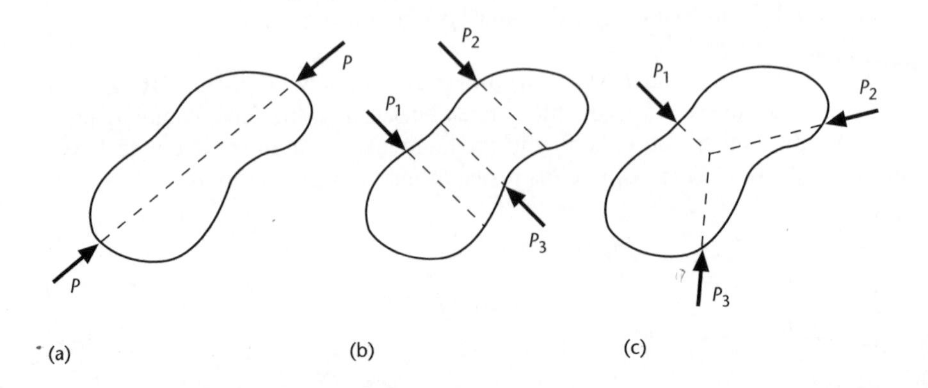

Figure 2.3
Equilibrium of forces in a plane.

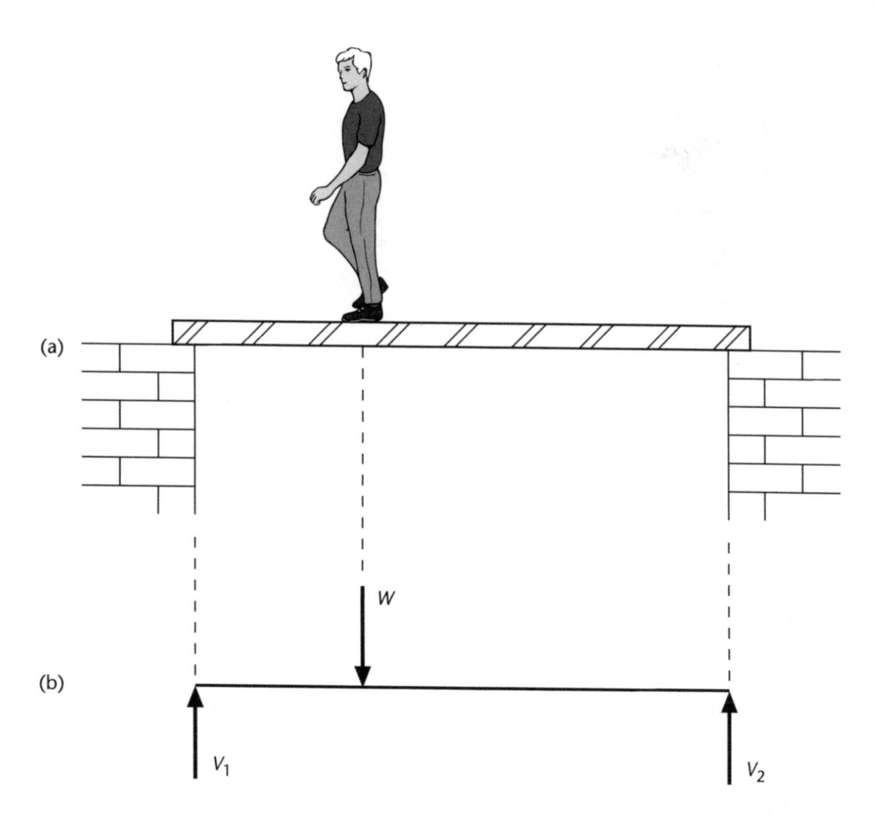

Figure 2.4
A simple free body diagram.

2.1.5 The force polygon

It has already been stated that the forces acting on a structure are vector quantities, that is, they have both magnitude and direction. Rather than resolving them into x, y and z components, we could simply say that, for equilibrium, the force vectors must sum to zero. This leads to the simple graphical method known as the force polygon. Figure 2.5(a) shows three forces acting on a body. A force polygon can be constructed by plotting end-to-end vectors parallel to the forces and with lengths proportional to the magnitudes of the forces. If the polygon closes, as in Figure 2.5 (b), then the forces are in equilibrium, since their vectors sum to zero. If it does not close (Figure 2.5(c)), then the force required to produce equilibrium is the vector $\mathbf{P}_4$ required to close the polygon. Alternatively, we can say that the force vectors $\mathbf{P}_1$ to $\mathbf{P}_3$ can be replaced by a single *resultant* force $\mathbf{P}_R$, which is equal to $-\mathbf{P}_4$.

2.1.6 Statical determinacy

One of the most important structural concepts is that of statical determinacy. A system of forces is said to be statically determinate if all of the unknown forces can be determined by the equations of statics alone. In the case of two-dimensional systems, as we have seen, it is normally possible to formulate three equations of static equilibrium, which can be solved simultaneously. This suggests that a statically determinate system will normally be limited to three unknown forces. In fact, as we will see later in this chapter, it is often possible to solve systems with more than three unknowns by making use of *equations of condition*. We shall return to the subject of statical determinacy several times over the next few chapters.

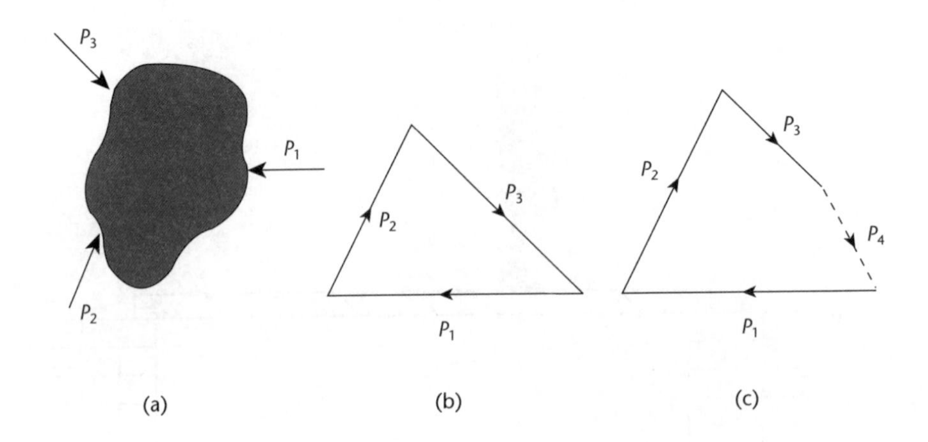

Figure 2.5
The force polygon.

2.1.7 The principle of superposition

This important principle is valid for linear elastic structures in which changes in geometry caused by the external loads are small. The principle states that the total effect of two different inputs to a system is equal to the sum of their effects when applied separately. Thus, if a spring extends a distance a under a load A and a distance b under a load B, then its extension under a load $(A + B)$ will be $(a + b)$. The principle may seem trivial at first glance, but it proves extremely useful when analysing more complex structures.

It is worth emphasising the restrictions on the use of superposition. The principle is not valid if the material of the structure is inelastic, or if its elastic limit is exceeded, or if the geometry of the structure changes appreciably when the load is applied. Obviously, the geometry of all structures changes slightly under load; individual members will extend, contract or bend, causing deflections of points on the structure. In most cases the resulting changes in geometry are small and can be neglected in applying the principle.

2.1.8 Compatibility of displacements

In addition to assessing the forces on structures, it is important to consider how they deform. In particular, in order for a structure to hold together under load, the displacements of its various elements must be compatible with each other.

Consider, for example, the simple structure shown in Figure 2.6. The frame consists of two bars AB and BC, connected to each other and to a rigid wall by frictionless pins, so that the bars are free to rotate at their ends. The vertical load at B will cause purely axial forces to be generated in the two members, which in turn will cause changes in length, whose magnitude depends on the elastic properties of the members. Obviously there must be a relationship between these changes in length and the movement of point B. For displacements that are small compared to the overall dimensions of the structure, this can be obtained by resolving the movement of each bar into a component along the bar axis (the extension or

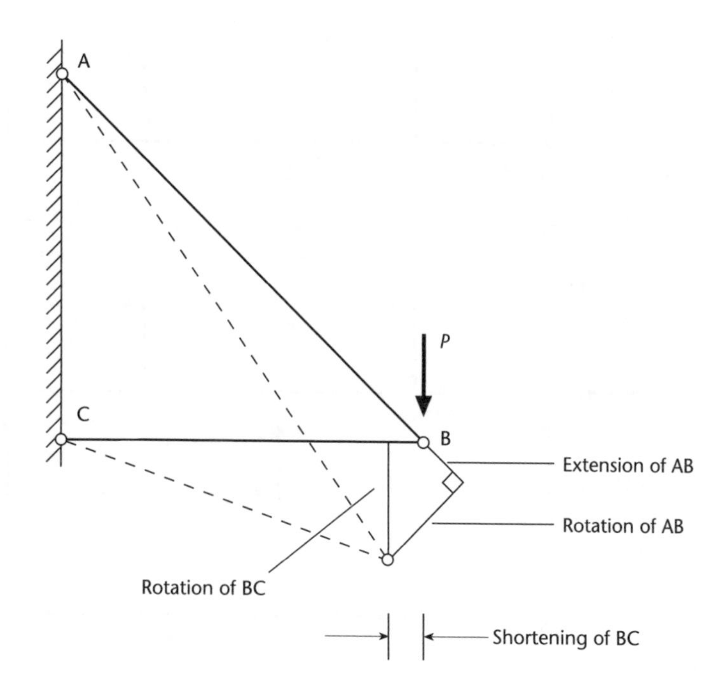

Figure 2.6
Displacement components in a simple frame.

compression caused by the axial force) and a perpendicular component due to the rotation of the bar about its pinned support, as shown in Figure 2.6. Strictly speaking, the rotational component should be a circular arc centred at the pinned support, but for small deformations the difference is negligible.

As well as the calculation of structural deformations, we shall see that displacement compatibility is also extremely useful for finding the forces in structures that are not statically determinate.

2.2 Determination of reactions

Structural supports play a vital role in determining how a structure carries loads and how it deforms. In Chapter 1 we introduced the most common idealisations of support conditions and discussed how they related to support conditions found in real structures. In order to analyse a structure, we need to consider what deformations can occur at these idealised supports, and what forces they exert on the structure. Consider the beam in Figure 2.7(a):

- The beam has a *pinned* support at end A, which prevents vertical and horizontal deflections but allows the beam to rotate. The reaction provided by this support will be a force R_A inclined at some angle θ_A to the vertical, but usually it is simpler to resolve it into vertical and horizontal components V_A and H_A.
- The central support at B is a *roller*, which provides only a vertical reaction, and therefore no restraint to rotation or horizontal movement. Note that the roller support prevents both upwards and downwards movements at B.

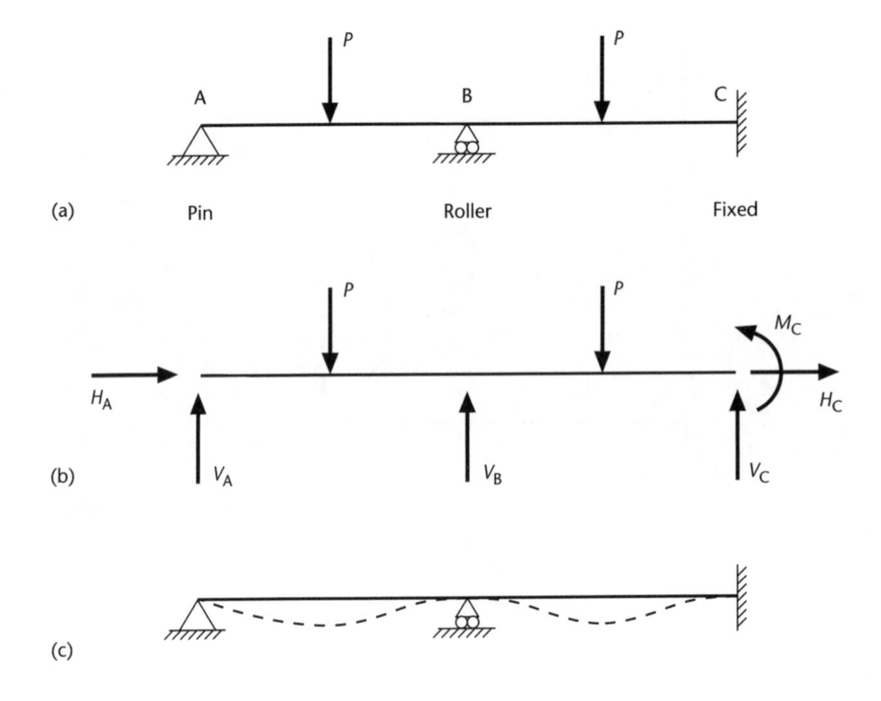

Figure 2.7
Idealised support types.

- At C the beam is held by a *fixed* or *built-in* support, which prevents horizontal and vertical deflections, and also rotation of the beam. In addition to the vertical and horizontal reactions, this requires a moment reaction M_c

The free body diagram for the beam is therefore as shown in Figure 2.7(b), and the forces acting on it will cause it to deform roughly as in Figure 2.7(c), in which the vertical scale of the deformations is, of course, exaggerated. Note that points A and C are fixed in space, and that the slope at C is zero. In theory, the point B can move horizontally, but in practice any such movement is likely to be negligibly small compared to the bending deflections in the beam away from the supports.

It must be emphasised that these support conditions are idealised, in that it is assumed that no friction exists in the roller or pin, and that the fixed support provides complete rigidity. In practice, conditions may well deviate from these ideal situations.

2.2.1 Stability and determinacy of reactions

In order for the support reactions of a structure to be calculated by simple statics, they must form a statically determinate set. That is, the number of independent equilibrium equations must equal the number of unknown reactions. For two-dimensional structures, we have already seen that three equilibrium equations are available, so that a statically determinate set of reactions will normally comprise three unknowns.

If there are less than three reactions, there will not be enough unknowns to satisfy the three equations, so the structure is *unstable* as far as the support conditions are concerned.

If there are more than three reactions, then the equations cannot be completely solved and the system is said to be statically indeterminate or *redundant*. For example, if there were five unknowns, two of them could be assigned any value.

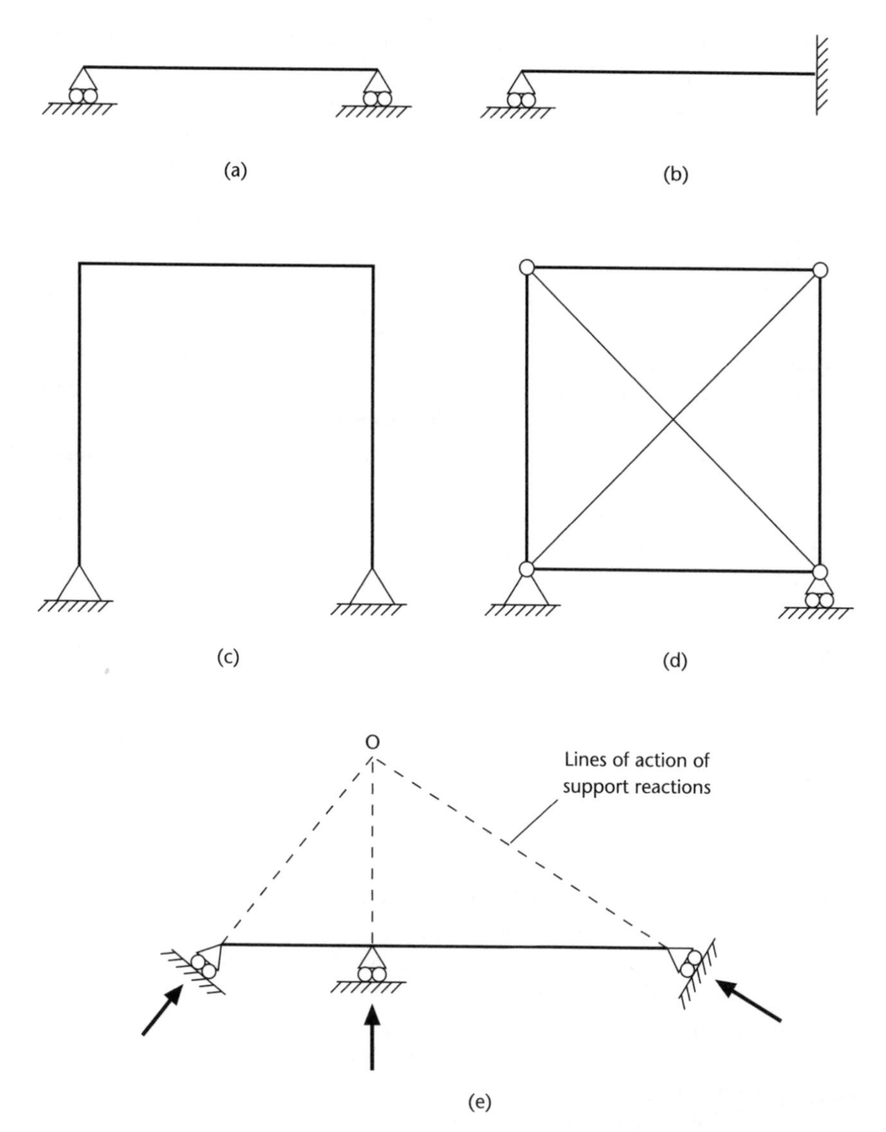

Figure 2.8
Determinacy of reactions.

The remaining three could then be found, their values depending entirely on the values chosen for the first two reactions. Redundant systems can be solved, but require quite sophisticated techniques that will not be covered until later in this book.

Figure 2.8 shows examples of several different structures. The beam (a) has only two vertical reactions and is therefore unstable. The beam (b) has four reactions (vertical at the left-hand end, vertical, horizontal and a moment at the right-hand end) and is therefore redundant to the first degree, that is, there is one more reaction than can be determined by statics alone. The portal frame (c) is also redundant, as there will be vertical and horizontal reactions at each support. The frame (d) is statically determinate in terms of reactions, but it will be seen later that it is redundant as far as the forces in the members are concerned.

It should be noted that in all cases the question of statical determinacy is independent of the loading applied to the system. Consider, for example, system (a): if only vertical loading is applied it might be tempting to say that there would be purely vertical reactions at each support, and since the support system is capable of sustaining these the structure is stable. However, an instability arises if a horizontal force, no matter how small, is applied. It can be seen that the stability or otherwise of the structure must be considered independently of any particular loading case; there would be little point in designing a building to resist only vertical loads if it collapsed as soon it was subjected to a small lateral wind load.

A slightly unusual case is shown in Figure 2.8(e). The beam has three reactions and so appears statically determinate at first glance. However, the lines of action of the three reactions, shown dotted, pass through a single point, O. Any load that did not also pass through O would create a moment which could not be resisted by the reactions, hence the beam would tend to rotate about O. The beam is therefore unstable. More generally, we can state that, for a system of reactions to be stable, it must be capable of resisting any small displacement or rotation that is applied to the structure.

In some cases two-dimensional systems having more than three reactions may be statically determinate, since an additional *equation of condition* can be generated. This concept will be introduced in the examples in the following section.

2.2.2 Calculation of reactions

For statically determinate structures reactions can be found by resolving forces, taking moments and sometimes using an equation of condition. Before illustrating the procedure, we need to give a little thought to the external loads on the system and the moments that they generate.

The simple idealisations of external loads that we use in structural analysis were briefly introduced in Chapter 1. In most cases we can treat a load either as a concentrated force acting at a point or as a uniformly distributed load (UDL) acting over all or part of the length of a member.

Often we wish to take moments of the forces about a point. For a concentrated load this is straightforward; the moment is simply given by the magnitude of the force multiplied by the perpendicular distance from the point to the line of action of the force. For a UDL, the calculation is slightly more complex. First, consider the loaded area split into short elements of length $\mathrm{d}x$, as shown in Figure 2.9. The force acting on a single element is then $w\,.\,\mathrm{d}x$ and the moment of this concentrated force about C is $wx\,.\,\mathrm{d}x$. The total moment is found by integrating this expression:

$$M_c = \int_a^b wx\,.\,\mathrm{d}x = \frac{w(b^2 - a^2)}{2} = w(b-a)\cdot\frac{(a+b)}{2} \tag{2.7}$$

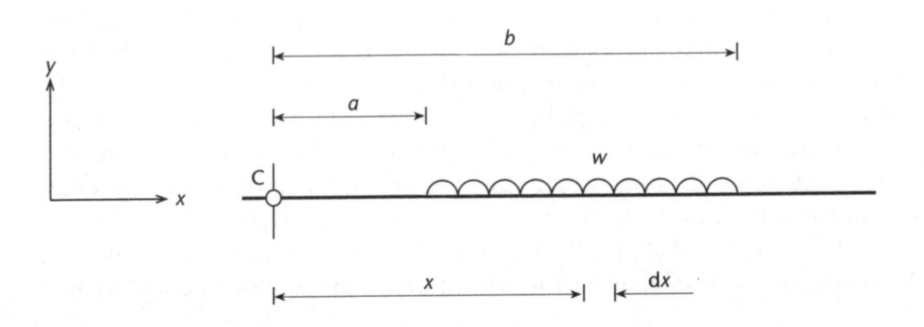

Figure 2.9
Moment calculation for a uniform load.

Here the term $w(b-a)$ represents the total load and $(a+b)/2$ is the distance from C to the centroid of the load. So equation (2.7) shows that the moment can be found by replacing the UDL by a point load equal in magnitude to its total value, and acting at its centroid. The same principle can be applied to all distributed loads, whether or not they are uniform; the load is replaced by a concentrated load equal to the total distributed load and acting at the centroid of the load system.

EXAMPLE 2.1

Calculate the support reactions for the beam in Figure 2.10(a), which rests on pinned and roller supports.

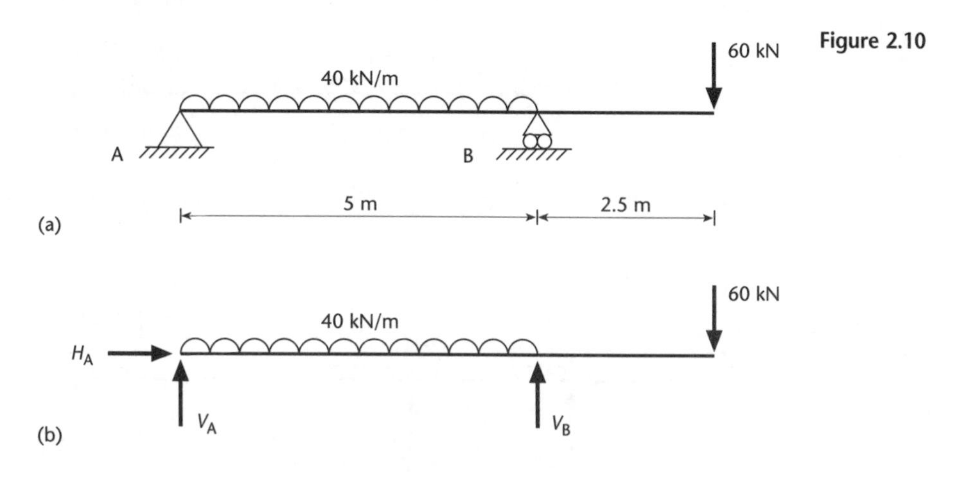

Figure 2.10

First, the free body diagram is drawn (Figure 2.10(b)). It can be seen that there are three unknown reactions, so the system is statically determinate. The reactions can therefore be found by resolving horizontally and vertically and taking moments about any point.

Resolving horizontally, Equation (2.1):

$$\sum P_x = 0 \qquad H_A = 0$$

Moments about A, Equation (2.6):

$$\sum M_z = 0 \qquad 5V_B - (40 \times 5 \times 2.5) - 60 \times 7.5 = 0 \qquad \rightarrow V_B = 190 \text{ kN}$$

Resolving vertically, Equation (2.2):

$$\sum P_y = 0 \qquad V_A + V_B - (40 \times 5) - 60 = 0 \qquad \rightarrow V_A = 70 \text{ kN}$$

These results could be checked by seeing whether the moments about some other point on the structure, such as B, sum to zero; this is always a good idea, as numerical mistakes are easy to make.

It can be seen from this example that solving a system of three reactions using the overall equilibrium of the structure is straightforward. In the following two examples we shall see that for certain structures it is possible to solve for more than three reactions. For each extra reaction, we need an additional piece of information over and above the three overall equilibrium equations. For instance, if a structure contains an internal pin joint then an additional equation can be obtained from the fact that the moment at the pin must be zero.

EXAMPLE 2.2

Calculate the support reactions for the beam in Figure 2.11(a).

Figure 2.11

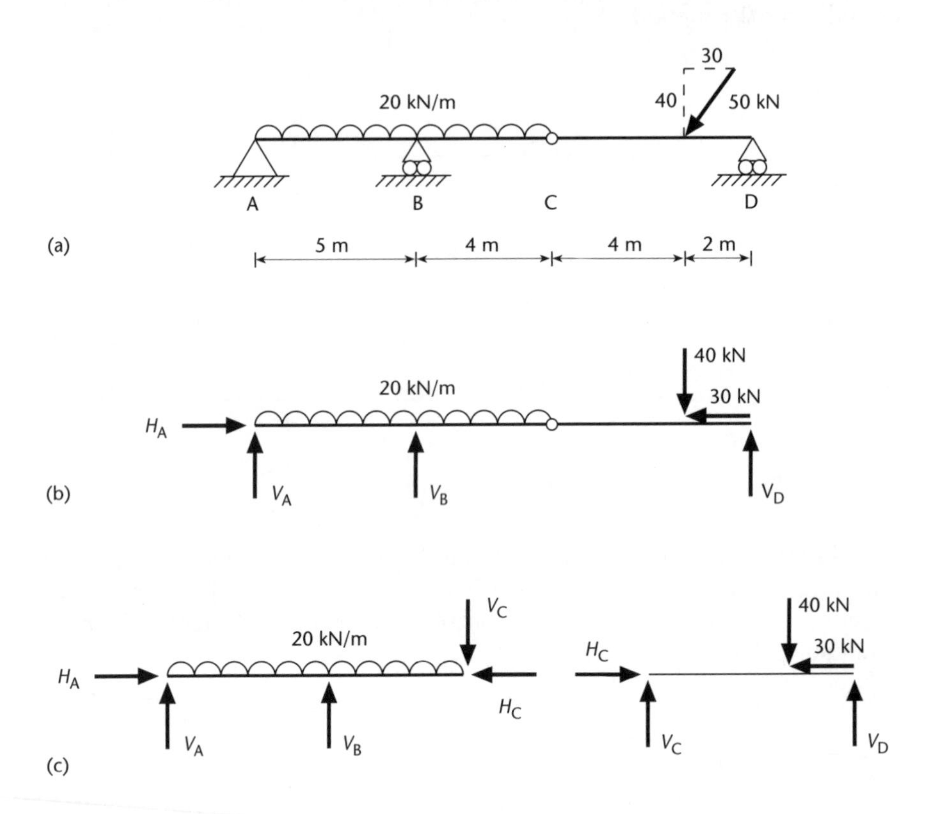

Again, we start by drawing the free body diagram (Figure 2.11(b)). The beam appears at first to be redundant, since there are four unknown reactions. However, there is a pin at C, allowing the two segments of beam either side of C to rotate relative to one another. The pin can transmit vertical and horizontal forces, but no moment.

For analysis purposes, we can split the structure at C and draw free body diagrams for each half, as shown in Figure 2.11(c). The internal forces exerted on each other by the two elements meeting at C are now included on the free body diagrams as though they were external forces. These forces must be equal and opposite, in order for the joint to be in equilibrium. Splitting the structure has thus introduced *two* additional unknowns, H_c and V_c, but has given us *three* extra equations, since we can now formulate separate equilibrium equations for ABC and for CD. So we have made a net gain of one equation, and the structure is therefore statically determinate.

Having split the structure, there are various ways to proceed; we can use the equilibrium equations for either half, or for the whole structure, or some combination. In this case the most direct solution is as follows:

For CD – Figure 2.11(c):
Moments about C:

$$6V_D - 40 \times 4 = 0 \qquad \rightarrow V_D = 26.7 \text{ kN}$$

For whole structure – Figure 2.11(b):
Moments about A:

$$5V_B + 26.7 \times 15 - 20 \times 9^2/2 - 40 \times 13 = 0 \qquad \rightarrow V_B = 186.0 \text{ kN}$$

Resolving vertically:

$$V_A + 186.0 + 26.7 - 220 = 0 \qquad \rightarrow V_A = 7.3 \text{ kN}$$

Resolving horizontally: $\rightarrow H_A = 30.0$ kN

Note that, when dealing with an inclined force, it is simplest to resolve it into vertical and horizontal components. This makes its moment about any point on the beam very easy to calculate, since the horizontal component runs along the beam axis and so produces no moment.

EXAMPLE 2.3

Calculate the support reactions for the three-pinned portal frame structure in Figure 2.12(a).

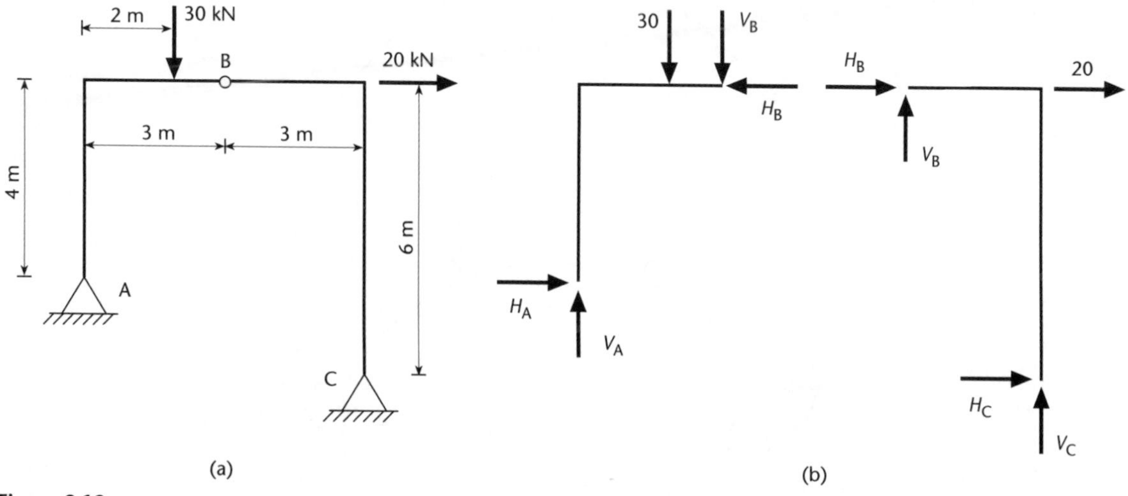

Figure 2.12

The solution procedure is similar to above. The structure is split at B and free body diagrams drawn for each side, Figure 2.12(b).

Moments about B for AB:	$4H_A + 30 \times 1 - 3V_A = 0$
Moments about B for BC:	$3V_C + 6H_C = 0$
Resolving vertically for ABC:	$30 - V_A - V_C = 0$
Resolving horizontally for ABC:	$H_A + H_C + 20 = 0$

These four equations can be solved simultaneously to give:

$$V_A = 2\text{ kN},\quad H_A = -6\text{ kN},\quad V_C = 28\text{ kN},\quad H_C = -14\text{ kN}$$

The negative signs of the horizontal reactions indicate that they act in the opposite directions to those indicated on the free body diagram.

The additional equation used to make the structures in Examples 2.2 and 2.3 statically determinate is known as an *equation of condition*. As well as making use of pin joints, an equation of condition can often be obtained from considerations of symmetry. The symmetrical arch rib in Figure 2.13(a) is fixed at both ends and has loads P symmetrically applied. From the free body diagram (Figure 2.13(b)), there are six unknown reactions, so apparently the arch is indeterminate to the third degree. However, if use is made of symmetry, then $H_A = H_B$, $M_A = M_B$ and $V_A = V_B$. Resolving vertically then shows that both V_A and V_B are equal to P. Symmetry has therefore enabled us to reduce the degree of indeterminacy from three to two.

Equations of condition of this type can only be used if there is complete symmetry of the structure, its supports and the loading.

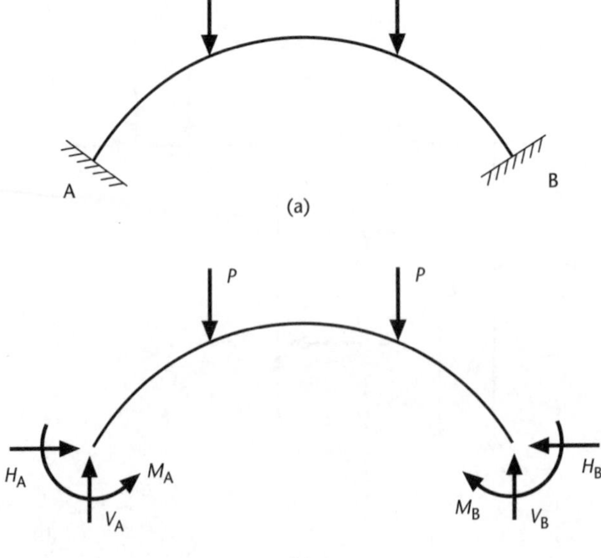

Figure 2.13
Use of symmetry.

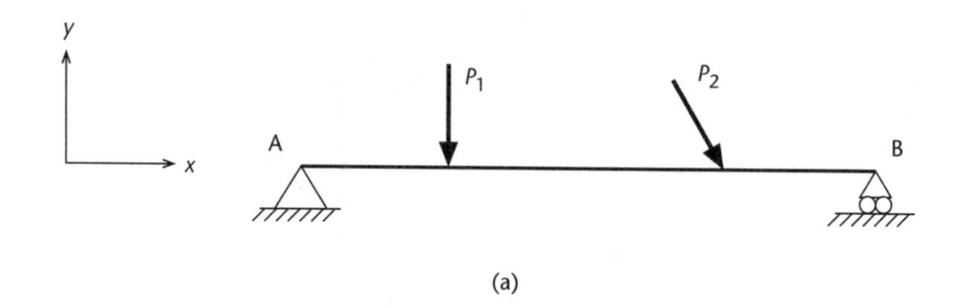

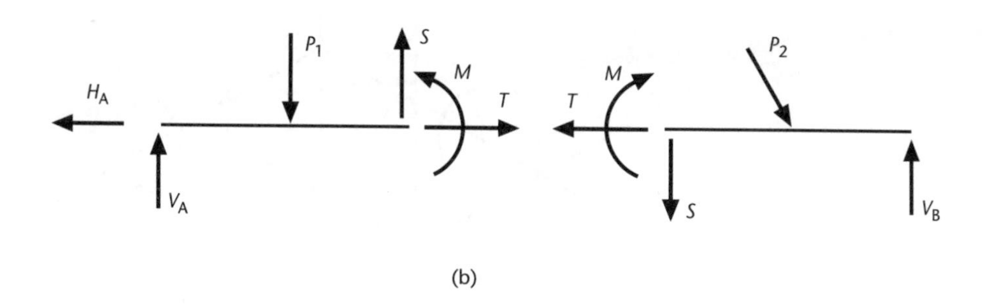

Figure 2.14
Internal forces in a 2D beam.

2.3 Internal forces in structures

So far, we have concentrated on the *external* forces applied to structures – the applied loads and the support reactions. In order for the structure to transmit the external loads to the ground, *internal* forces must be developed within the individual members. The aim of the design process is to produce a structure that is capable of carrying all these internal forces, which may take the form of axial forces, shear forces, bending moments or torques.

Consider first a two-dimensional beam where the applied forces and reactions all lie in a single plane (Figure 2.14(a)). The internal forces at a point in the structure can be found by splitting it at that point and drawing free body diagrams for the two sides (Figure 2.14(b)). The requirements of equilibrium state that not only must the resultant force on the entire structure be zero, but the resultant on any segment of it must also be zero. It is therefore clear that there must be forces acting at the cut point, as shown. These are drawn on the free body diagrams of the segmented structure as though they were external loads, but they are in fact the internal forces in the beam. The forces can be thought of as the external forces that would have to be applied to the cut beam in order to produce the same deformations as in the original, uncut beam. The forces shown are an axial force T, a transverse force S, known as a shear force, and a bending moment M.

For equilibrium at the cut point, the forces acting on the faces either side of the cut must be equal and opposite; this means that, when the two segments are put together to form the complete structure, there is no resultant external load at that point.

For a member in three-dimensional space, a total of six internal forces must be considered, as shown in Figure 2.15. Here, there is again an axial force T in the x direction, and the resultant shear force has been resolved into components S_y and S_z parallel to the y and z axes respectively. There are also moments about each of the three axes: M_y tends to cause the structure to bend in the horizontal $(x-z)$

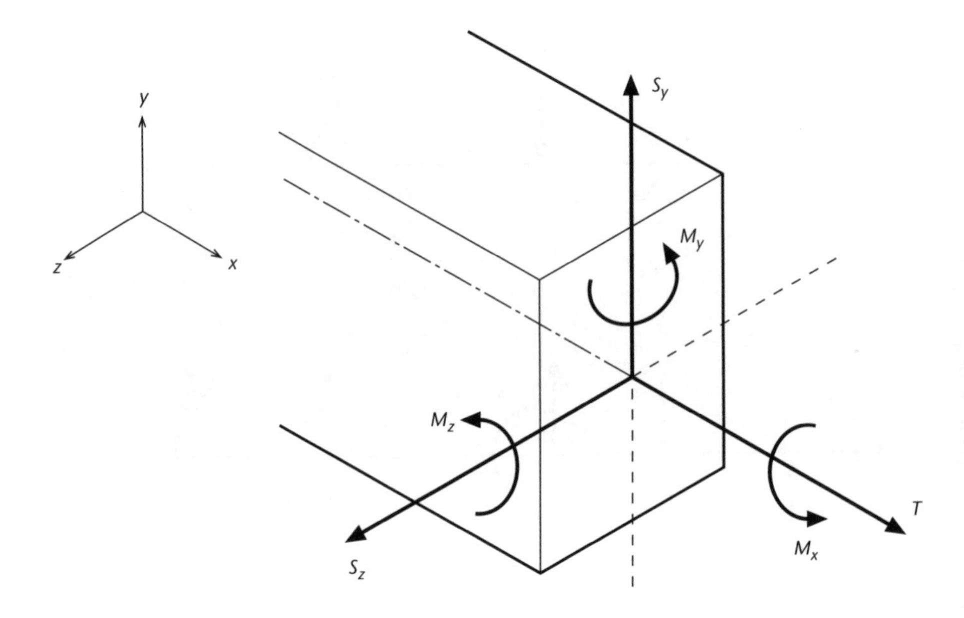

Figure 2.15
Internal forces in a 3D beam.

plane; M_z causes bending in the vertical $(x - y)$ plane; M_x causes the member to twist about its longitudinal axis, and is called a *torque*. For the time being, however, we will restrict ourselves to two-dimensional systems.

2.3.1 Sign convention for internal forces

Before defining the internal forces more fully, we need to extend the sign convention introduced earlier. For a two-dimensional system, positive forces act in the positive x and y directions, and a positive moment about the z-axis acts from the positive x towards the positive y-axis, that is anti-clockwise. We can also define a *positive face* of a member as one whose outward normal is in a positive axis direction. Thus, for the beam segment in Figure 2.16(a), the right-hand face is a positive x-face, the top surface is a positive y-face and the other two faces are negative.

We can now define a positive internal force as one which acts *either* in a positive direction on a positive face *or* in a negative direction on a negative face. Conversely, a negative force either acts in a negative direction on a positive face or *vice versa*.

Axial force

Axial forces may be either tensile, tending to stretch the member, or compressive, causing it to shorten. Figure 2.16(b) shows a bar under external loads which put it into tension. If the bar is cut at some point along its length, then for equilibrium of the cut sections, the forces at the cut must be in the directions shown. Both of these are positive, since the force on the left-hand segment acts in the positive x-direction on a positive face and that on the right-hand segment acts in the negative x direction on a negative face. Thus, tensile internal forces always pull inwards from the member ends and are taken as positive. Compressive forces are opposite to this; they push outwards and are taken as negative.

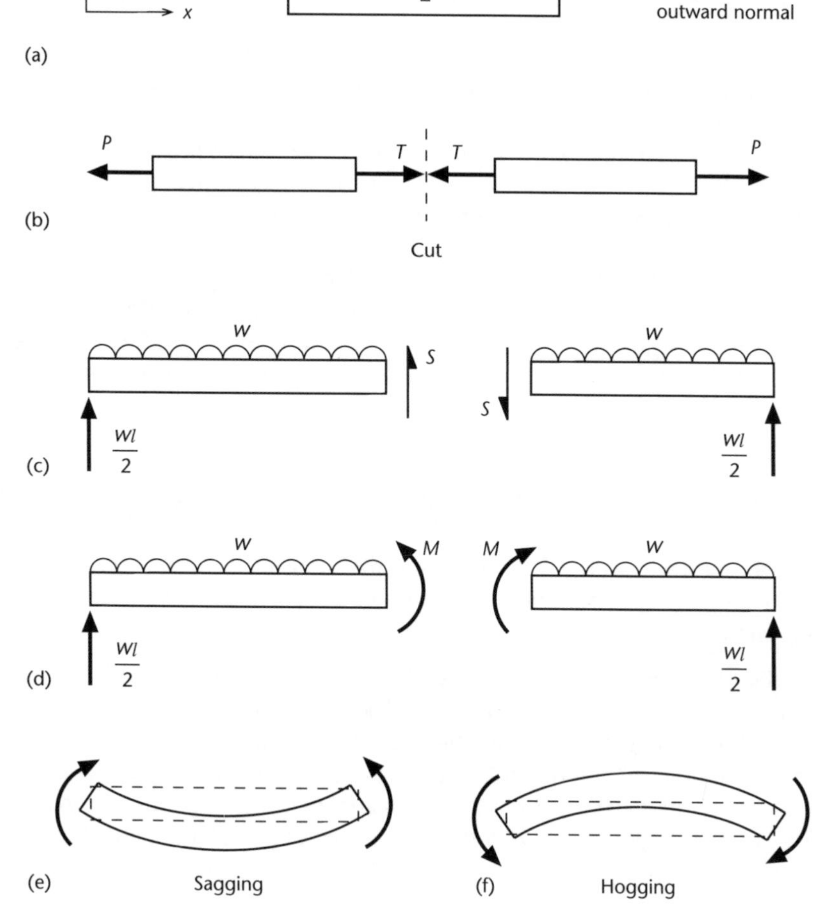

Figure 2.16
Sign convention for internal forces.

Shear force

If we take a vertical section through a beam in equilibrium, it will be necessary to apply forces normal to the axis of the beam to maintain equilibrium. These forces S are known as *shear forces*. Figure 2.16(c) shows a simply supported beam subjected to a uniform external load. If we split the beam somewhere to the right of its midpoint, the resultant of the external forces on the left-hand portion is downwards, and equilibrium therefore requires an upwards shear force at the cut. For the right-hand portion, the resultant external force and the shear force each act in opposite senses. Again, both the shear forces shown are positive, since the one on the left-hand segment acts in the positive y direction on a positive x-face, and so on.

Bending moment

If we again consider a vertical section through a beam we can see that, in addition to a shear force, equilibrium implies the existence of a moment about a horizontal axis through the centroid of the section. This is known as the bending moment M. Returning to the uniformly loaded beam, we again split the beam to the right of its midpoint (Figure 2.16(d)). Considering the left-hand segment, if we take moments of the external forces about the cut end, then the clockwise moment due to the support reaction will be larger than the anti-clockwise moment due to the uniform load. For equilibrium there must therefore be an anti-clockwise moment at the cut end. The opposite is true for the right-hand segment. Once again, the bending moments shown are both positive, since one acts anti-clockwise on a positive face, the other clockwise on a negative face.

Note that a positive bending moment is one that causes *sagging*, that is, the beam bends downwards, so that the upper surface is concave (Figure 2.16(e)). A negative bending moment is termed *hogging*, and causes an opposite curvature (Figure 2.16(f)).

When considering equilibrium of part of a member or structure, it can be helpful to think of the internal forces at the cut point as the reactions of the structure to the external loads. It follows that the internal forces must be equal in magnitude but opposite in direction to the resultant of all the external loads on the section.

EXAMPLE 2.4

For the beam of Figure 2.11, calculate the axial force, shear force and bending moment 6 m to the right of A. (The support reactions have already been calculated in Example 2.2.)

To find the internal forces at a point, we imagine the structure split into two at that point. We then consider the free body diagram of the structure to one side of the cut, with the internal forces at the cut drawn on as though they were external forces (Figure 2.17). In doing this, it is always a good idea to draw all the internal forces as positive even if, as here, it may sometimes seem obvious that they are not. The internal forces can now be found using simple equilibrium equations.

Figure 2.17

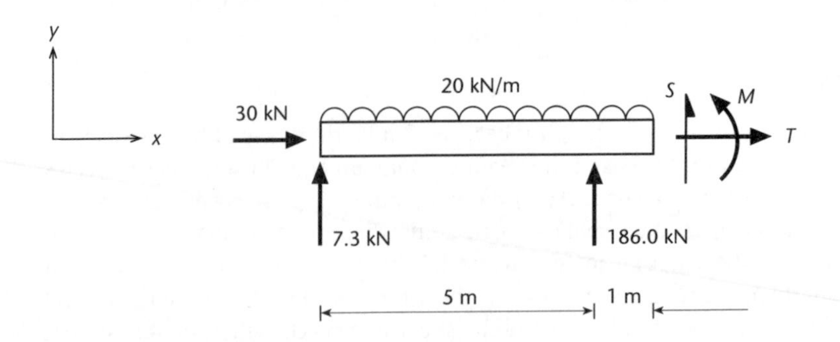

Resolving horizontally:

$$T + 30 = 0 \qquad \rightarrow T = -30.0 \text{ kN}$$

Resolving vertically:

$$S - (20 \times 6) + 7.3 + 186.0 = 0 \qquad \rightarrow S = -73.3 \text{ kN}$$

Moments about right end:

$$M + (20 \times 6^2/2) - (7.3 \times 6) - (186.0 \times 1) = 0 \qquad \rightarrow M = -130.2 \text{ kNm}$$

The negative values mean that the axial force in the beam is compressive, the resultant shear force acts downwards on the face shown and the beam is hogging at the point under consideration.

2.3.2 Relations between load, shear force and bending moment

Figure 2.18 shows a small element of length dx cut from a beam. It is loaded by a force on the top surface of average intensity w per unit length, acting downwards. To keep the element in equilibrium it is necessary to apply both shear forces and bending moments, which change in value from one side of the element to the other. It is worth emphasising a couple of points about the sign convention here:

- The internal forces shown are all positive. Note that the positive shear forces create an *anti-clockwise* couple about the centre of the element – this can provide a convenient way of remembering the correct convention, and one that gives more of a physical feel for the behaviour.
- The downwards external load w is taken as positive even though it acts in the negative y direction. This is not strictly in accordance with our sign convention, but it is worth introducing this small inconsistency for the sake of convenience – otherwise we would need to incorporate a negative sign every time we applied a gravity load to a structure.

Resolving vertically:

$$S + \mathrm{d}S - S - w\,.\,\mathrm{d}x = 0$$

so:

$$\frac{\mathrm{d}S}{\mathrm{d}x} = w \qquad (2.8)$$

Taking moments about the right-hand edge:

$$M + \mathrm{d}M - M + S\,.\,\mathrm{d}x + \frac{w\,.\,\mathrm{d}x^2}{2} = 0$$

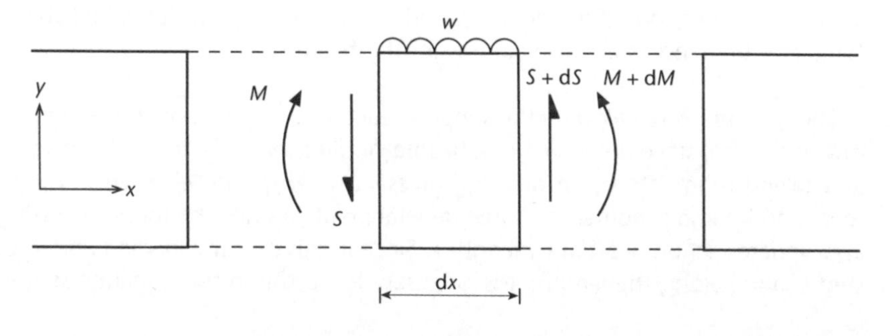

Figure 2.18
Forces on an elemental length of beam.

Now dx, dM etc. are very small quantities. Products or higher powers of these terms will be far smaller than the other terms and so can be neglected, leaving

$$\frac{dM}{dx} = -S \tag{2.9}$$

and substituting from Equation (2.8) gives

$$\frac{d^2M}{dx^2} = -w \tag{2.10}$$

Of these useful relationships, the one given by equation (2.9) is particularly important. It tells us that the rate of change of bending moment along a member is equal in magnitude but opposite in sign to the shear force. Therefore, *the value of the bending moment will be a maximum or minimum when the shear force is zero.* Exceptions to this rule may arise if the maximum value occurs at the end of a beam, or if there is a sudden change in the bending moment caused by a concentrated moment.

Equation (2.9) can be integrated to give

$$M_2 - M_1 = \int_{x_1}^{x_2} -S dx$$

That is, the change in bending moment between any two points is given by minus the integral of the shear force between those points. Again, an exception must be made for the case where a couple is applied between the two points considered.

2.4 Shear force and bending moment diagrams for beams

As we have already seen, a beam is a structural element which supports loads applied normal to its axis, so that bending and shear are the dominant internal forces. One of the early steps in beam design is a plot of the distribution of shear forces and bending moments along the length of the member. This can be done by considering equilibrium of a section, as in Example 2.4. However, instead of using a fixed length, we now analyse a section of variable length x. The procedure is best illustrated by an example.

EXAMPLE 2.5

Figure 2.19(a) shows a simply supported beam (that is, one having a pinned support at one end and a roller at the other) carrying a point load. Plot the shear force and bending moment diagrams for the beam.

The first step is to calculate the support reactions. The horizontal reaction at A is obviously zero, since there are no horizontal applied loads. Then resolving vertically and taking moments about one end gives $V_A = Wb/l$ and $V_B = Wa/l$. The shear force and bending moment at some general point can now be found by taking an appropriate section of arbitrary length x. Since in this case the loading undergoes a step change along the length, it is necessary to do this in two separate stages.

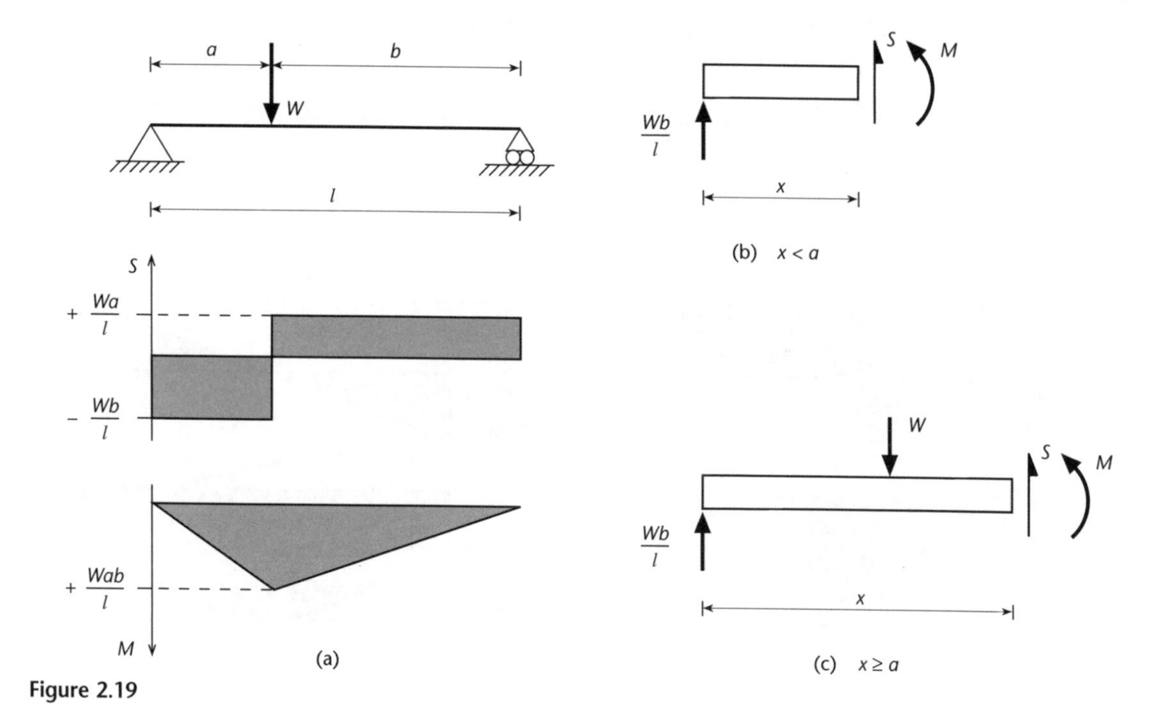

Figure 2.19

$x < a$: The free-body diagram in Figure 2.19(b) shows the forces acting on the beam.

Resolving vertically: $S + Wb/l = 0 \quad \rightarrow S = -Wb/l$

Taking moments about the cut point: $M - Wbx/l = 0 \quad \rightarrow M = Wbx/l$

$x \geq a$: The free-body diagram is shown in Figure 2.19(c). Proceeding as above gives:

$S = W - Wb/l = Wa/l$

$M = Wbx/l - W(x - a)$

The shear force and bending moment diagrams can now be plotted. It is most convenient to draw them directly under a sketch of the beam, as shown in Figure 2.19(a). It is a good practice to mark the significant values and their locations clearly on the diagrams.

Note that, while shear forces are plotted with the positive axis upwards in the conventional way, bending moments are plotted positive downwards. This means that, if superimposed on the axis of the beam, a sagging (positive) moment plots below the beam and a hogging moment above. Now, referring to Figure 2.16(e) and (f), we can see that a sagging curvature causes tension on the bottom face of the beam, while hogging causes tension at the top. So, using our convention, *the bending moment always plots on the side of the beam that it is tension*. This convention can provide an extremely useful check on results, and also simplifies the plotting of bending moment diagrams for more complex structures, as we shall see in the next chapter.

Consider the form of the bending moment and shear force diagrams in Figure 2.19. The shear force is constant and negative to the left of the point load, but undergoes a step change at the point of load application. The bending moment

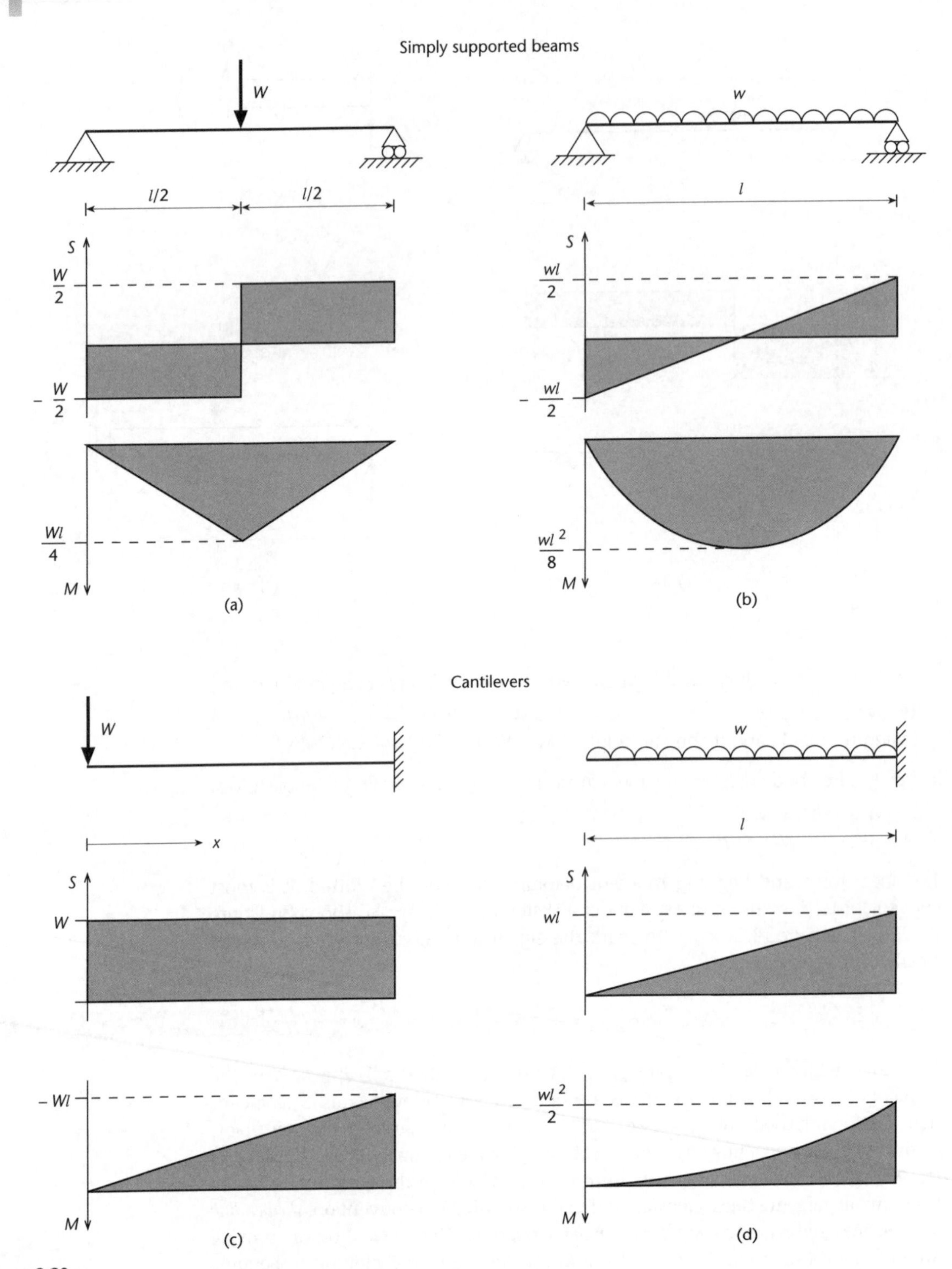

Figure 2.20
Shear force and bending moment diagrams for some common loadcases.

diagram is positive (sagging) throughout, with the maximum value occurring under the load, where there is also a sudden change from a positive to a negative slope. It can be seen that the shear force is always equal to minus the differential of the bending moment; this is in agreement with the relationship derived in the previous section, increasing our confidence that the diagrams have been drawn correctly.

The shear force and bending moment diagrams for some other simple cases are shown in Figure 2.20. These can be regarded as standard results, which are worth committing to memory as they come up so often in structural analysis. Case (a) is simply a special case of the more general result proved in the preceding example, with the point load applied at the centre of the simply supported span. This results in a symmetrical bending moment diagram and an anti-symmetric shear force diagram.

In (b) the simply supported span is subjected to a UDL of w per unit length. This results in equal vertical reactions of $wl/2$. The shear force at a distance x from the left-hand end is then $-w(l/2 - x)$ and the bending moment is $w(lx - x^2)/2$. Thus there is a linear variation of shear force across the structure, and a parabolic distribution of bending moment.

Diagrams (c) and (d) show the results for cantilevers, that is, beams that are rigidly fixed at one end and free at the other. The cantilevers could be drawn either way round, but it is probably simplest to draw them as shown, with the variable distance x measured from the free end. The bending moment and shear force calculations for the arbitrary section of length x then involve only the applied loads, not the support reactions. As with the simply supported beam, the point load case (c) gives constant shear force and linearly varying bending moment, though this time the beam is hogging throughout. The uniform load (d) results in a linearly varying shear force and a parabolic bending moment diagram, again hogging. Note that, unlike a pinned or roller support, a fixed support is capable of sustaining a moment, so that the bending moment does not reduce to zero at the right-hand end.

EXAMPLE 2.6

Figure 2.21(a) shows the beam whose reactions we determined in Example 2.1. Calculate the shear force and bending moment diagrams.

Again we must split the analysis into two stages:

$x < 5$ m (Figure 2.21(b)): Shear force $S = 40x - 70$

Bending moment $M = 70x - 40x^2/2$

$x \geq 5$ m (Figure 2.21(c)): $S = 40 \times 5 - 70 - 190 = -60$

$M = 70x + 190(x - 5) - 200(x - 2.5) = 60x - 450$

The shear force and bending moment diagrams can now be drawn, as shown in Figure 2.21(a). To complete the solution, it is necessary to identify all the key points on the diagrams, such as maxima and zeroes.

The shear force is zero when $40x - 70 = 0$, that is at $x = 1.75$ m. Since the shear force is equal to the differential of the bending moment, this is also the location of the maximum bending moment, hence $M_{max} = 70 \times 1.75 - 40 \times 1.75^2/2 = 61.25$ kNm. Lastly, the bending moment is zero at the beam ends, neither of which can support a moment, and also when $70x - 40x^2/2 = 0$, that is at $x = 3.5$ m. This point is known as a *point of contraflexure*.

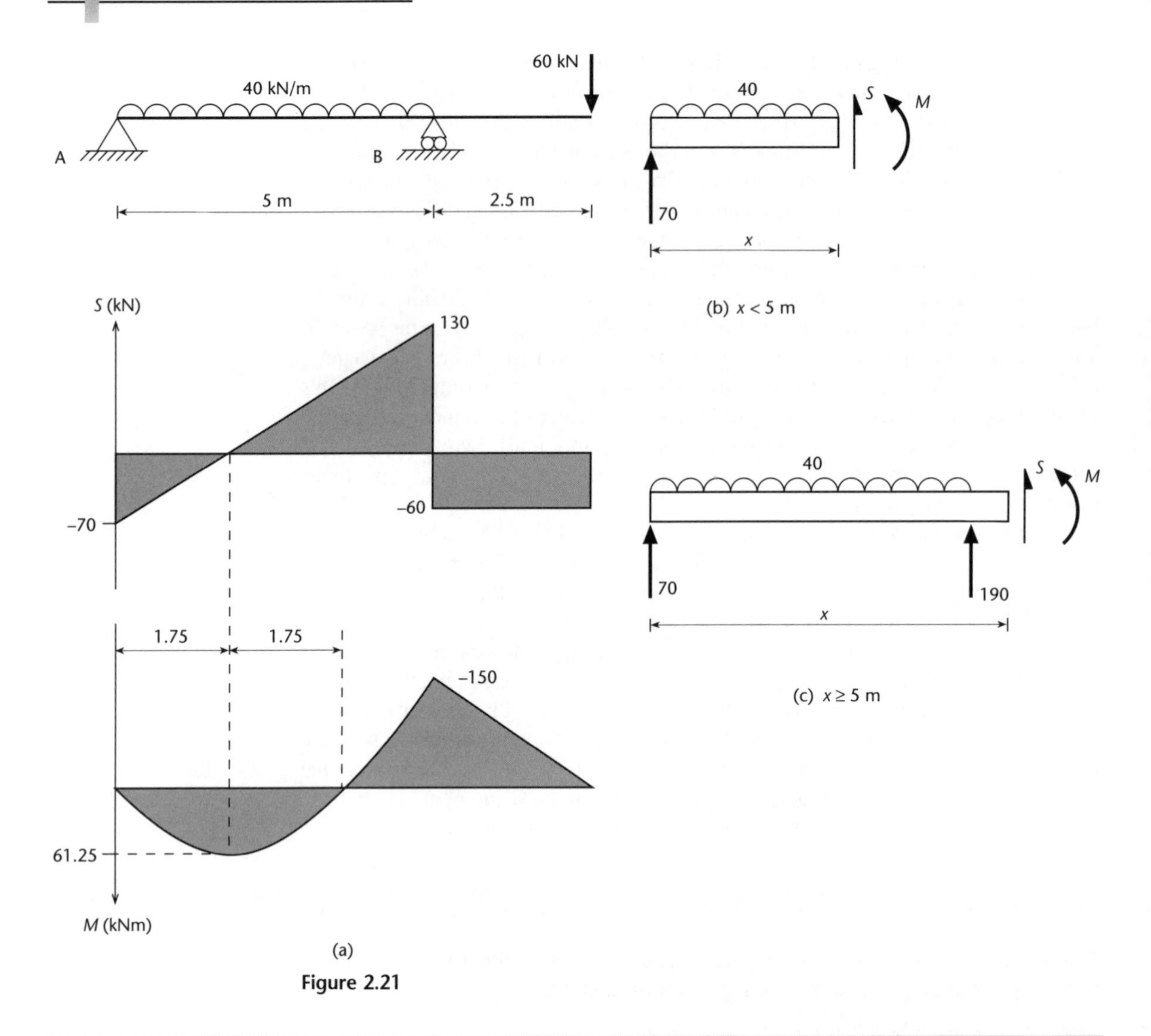

Figure 2.21

The ways in which shear forces and bending moments vary with loading should now be clear. For example, in Figure 2.21 the shear force starts as negative and equal in magnitude to the vertical reaction at A. It then increases linearly under the UDL, its gradient equal to the magnitude of the load. At B it undergoes a sudden decrease equal in magnitude to the reaction force applied there. It then remains constant until another point load is encountered at C, which returns the shear force to zero. The bending moment starts at zero, since a pinned end can carry no moment, varies parabolically under the UDL, undergoes a sudden reversal in slope at B, then goes linearly to zero at C.

2.4.1 Use of superposition

The principle of superposition, introduced earlier, can often be used to good effect in the production of shear force and bending moment diagrams. For any case, it would be possible to draw individual diagrams for each load applied to a beam separately, then sum the diagrams to get the total effect when all the loads are applied simultaneously. For the majority of cases, this would be a rather tedious approach, so it is necessary to be selective.

EXAMPLE 2.7

Find the bending moment diagram for the simply supported beam shown in Figure 2.22(a).

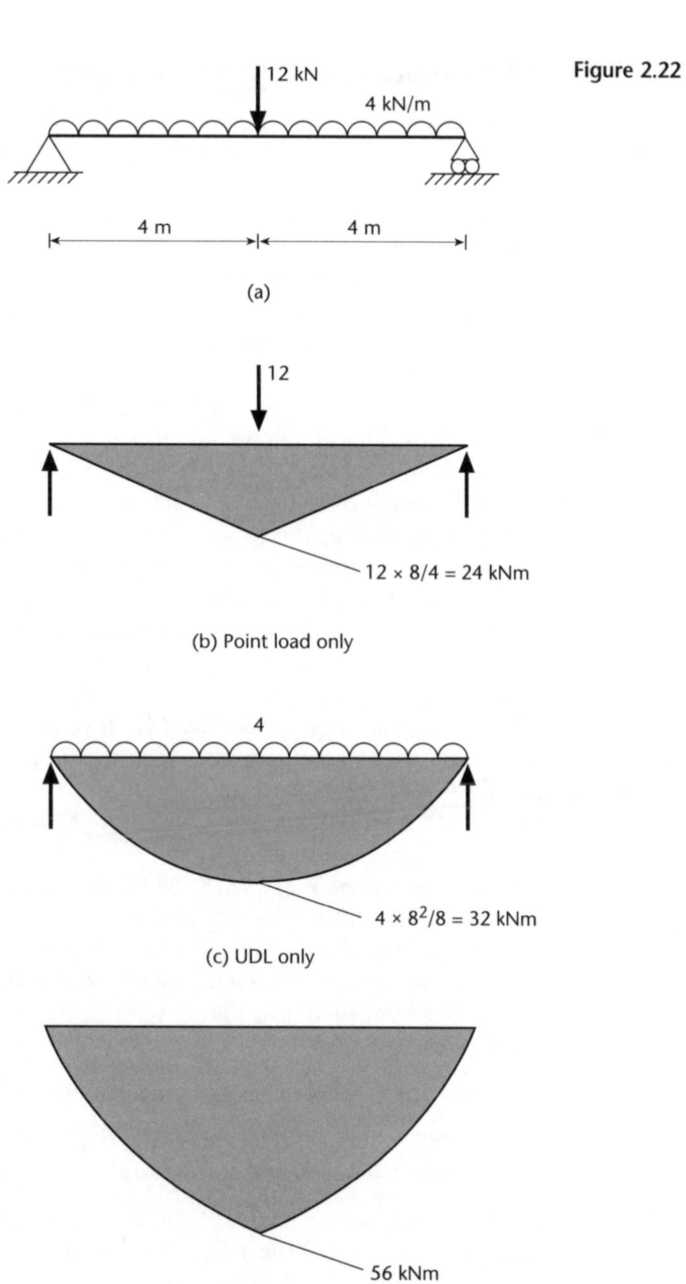

Figure 2.22

The solution using superposition is shown in the figure. The loadcase is split into two parts, a central point load and a uniform load. For each part, the bending moment diagram can be plotted using the standard results of Figure 2.20; these are shown superimposed on the loading diagrams. The total bending moment diagram due to the combined loading is then found by simply summing the two effects, and is plotted at the bottom.

2.4.2 Relating bending moment diagrams to deflected shapes

While we are not yet in a position to calculate beam deflections, it is useful to have a qualitative idea of how the deformed shape of a beam corresponds to its bending moment diagram. The most important rule here is that the bending moment is proportional to the beam curvature, that is, the second differential of its deflection. This means that the beam will curve most where the bending moment is a maximum, and will be straight at a point of contraflexure. An ability to sketch bending moment diagrams and deflected shapes is extremely valuable, as it gives a good feel for structural behaviour and is an excellent way of checking that calculated values are reasonable.

Some examples are shown in Figure 2.23, in which dimensions and load values have deliberately been omitted in order to emphasise the qualitative nature of the exercise. Figure 2.23(a) shows a simply supported beam with a central point load. We have already seen that the bending moment diagram for this structure varies linearly from zero at the two supports to a maximum at the load point; this is shown below the structure. Since the two are directly proportional, it follows that the curvature varies in the same way – at the supports it is zero, that is the beam is straight, and at midspan the curvature is maximum. From symmetry, the deflection is also maximum at the centre. The deflected shape is therefore as indicated by the dashed line.

In case (b) the concentrated load is off-centre. The bending moment diagram again consists of straight lines between the supports and the load point, and the curvature must again vary in proportion to the bending moments. However, because of the lack of symmetry, we can no longer identify the point of maximum deflection without detailed analysis.

Case (c) is a *propped cantilever*, loaded only on the overhanging portion at the right-hand end. This is a redundant structure which we are not yet in a position to analyse mathematically, but we can still perform a qualitative analysis. With redundant structures, it is often easier to start with the deflected shape and then deduce the form of the bending moment diagram, rather than the other way round. We know that:

- there must be no deflection or rotation at the fixed end;
- no vertical deflection is possible at the roller support;
- the beam is bound to bend downwards under the load.

It is then a fairly simple matter to sketch the deflected shape, shown dashed on the figure. We can now use the deflected shape to help us sketch the bending moment diagram. Starting from the right:

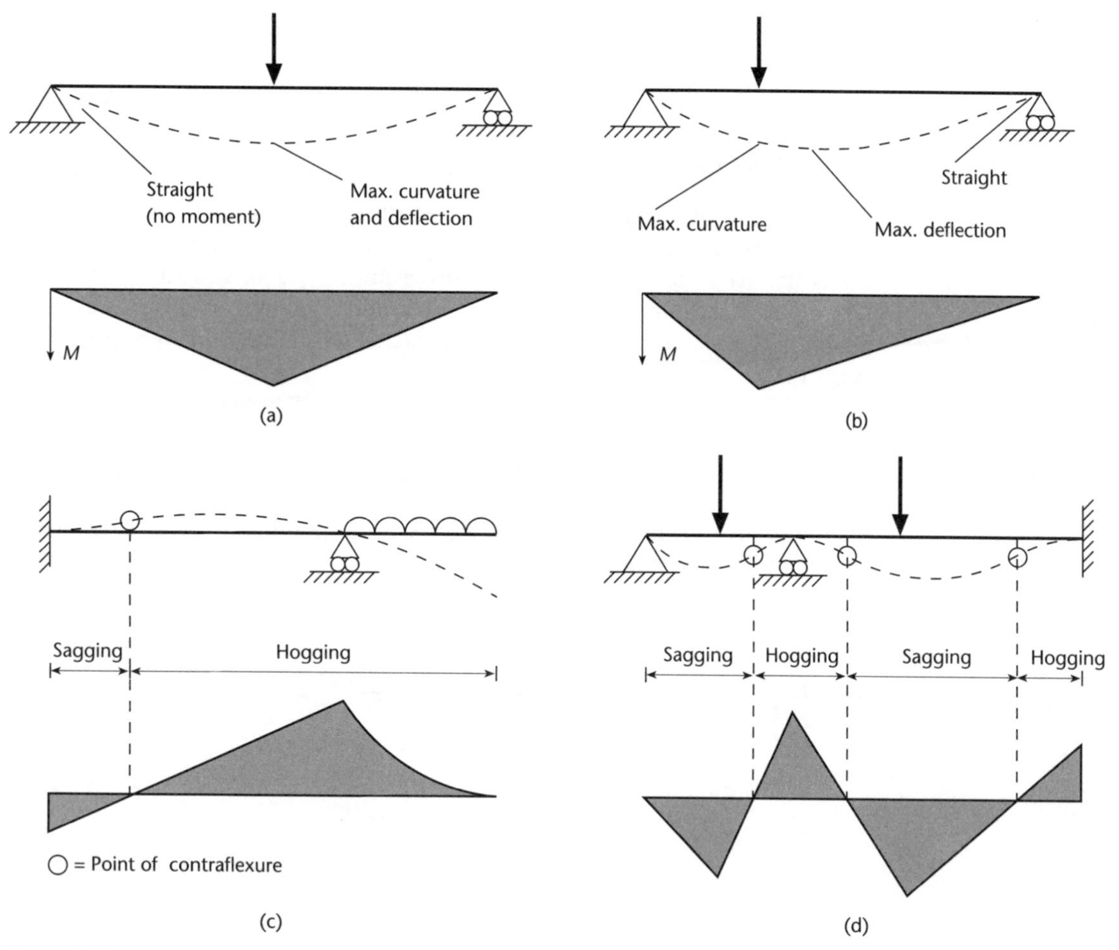

Figure 2.23
Sketched bending moment diagrams and deflected shapes.

- the bending moment must be zero at the free end;
- it must become increasingly negative towards the support, as the beam is hogging;
- the bending moment distribution caused by the uniform load must be parabolic;
- to the left of the roller the bending moment can only vary linearly, as the beam is not loaded;
- it must be zero at the point of contraflexure – the point where the curvature of the beam changes from hogging to sagging.

The bending moment diagram is therefore as shown, with a small sagging moment at the fixed support.

Case (d) is a rather more complex structure, but can be tackled in much the same way. The deflected shape can be sketched by noting that there is zero deflection at the pin and roller, and complete fixity at the right-hand end. It is then obvious that the beam is sagging over most of its length and hogging near the roller support and the fixed end. With only point loads applied to the structure, the bending moment diagram must consist of straight lines between the supports and the load points, and can therefore be drawn as shown. We can see that there are three points of contraflexure in the beam, where the bending moment is zero and the curvature changes between sagging and hogging.

2.4.3 Minimisation of bending moments

If the designer of a structure has some flexibility over where the supports can be located, then this can be used to advantage to reduce the bending moments that the beams have to carry, and so produce a more efficient structure. Since both hogging and sagging moments can cause failure, the aim of this process is to minimise the magnitude of the bending moments, whether they are positive or negative.

EXAMPLE 2.8

For the beam shown in Figure 2.24, position the supports so as to minimise the maximum bending moment in the beam.

Figure 2.24

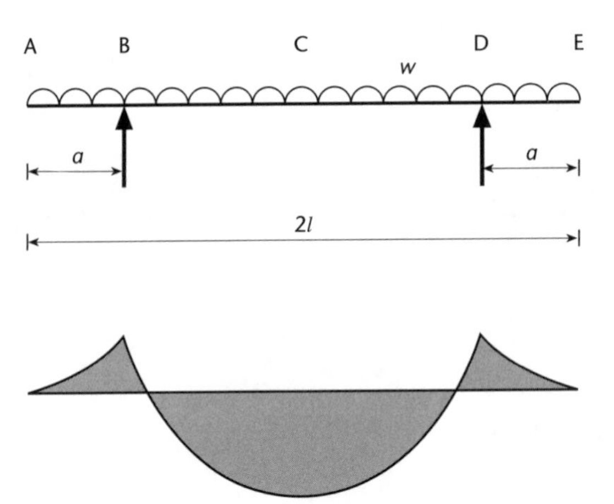

The general shape of the bending moment diagram is shown in the figure, the largest hogging moment occurring at the supports, B and D, and the maximum sagging at midspan, C. The values are:

$$M_B = -wa^2/2$$
$$M_c = -wl^2/2 + wl(l - a) = wl^2/2 + wla$$

Now to minimise the maximum bending moment in the beam, the supports must be positioned so that the largest hogging and sagging moments are of equal magnitude. Moving the supports outwards from these optimum positions would increase the sagging moment at midspan, while moving them inwards would increase the hogging moment over the supports. Either way, the largest bending moment in the structure would increase. Therefore, equating M_c and $-M_B$:

$$wl^2/2 + wla = wa^2/2$$

Rearranging:

$$1 + 2(a/l) - (a/l)^2 = 0$$

This quadratic equation can be solved to give:

$$a = l(\sqrt{2} - 1) = 0.414l$$

Substituting back into the moment expressions gives:

$$M_C = -M_B = 0.0858wl^2$$

2.5 Problems

Where appropriate, answers are given at the end of the book. The questions marked with an asterisk are a little more challenging.

2.1. For the beams in Figure 2.25, state whether each is statically determinate, and either calculate the reactions or state the degree of redundancy, as appropriate.

2.2. For each of the plane frames in Figure 2.26, calculate the reactions at A. Take $g = 9.81\ \mathrm{m/s^2}$.

2.3. Determine the support reactions for the beam ABC in Figure 2.27. Draw the shear force and bending moment diagrams, indicating function values and positions of key points such as maxima, zeros etc.

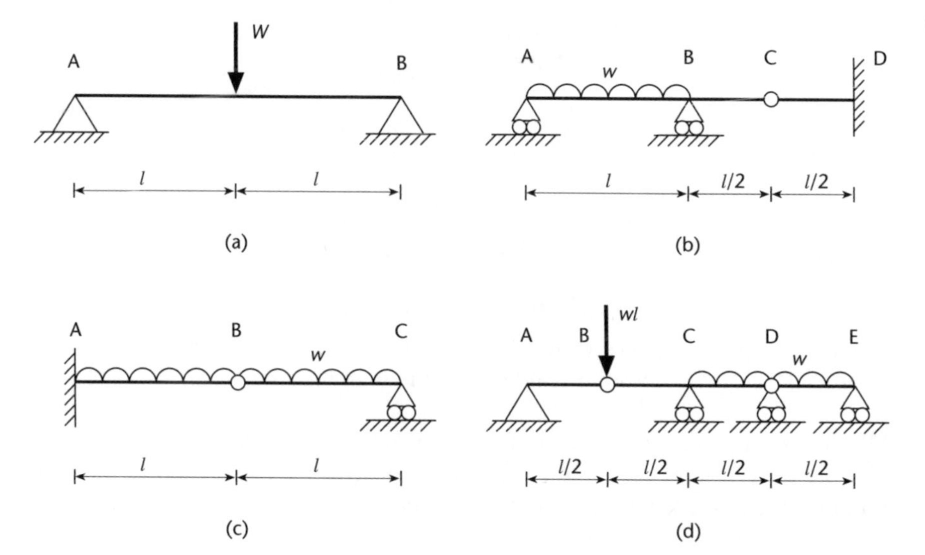

Figure 2.25

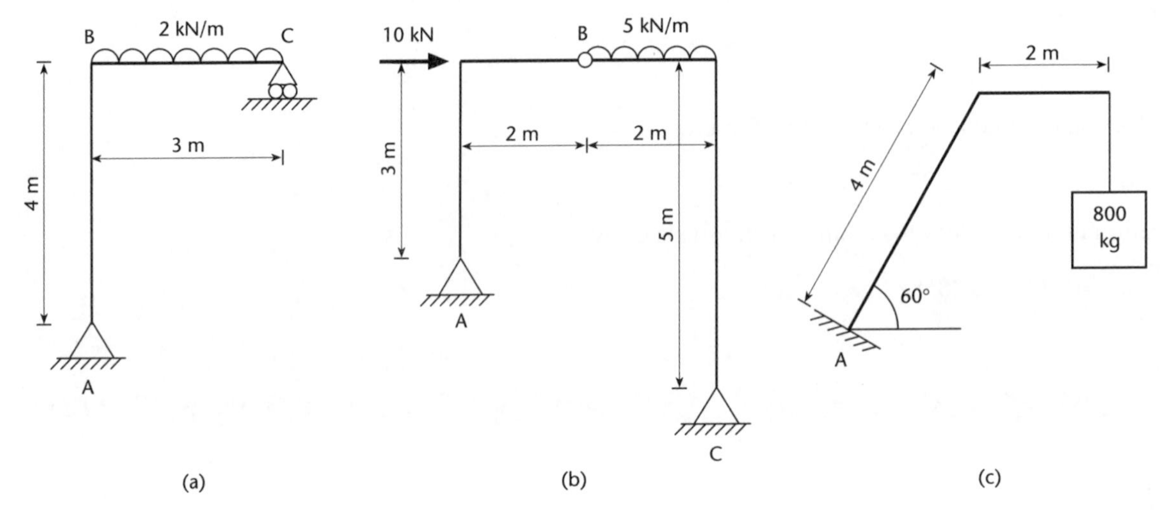

Figure 2.26

Figure 2.27

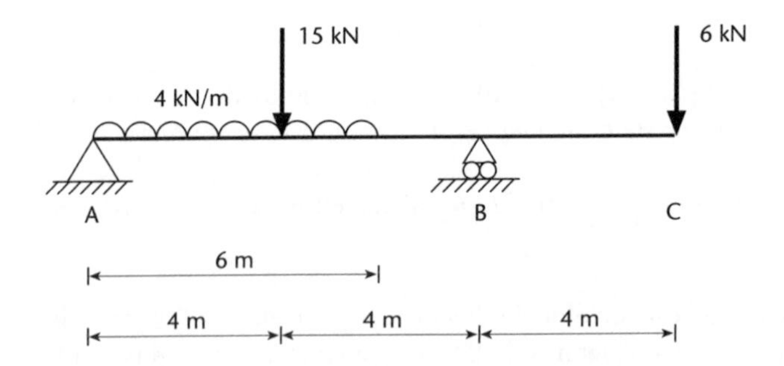

2.4. A cantilever (one end free, one end fixed) of length l carries a uniform downwards load w over its length and an upwards point load P at its free end. Use superposition of standard solutions to find an expression for the bending moment at any point. If the bending moment is zero at the midpoint of the beam, find a relationship between wl and P and sketch the bending moment diagram, indicating maximum values and their locations.

2.5. For each of the beams shown in Figure 2.28, sketch the deflected shape and the shear force and bending moment diagrams, identifying any points of contraflexure.

2.6.* Four bending moment diagrams are shown in Figure 2.29. Sketch the corresponding shear force and free body diagrams.

2.7.* A simply supported beam of length l is symmetrically loaded with a distributed load whose intensity increases from zero at the supports to w per unit length at midspan. Draw the shear force and bending moment diagrams.

2.8. Show that the beam ABCD in Figure 2.30 is statically determinate, and calculate the reactions at the supports. Draw the shear force and bending moment diagrams, indicating key values and their locations.

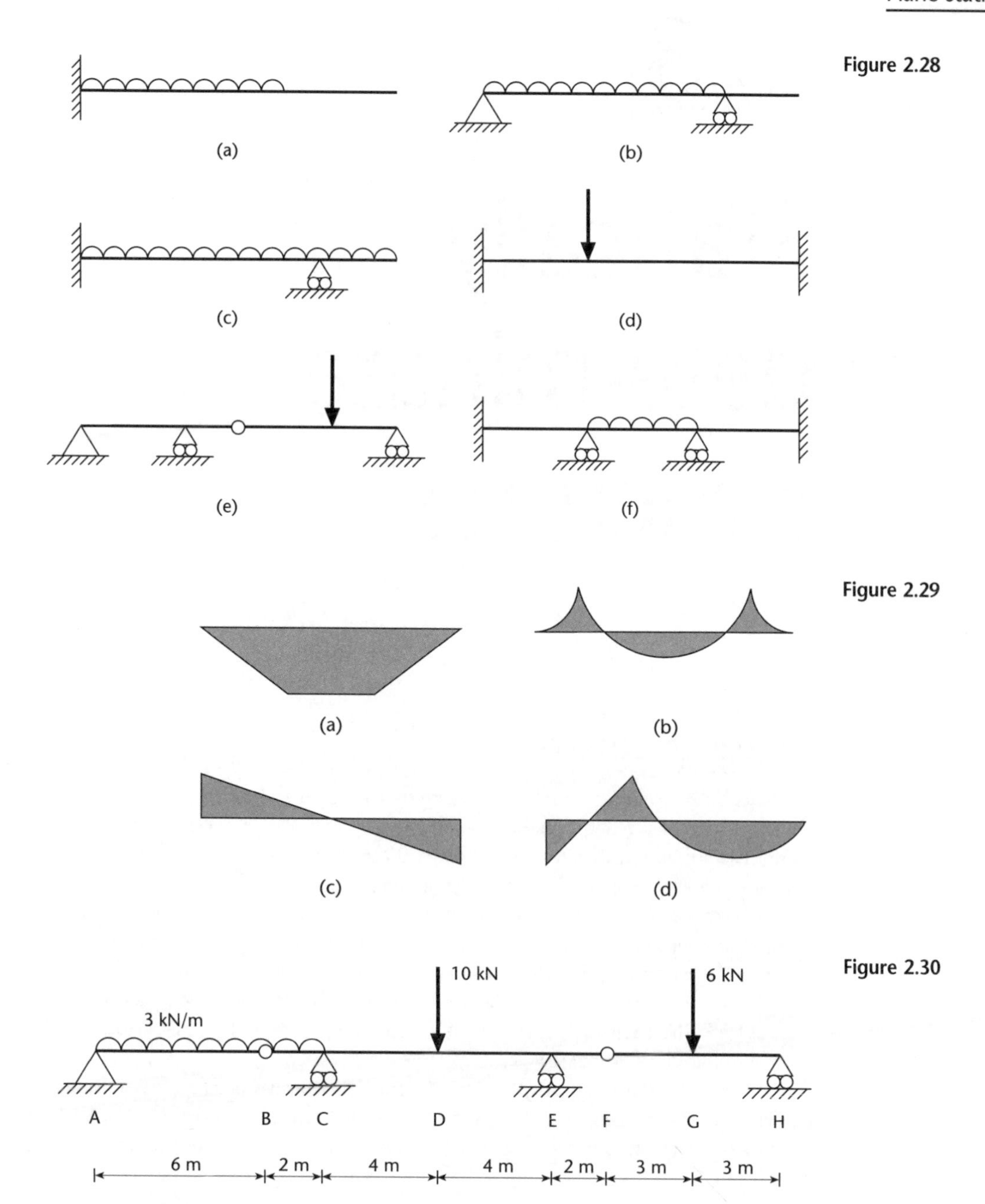

Figure 2.28

Figure 2.29

Figure 2.30

2.9.* A simply supported beam of span l carries a uniformly distributed load w. To reduce the bending moments in the beam, an upwards point load P is applied at the centre. Find the value of P required to reduce the midspan moment to zero, and draw the shear force and bending moment diagrams for this case.

2.10.* For the beam in Problem 2.9, find the value of P required to minimise the maximum bending moment in the beam. Draw the corresponding shear force and bending moment diagrams.

CHAPTER 3

Statically determinate structures

CHAPTER OUTLINE

In Chapter 2 we introduced the basics of statics: the concepts of force and moment equilibrium; statical determinacy; the calculation of support reactions; and the internal forces within structural members (axial forces, shear forces and bending moments). We can now use these simple principles to calculate the internal forces within statically determinate structures.

A statically determinate structure is one for which it is possible to determine all the member forces and support reactions using simple statics. All such structures possess the minimum number of members and reactions required to enable them to carry external loads. If one member or reaction is removed, they become unstable, that is, unable to support even arbitrarily small loads. This makes them rather vulnerable to localised damage, and is the main reason why most real structures are not statically determinate.

We will look at three principal types of structure: pin-jointed frames, or *trusses*, in which all the joints between members are pinned; *moment frames*, in which at least some of the joints are capable of transmitting moments; and cable and arch structures. This chapter will enable the reader to:

- assess the statical determinacy of frame structures;
- calculate axial forces in 2D and 3D trusses;
- understand the basic concepts underlying the computer analysis of statically determinate structures;
- calculate axial forces, shear forces and bending moments in 2D moment frames;
- plot bending moment diagrams for 2D moment frames;
- analyse simple cables and arches.

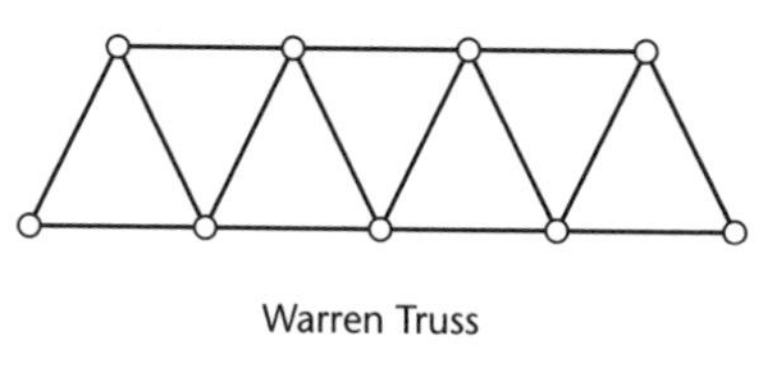

Warren Truss

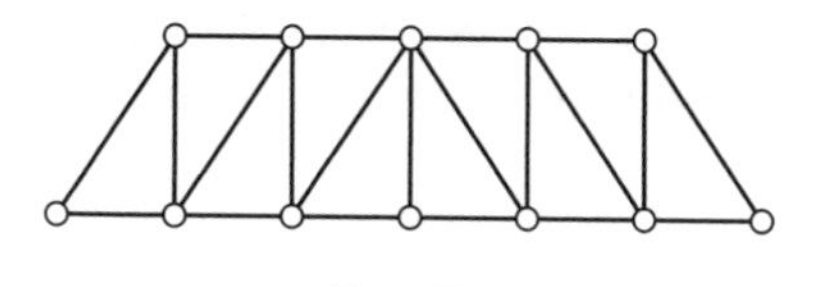

Howe Truss

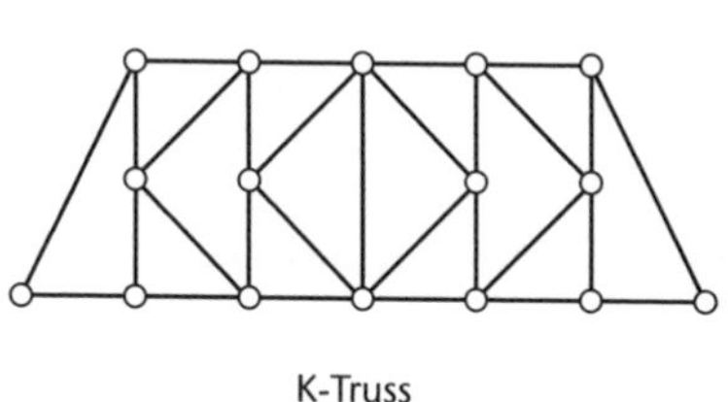

K-Truss

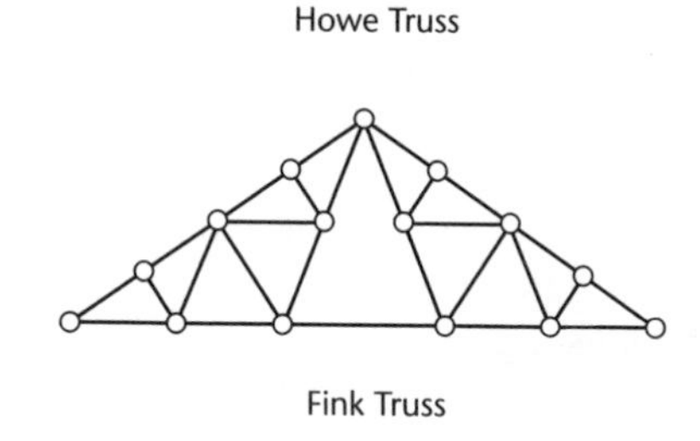

Fink Truss

Figure 3.1
Plane trusses.

3.1 Plane pin-jointed trusses

A structure of this type consists of a number of members, usually called *bars*, connected by pins at their ends and all lying in a single plane. Of course, all structures are in reality three dimensional, but in many cases it is possible to treat the problem from a two-dimensional point of view. For instance, many bridges consist of two parallel trusses connected by transverse members carrying the roadway. If the loads transmitted by the transverse members are known, then each truss can be analysed as a 2D system.

In this section we shall consider only pin-jointed frames in which all the applied loads act through the joints. Since the pins can carry no moment, only axial tensile and compressive forces can be developed in the members. If loads were applied between the joints of a particular member, then these would cause shear forces and bending moments to be set up in that member in addition to the axial force.

Figure 3.1 shows some typical pin-jointed trusses. The first three are commonly used in bridges, while the fourth is a roof truss.

3.1.1 Stability and determinacy

A system of forces is statically determinate if the number of unknown forces is exactly equal to the number of equilibrium equations available. We introduced this concept in Chapter 2 and applied it to the support reactions acting on a structure. We now need to extend it to include the internal forces; if all the unknown forces (reactions and internal bar forces) are to be determined by simple statics, then they must together form a statically determinate set.

The number of unknown forces is easy to assess, since each bar carries only an axial force. Therefore, if the frame comprises b bars and is supported by r reactions, the number of forces to be found is $b + r$.

How many independent equations can we formulate to solve for these forces? We saw in Chapter 2 that internal forces can be found by making imaginary cuts through the structure and considering equilibrium of the sections so formed. For a plane truss we can cut out each joint in turn and resolve the forces on the joint in two perpendicular directions. While other equilibrium equations can be formulated, they provide no extra information. So the total number of independent equations is $2j$, where j is the number of joints. This leads to the simple equation that must be satisfied by a statically determinate 2D truss:

$$b + r = 2j \tag{3.1}$$

If $b + r < 2j$ the structure is *unstable* or a *mechanism* – there are insufficient forces to hold the members in place, so that the structure is free to collapse or move uncontrollably.

If $b + r > 2j$ the structure is *statically indeterminate* or *redundant* – the structure is stable but there are not enough equations to enable us to find all the forces by statics alone. Many real structures are redundant; we shall deal with these in later chapters.

EXAMPLE 3.1

Assess the statical determinacy of the three plane trusses shown in Figure 3.2.

Figure 3.2

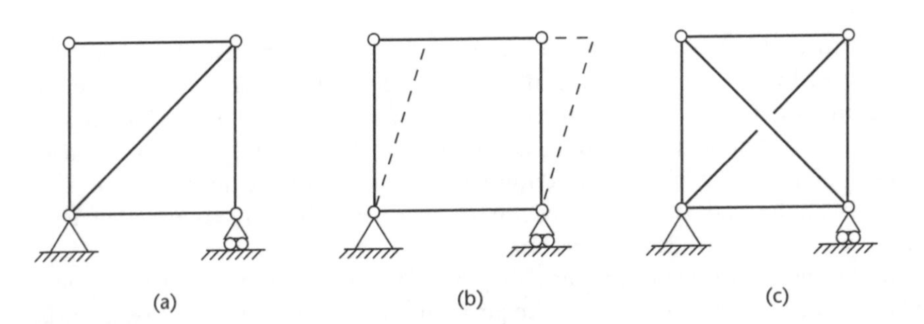

(a) The structure has four joints, comprises five bars and is supported by three reactions (two at the pinned support, one at the roller). So:

$$b + r = 5 + 3 = 8, \quad \text{and} \quad 2j = 2 \times 4 = 8$$

The structure satisfies equation (3.1) and so *may* be statically determinate. However, we need a more rigorous check to make sure; this is discussed below.

(b) The diagonal bar has been removed, leaving:

$$b + r = 4 + 3 = 7, \quad \text{and} \quad 2j = 2 \times 4 = 8$$

The frame is a mechanism which, with the application of a very small lateral force or displacement, could collapse as shown by the broken lines.

(c) The frame has two diagonals, so that:

$$b + r = 6 + 3 = 9, \quad \text{and} \quad 2j = 2 \times 4 = 8$$

It is therefore redundant and cannot be analysed using statics alone.

It is important to be aware of the limitations of equation (3.1). All statically determinate 2D frames will satisfy this equation, but so will some structures which

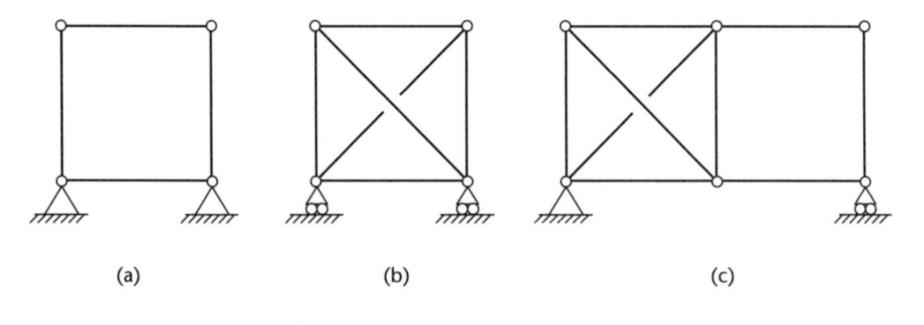

Figure 3.3
Frames which are part-redundant, part-mechanism (all satisfying $b + r = 2j$*).*

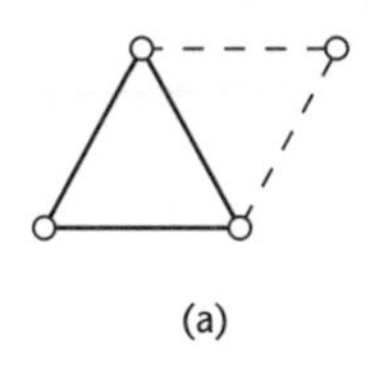

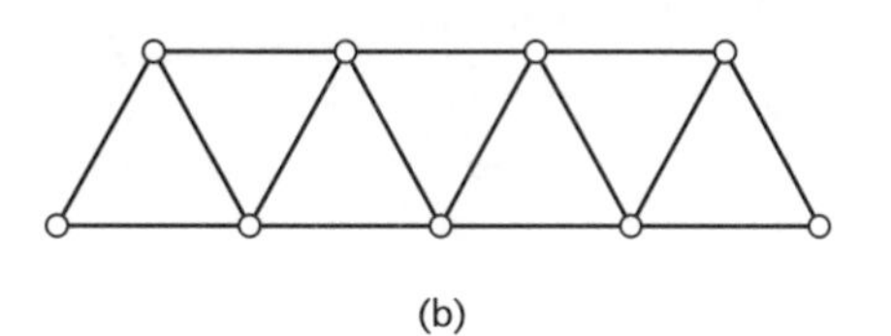

Figure 3.4
Determinacy of a frame.

are not statically determinate. It is easy to see that the three plane trusses in Figure 3.3 all satisfy $b + r = 2j$, but none of them is statically determinate! In (a) the reactions are redundant but the frame is a mechanism, which could fold up if a small horizontal disturbance was applied to the top bar. In (b) the frame is redundant but the supports are insufficient to prevent large horizontal movements of the whole structure. In (c) one side of the structure is a mechanism and the other is redundant. Thus, when assessing statical determinacy it is necessary to consider the determinacy not only of the structure as a whole, but also of the reactions alone (see Chapter 2), the frame alone, and even of small parts of the frame.

Various authors have suggested further formulae to deal with the determinacy of the frame alone. However, these tend to cause more confusion rather than less, so it is generally better to use an intuitive approach. Most trusses are based on a triangular system. Suppose we start with a single triangle ABC (Figure 3.4(a)). Given appropriate support conditions, this is statically determinate. If we wish to connect a fourth joint D to the basic triangle then, to maintain statical determinacy, we must do so using two bars. We can go on adding further joints, so long as each new joint is accompanied by two additional bars. We can therefore see that the frame in Figure 3.4(b) is statically determinate, and the full structure will therefore be determinate so long as a suitable set of reactions is provided.

EXAMPLE 3.2

Suggest alterations to the two frames shown in Figure 3.5 in order to make them statically determinate.

(a) $b = 8$, $r = 4$, $j = 6$, so $b + r = 2j$ is satisfied. However, only three equations can be found in terms of the four reactions, so the reactions are indeterminate. Also, working through the frame starting from the triangle ABD, we immediately find that joint E is inadequately supported. So to make the structure statically determinate we must change either of the supports to a roller and insert a new bar between B and E.

Figure 3.5

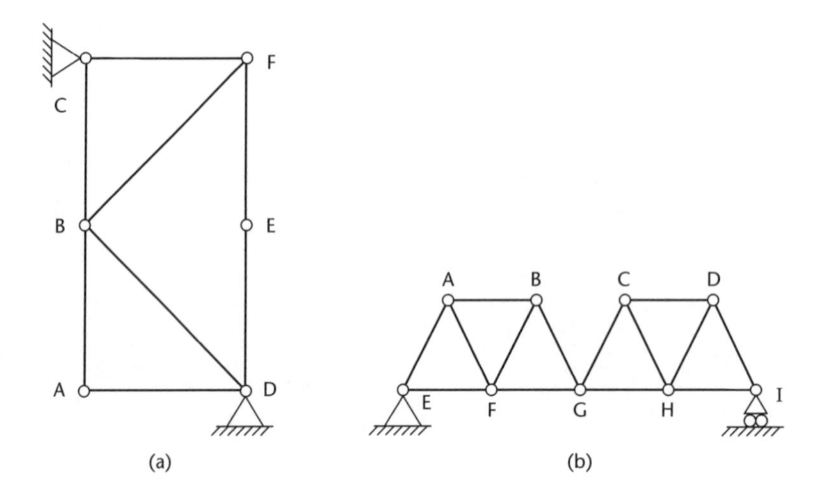

(b) $b = 14$, $r = 3$, $j = 9$, so the structure is a mechanism, requiring one additional member or reaction. At first glance, the three reactions appear to form a statically determinate set. However, an equation of condition can be generated by splitting the frame at G, so an additional reaction is required. Looking at the frame alone, if we work through starting from the triangle AEF we can get as far as G by adding two bars for each joint, but then this approach fails.

There are two possible remedial approaches. *Either* add a bar between B and C – this makes the frame determinate and also prevents the structure from being split at G, thus eliminating the equation of condition. *Or* add an additional support, such as a roller at G. This effectively performs the same function as the bar BC – both serve to prevent the two halves of the frame from rotating about the outer supports, with a large vertical deflection at G.

3.1.2 Resolution at joints

Having established that a frame is statically determinate, there are then various ways of going about calculating the member forces. Probably the simplest of these is resolving at joints. We have already introduced the idea of making imaginary cuts through members or structures and considering equilibrium of the resulting segments; we can do this so long as the internal member forces at the cuts are treated as external loads. We shall now apply this approach by cutting out a single joint at a time from the structure, and considering equilibrium of the forces exerted on it by the bars meeting at the joint. We can resolve in two perpendicular directions (usually, but not always, horizontally and vertically) at each joint. The method is best demonstrated by some examples.

EXAMPLE 3.3

Find the forces in the Warren truss in Figure 3.6(a). All members have equal length, so that the inclined bars make an angle of 60° to the horizontal.

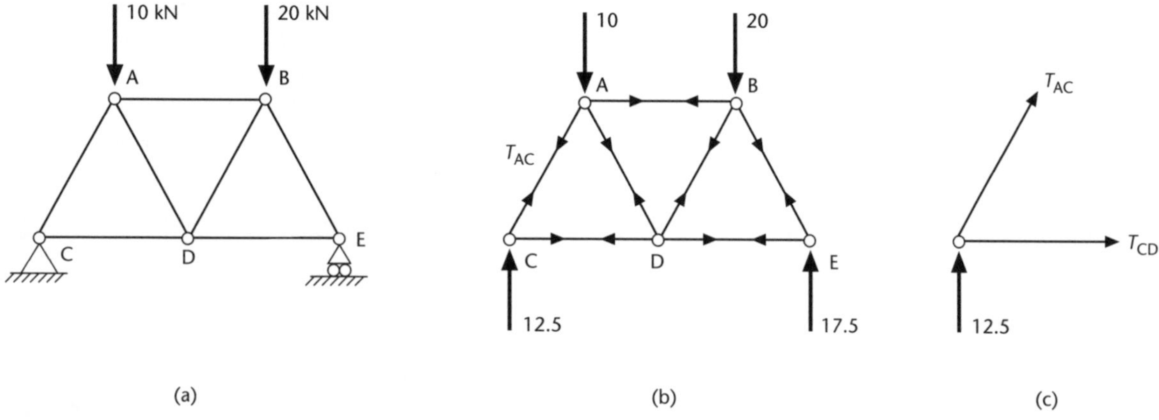

Figure 3.6

First find the support reactions:

Resolving horizontally: $H_C = 0$

Moments about E: $20 \times 0.5 + 10 \times 1.5 - V_C \times 2 = 0 \quad \rightarrow V_C = 12.5$ kN

Resolving vertically: $10 + 20 - 12.5 - V_E = 0 \quad \rightarrow V_E = 17.5$ kN

We can now go on to find the axial forces in the bars. Since we do not know which are tensile and which compressive, we initially assume they are all tensile (positive). If any force is compressive, this will simply result in a negative sign in the calculations. We start by marking the internal forces as arrows on the members (Figure 3.6(b)). Remember that a tensile internal force pulls inwards, reacting against the external loads which are causing it to stretch. Therefore all of our force arrows are directed *away* from the joints. We can now work round the structure resolving in two directions at each joint, and so find the internal forces. In doing this it can be helpful to draw the free body diagram for each joint, as shown for joint C in Figure 3.6(c). In practice, however, it is usually just as easy to work directly from the diagram for the full frame. The process can be conveniently tabulated, as shown below.

Joint and direction	Equation	Result
C vertical	$12.5 + T_{AC} \sin 60° = 0$	$T_{AC} = -14.43$ kN
C horizontal	$T_{CD} + T_{AC} \cos 60° = 0$	$T_{CD} = 7.22$ kN
A vertical	$T_{AC} \sin 60° + T_{AD} \sin 60° + 10 = 0$	$T_{AD} = 2.88$ kN
A horizontal	$-T_{AC} \cos 60° + T_{AD} \cos 60° + T_{AB} = 0$	$T_{AB} = -8.66$ kN
E vertical	$17.5 + T_{BE} \sin 60° = 0$	$T_{BE} = -20.21$ kN
E horizontal	$T_{DE} + T_{BE} \cos 60° = 0$	$T_{DE} = 10.10$ kN
D vertical	$T_{AD} \sin 60° + T_{BD} \sin 60° = 0$	$T_{BD} = -2.88$ kN

It can be seen that all the forces around the outside of the frame above the support level are compressive, while those along the bottom chord CDE are tensile. This pattern can be related to the deflected shape – as the truss deflects downwards the bottom chord must extend and the top part of the structure will tend to compress.

Note that not all joints and directions have been utilised in the above process. We kept the equations simple by avoiding the more complex joints as far as possible.

For instance, horizontal resolution at D would involve four bar forces. We could use resolution at some of these additional locations to check the results, as errors are easy to make. This is always worth doing, but especially so if the forces calculated cannot easily be related to the likely deflected shape of the structure.

EXAMPLE 3.4

Find the forces in the frame of Figure 3.7(a).

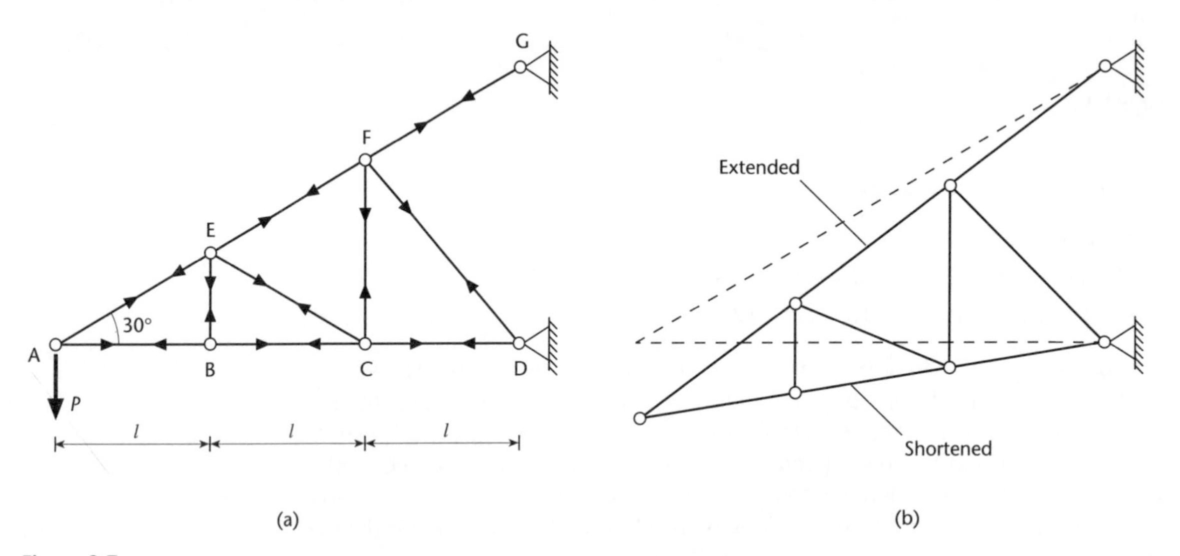

Figure 3.7

It is always worth spending a few moments inspecting the structure prior to starting an analysis. Sometimes it is possible to perform the analysis without calculating the support reactions, or to simplify it by recognising that some members are carrying no load. For this frame we can see at a glance that there is no force in BE (imagine vertical resolution at B). If we then resolve perpendicular to EF at E, it follows that the force in CE is zero. Continuing in this way, we find that all the interior members carry zero load.

It then follows that the forces in AB, BC and CD must all be equal, since all the bars joining these at B and C carry no load. The same is true for AE, EF and FG. There are therefore only two unknown forces, which can be found by resolving at A:

Vertically: $T_{AE} \sin 30° - P = 0 \quad \rightarrow T_{AE} = 2P \quad (= T_{EF} = T_{FG})$

Horizontally: $T_{AE} \cos 30° + T_{AB} = 0 \quad \rightarrow T_{AB} = -P\sqrt{3} \quad (= T_{BC} = T_{CD})$

This time, as the frame is cantilevered out from a wall, the downwards load causes tension along the top chord and compression along the bottom. Again, this accords with the likely deflected shape; as the frame bends downwards the top part is stretched while the bottom chord shortens (Figure 3.7(b)).

It might be wondered why it is necessary to include so many members which carry no force. There are two reasons. First, without them the frame would become unstable, since there would be too few bars to hold all the joints in place – the issue of stability is, of course, entirely independent of the loading. We could get round this problem by making AD and AG into single, continuous members. The second reason is that the forces are zero only under this particular load configuration. Most structures have to withstand many different loads, and certain members may be more useful in resisting some loadcases than others.

3.1.3 Method of sections

The forces in most 2D frames can be found quite simply using resolution at joints, but for some of the more complex structures the method proves fiddly, requiring the solution of many simultaneous equations. In these cases a lot of time and effort can often be saved by switching to another approach. By far the most powerful of the many alternatives is the method of sections.

In this method, rather than looking at the equilibrium of a single joint at a time, the structure is split into two sections by an imaginary cut, with the internal forces in any cut members treated as external loads. These forces are then found by considering the equilibrium of either section.

With a well-chosen section, a very direct solution can often be obtained. The choice is largely a matter of common sense combined with experience. In general, *the section should aim to cut through no more than three bars,* since only three equilibrium equations are available. However, Example 3.6 shows an exception to this rule.

EXAMPLE 3.5

For the Pratt truss in Figure 3.8(a), find the forces in members AB, CD and AD.

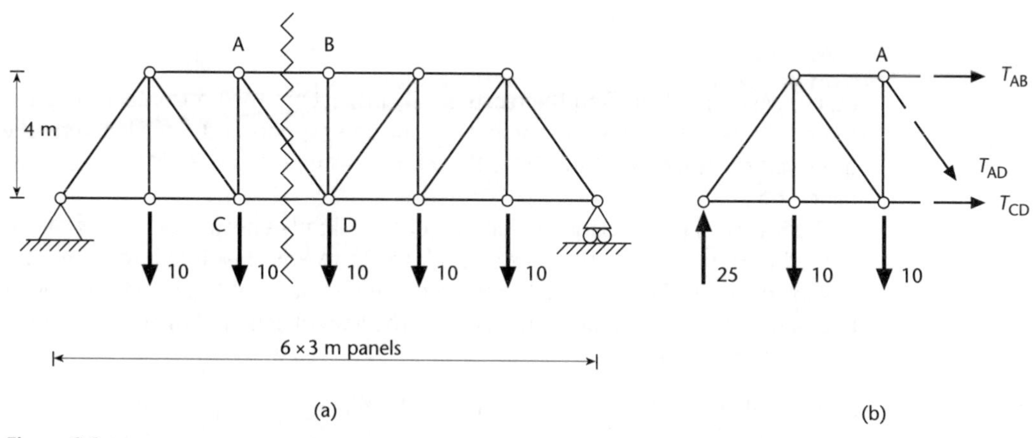

Figure 3.8

We first find the reactions. Since all the external loads are vertical, the horizontal reaction at the pin is zero and, using symmetry, both vertical reactions are equal to 25 kN.

We could then solve the problem using resolution at joints, but this would require us to work through the structure from one end in order to find the three required bar forces. It is far quicker to use the method of sections, making a vertical cut as indicated by the wavy line in the diagram. If we now consider equilibrium of the left-hand section of the frame, the free body diagram is as shown in Figure 3.8(b). Note here that we have again assumed that all the member forces are tensile, so that the forces in the cut bars all act away from the joints.

Having made the section, the calculations are simple. Note that each diagonal member makes an angle θ with the vertical where $\sin\theta = 0.6$, $\cos\theta = 0.8$.

Moments about A: $4T_{CD} + 3 \times 10 - 6 \times 25 = 0 \quad \rightarrow T_{CD} = 30.0$ kN

Resolve vertically: $25 - T_{AD} \times 0.8 - 10 - 10 = 0 \quad \rightarrow T_{AD} = 6.25$ kN

Resolve horizontally: $T_{AB} + T_{CD} + T_{AD} \times 0.6 = 0 \quad \rightarrow T_{AB} = -33.75$ kN

EXAMPLE 3.6

Find the forces in members AB, CD, BE and DE of the K-truss shown in Figure 3.9(a).

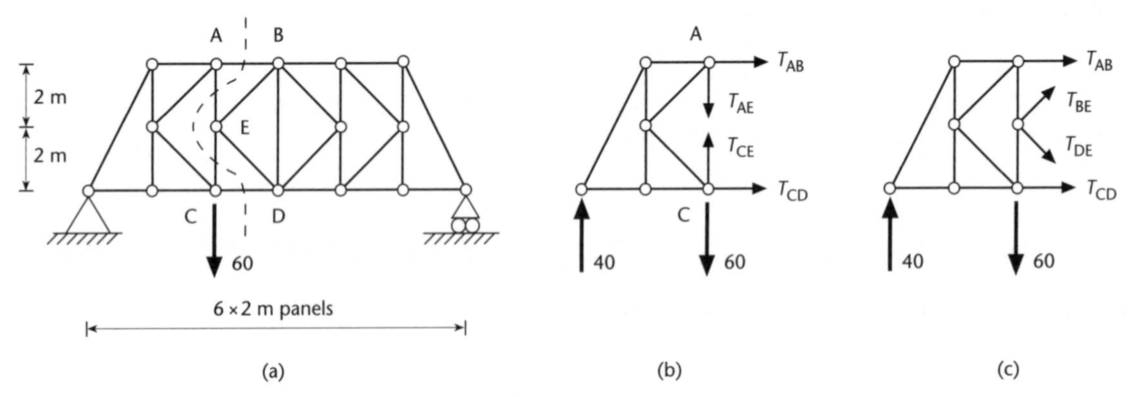

Figure 3.9

Again, we begin by finding the reactions. By inspection, the horizontal reaction at the pin is zero. Then taking moments about one support and resolving vertically gives vertical reactions of 40 kN at the left-hand support and 20 kN at the right-hand end.

The K-truss is rather elaborate, and requires a slightly unusual section, as shown dashed in Figure 3.9(a). This cuts through four bars and therefore introduces four unknown forces (Figure 3.9(b)). However, the geometry of the structure enables two of the four forces to be found directly. The lines of action of three of the forces pass through C, so:

Moments about C: $-4T_{AB} - 4 \times 40 = 0 \quad \rightarrow T_{AB} = -40.0$ kN

Similarly, moments about A: $4T_{CD} - 4 \times 40 = 0 \quad \rightarrow T_{CD} = 40.0$ kN

To find the forces in the diagonals we need a second section, this time obtained by taking a simple vertical cut through the area of interest, Figure 3.9(c). Then:

Resolve horizontally: $(T_{DE} + T_{BE}) \cos 45^\circ = 0$

Resolve vertically: $(T_{DE} - T_{BE})\cos 45° - 40 + 60 = 0$

Solving simultaneously: $T_{DE} = -14.14$ kN, $T_{BE} = 14.14$ kN

3.1.4 Introduction to matrix methods

We introduced the method of sections as a way of avoiding some of the more tedious and error-prone aspects of resolving at joints, particularly the solution of many simultaneous equations. Another way of speeding up the analysis is by using a computer. In fact, we can use the computer not only to solve the simultaneous equations, but also to set them up. However, this does require that the problem be approached in an entirely methodical way.

Returning to the method of joints, suppose we have a statically determinate structure consisting of N bars and R reactions. We know that we can generate $N + R$ equations in these unknowns by resolving in two perpendicular directions at each joint. Alternatively, we can generate N equations in terms of the bar forces alone by steering clear of joints and directions where there are reactions.

Consider a typical joint I, where several bars IJ, IK, IL etc. meet, and external loads are applied, as shown in Figure 3.10. We can write the equilibrium equations for this joint in a rather more methodical way than we have done previously by measuring all bar inclinations anti-clockwise from the positive x-axis. We must also be more rigorous about the way we define the external loads; a positive vertical load must act in the positive y-direction, that is, upwards.

Resolving horizontally:

$$T_{IJ}\cos\theta_{IJ} + T_{IK}\cos\theta_{IK} + T_{IL}\cos\theta_{IL} + \ldots + H_I = 0 \tag{3.2}$$

and vertically:

$$T_{IJ}\sin\theta_{IJ} + T_{IK}\sin\theta_{IK} + T_{IL}\sin\theta_{IL} + \ldots + V_I = 0 \tag{3.3}$$

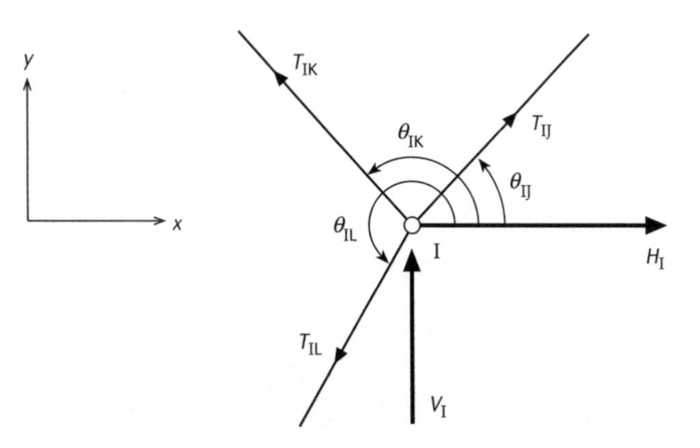

Figure 3.10
Equilibrium of a typical joint.

If we repeat this for all joints and directions where there are no reactions we will then have N equations for the unknown bar forces. We can write these in matrix form as:

$$\begin{bmatrix} \cdots & \cdots & \cdots & \cdots & \cdots & \cdots & \cdots & \cdots \\ \cdots & \cdots & \cos\theta_{\mathrm{IJ}} & \cos\theta_{\mathrm{IK}} & \cos\theta_{\mathrm{IL}} & \cdots & \cdots & \cdots \\ \cdots & \cdots & \cdots & \cdots & \cdots & \cdots & \cdots & \cdots \\ \cdots & \cdots & \cdots & \cdots & \cdots & \cdots & \cdots & \cdots \\ \cdots & \cdots & \cdots & \cdots & \cdots & \cdots & \cdots & \cdots \\ \cdots & \cdots & \sin\theta_{\mathrm{IJ}} & \sin\theta_{\mathrm{IK}} & \sin\theta_{\mathrm{IL}} & \cdots & \cdots & \cdots \\ \cdots & \cdots & \cdots & \cdots & \cdots & \cdots & \cdots & \cdots \\ \cdots & \cdots & \cdots & \cdots & \cdots & \cdots & \cdots & \cdots \end{bmatrix} \begin{bmatrix} \cdots \\ \cdots \\ T_{\mathrm{IJ}} \\ T_{\mathrm{IK}} \\ T_{\mathrm{IL}} \\ \cdots \\ \cdots \\ \cdots \end{bmatrix} = \begin{bmatrix} \cdots \\ -H_{\mathrm{I}} \\ \cdots \\ \cdots \\ \cdots \\ -V_{\mathrm{I}} \\ \cdots \\ \cdots \end{bmatrix} \tag{3.4}$$

or

$$\mathbf{Et} = \mathbf{f} \tag{3.5}$$

Here $\mathbf{E}$, the *equilibrium matrix*, is an $N \times N$ matrix containing sine and cosine terms, $\mathbf{t}$ is an $N \times 1$ vector of bar tensions and $\mathbf{f}$ is an $N \times 1$ vector of external forces. The dots represent other terms that we have not written out in full above. At this stage, it is worth mentioning the conventions used for the various types of quantity: we use *italics* to denote ordinary variables and upright, **bold** characters for matrices and vectors.

Multiplying equation (3.5) by the inverse of $\mathbf{E}$ gives:

$$\mathbf{t} = \mathbf{E}^{-1}\mathbf{f} \tag{3.6}$$

So if we can invert $\mathbf{E}$ then we can immediately find all the unknown bar tensions. Of course, if we cannot invert $\mathbf{E}$ it follows that the structure is not statically determinate.

Writing the equilibrium equations in matrix form makes them amenable to computational solution – there are many numerical algorithms which can solve large matrix equations very efficiently. A computer program to analyse statically determinate frames would normally set up the equations as well as solving them. Typically, the procedure would be as follows:

1 Read in the data – we require:
   - coordinates of all the joints in the structure;
   - a list of bars and the joints they span between;
   - the support conditions – which joints are supported and in which directions the reactions act;
   - the external loads – magnitudes, directions and the joints at which they act.
2 Assemble the vector of unknown bar forces $\mathbf{t}$.
3 Assemble the vector of external forces $\mathbf{f}$ by resolving the external loads in two perpendicular directions at each joint. Omit joints at which there is a pinned support. At roller-supported joints, resolve only perpendicular to the reaction.
4 From the joint coordinates, calculate the angle θ of each bar to the x-axis. Hence assemble the equilibrium matrix, by inserting $\sin\theta$ and $\cos\theta$ terms in the appropriate positions. The row in which a term is positioned corresponds to the joint and direction being considered, and the column corresponds to the bar

force being resolved. For example, the term $\cos\theta_{IK}$ in equation (3.4) arises when resolving the force T_{IK} (row 4 in the bar force vector) in the horizontal direction at joint I (row 2 in the load vector). Its position in the equilibrium matrix is therefore row 2, column 4.

5. Check whether the structure is statically determinate. This requires that the equilibrium matrix must have an inverse, which in turn implies that:

   - **E** must be a square matrix, i.e. the vectors **t** and **f** must have the same dimensions. If **E** has more rows than columns then there are not enough bar forces (the dimension of **t** is less than that of **f**) and the frame is a mechanism. If **E** has fewer rows than columns then there are not enough equilibrium equations and the frame is redundant.
   - The determinant of **E** must not be zero. If it is zero then this implies a local indeterminacy within the frame. For example, the determinant of the equilibrium matrix for the frame in Figure 3.3(c) would be zero.

6 Invert **E** and pre-multiply **f** by $\mathbf{E}^{-1}$ to give the unknown bar forces (equation (3.6)).

7 Having found the bar forces, use resolution at the supported joints to find the reactions, if required.

An important advantage of this approach is that the equilibrium matrix is dependent on the layout of the bars in the structure and on the support conditions, but not on the magnitudes of the applied loads. This means that if, as is often the case, we wish to analyse a structure under several different load configurations we need only set up and invert the equilibrium matrix once. It is then a simple matter to calculate a new external load vector, and hence a new set of bar forces, for each loadcase.

EXAMPLE 3.7

Figures 3.11(a) and (b) show a simple frame under two different sets of forces. Find all the member forces for each loadcase.

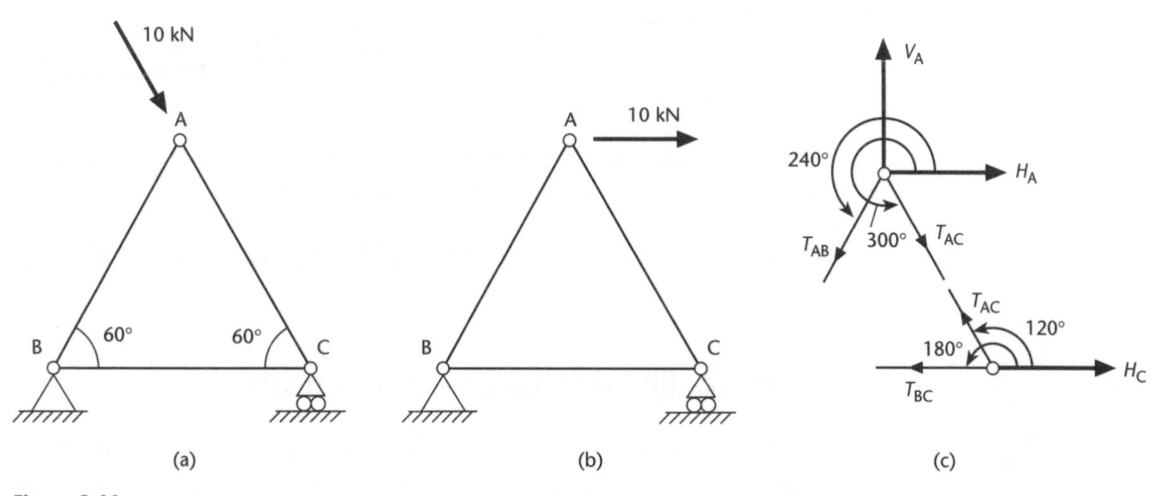

Figure 3.11

There are three unknown bar forces and, avoiding joints and directions where there are reactions, we can generate three equations: horizontal and vertical equilibrium at A, and horizontal equilibrium at C. For the moment, let the external forces at these locations be H_A, V_A and H_C respectively. The relevant bar angles from the positive x-axis are shown in Figure 3.11(c). We can write the equilibrium equations out in full as follows.

Horizontally at A: $T_{AB}\cos 240^\circ + T_{AC}\cos 300^\circ + H_A = 0$

Vertically at A: $T_{AB}\sin 240^\circ + T_{AC}\sin 300^\circ + V_A = 0$

Horizontally at C: $T_{AC}\cos 120^\circ + T_{BC}\cos 180^\circ + H_C = 0$

These can then be written in matrix form as:

$$\begin{bmatrix} \cos 240^\circ & \cos 300^\circ & 0 \\ \sin 240^\circ & \sin 300^\circ & 0 \\ 0 & \cos 120^\circ & \cos 180^\circ \end{bmatrix} \begin{bmatrix} T_{AB} \\ T_{AC} \\ T_{BC} \end{bmatrix} = \begin{bmatrix} -H_A \\ -V_A \\ -H_C \end{bmatrix}$$

or, substituting numbers (to three decimal places) for the sine and cosine terms:

$$\begin{bmatrix} -0.5 & 0.5 & 0 \\ -0.866 & -0.866 & 0 \\ 0 & -0.5 & -1 \end{bmatrix} \begin{bmatrix} T_{AB} \\ T_{AC} \\ T_{BC} \end{bmatrix} = \begin{bmatrix} -H_A \\ -V_A \\ -H_C \end{bmatrix}$$

Note that, when setting up the equilibrium matrix by hand, it is generally easier to start by writing out the equations in full, whereas if we were doing it computationally the more mechanistic approach outlined earlier would be preferable. If we now invert the equilibrium matrix we obtain:

$$\mathbf{E}^{-1} = \begin{bmatrix} -1 & -0.577 & 0 \\ 1 & -0.577 & 0 \\ -0.5 & 0.289 & -1 \end{bmatrix}$$

The solutions to the two loadcases can now be found:

$$\text{(a)}\quad \mathbf{f} = \begin{bmatrix} -H_A \\ -V_A \\ -H_C \end{bmatrix} = \begin{bmatrix} -5.0 \\ 8.66 \\ 0 \end{bmatrix} \qquad \therefore \mathbf{t} = \begin{bmatrix} T_{AB} \\ T_{AC} \\ T_{BC} \end{bmatrix} = \begin{bmatrix} 0 \\ -10 \\ 5 \end{bmatrix} \text{kN}$$

$$\text{(b)}\quad \mathbf{f} = \begin{bmatrix} -H_A \\ -V_A \\ -H_C \end{bmatrix} = \begin{bmatrix} -10.0 \\ 0 \\ 0 \end{bmatrix} \qquad \therefore \mathbf{t} = \begin{bmatrix} T_{AB} \\ T_{AC} \\ T_{BC} \end{bmatrix} = \begin{bmatrix} 10 \\ -10 \\ 5 \end{bmatrix} \text{kN}$$

Of course, we would not normally resort to a matrix approach to analyse such a simple structure, but it serves to illustrate the procedure.

3.2 Three-dimensional pin-jointed trusses

While the ability to analyse plane trusses is extremely useful, most real structures are three-dimensional, or *space* structures. Moving to three dimensions does not alter the basic structural behaviour – the truss still comprises bars carrying axial forces only, connected by pins which offer no resistance to rotation. However, the visualisation and analysis now become rather more complex.

The first issue to be addressed is statical determinacy. Since we can now resolve in three perpendicular directions at each joint, equation (3.1) is no longer valid and a statically determinate 3D structure must instead satisfy:

$$b + r = 3j \tag{3.7}$$

If $b + r < 3j$ the structure is unstable, whereas if $b + r > 3j$ then it is redundant. As with plane frames, equation (3.7) is not alone sufficient to check for determinacy. It is also necessary to check the structure for localised instabilities or redundancies. This is rather more difficult to do by inspection than for plane structures, though the underlying principle is very similar. If we again start from the basic, stable unit of a triangle, then each time we add another joint to the structure it must be held in place by *three* additional members. We can often make use of this rule when checking through a framework.

When calculating the forces in a 3D frame, we can make use of all the methods outlined above. The equilibrium matrix approach proves particularly useful, since the algebra associated with the analysis of 3D structures quickly becomes complex, so that hand calculations are highly error-prone. One additional approach that is particularly suited to 3D structures is the method of tension coefficients.

3.2.1 Tension coefficients

If we have an inclined bar in 3D space whose axial force we wish to resolve into x, y and z components, then we must find the angle the bar makes with each of the three axes. This can be difficult and time consuming. The strength of the tension coefficients approach is that it eliminates the need to calculate these angles.

Consider the bar AB in Figure 3.12, which makes angles of α, β and γ with the x, y and z axes respectively. The x, y and z components of the bar tension are:

$$\begin{aligned} T_x &= T_{AB} \cos\alpha \\ T_y &= T_{AB} \cos\beta \\ T_z &= T_{AB} \cos\gamma \end{aligned} \tag{3.8}$$

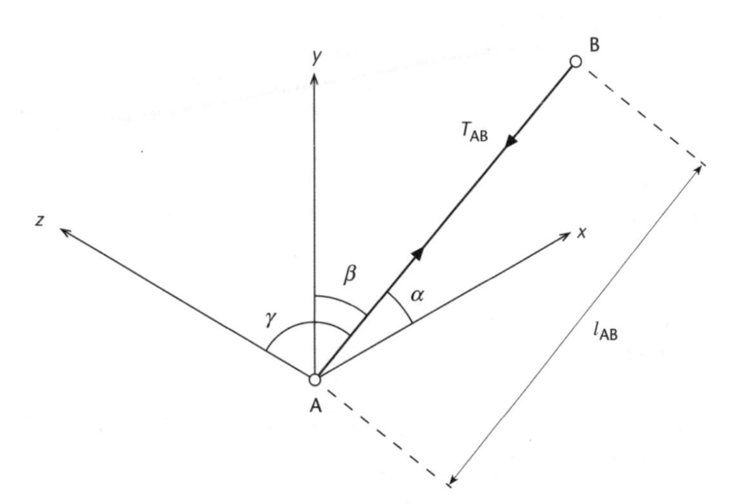

Figure 3.12
Bar in 3D space.

Similarly, the projected lengths of the bar in these directions are:

$$\begin{aligned} l_x &= l_{AB}\cos\alpha \\ l_y &= l_{AB}\cos\beta \\ l_z &= l_{AB}\cos\gamma \end{aligned} \tag{3.9}$$

Suppose that, instead of the bar force, we work in terms of the ratio of force to length, that is:

$$t_{AB} = \frac{T_{AB}}{l_{AB}} \tag{3.10}$$

t_{AB} is called the *tension coefficient* for bar AB. By substituting from equations (3.9) and (3.10), the force components given in equation (3.8) can be written as:

$$\begin{aligned} T_x &= t_{AB}l_{AB}\cdot\cos\alpha = t_{AB}l_x \\ T_y &= t_{AB}l_{AB}\cdot\cos\beta = t_{AB}l_y \\ T_z &= t_{AB}l_{AB}\cdot\cos\gamma = t_{AB}l_z \end{aligned} \tag{3.11}$$

So the resolved component of the bar force in any direction is simply the tension coefficient multiplied by the projected length of the bar in that direction. This is extremely useful since, if the plan and elevation of a space structure are given, the projected bar lengths can be found much more easily than the angles the bar makes with the various axes.

The analysis can now be performed by resolving at joints, but working in terms of tension coefficients rather than bar forces. Once the tension coefficients have been found, these can be converted to forces using equation (3.10).

EXAMPLE 3.8

Find the forces in the structure shown in Figure 3.13.

The structure is shown in two elevations and a plan view, and from these an isometric view has been sketched. This enables us to visualise the structural form but does not help with the analysis.

We first check the statical determinacy. There are 12 bars, 7 joints and 9 reactions (since a pinned support in 3D space implies three perpendicular reactions). Thus $b + r = 3j$ is satisfied. Inspection of the structure shows that no part is a mechanism, so the frame is statically determinate.

We can now use resolution at joints to determine the tension coefficients. These can all be found from joints D, E, F and G, so that it is not necessary to calculate the support reactions.

Joint and direction	Equation	Results
G x	$-3t_{EG} - 3t_{FG} - 6t_{DG} = 0$	
G y	$-4t_{EG} - 4t_{FG} - 4t_{DG} - 100 = 0$	Solve simultaneously:
G z	$-3t_{EG} + 3t_{FG} = 0$	$t_{EG} = t_{FG} = -25$, $t_{DG} = 25$

E x	$-3t_{DE} + 3t_{EG} = 0$	$t_{DE} = -25$
E y	$-6t_{BE} + 4t_{EG} = 0$	$t_{BE} = -50/3$
E z	$3t_{DE} + 3t_{EG} + 6t_{EF} = 0$	$t_{EF} = 25$
F x	$-3t_{DF} + 3t_{FG} = 0$	$t_{DF} = -25$
F z	$-3t_{DF} - 3t_{FG} - 6t_{EF} - 6t_{BF} = 0$	$t_{BF} = 0$
F y	$4t_{FG} - 3t_{BF} - 6t_{CF} = 0$	$t_{CF} = -50/3$
D x	$3t_{DE} + 3t_{DF} + 6t_{DG} + 3t_{BD} + 3t_{CD} - 2t_{AD} = 0$	
D y	$4t_{DG} - 6t_{BD} - 6t_{CD} - 6t_{AD} = 0$	Solve simultaneously:
D z	$-3t_{DE} + 3t_{DF} - 3t_{BD} + 3t_{CD} = 0$	$t_{BD} = t_{CD} = 20/6, t_{AD} = 10$

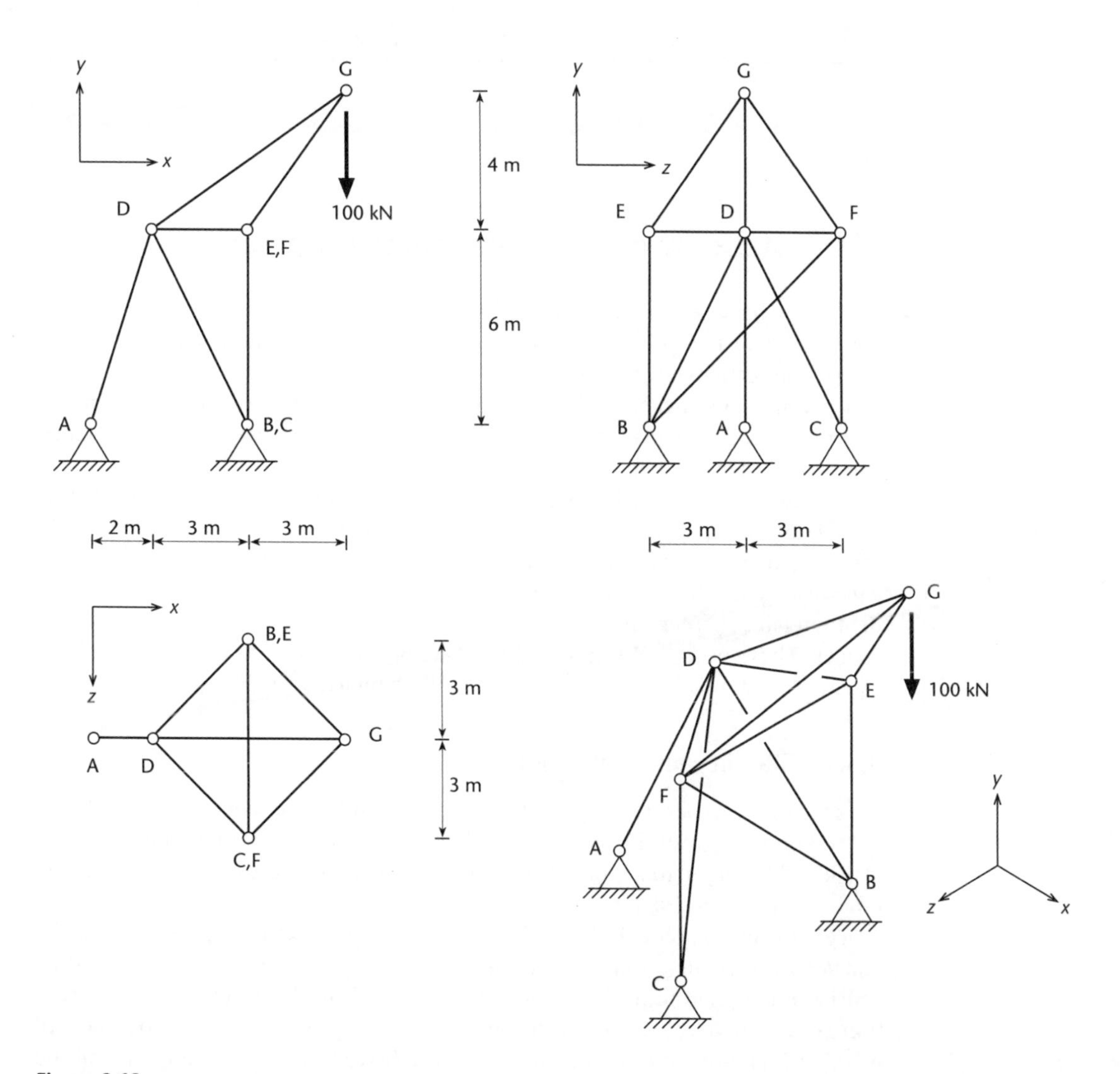

Figure 3.13

Having found the tension coefficients, it is then a simple matter to convert these to bar forces. Again, the procedure can conveniently be tabulated:

Bar	t (kN/m)	l (m)	$T = tl$ (kN)
AD	10	$\sqrt{(2^2+6^2)} = 6.32$	63.2
BD	20/6	$\sqrt{(3^2+3^2+6^2)} = 7.35$	24.5
BE	−50/3	6.0	−100.0
BF	0	$\sqrt{(6^2+6^2)} = 8.49$	0
CD	20/6	$\sqrt{(3^2+3^2+6^2)} = 7.35$	24.5
CF	−50/3	6.0	−100.0
DE	−25	$\sqrt{(3^2+3^2)} = 4.24$	−106.1
DF	−25	$\sqrt{(3^2+3^2)} = 4.24$	−106.1
DG	25	$\sqrt{(6^2+4^2)} = 7.21$	180.3
EF	25	6.0	150.0
EG	−25	$\sqrt{(3^2+3^2+4^2)} = 5.83$	−145.8
FG	−25	$\sqrt{(3^2+3^2+4^2)} = 5.83$	−145.8

3.3 Statically determinate moment frames

So far we have looked at frames in which all the joints are pinned. In reality, comparatively few joints allow complete freedom of rotation. Certain types of bolted connections in steel frames can reasonably be approximated as pinned, as can many of the mechanical connections used to join the elements of a precast concrete frame. However, welded steel frames or cast-in-place concrete structures can be regarded as having effectively rigid joints.

If at least some of the joints between the elements of a frame are rigid (that is, capable of transmitting moments) then bending moments and shear forces can be set up in the members, in addition to the axial forces. Such a structure is known as a *moment frame*. Most moment frames are redundant and will be dealt with in some depth in later chapters. In this section we shall look at plane, statically determinate frames. These occur quite rarely in practice, but the ability to analyse them is an essential prerequisite to the analysis of more complex structures.

3.3.1 Stability and determinacy

Because the members of a moment frame carry bending moments and shear forces in addition to axial loads, the number of unknowns to be determined is considerably larger than for a truss. We therefore cannot use equation (3.1) to establish the statical determinacy.

We saw in Chapter 2 that a beam is statically determinate so long as the reactions can be determined, since all the internal forces can then be found by considering equilibrium of a section. The same is true for single-storey moment frames. The issue of determinacy can therefore be reduced to consideration of whether it is possible to find all the reactions. If we start by assuming that all the joints are rigid, then there are just three equilibrium equations (horizontal and vertical resolution, and moments about any point). We know that we can also generate an equation of condition by splitting the structure at an internal pin joint. Therefore we can generate an additional equation for each internal pin. If the total number of equations thus generated equals the number of unknown reactions, then the structure may be statically determinate. We can express this as a formula:

$$r = j + 3 \tag{3.12}$$

where j is the number of internal pin joints. If $r < j + 3$ then there are insufficient reactions and the structure is unstable. If $r > j + 3$ then the structure is redundant.

For multi-storey frames the issue of statical determinacy is rather more complex and equation (3.12) cannot be used; these structures will not be considered in this chapter.

EXAMPLE 3.9

Assess the statical determinacy of the frames shown in Figure 3.14.

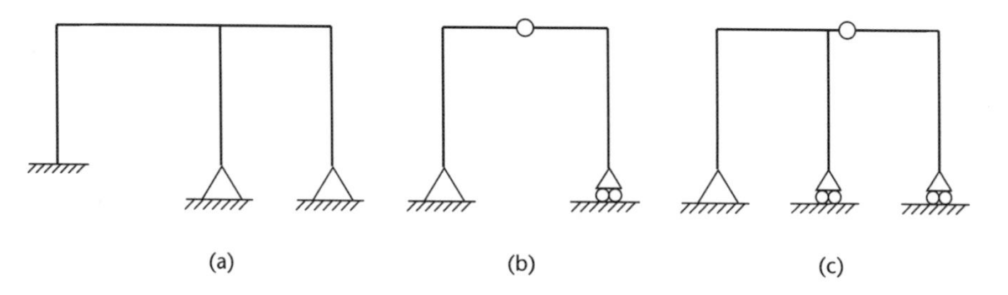

Figure 3.14

(a) There are seven reactions (three at the built-in support, two at each of the pinned supports). Since there are no internal pins, only three equations are available. The structure therefore has four redundancies.

(b) There are three reactions, two at the pin, one at the roller. There is one pin joint in the structure, so a total of four equations is available. The frame is therefore a mechanism, requiring one additional reaction to make it stable.

(c) Since there are four reactions and four equations available, and no part of the structure is unstable, this frame is statically determinate.

3.3.2 Calculation of internal forces

If a frame has been found to be statically determinate, we can use simple statics to determine the axial forces, shear forces and bending moments in the structure on a member-by-member basis. This does not introduce any new methods beyond those used for beams in Chapter 2, but confusion can arise if the analysis is not performed in a careful, methodical way.

EXAMPLE 3.10:

Find the distribution of axial force, shear force and bending moment in the frame of Figure 3.15(a).

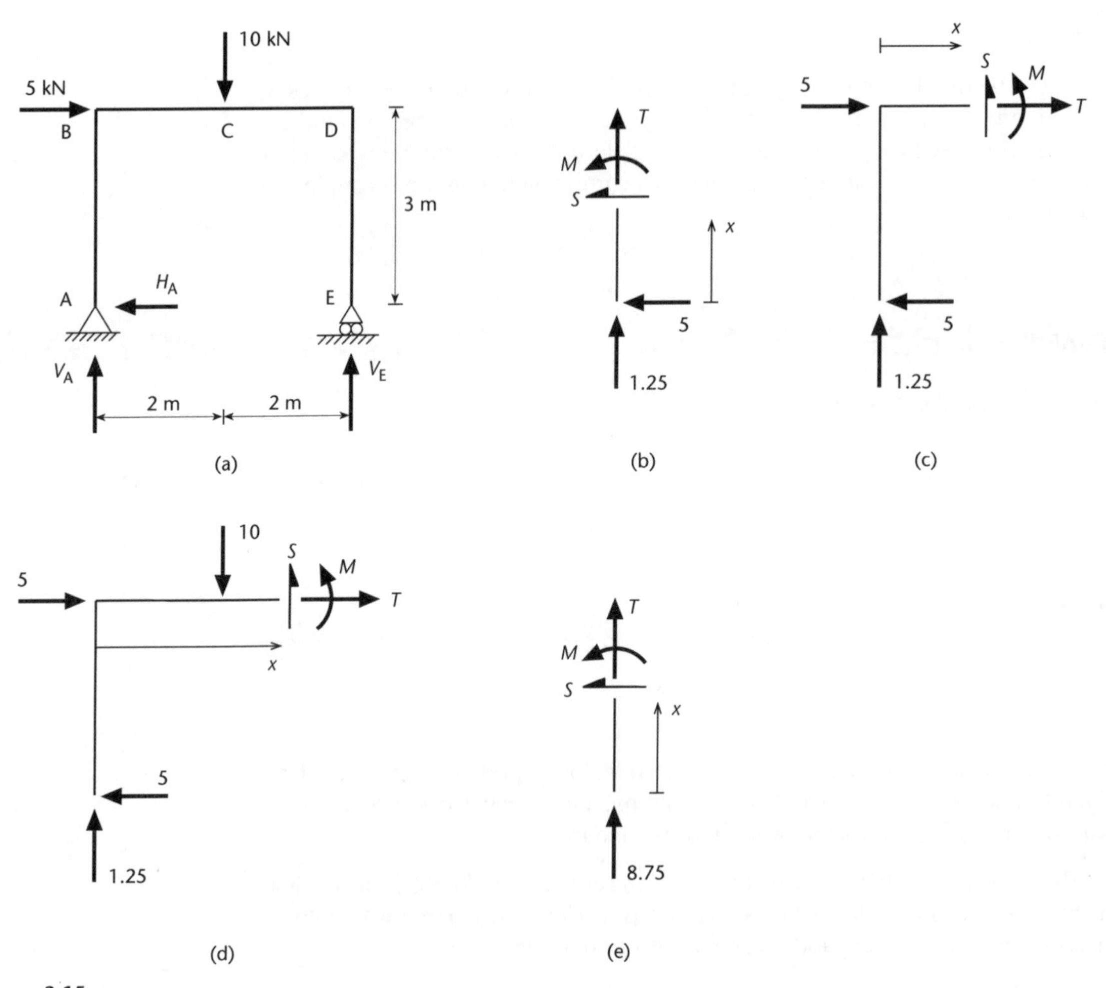

Figure 3.15

First find the reactions:

Resolve horizontally: $H_A = 5\text{ kN} \leftarrow$

Moments about E: $10 \times 2 - 5 \times 3 - 4V_A = 0 \quad \therefore V_A = 1.25\text{ kN} \uparrow$

Resolve vertically: $V_E = 8.75\text{ kN} \uparrow$

We can now work our way around the structure calculating internal forces by taking sections. We must use a different section each time we come to a joint, corner or concentrated load. Consider first a section of the column AB of length x (Figure 3.15(b)). Resolving in each direction and taking moments about the cut end gives:

$$T = -1.25$$
$$S = -5$$
$$M = 5x$$

These expressions are valid over the whole length of AB. Next consider the section (c) made by cutting through the beam BCD to the left of C. We now let the distance from B be x. (Note that we always measure the variable distance x along the member of interest, regardless of its orientation.)

$$T = 0$$
$$S = -1.25$$
$$M = 5 \times 3 + 1.25x = 15 + 1.25x$$

Next make a cut between C and D (Figure 3.15(d)):

$$T = 0$$
$$S = 10 - 1.25 = 8.75$$
$$M = 15 + 1.25x - 10(x - 2) = 35 - 8.75x$$

As we get further round the frame, the internal force expressions steadily accumulate more terms. At this point it is probably simpler to start working back around the structure from the opposite end. We therefore take the section shown in (e):

$$T = -8.75$$
$$S = 0$$
$$M = 0$$

We now have expressions for the axial force, shear force and bending moment in each part of the frame.

Example 3.10 raises several important points about moment frames:

- For the axial forces, positive values are always tensile, negative compressive. However, the signs of the shear force and bending moment terms are dependent on the way we choose to approach the problem. For example, for the column AB we chose to measure the variable length x from A towards B, and this resulted in a positive shear force. If we instead took x as positive going downwards from B, the sign of the shear force would reverse.
- The axial and shear forces in members meeting at right angles are related. Consider, for example, the members meeting at D (Figure 3.16). DE carries an axial (vertical) compressive force, and this is transmitted into CD as a vertical shear force having the same magnitude. Similarly, since there is no horizontal axial force in CD, it follows that there is no horizontal shear force in DE. The same relationships could be established at B, though here the situation is complicated slightly by the presence of an external load acting at the joint.
- Bending moments in members meeting at a rigid joint must be in equilibrium, so that they do not impose a resultant moment on the joint. For a joint between just two

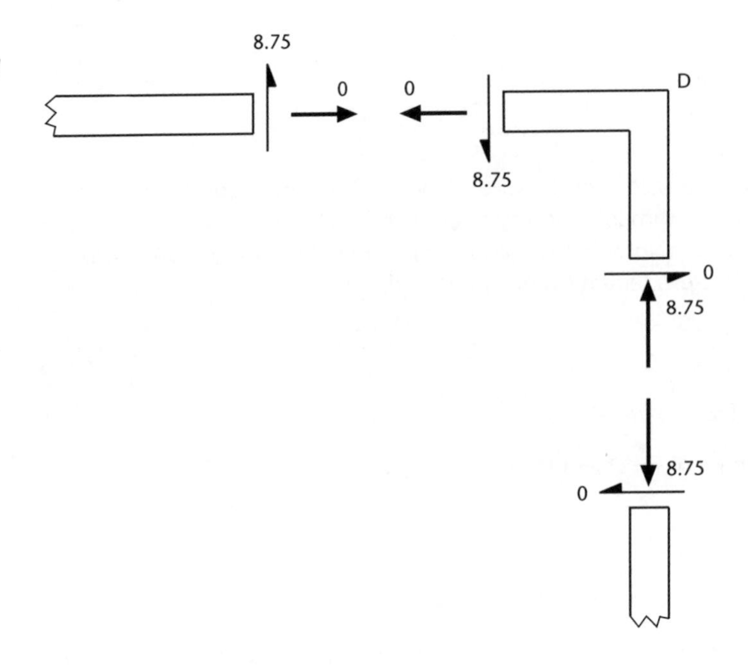

Figure 3.16
Force equilibrium at a rigid joint.

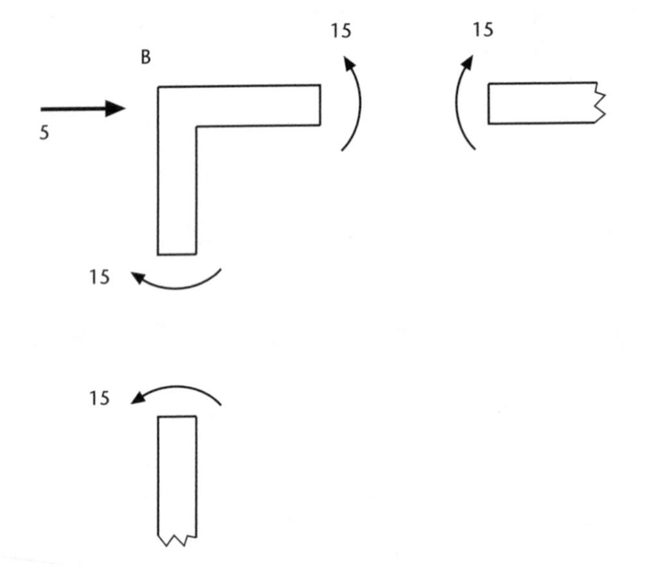

Figure 3.17
Moment equilibrium at a rigid joint.

members, this means that the moment in each must be the same. The example of joint B is shown in Figure 3.17.

- While we do not normally plot axial and shear force distributions for frames, we very often wish to plot bending moment diagrams. This can most conveniently be done by taking an outline of the frame itself as the baseline. The sign convention we used when plotting bending moment diagrams for beams in Chapter 2 was to plot positive values downwards, so that the lines were always positioned on the tension sides of the beams. As was mentioned earlier, the concept of positive and negative moments in frames is rather ambiguous. Therefore, when plotting the bending moment diagram we simply follow the convention that *the bending moment is always plotted on the tension side of the members.*

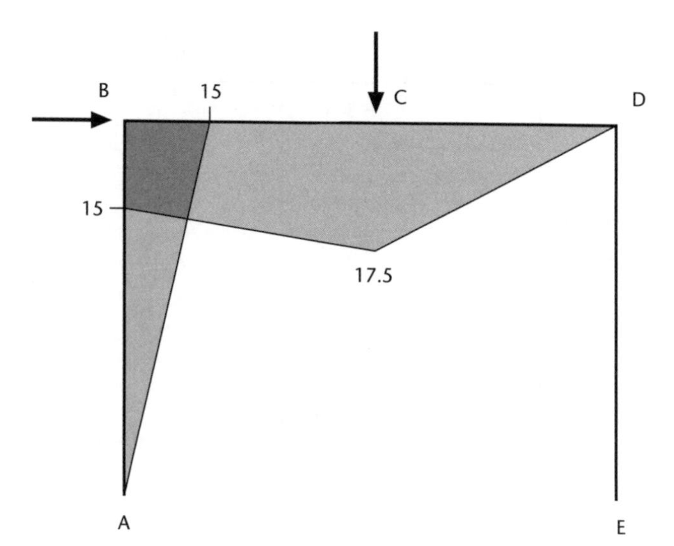

Figure 3.18
Bending moment diagram for a frame.

Figure 3.18 shows the bending moment diagram for the frame of Example 3.10. Consider the member AB. Referring to Figure 3.15(b), a positive value of bending moment as drawn implies tension on the inside face of the member, hence the bending moment is plotted inside the frame. A similar deduction can be made for BCD. Note that the bending moments at B are the same in members AB and BCD.

EXAMPLE 3.11

Plot the bending moment diagram for the frame of Figure 3.19(a).
This frame is slightly more complex and carries uniform loads, which will cause a parabolic distribution of bending moments in the beam ABCD. Remember that a full solution requires the clear identification of all maxima, minima and zeros.

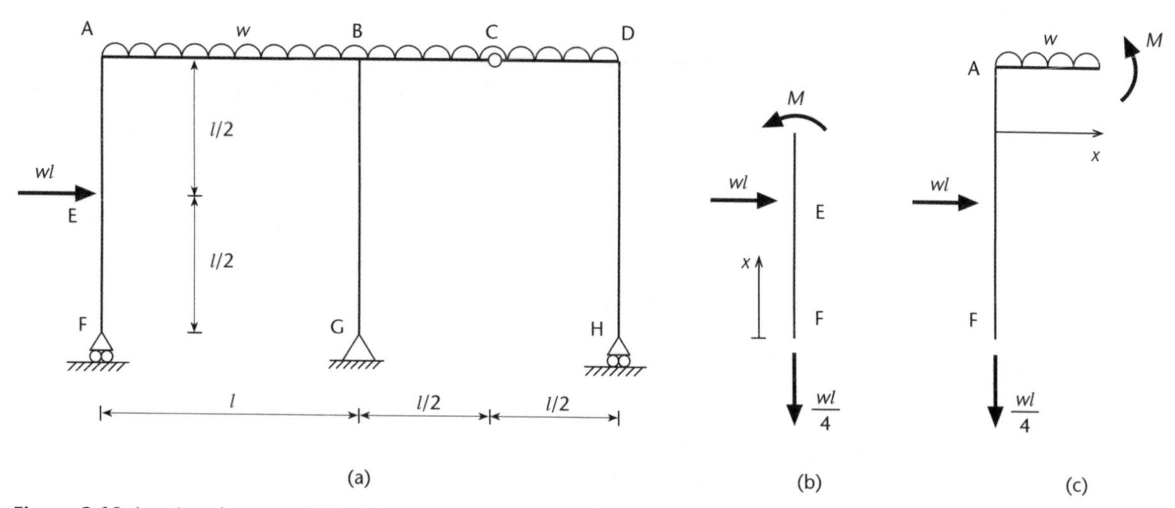

Figure 3.19 *(continued on page 84)*

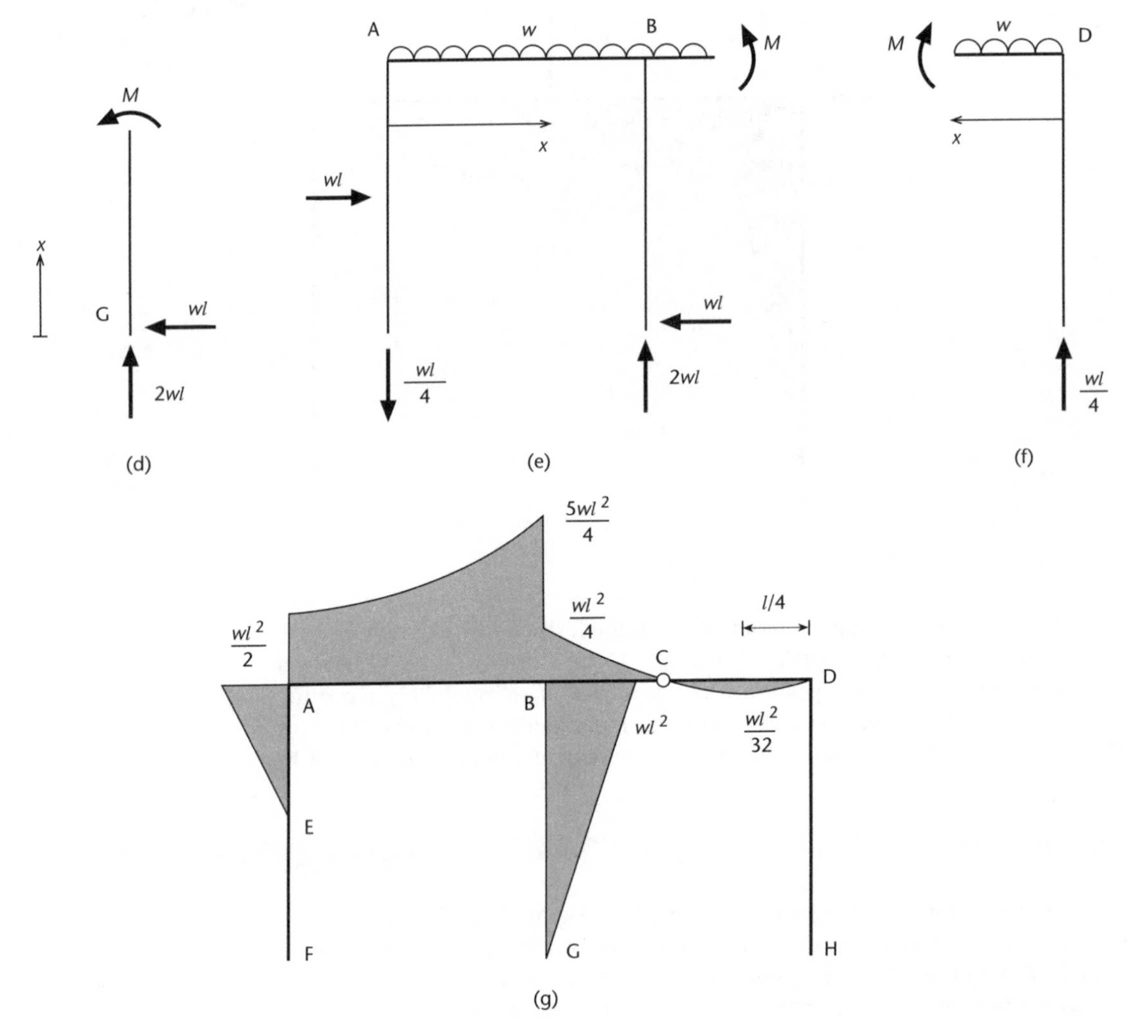

Figure 3.19 *(continued)*

There are four reactions, which can be found using the three static equilibrium equations and the equation of condition formed by splitting the structure at the pin C:

$$V_F = wl/4 \downarrow, \quad V_G = 2wl \uparrow, \quad V_H = wl/4 \uparrow, \quad H_G = wl \leftarrow$$

Having found the reactions, we proceed in the same way as before, drawing free body diagrams for successive sections of the frame, as shown Figure 3.19(b)–(f).

(b) Column FEA – below E: $M = 0$, above E: $M = -wl(x - l/2)$

M therefore varies linearly from 0 at E to $-wl^2/2$ at A.

(c) Beam AB: $M = -wl^2/2 - wlx/4 - wx^2/2$

M varies parabolically from $-wl^2/2$ at A to $-5wl^2/4$ at B. It is easy to show that there is no value of x between 0 and l for which either $M = 0$ or $dM/dx = 0$. Therefore there are no zeros or turning points in AB.

(d) Column GB: $M = wlx$

M varies linearly from 0 at G to wl^2 at B.

(e) Beam BC: $M = -wl^2/2 - wlx/4 - wx^2/2 + wl^2 + 2wl(x - l)$

M varies parabolically from $-wl^2/4$ at B to zero at the pin at C. Again, examination of M and its differential shows that there are no zeros or turning points between B and C.

(f) Beam CD: $M = wlx/4 - wx^2/2$

(This has been tackled by working back from the right-hand end of the structure.) M varies parabolically and is zero at both C and D. There must therefore be a turning point along CD. Putting $dM/dx = 0$ gives $wl/4 - wx = 0$, so $x = l/4$, and at this point $M = wl^2/32$.

We can now plot the bending moment diagram for the frame (Figure 3.19(g)), in which the lines are again plotted on the tension sides of the members. Note that the requirements of moment equilibrium at the rigid joints are again satisfied. This is easy to see at A and D, where only two members meet. At B, the moment in AB is balanced by the opposing moments in BC and BG. An alternative way of thinking about this is to say that the column BG imposes a concentrated moment on the beam at B, causing a sudden reduction in bending moment in the beam.

3.3.3 Qualitative analysis

An understanding of structural behaviour and an ability to check that numerical results make sense are essential skills for an engineer. For moment frames, useful checks include:

- the shear and axial forces in members meeting at a joint must be in equilibrium;
- the bending moments in members meeting at a joint must be in equilibrium;
- the bending moment diagram must correspond to a credible deflected shape.

We have already discussed some of these aspects in the previous section, but since they are so important we shall emphasise them here using some simple examples.

EXAMPLE 3.12

Make rough estimates of the distributions of axial force, shear force and bending moment in the frame of Figure 3.20(a).

First, consider the reactions. With only the vertical load, the vertical reactions at A and D would be equal, from symmetry. However, if we take moments about one of the supports, it is clear that the horizontal load will tend to increase the upwards reaction at D and reduce the one at A. There is therefore a small upwards reaction at A, while at D there is a large upwards reaction and a horizontal reaction of magnitude H to the left. (In fact, the vertical reaction at A might act downwards, depending on the relative magnitudes of V and H, but we shall assume here that it acts upwards.)

Figure 3.20

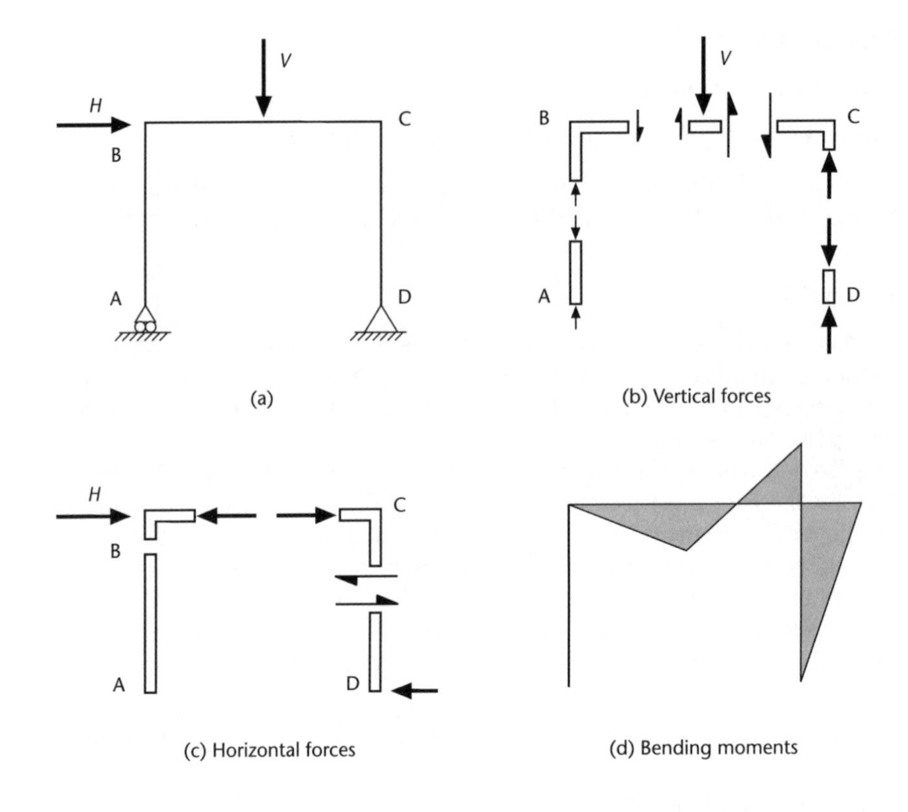

We can now establish the flow of axial and shear forces through the frame. Consider first the vertical forces (Figure 3.20 (b)). The vertical force in each column is equal to the reaction at its base. Hence there is a small compression in AB and a much larger one in CD. These are transmitted into the beam BC as shear forces; there is therefore a small shear force to the left of the load and a much larger one, of opposite sign, to the right. The change in shear force at the load point must be equal in magnitude to V.

For the horizontal forces (Figure 3.20(c)), we shall start at B. The column AB cannot support a shear force because of the roller support at A, which provides no horizontal reaction. Therefore the entire force H is carried as compression in BC; the column AB makes no contribution to carrying this load. At C the compressive force becomes a horizontal shear force of magnitude H in CD and this is transmitted into the ground as a reaction of magnitude H at the base.

Last, consider the bending moments. Since the frame carries only point loads, the bending moment diagram will consist entirely of straight lines. Again the roller at A causes there to be no moments in AB. The moment in the left-hand half of BC is equal to the vertical reaction at A multiplied by the distance from the face of AB. The moment therefore increases quite slowly, as the reaction at A is small. Alternatively, we could reach the same result by remembering that the slope of the bending moment is equal to the magnitude of the shear force. The bending moment reverses its slope at the load point, where the shear force switches sign, and now takes a much steeper gradient. There is therefore a moment of opposite sign at C. Finally, the bending moment reduces linearly to zero at the pin D.

EXAMPLE 3.13

Consider the frames of Examples 3.10 and 3.11. Working from the bending moment diagrams, sketch the deflected shapes.

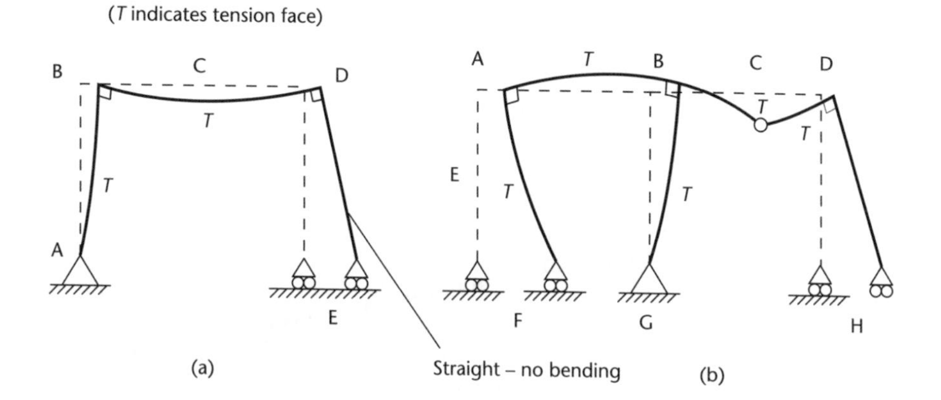

Figure 3.21

The frame of Example 3.10 is shown in Figure 3.15(a), and its bending moment diagram in Figure 3.18. The column AB experiences tension on its inside face, and must therefore bend as shown in Figure 3.21(a), with B deflecting to the right due to the lateral load. BCD is also in tension on the inside and must therefore bend downwards. DE carries no moment and therefore remains straight, its inclination determined by the rotation of BCD at D. Note that the members meeting at joints B and D remain at right angles to each other.

For the frame of Example 3.11 (see Figure 3.19), ABC is in tension on the top face and therefore must curve upwards (Figure 3.21(b)). AF is in tension on the outside above E and is straight below E, so that the roller support F moves inwards. To the right of the pin, CD bends down, giving tension on the bottom face, and DH remains straight.

3.4 Cable and arch structures

3.4.1 Suspension cables

Probably the best known use of cables in structural engineering is in suspension bridges, though they can also be used in other applications, such as lightweight roof structures. The structural behaviour of a cable is governed by two important characteristics:

- since it can support neither bending nor compression, a cable carries loads solely in tension;
- because it is so flexible, the shape taken up by the cable will vary with the loading applied to it.

Figure 3.22(a) shows a cable pinned to two supports and carrying a single point load. As the cable has no flexural stiffness, it deforms into two straight-line

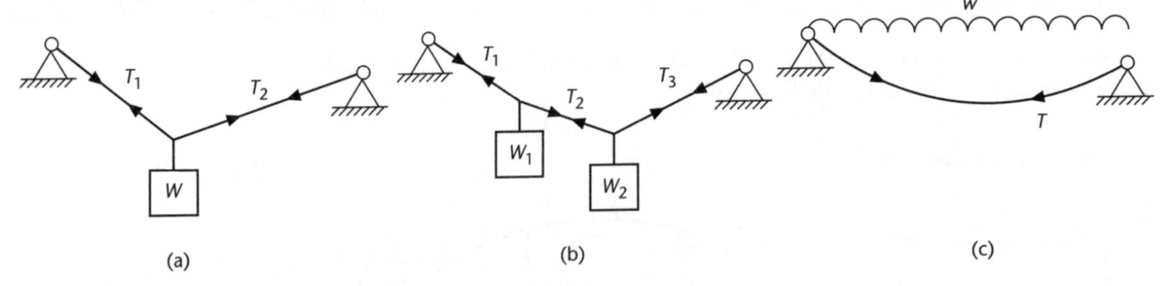

Figure 3.22
Simple cables.

segments. Given the dimensions of the problem, the inclination of these can easily be found and hence the tensions in the two parts can be determined by resolution at the load point. If we add a second point load (Figure 3.22(b)), the cable forms into three straight segments. We can go on adding loads in this way, until eventually we arrive at the case of a uniform load, which causes the cable to take up a parabolic shape (Figure 3.22(c)). We can easily prove this using simple statics.

Figure 3.23(a) shows a cable subjected to a variable loading $w(x)$ per unit horizontal span (note: *not* per unit length of cable). Since all the applied loads are vertical, the horizontal reactions at either end must be equal and opposite; let these be H. If we now imagine splitting the cable at any point and resolving horizontally, it is obvious that the horizontal component of the cable tension is H throughout.

Next consider the forces acting on a short length dx of the cable at a distance x from the left-hand support (Figure 3.23(b)). If dx is arbitrarily small, we can treat the external load on this section as uniform. Additionally, we can resolve the tension at each end of the section into vertical and horizontal components as follows:

- If the tension at x from the left-hand end is T and the angle of the cable to the horizontal is θ, then the horizontal component of the tension is $H = T\cos\theta$ and the

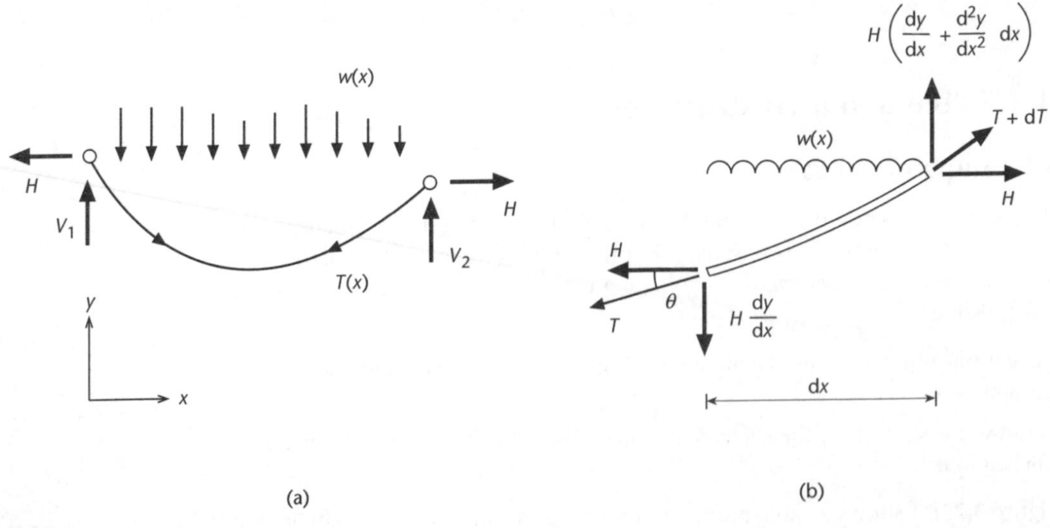

Figure 3.23
Derivation of cable governing equation.

vertical component is $V = T\sin\theta$. Eliminating T from these two expressions gives $V = H\tan\theta$;

- But $\tan\theta$ is equal to the slope of the cable, $\mathrm{d}y/\mathrm{d}x$, so $V = H\mathrm{d}y/\mathrm{d}x$;
- The rate of change of the slope with x is $\mathrm{d}^2y/\mathrm{d}x^2$. So at $(x + \mathrm{d}x)$ from the left-hand end the slope has increased by $(\mathrm{d}^2y/\mathrm{d}x^2)\cdot\mathrm{d}x$ and the vertical component of the tension has changed accordingly.

Resolving vertically:

$$H\left(\frac{\mathrm{d}y}{\mathrm{d}x} + \frac{\mathrm{d}^2y}{\mathrm{d}x}\mathrm{d}x\right) - H\frac{\mathrm{d}y}{\mathrm{d}x} - w(x)\mathrm{d}x = 0$$

Hence:

$$\frac{\mathrm{d}^2y}{\mathrm{d}x^2} = \frac{w(x)}{H} \tag{3.13}$$

This is the governing equation relating the shape of a cable to the transverse load distribution. If we now consider the special case of a uniform load over the horizontal span of the cable, then w is now a constant rather than a function of x and we can integrate (3.13) to give an expression for the shape of the cable:

$$y = \frac{wx^2}{2H} + Bx + C \tag{3.14}$$

which is the equation of a parabola. The constants B and C for a particular cable must be determined from the boundary conditions. Cable problems generally involve relating the applied loads and support reactions to the cable shape, and can be tackled by two principal approaches, as illustrated by the following example.

EXAMPLE 3.14

Find the support reactions for the cable in Figure 3.24(a).

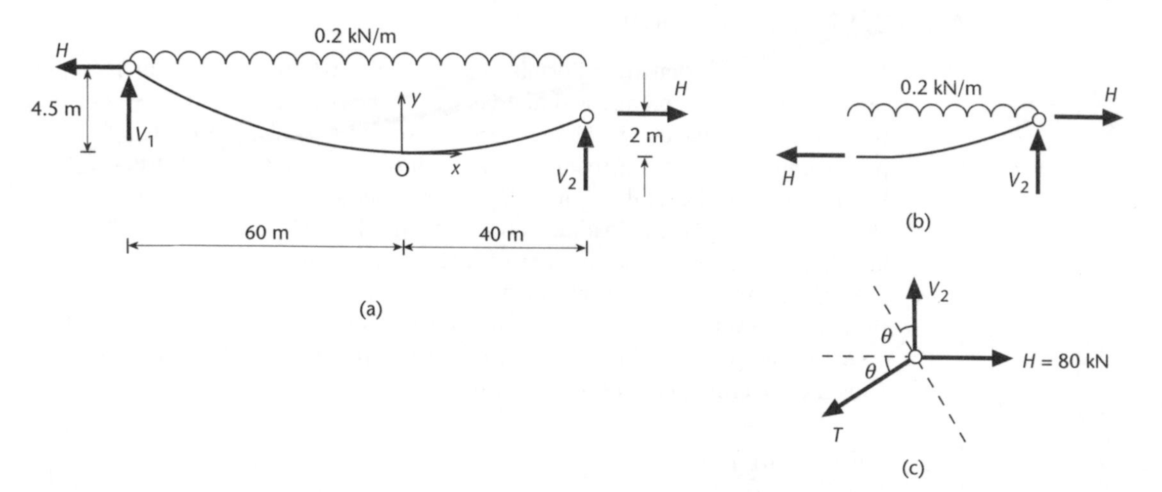

Figure 3.24

The simplest approach is to take sections, just as we have done previously for trusses and moment frames. We can cut the cable at any point, knowing that only an axial tension will exist at the cut. If we cut at the lowest point O, we know that the cable is horizontal and, since it carries only an axial force, the line of action of the force is also horizontal. Consider the right hand section (Figure 3.24(b)).

Moments about the support:	$0.2 \times 40^2/2 - 2H = 0$	$\rightarrow H = 80$ kN
Resolve vertically:	$V_2 - 0.2 \times 40 = 0$	$\rightarrow V_2 = 8$ kN
Resolve vertically (whole structure):	$V_1 + V_2 - 0.2 \times 100 = 0$	$\rightarrow V_1 = 12$ kN

The problem could also be solved by making use of the governing equation derived earlier. We know that the uniform loading results in a parabolic cable shape of the form:

$$y = Ax^2 + Bx + C$$

We can find the coefficients from the coordinates given in the diagram:

At $x = 0, y = 0$	$\rightarrow C = 0$	
At $x = 40, y = 2$	$2 = 1600A + 40B$	
At $x = -60, y = 4.5$	$4.5 = 3600A - 60B$	$\rightarrow A = 0.00125, B = 0$

But, comparing with equation (3.14), $A = w/2H$, hence $H = 80$ kN.

Now the tangent of the cable slope at any point is $dy/dx = 0.0025x$, and the value at the right-hand end is $\tan\theta = 0.0025 \times 40 = 0.1$. Consider equilibrium at the left-hand support (Figure 3.24(c)):

Resolving at right angles to the cable: $V_2 \cos\theta = 80 \sin\theta$

$$\rightarrow V_2 = 80 \tan\theta = 8 \text{ kN}$$

Repeating at the left hand end gives: $V_1 = 12$ kN

3.4.2 Parabolic arches

We have shown that a uniformly loaded cable, whose internal force is pure tension, *must* have a parabolic shape. As we saw in section 1.3.4, by simply turning the structure upside down (Figure 3.25), we obtain a parabolic arch, which acts in pure compression. Of course, an arch is not just an inverted cable. Normally it will be made from rather stiffer material, so that its bending resistance is much higher. However, so long as the loading is uniform and the arch shape is a parabola, this bending resistance will not be utilised and the inverted cable analogy holds.

Parabolic arches can be analysed by the methods outlined above for cables. However, because the curvature and the direction of the internal force have reversed, the relationship between the loading and the cable shape changes sign. Thus the governing equation, (3.13), becomes:

$$\frac{d^2y}{dx^2} = -\frac{w(x)}{H} \tag{3.15}$$

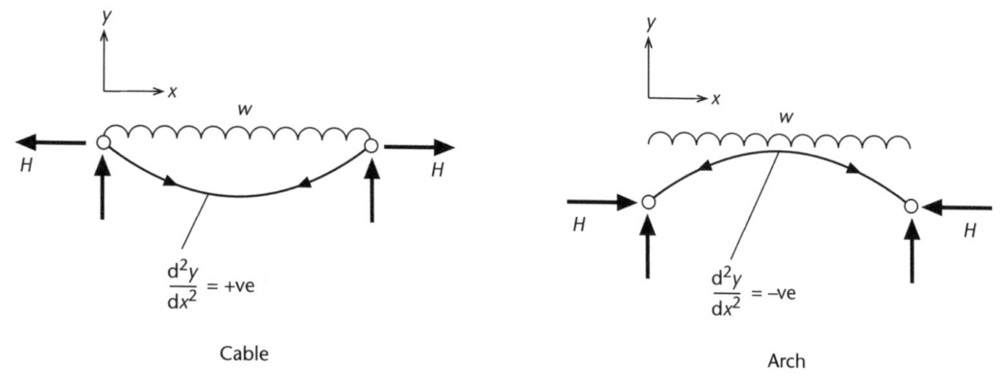

Figure 3.25
The parabolic arch as an inverted cable.

and if w is constant this can be integrated to give the arch shape:

$$y = -\frac{wx^2}{2H} + Bx + C \tag{3.16}$$

3.4.3 Three-pinned arches

Parabolic arches, carrying loads in pure compression, are common. However, other arch shapes are also quite frequently used. While these still carry a large proportion of their load in compression, they also sustain bending moments and shear forces. Non-parabolic arches with pinned ends are redundant, since they possess four reactions, for which only three equations can be formulated. But if we introduce a third pin within the arch then an equation of condition is available, allowing the structure to be analysed. A three-pinned arch is therefore statically determinate.

EXAMPLE 3.15

Find the distributions of internal forces in the three-pinned, semi-circular arch shown in Figure 3.26(a).

Using symmetry, the vertical reactions at the supports are both 10 kN upwards. Splitting at B and taking moments about B for either side, we find that the horizontal reactions are both 17.3 kN inwards.

Since the structure is symmetrical, we need only analyse one half. Consider the section of AB shown in Figure 3.26(b). Note that, because of the circular form, it is easiest to work in terms of the angle θ measured from the centre, rather than the distance along the arch.

Resolve vertically: $S\sin\theta + T\cos\theta + 10 = 0$

Resolve horizontally: $T\sin\theta - S\cos\theta + 17.3 = 0$

Solving simultaneously: $S = -10\sin\theta + 17.3\cos\theta,\ T = -10\cos\theta - 17.3\sin\theta$

Figure 3.26

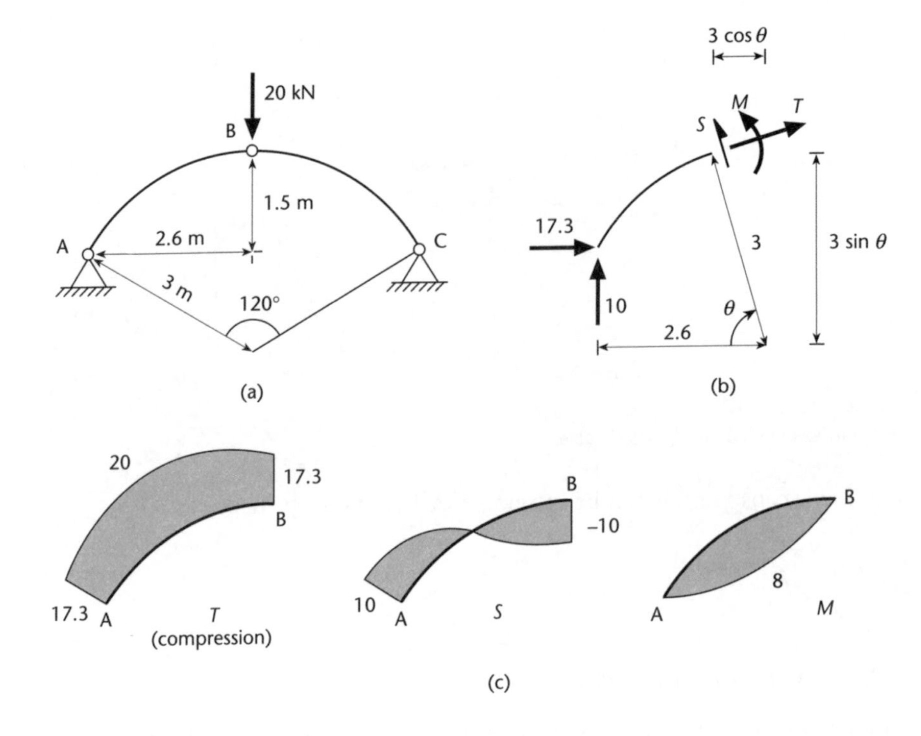

Moments about the cut: $M + 17.3(3\sin\theta - 1.5) - 10(2.6 - 3\cos\theta) = 0$

Simplifying: $M = 30\cos\theta - 52\sin\theta$

These solutions are valid for the whole of AB, that is for $30° \leq \theta \leq 90°$. It can be shown by differentiation that both T and M take maximum values at $\theta = 60°$. We can plot the variations of T, S and M around the arch as shown in Figure 3.26(c). Clearly there is a large compression throughout the arch, taking a maximum value midway between the pins, where the shear force is zero. At other locations the presence of a shear force tends slightly to reduce the axial thrust. The non-parabolic shape of the arch also results in significant bending between the pins, with the maximum again occurring midway between A and B.

3.5 Problems

Where appropriate, answers are given at the end of the book. The questions marked with an asterisk are a little more challenging.

3.1. Assess the statical determinacy of the four plane frames shown in Figure 3.27. For frames that are not statically determinate, suggest modifications to make them so.

3.2. Find the reactions and the forces in all the bars of Figure 3.27(a) when a horizontal load of 5 kN to the right is applied at A. All members have equal length.

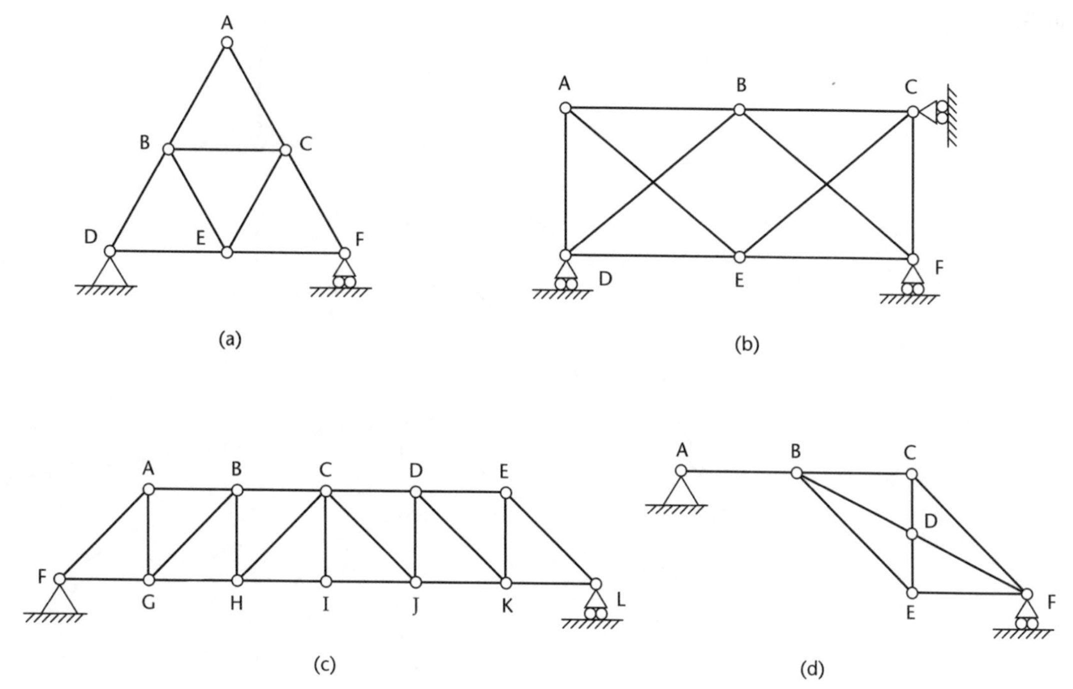

Figure 3.27

3.3. Vertical point loads of 2 kN are applied at each joint along the bottom chord of the frame in Figure 3.27(c). Find the forces in members BC, CH, HI and CI. All horizontal and vertical members have equal length.

3.4.* Write out the equilibrium matrix equations for the two structures shown in Figure 3.28. Use these equations to assess the statical determinacy of the structures.

3.5.* Figure 3.29 shows the plan and elevation of a derrick-type crane, in which all members are pin jointed. ABC and DEF are both equilateral triangles, of side 10 m and 5 m respectively. Find all the member forces when the 10 kN load is applied at G.

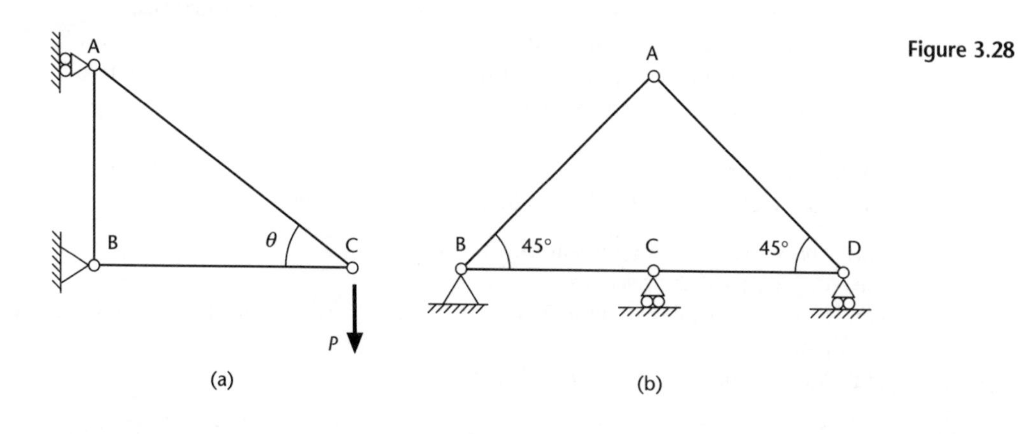

Figure 3.28

Figure 3.29

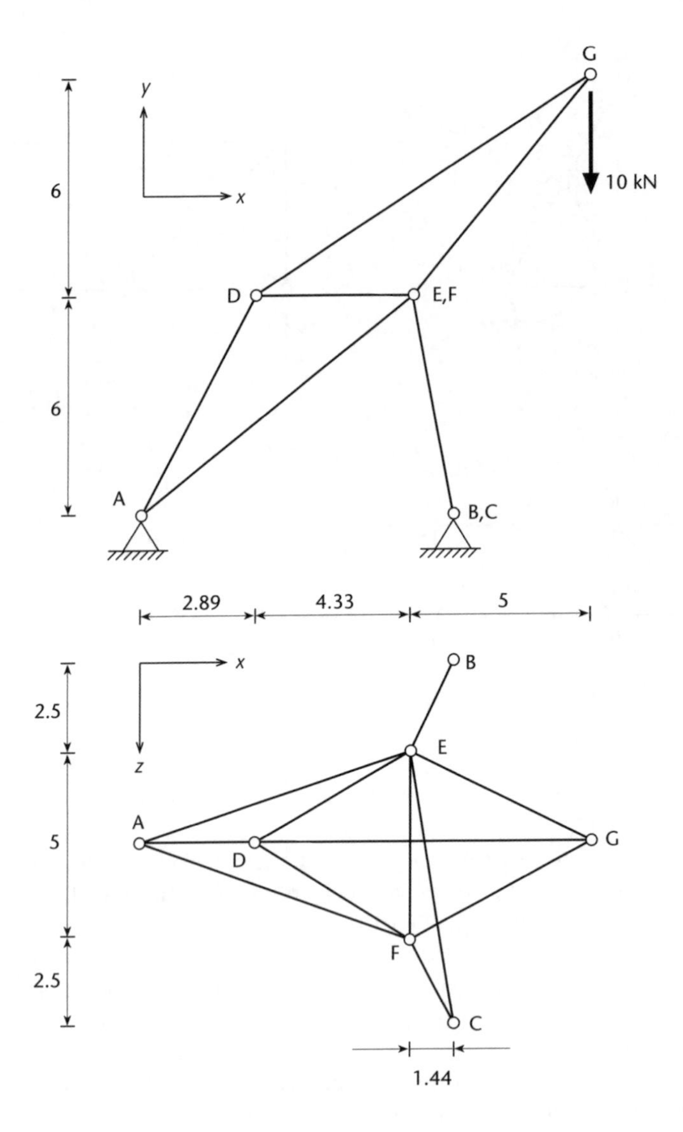

3.6. For the frame in Figure 3.30:

(a) Calculate the support reactions.
(b) Find the distributions of axial and shear force across the structure.
(c) Draw the bending moment diagram, indicating key values and their locations.
(d)* Draw an exploded view of joint B, showing the directions of the internal forces at the joint. Hence check that the joint is in equilibrium.

3.7. The framework in Figure 3.31 has rigid joints at A and C. At B there is a rigid joint between AB and BE, but a pin at the left-hand end of BC.

(a) Show that the frame is statically determinate.
(b) Calculate the support reactions.
(c) Draw the bending moment diagram, clearly identifying all zeros and turning points.
(d)* Make a rough sketch of the deflected shape.

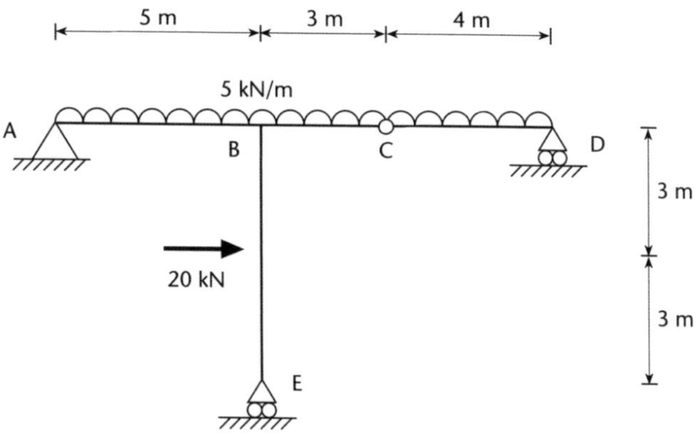

Figure 3.30

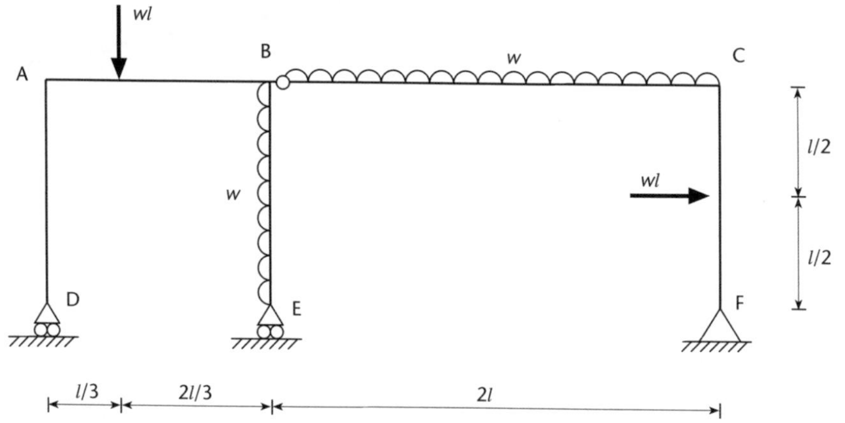

Figure 3.31

3.8.* Without calculation, sketch the deflected shapes and the bending moment diagrams for the frames shown in Figure 3.32.

3.9. A symmetrical parabolic arch bridge has a horizontal span between the supports of 16 m and a central rise of 5 m. It carries a uniform load of 3 kN/m of horizontal span. Calculate the reactions at the supports and find an equation for the shape of the arch.

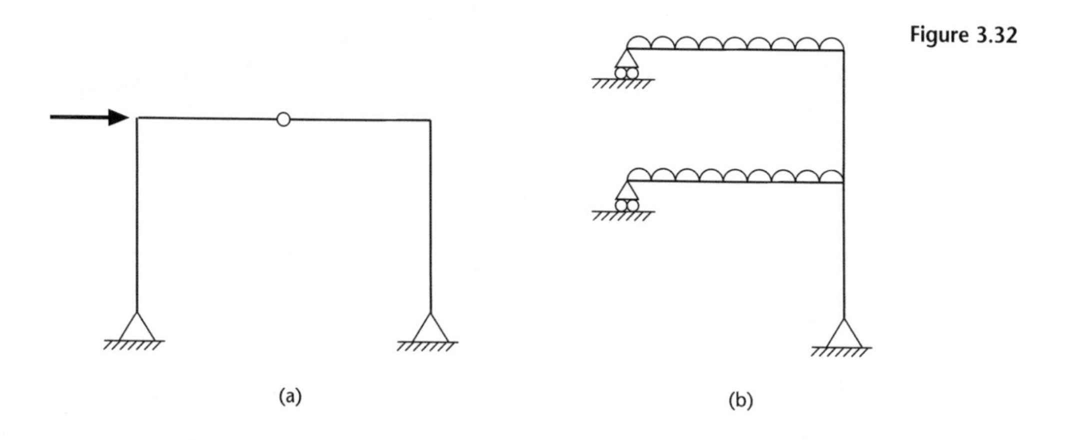

Figure 3.32

3.10* A semi-circular arch rib of radius R has level, pinned supports and an internal pin at the crown. It carries a uniformly distributed load w per unit horizontal span. Find the support reactions and the distributions of axial force, shear force and bending moment in one half of the arch.

CHAPTER 4

Stress and strain

CHAPTER OUTLINE

We have so far looked at structures only at the macro scale, determining the reactions and internal forces caused by sets of applied loads. In Chapters 4–6 we shall go down to a more detailed level and look at the effects that these internal forces have on the members. This field of study is known as *solid mechanics* or *strength of materials*, and is important to structural analysts and designers for two main reasons:

- Having calculated the internal forces to which a structure will be subjected, a designer must select suitable member dimensions and materials for the structural members. This requires an understanding of how the internal forces are carried within the members and what deformations they cause.
- For statically indeterminate structures, it is not possible to calculate the internal forces from the geometry of the structure alone. For these structures, the distribution of internal forces is also dependent on the relative stiffnesses of the members, that is, how easily they can be deformed. This again requires an understanding of solid mechanics.

In this chapter we shall restrict ourselves to the analysis of simple one- and two-dimensional systems of stresses within members. Chapters 5 and 6 will then apply these principles to the analysis of elements loaded in bending, shear and torsion. This chapter should enable the reader to:

- understand the concepts of stress and strain;
- calculate stresses and strains in simple one-dimensional elastic systems;
- analyse simple structures in which yielding occurs;
- resolve stresses and strains in two dimensions;
- appreciate the significance of the principal elastic constants and the relationships between them.

4.1 Stress

The effect that an internal force has on a member will depend on the properties of the material from which the member is made, and on its dimensions; obviously a very large member will be able to sustain greater forces than a smaller one made from the same material. This dependence on element size can be conveniently accounted for by performing calculations in terms of *stress* rather than force. In simple terms, stress can be defined as the force acting on a member or part of a member divided by the area over which it acts. Stress is a useful quantity because:

- it provides a measure of how the internal force is distributed through a member;
- the ways in which a member responds to a certain amount of stress are functions solely of the material properties, and are independent of the member dimensions.

The SI unit of stress is the Pascal (Pa). One Pascal is equal to one Newton per square metre. However, in many applications we find that the Pascal is rather a small quantity, so it is quite common to work in terms of mega-Pascals ($1\ \text{MPa} = 1\ \text{N/mm}^2$).

We have already seen that elements within structures can be subjected to combinations of axial force, shear force and bending moment. These in turn can give rise to quite complex stress patterns within the members. However, it is always possible to resolve the stresses acting on a surface or cross-section into just two types: *direct* or *normal* stresses acting at right angles to the surface, and *shear* or *tangential* stresses acting in the plane of the surface.

4.1.1 Direct stress

Figure 4.1(a) shows a short length of bar carrying an axial tensile force P_x. If we make an imaginary cut in the bar at right angles to its axis we know that, in order

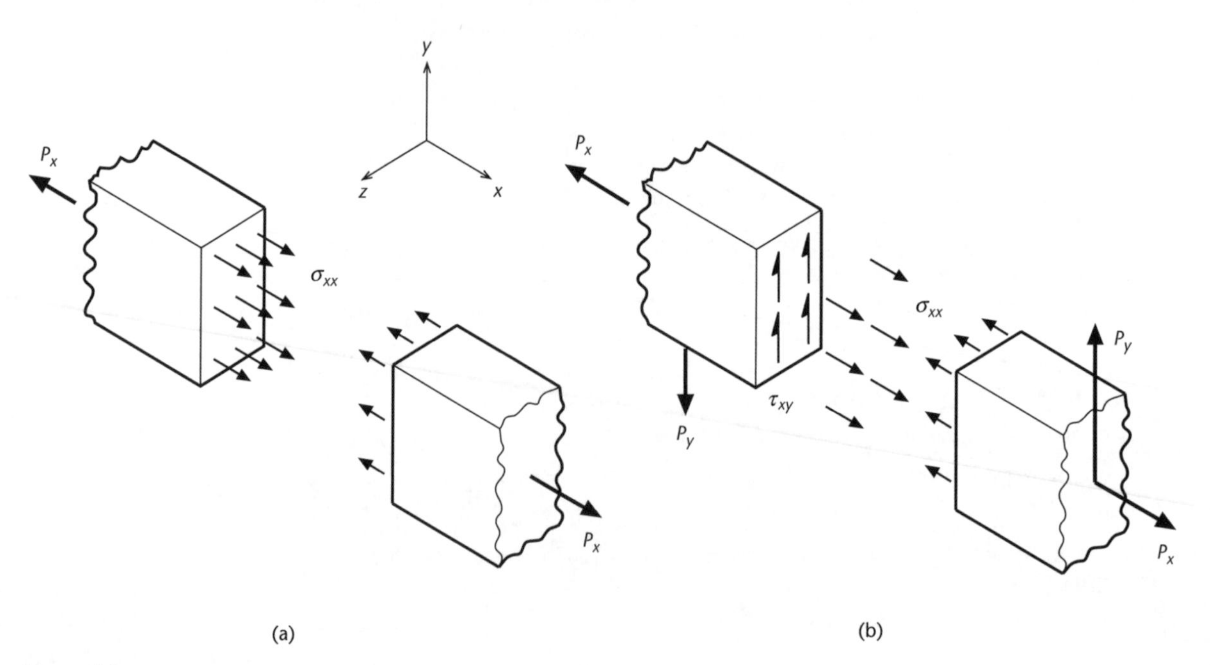

Figure 4.1
Direct and shear stresses.

to maintain equilibrium, there must be an internal force at the cut. Up to now we have considered the bar as a line element, but in reality it must have a finite cross-sectional area A. The internal force will create a stress acting over this area. In this case, since the resultant force acts along the member axis, the stress will be uniform and equal to:

$$\sigma_{xx} = \frac{P_x}{A} \tag{4.1}$$

Here the first subscript refers to the face on which the stress acts, and the second to its direction. So σ_{xx} is the direct stress acting in the x direction on a face whose outward normal is also in the x direction.

More generally, it is possible for the stress to vary over the area being considered. In this case we take an elemental area $\mathrm{d}A$. If the force acting on the area is $\mathrm{d}P_x$ then the direct stress is:

$$\sigma_{xx} = \frac{\mathrm{d}P_x}{\mathrm{d}A} \tag{4.2}$$

and integrating (4.2) gives the total force acting on the area A:

$$P_x = \int_A \sigma_{xx}\mathrm{d}A \tag{4.3}$$

4.1.2 Shear stress

Returning to our simple bar, suppose that, instead of acting axially, the resultant force at the cut point acts at an angle to the x-axis, in the x–y plane. The force can be resolved into components P_x and P_y acting in the two axis directions (Figure 4.1 (b)). The P_x component produces a direct stress σ_{xx} as before, but the P_y component acts in the plane of the cross-section, and so produces a shear stress. If this is assumed constant over the area, then it is given by:

$$\tau_{xy} = \frac{P_y}{A} \tag{4.4a}$$

Similarly, if there were a z component of internal force at the cut, this would produce a horizontal shear stress τ_{xz}:

$$\tau_{xx} = \frac{P_x}{A} \tag{4.4b}$$

The subscript convention is the same as for the direct stresses, with the first subscript referring to the face and the second to the direction of the force, so that τ_{xy} is the shear stress in the y direction on an x face, and so on.

Again we must consider the cases where τ_{xy} and τ_{xz} are not constant over the area. As before, the stresses are found by considering the forces acting on an elemental area $\mathrm{d}A$:

$$\tau_{xy} = \frac{\mathrm{d}P_y}{\mathrm{d}A}, \quad \tau_{xz} = \frac{\mathrm{d}P_z}{\mathrm{d}A} \tag{4.5}$$

So far we have considered only stresses acting on the x-faces of a member. If we considered a small three-dimensional element cut from a solid body, then there

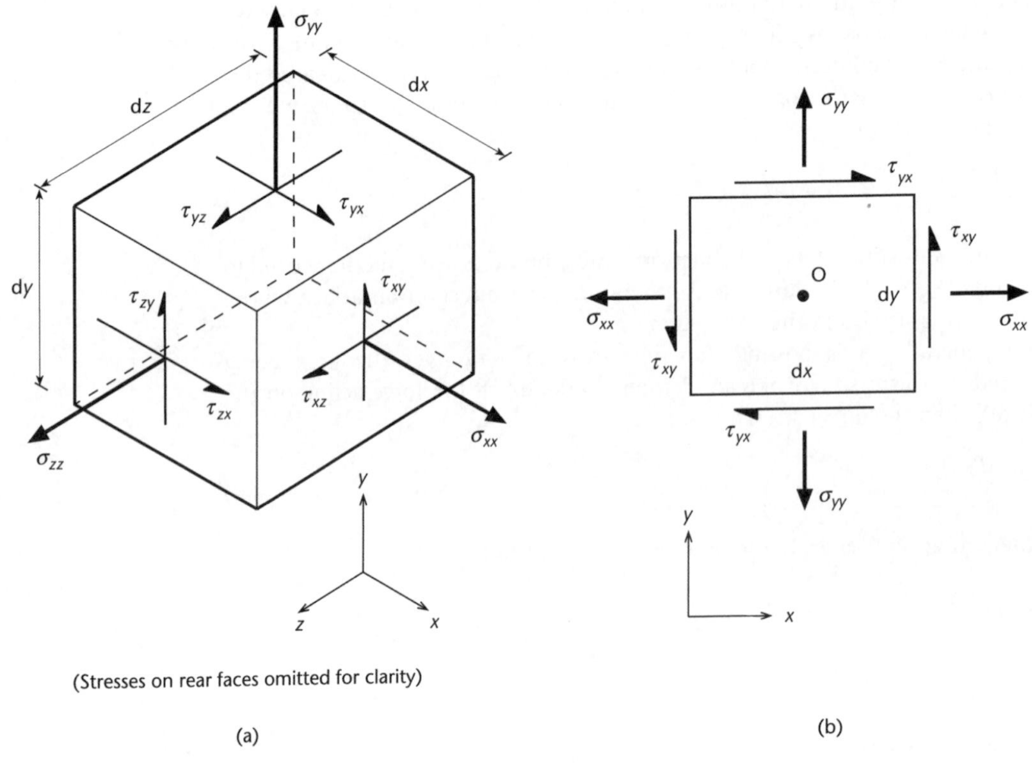

Figure 4.2
Stress components in three dimensions.

may be sets of direct and shear stresses on each pair of opposite faces. The possible stress components can be seen in Figure 4.2(a) – for clarity, only the stresses on the front three faces are shown but there are, of course, equal and opposite stresses acting on the opposite, hidden faces. There is a total of nine different stress components; three direct and six shear.

If we look directly at, for example, the z-face of the elemental volume, then we see the stresses indicated in Figure 4.2(b). Now the element must be in equilibrium, so taking moments about an axis through O and normal to the page:

- Moment = (stress × area) × distance
- Anti-clockwise moment due to $\tau_{xy} = (\tau_{xy} \cdot \mathrm{d}y \cdot \mathrm{d}z) \cdot \mathrm{d}x$
- Clockwise moment due to $\tau_{yx} = (\tau_{yx} \cdot \mathrm{d}x \cdot \mathrm{d}z) \cdot \mathrm{d}y$.
- For the element to be in equilibrium, the anti-clockwise moment due to the vertical shear stresses τ_{xy} must be cancelled out by the clockwise moment due to τ_{yx}, so:

$$\tau_{xy} = \tau_{yx} \tag{4.6}$$

τ_{xy} and τ_{yx} are called *complementary* shear stresses. Using a similar argument, we can also see that the shear stresses acting around the perimeters of the x- and y-faces are complementary, that is $\tau_{yz} = \tau_{zy}$ and $\tau_{xz} = \tau_{zx}$. Thus, instead of the nine different stresses in Figure 4.2(a), we have only six, since each pair of shear stresses must be equal. Also, it is important to note that it is impossible for τ_{xy}, say, to exist without τ_{yx} a shear stress is *always* accompanied by a complementary shear stress on a perpendicular face.

EXAMPLE 4.1

A circular bar of cross-sectional area A carries an axial tensile force P. Calculate the direct and shear stresses on planes making angles of 90°, 60°, 45° and 30°, with the bar axis.

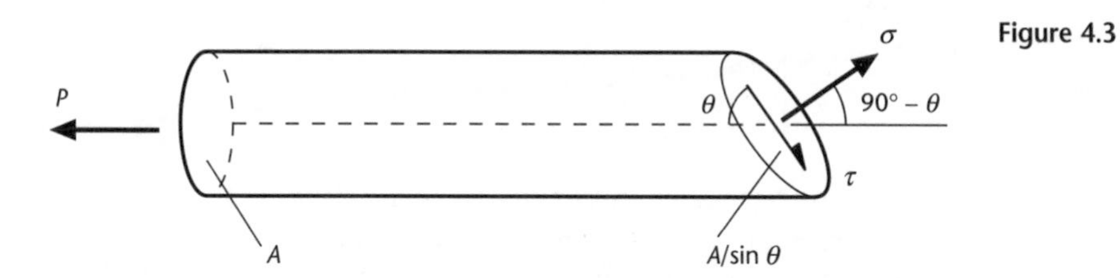

Figure 4.3

Figure 4.3 shows the free body diagram created by cutting the bar at an angle θ to its axis. Since the load is applied axially, the direct stress σ and shear stress τ will both be constant over the cut surface.

Resolving normal to the surface: $\sigma \cdot (A/\sin\theta) = P\sin\theta \;\rightarrow \sigma = (P/A) \cdot \sin^2\theta$

Resolving parallel to the surface: $\tau \cdot (A/\sin\theta) = P\cos\theta \;\rightarrow \tau = (P/A) \cdot \sin\theta\cos\theta$

Substituting numerical values for θ, we get:

$\theta = 90°$	$\sigma = P/A$	$\tau = 0$
$\theta = 60°$	$\sigma = 0.75P/A$	$\tau = 0.433P/A$
$\theta = 45°$	$\sigma = 0.5P/A$	$\tau = 0.5P/A$
$\theta = 30°$	$\sigma = 0.25P/A$	$\tau = 0.433P/A$

It can be seen that the direct stress decreases as θ decreases. This is both because the stress is becoming increasingly close to perpendicular to the applied force, and because it is acting over an increasingly large area. For the shear stress, these two effects oppose each other, so that the stress initially increases with angle, but at large values of θ the inclined area becomes very large and τ therefore reduces.

4.2 Strain

In addition to the forces they sustain, we are concerned with the deformations that loaded structures undergo. As with force, it is useful to have a measure of how deformation is distributed throughout a member. The quantity we use is the *strain*, defined as the change in an element dimension caused by a loading, divided by the original dimension. As with stress, both direct and shear strains can be developed, depending on the applied loading.

Before deriving general mathematical expressions for direct and shear strains, we shall explain their physical meanings in simple terms. Consider again the case

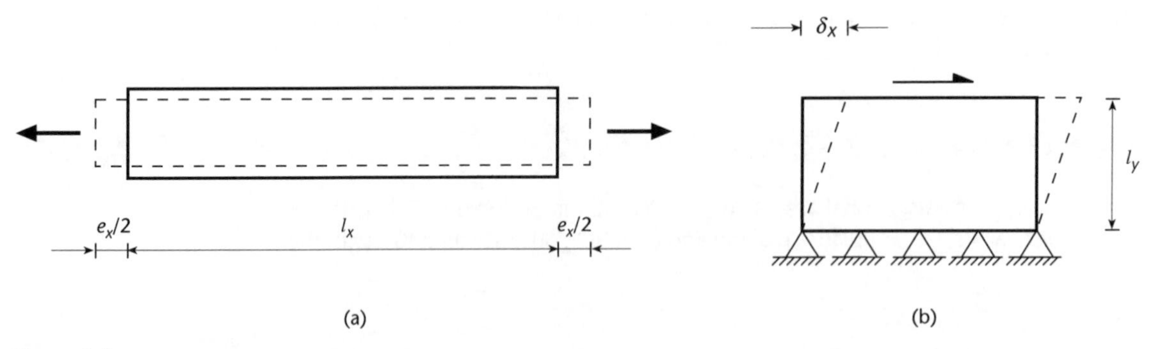

Figure 4.4
Direct and shear strains.

of a uniform bar under an axial tensile load (Figure 4.4(a)). The bar extends by an amount e_x – it also undergoes a lateral contraction as shown, but we will ignore this for the moment. The *direct strain* is constant throughout the bar and is equal to the extension divided by the original length l_x, that is:

$$\epsilon_{xx} = \frac{e_x}{l_x} \tag{4.7}$$

While a direct strain implies a change in the dimensions of the member, its basic shape remains unaltered, that is none of the angles between the edges of the member changes.

Now consider a thin rectangular plate. If the plate is held in place by pinned supports along one edge and a shear load is applied at the opposite edge, it will deform as shown in Figure 4.4(b). In this case, the dimensions of the member remain unchanged, but the angle between the edges alters. We describe this type of deformation as a *distortion*, and we quantify it by the *shear strain*, defined as the change in angle between two initially perpendicular lines. Since the deformations we are dealing with are generally small, we can express the shear strain using small angle approximations as:

$$\gamma_{xy} \approx \tan \gamma_{xy} = \frac{\delta_x}{l_y} \tag{4.8}$$

In general, structural deformations are very small compared to the overall dimensions, so that both direct and shear strains take values much less than one.

The above definitions are adequate for very simple cases, but more generally the deformations within a member may vary with position in 3D space. The displacement of a point can be described in terms of three perpendicular components u, v and w in the x, y and z directions respectively. Since displacement is a function of position each of these components is a function of x, y and z, that is:

$$u = u(x, y, z), \quad v = v(x, y, z), \quad w = w(x, y, z)$$

For clarity, we shall limit the ensuing proof to two dimensions. Suppose we marked a small rectangle with dimensions $dx \times dy$ on a thin plate lying in the x–y plane. Under the combined actions of direct and shear stresses acting in the plane of the plate, the rectangle would deform as shown in Figure 4.5. This has been drawn making the assumption that all the deformations are positive; direct strains

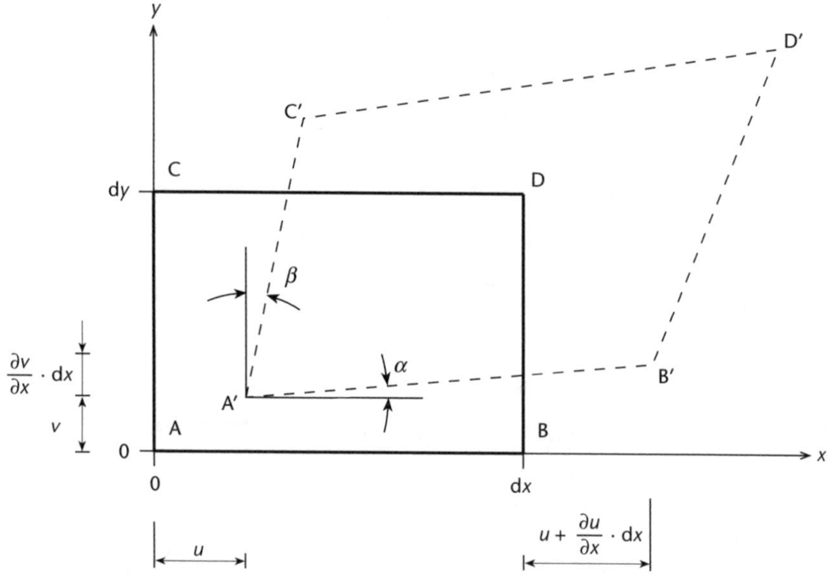

Figure 4.5
Deformation of a 2D element under combined direct and shear loading.

are extensional, while a positive shear strain implies that, for example, the y displacement of edge AB increases with increasing x.

The direct strain is defined as the change in length divided by the original length. Consider the extension of edge AB. A has displaced in the x direction by u, B by $u + (\partial u/\partial x) \cdot \mathrm{d}x$, where $\partial u/\partial x$ is the rate of change of u in the x direction. The partial derivative is used as u is also a function of y and z. The strain in the x direction is the change in x dimension divided by the original value:

$$\epsilon_{xx} = \frac{u + \dfrac{\partial u}{\partial x}\mathrm{d}x - u}{\mathrm{d}x} = \frac{\partial u}{\partial x}$$

and a similar result can be obtained for ε_{yy} by consideration of edge AC.

The shear strain is given by the change in the angle CAB, which is the sum of the two angles α and β. The tangent of α is given by the vertical component of A′B′ divided by its horizontal component:

$$\tan\alpha = \frac{\dfrac{\partial v}{\partial x}\mathrm{d}x}{\mathrm{d}x + \dfrac{\partial u}{\partial x}\mathrm{d}x} = \frac{\dfrac{\partial v}{\partial x}}{1 + \epsilon_{xx}} \approx \frac{\partial v}{\partial x}$$

since $\varepsilon_{xx} \ll 1$. Similarly,

$$\tan\beta = \frac{\partial u}{\partial y}$$

Now if the angles are small then there is negligible difference between the angles themselves and their tangents, so the shear strain is given by:

$$\gamma_{xy} \approx \tan\alpha + \tan\beta = \frac{\partial v}{\partial x} + \frac{\partial u}{\partial y}$$

To summarise, we shall state the general results for the three-dimensional case. If an arbitrarily small element $dx \times dy \times dz$ experiences displacement components u, v and w in the x, y and z directions, then its deformation can be described by three direct strains in those directions:

$$\epsilon_{xx} = \frac{\partial u}{\partial x} \qquad \epsilon_{yy} = \frac{\partial v}{\partial y} \qquad \epsilon_{zz} = \frac{\partial w}{\partial z} \tag{4.9}$$

and three shear strains in the x–y, y–z and z–x planes:

$$\gamma_{xy} = \frac{\partial v}{\partial x} + \frac{\partial u}{\partial y} \qquad \gamma_{yz} = \frac{\partial w}{\partial y} + \frac{\partial y}{\partial z} \qquad \gamma_{zx} = \frac{\partial u}{\partial z} + \frac{\partial w}{\partial x} \tag{4.10}$$

4.3 One-dimensional elasticity

The science of elasticity deals with the behaviour of solids under loads which cause *elastic* deformations. A loaded body is said to deform elastically if it returns to its original size and shape when the load is removed. Most materials used in engineering, such as metals, polymers and ceramics, behave elastically under low or moderate loads, only becoming inelastic if subjected to very high stress levels. In general, engineers aim to design structures so as to avoid inelastic behaviour, since this implies permanent, irrecoverable deformations. However, in some cases structures are permitted to respond inelastically to extreme loads.

In this section we shall introduce the basic stress–strain relations and apply them to a range of one-dimensional elasticity problems. Two- and three-dimensional problems and inelastic behaviour will be introduced in subsequent sections.

4.3.1 Stress–strain relations

Elasticity theory is based on an experimentally determined law first published by Robert Hooke in 1678. Hooke's law states that, for certain *linear elastic* materials under uniaxial loading, the direct strain is proportional to the direct stress. The constant of proportionality between the stress and strain is known as Young's modulus, given the symbol E. Hooke's law for a uniaxially loaded elastic material can therefore be written as:

$$\sigma_{xx} = E\varepsilon_{xx} \tag{4.11}$$

Young's modulus is a property solely of the material from which a member is made, and is independent of its size and shape. As strain is dimensionless, E has the units of stress, that is, Pascals.

Figure 4.6 shows typical graphs of stress against strain determined by performing tensile loading tests on different types of material. In (a) the material is linear elastic; the plot of stress against strain is a straight line of gradient E, and when the load on the specimen is removed the stress and strain return to zero along the same line. In (b) the material is elastic since the loading and unloading curves are identical, but the gradient varies, so the material is non-linear. The material in (c) is both non-linear and inelastic; the unloading curve follows a different path to the loading curve, so that the strain does not return to zero.

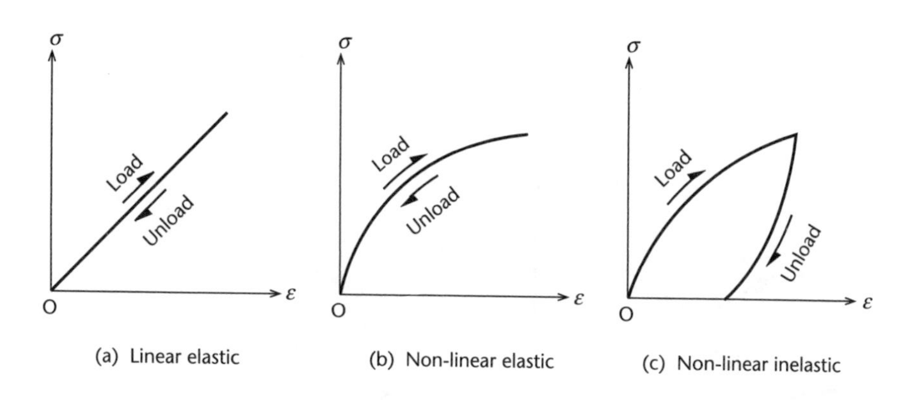

Figure 4.6
Stress–strain curves for different material types.

If a specimen is loaded uniaxially it will be found that, as the length changes under the load, so do the cross-sectional dimensions. Imagine stretching a rubber band; as its length increases, the band becomes thinner. The strains in the y and z directions are proportional to the axial strain, but opposite in sign. This can be expressed mathematically as:

$$\varepsilon_{yy} = \varepsilon_{zz} = -\nu\varepsilon_{xx} \tag{4.12}$$

ν is known as Poisson's ratio and, like Young's modulus, it is a property of the material from which the member is made. Poisson's ratio is dimensionless, and for most materials takes a value of less than 0.5. Values of Young's modulus and Poisson's ratio for some common engineering materials are shown in Table 4.1. In this book we shall deal only with *isotropic* materials, that is, ones whose properties are the same in every direction. Most metals are isotropic. An example of an *anisotropic* material is timber, which exhibits quite different properties when loaded parallel to the grain and normal to the grain.

In practice we often encounter more complex stress systems than the simple uniaxial case. The 3D element in Figure 4.7 is subjected to direct stresses in the three axis directions, resulting in corresponding strains. We can calculate the total strain in each direction using the principle of superposition introduced in Chapter 2; we consider the effects of each stress acting independently, then simply add the results to give the overall effect.

With σ_{xx} only: $\varepsilon_{xx} = \sigma_{xx}/E, \quad \varepsilon_{yy} = \varepsilon_{zz} = -\nu\varepsilon_{xx} = -\nu\sigma_{xx}/E$

With σ_{yy} only: $\varepsilon_{yy} = \sigma_{yy}/E, \quad \varepsilon_{xx} = \varepsilon_{zz} = -\nu\sigma_{yy}/E$

With σ_{zz} only: $\varepsilon_{zz} = \sigma_{zz}/E, \quad \varepsilon_{xx} = \varepsilon_{yy} = -\nu\sigma_{zz}/E$

Table 4.1 *Mechanical properties of structural materials*

Material	Young's modulus, *GPa* ($= Pa \times 10^9$)	Poisson's ratio
Structural steel	205	0.3
Aluminium alloy	70	0.33
Concrete	20–25	0.2
Rubber	0.001–1	0.49

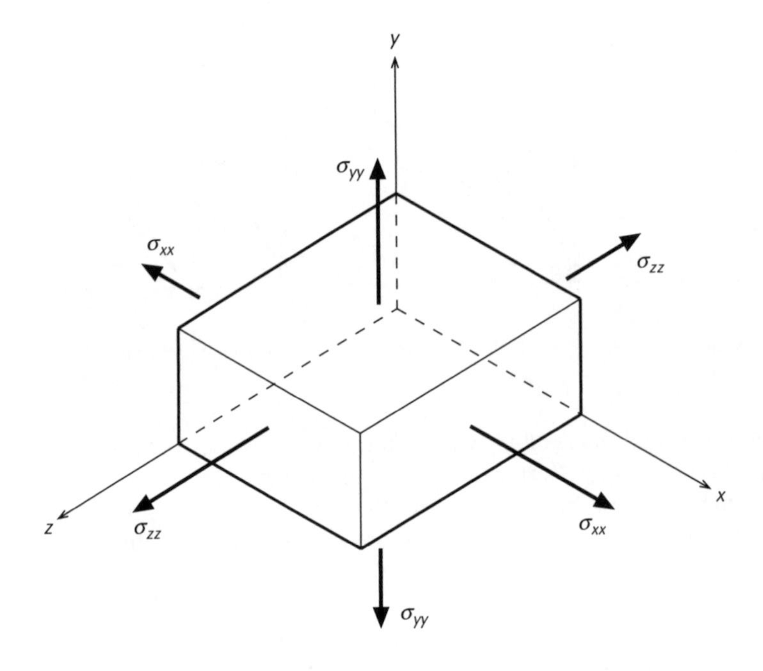

Figure 4.7
3D element subjected to three direct stresses.

Combining these, we get the *generalised form* of Hooke's law:

$$
\begin{aligned}
\varepsilon_{xx} &= \frac{1}{E}\left[\sigma_{xx} - \nu(\sigma_{yy} + \sigma_{zz})\right] \\
\varepsilon_{yy} &= \frac{1}{E}\left[\sigma_{yy} - \nu(\sigma_{zz} + \sigma_{xx})\right] \\
\varepsilon_{zz} &= \frac{1}{E}\left[\sigma_{zz} - \nu(\sigma_{xx} + \sigma_{yy})\right]
\end{aligned}
\tag{4.13}
$$

As with direct stresses and strains, we can show experimentally that shear stresses and strains are proportional for a linear elastic material:

$$
\tau_{xy} = G\gamma_{xy} \qquad \tau_{yz} = G\gamma_{yz} \qquad \tau_{zx} = G\gamma_{zx} \tag{4.14}
$$

G is called the *shear modulus* or *modulus of rigidity*, and can be thought of as the shear equivalent of Young's modulus. For shear, a strain in one plane has no effect on the perpendicular planes, so there is no need to introduce a shear equivalent for Poisson's ratio.

4.3.2 Statically determinate systems of parallel bars

If a system of bars is statically determinate, we can find all the bar forces using the methods described in Chapters 2 and 3. We can then use the stress–strain relationships to calculate how the bars deform. Consider a single bar of length l and cross-sectional area A, made from a material with Young's modulus E and Poisson's ratio ν. The bar sustains an axial tension T, causing it to extend by an amount e. From equations (4.1) and (4.7), the axial stress and strain are:

$$
\sigma_{xx} = \frac{T}{A} \qquad \varepsilon_{xx} = \frac{e}{l}
$$

For uniaxial loading $\sigma_{yy} = \sigma_{zz} = 0$, so Hooke's law (equation (4.11)) gives $(T/A) = E \cdot (e/l)$, or:

$$e = \frac{Tl}{AE} \qquad (4.15)$$

Over the next few chapters we shall frequently make use of the concept of the *stiffness* of a member or structure. Stiffness is defined as the force (or moment) required to produce a unit displacement (or rotation). Many different kinds of stiffness can be calculated for a member, depending on whether it is acting in bending, shear, torsion or whatever, so when we refer to stiffness we must always be clear about what type of structural action we are dealing with. For a simple bar in tension, we are concerned only with the axial stiffness k_A, which can be found by rearranging equation (4.15):

$$k_A = \frac{T}{e} = \frac{AE}{l} \qquad (4.16)$$

EXAMPLE 4.2

Figure 4.8 shows a simple structure comprising two steel bars joined coaxially. The structure is rigidly fixed at A and loaded axially at C. Determine the stress in each bar, the displacement of C and the axial stiffness of the structure.

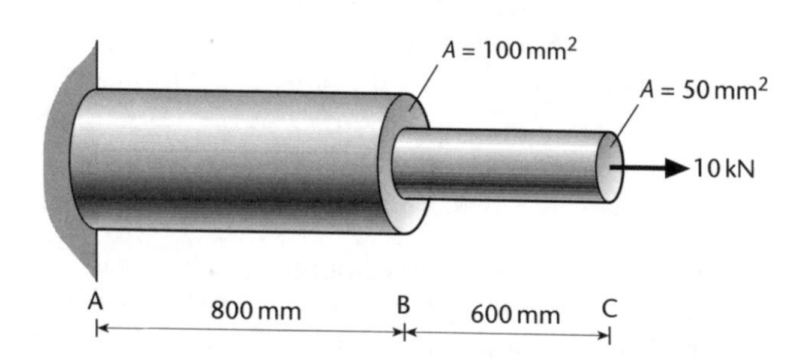

Figure 4.8

By inspection, both members carry only axial forces. Resolving horizontally at joints C and B gives $T_{AB} = T_{BC} = 10$ kN. So, working in units of N, mm and MPa, the stresses are given by equation (4.1):

$$\sigma_{AB} = \frac{10,000}{100} = 100 \text{ MPa}, \qquad \sigma_{BC} = \frac{10,000}{50} = 200 \text{ MPa}$$

The extensions are found from equation (4.15), taking the steel properties from Table 4.1:

$$e_{AB} = \frac{10,000 \times 800}{100 \times 205,000} = 0.39 \text{ mm}, \qquad e_{BC} = \frac{10,000 \times 600}{50 \times 205,000} = 0.59 \text{ mm}$$

So the total displacement of C = 0.39 + 0.59 = 0.98 mm

The axial stiffness of the structure is the force required to produce a unit displacement. We know that a force of 10 kN results in a displacement of 0.98 mm. So the stiffness is:

$$k_{ABC} = \frac{10,000}{0.98} = 10,200 \text{ N/mm}$$

Note that the individual stiffnesses of the two bars are: $k_{AB} = 10,000/0.39 = 25,640$ N/mm, and $k_{BC} = 10,000/0.59 = 16,950$ N/mm, so that the member and structure stiffnesses satisfy the relation:

$$k_{ABC}^{-1} = k_{AB}^{-1} + k_{BC}^{-1}$$

In fact, the coaxial bars behave just like two springs in series:

- the force in each bar is the same, and equal to the applied load;
- the total extension is the sum of the two individual extensions;
- the reciprocal of the total stiffness is the sum of the reciprocals of the two individual stiffnesses.

4.3.3 Redundant systems of parallel bars

For a redundant structure, statics alone will not enable us to calculate the internal forces. We need to generate some additional equations in order to solve the problem. These equations can be obtained using elasticity theory; we use Hooke's law to relate the unknown forces to the deformations, then apply the principle of *compatibility of displacements*, which states that, for a structure to hold together, the displacements of its various parts must be compatible with each other.

EXAMPLE 4.3

The structure in Figure 4.9(a) comprises two bars, each of length l, and with areas A_1 and A_2 and Young's moduli E_1 and E_2. The bars are connected to a rigid support at A and loaded via a very stiff endplate BC, in such a way that the endplate always remains parallel to the support. Find the forces in the bars and the displacement of the endplate BC.

Figure 4.9

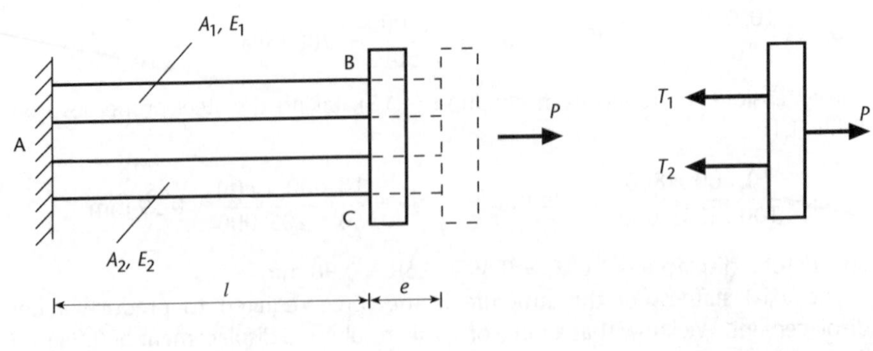

Although there are only two unknown bar forces, they cannot be found by statics alone as only horizontal resolution provides any useful information. To solve the problem we need one additional piece of information. The solution procedure is as follows:

First apply horizontal equilibrium – from the free body diagram for the plate BC (Figure 4.9(b)):

$$T_1 + T_2 = P$$

Now consider displacement compatibility – since the endplate is rigid and does not change its orientation during the loading, both bars must undergo the same extension e. The extension can be related to the bar forces using equation (4.15):

$$e = \frac{T_1 l}{A_1 E_1} = \frac{T_2 l}{A_2 E_2} \quad \text{hence } T_1 = \frac{k_1}{k_2} T_2 \quad \text{where } k_1 = \frac{A_1 E_1}{l}, k_2 = \frac{A_2 E_2}{l}$$

We now have two simultaneous equations for T_1 and T_2, which can be solved to give:

$$T_1 = \frac{k_1}{k_1 + k_2} P \qquad T_2 = \frac{k_2}{k_1 + k_2} P$$

And substituting back into the displacement equation gives:

$$e = \frac{P}{k_1 + k_2}$$

The bars thus behave like two springs in parallel:

- the extension in each bar is the same;
- the sum of the two bar forces equals the applied load;
- the internal force in each bar is proportional to its stiffness;
- the total stiffness is the sum of the two individual stiffnesses.

Example 4.3 has introduced an extremely important, fundamental principle, which will be used regularly in the remainder of this book. It is worth restating:

- *Statically determinate* structures can be analysed using *static equilibrium only*.
- *Redundant* structures require both *equilibrium* **and** *displacement compatibility*.

4.3.4 Thermal stresses and strains

Other effects besides external forces can cause internal stresses and deformations, one of the most common causes being changes in temperature. If a body free from restraints is subjected to a temperature rise θ it will expand in all directions, the strain in any direction given by:

$$\varepsilon = \alpha\theta \tag{4.17}$$

α is the *coefficient of linear thermal expansion*, which takes a constant value for a linear elastic material. Since strain is non-dimensional, α has units of $^\circ C^{-1}$.

Consider, for example, a circular steel bar 2.0 m long and 40 mm in diameter, subjected to a temperature rise of 30°C. For steel $\alpha = 12 \times 10^{-6}\,{}^\circ\mathrm{C}^{-1}$, so the strain in any direction due to the temperature change is:

$$\varepsilon = 30 \times 12 \times 10^{-6} = 3.6 \times 10^{-4}$$

The bar therefore extends by: $2000 \times 3.6 \times 10^{-4} = 0.72$ mm

and its diameter increases by: $40 \times 3.6 \times 10^{-4} = 0.0144$ mm.

Statically determinate structures provide no restraint to thermal movements; the members are therefore free to expand and contract, with no stresses induced. In redundant structures, however, the geometry of the structure can severely restrict the thermal expansion, causing large internal stresses.

Returning to the steel bar, suppose that, instead of being unrestrained, the bar is fixed between rigid endplates which prevent any axial movement. If unrestrained, we know that the bar would extend by 0.72 mm. The effect of the restraints must therefore be to impose a compressive load on the bar, preventing the extension. We can calculate the magnitude of the load using superposition:

- Consider first the temperature rise alone, without the restraint effect. As before $\varepsilon_{xx} = 3.6 \times 10^{-4}$.
- Now consider the restraint effect alone, with no temperature change. The restraint causes an axial stress σ_{xx} which in turn creates a strain $\varepsilon_{xx} = \sigma_{xx}/E$.
- We know that no extension is allowed, so the total strain is zero. Therefore, combining the two effects using superposition gives:

$$3.6 \times 10^{-4} + \sigma_{xx}/E = 0$$

For steel $E = 205$ GPa, so $\sigma_{xx} = -73.8$ MPa. Multiplying by the cross-sectional area of the bar, this stress corresponds to an axial compressive force of 92.7 kN.

EXAMPLE 4.4

The structure of Example 4.3 undergoes a temperature rise θ with no external load applied. If the bars have thermal expansion coefficients α_1 and α_2, find the forces in each bar.

If not joined together, the bars would extend by different amounts (unless $\alpha_1 = \alpha_2$). The endplate forces the bars to extend equally and therefore must set up axial forces in the bars. Again we must use both equilibrium and compatibility to find the forces in this redundant structure.

Equilibrium – the horizontal forces on BC are as in the previous example, except that $P = 0$, so:

$$T_1 + T_2 = 0$$

Displacement compatibility – the extensions of bars 1 and 2 can be found by superposition of the temperature and axial force effects, and must be equal.

Extensions due to temperature rise alone: $e_1 = l\alpha_1\theta, \quad e_2 = l\alpha_2\theta$

Due to axial forces alone: $e_1 = T_1 l/A_1 E_1 \quad e_2 = T_2 l/A_2 E_2$

Equating the total extensions: $l(\alpha_1\theta + T_1/A_1E_1) = l(\alpha_2\theta + T_2/A_2E_2)$

Solving the equilibrium and compatibility equations simultaneously:

$$T_1 = \frac{A_1E_1 \cdot A_2E_2(\alpha_2 - \alpha_1)\theta}{A_1E_1 + A_2E_2} = -T_2$$

Therefore if $\alpha_2 > \alpha_1$ then bar 1 is in tension and 2 in compression. If $\alpha_1 > \alpha_2$ then the reverse is true.

4.3.5 Strain energy

When an axial force is applied to a bar it causes an extension. The force thus moves through a certain distance and work is done on the bar. Assuming that there is no heat generated, and that the bar is stationary at the start and end of the loading, then the work done is all stored in the bar as potential energy. This particular form of potential energy is known as strain energy, given the symbol U.

Of course, a load cannot be applied instantaneously. Its magnitude must increase from zero over a period of time, and if the material remains linear then the extension will increase proportionately. Thus a plot of load against extension will be a straight line (Figure 4.10(a)). For a small displacement increment de starting from some general point (P, e) on the graph, the strain energy imparted to the bar is:

$$\mathrm{d}U = P\,.\,\mathrm{d}e = (AE/l)e \cdot \mathrm{d}e$$

The total strain energy caused by the load P_0 is found by integrating this expression from 0 to e_0:

$$U = \frac{AEe_0^2}{2l} = \frac{P_0e_0}{2} = \frac{P_0^2l}{2AE} \tag{4.18}$$

Thus the strain energy in an axially loaded member is simply the area under the load–extension curve.

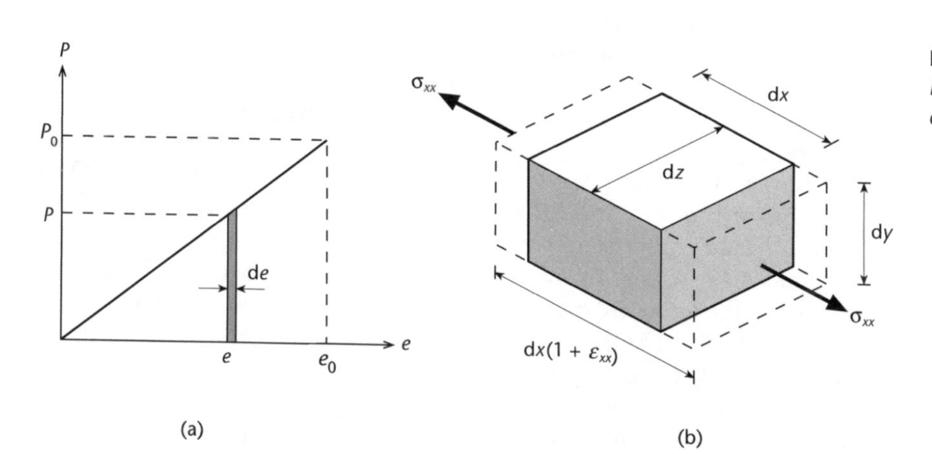

Figure 4.10
Derivation of strain energy in an element under axial load.

It is useful also to have a definition of strain energy in terms of stress and strain. Consider the small 3D element loaded by a single stress σ_{xx} (Figure 4.10(b)). The force on the element is $\sigma_{xx}\mathrm{d}y\mathrm{d}z$. If the element undergoes a small strain increment $\mathrm{d}\varepsilon_{xx}$ then the extension is $\mathrm{d}e = \mathrm{d}\varepsilon_{xx}\mathrm{d}x$. So, using the same approach as above, the increment in strain energy is:

$$\mathrm{d}U = (\sigma_{xx}\mathrm{d}y\mathrm{d}z)\cdot(\mathrm{d}\varepsilon_{xx}\mathrm{d}x)$$

Now a small element of volume $\mathrm{d}V = \mathrm{d}x\mathrm{d}y\mathrm{d}z$ and $\sigma_{xx} = E\varepsilon_{xx}$. Substituting in the above equation and integrating from 0 to ε_{xx}:

$$\frac{\mathrm{d}U}{\mathrm{d}V} = \frac{E\varepsilon_{xx}^2}{2} = \frac{\sigma_{xx}\varepsilon_{xx}}{2} = \frac{\sigma_{xx}^2}{2E} \tag{4.19}$$

Equation (4.19) states that the strain energy per unit volume is equal to the area under the stress–strain curve. A similar result can be obtained for an element subjected to a shear stress:

$$\frac{\mathrm{d}U}{\mathrm{d}V} = \frac{G\gamma_{xy}^2}{2} = \frac{\tau_{xy}\gamma_{xy}}{2} = \frac{\tau_{xy}^2}{2G} \tag{4.20}$$

4.4 Two-dimensional elasticity

There are many situations in which the elastic behaviour of a solid body is, or can be approximated as, essentially two-dimensional. In this section we will deal with two important types of 2D system:

- We shall concentrate on *plane stress* systems. As the name implies, all the stresses act in a 2D plane, normally assumed to be the x–y plane, so that $\sigma_{zz} = \tau_{yz} = \tau_{zx} = 0$. However, the loaded body may still deform in all three dimensions, as there may be a strain in the z direction due to the Poisson's ratio effect. The most common application of plane stress theory is to the analysis of thin plates.
- We shall also consider *plane strain* systems. In this case all the strain components lie within a 2D plane so that $\varepsilon_{zz} = \gamma_{yz} = \gamma_{zx} = 0$.

4.4.1 Resolution of plane stresses

We shall first consider the analysis of plane stress systems. Suppose we took a thin, square sheet of rubber and applied equal shear forces along the four edges, as shown in Figure 4.11(a). The two vertical shear forces will cause a relative vertical displacement between edges AB and CD. Similarly, the horizontal shear forces will cause edge BC to deflect horizontally relative to AD. As a result, the square will deform into a rhombus (Figure 4.11(b)). Obviously the deformation has caused the diagonal AC to lengthen and BD to shorten and we can see intuitively that a similar deformation could be achieved by pulling outwards on the corners A and C while simultaneously pushing inwards along BD (Figure 4.11(c)). So long as the magnitudes of the forces are chosen correctly, the two force systems in (a) and (c) are exactly equivalent. In fact, the same force system can be represented in many different ways, depending on the directions in which we choose to resolve.

Now consider a more general case. Figure 4.12(a) shows a set of positive stresses acting on the faces of a small rectangular element in the x–y plane. Since this is a

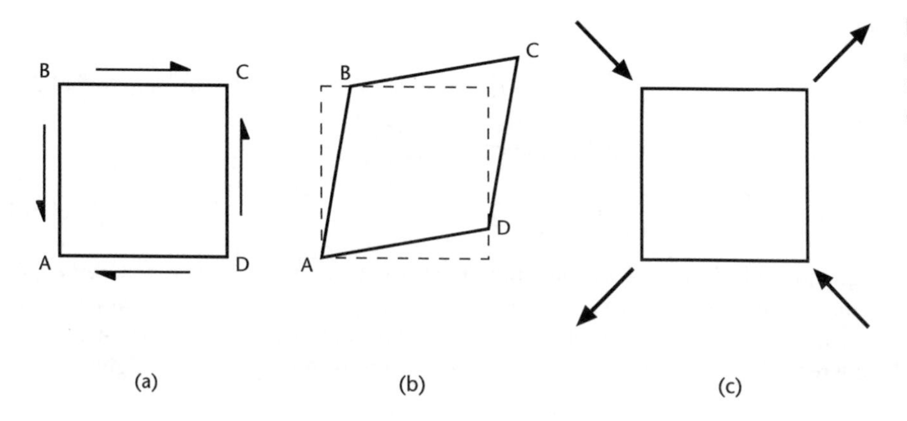

Figure 4.11
Equivalence of a shear force to perpendicular tensile and compressive forces.

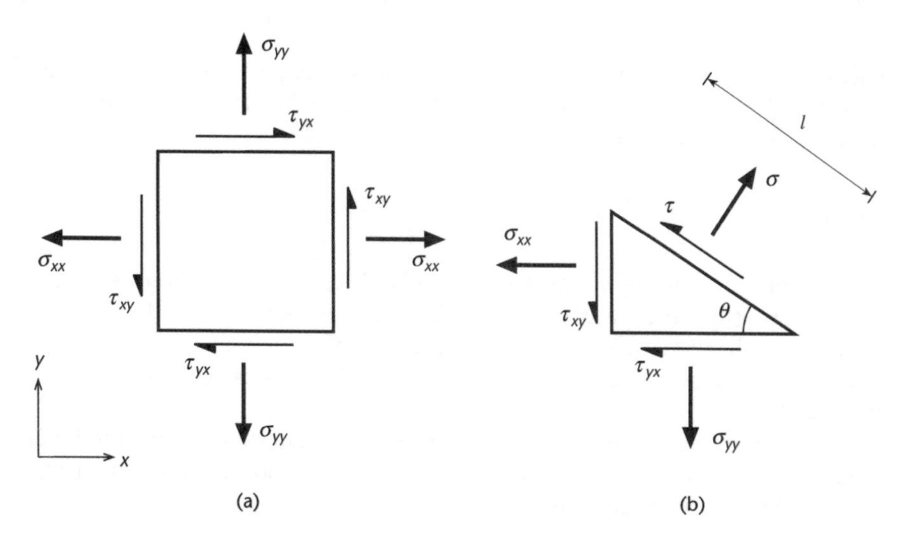

Figure 4.12
Elements under plane stress.

plane stress system, all stress components in the z direction are zero. The stresses shown follow the same naming and sign convention as used previously. For example σ_{xx} is the direct stress acting in a positive x direction on a positive x-face, or in a negative x direction on a negative x-face.

When designing a member to withstand a set of stresses we need to determine the largest stress present. For a system such as that shown in Figure 4.12(a) the value of the largest stress is far from obvious; the stress system has been resolved into x and y components, but we know that it could be resolved in any direction, and it is possible that a direction could be found in which a larger stress exists. Suppose we make an imaginary cut through the element at some angle θ (Figure 4.12(b)). We can find the stresses acting on the cut face by resolution, though we must take care to resolve forces *not* stresses, by multiplying by the area over which the stresses act. If the element has thickness t in the z direction then resolving normal to the cut gives:

$$\sigma(lt) = \sigma_{xx}(lt\sin\theta)\sin\theta + \sigma_{yy}(lt\cos\theta)\cos\theta + \tau_{xy}(lt\sin\theta)\cos\theta + \tau_{yx}(lt\cos\theta)\sin\theta$$

However, we know that $\tau_{xy} = \tau_{yx}$ so this simplifies to:

$$\sigma = \sigma_{xx}\sin^2\theta + \sigma_{yy}\cos^2\theta + 2\tau_{xy}\sin\theta\cos\theta$$

or

$$\sigma = \frac{\sigma_{xx} + \sigma_{yy}}{2} + \frac{\sigma_{xx} - \sigma_{yy}}{2}\cos 2\theta + \tau_{xy}\sin 2\theta \quad (4.21)$$

Similarly, resolving parallel to the cut gives:

$$\tau = \frac{\sigma_{xx} - \sigma_{yy}}{2}\sin 2\theta - \tau_{xy}\cos 2\theta \quad (4.22)$$

Equations (4.21) and (4.22) give the direct stress σ and shear stress τ at any angle from the original stresses, and are often referred to as the *stress transformation equations*. To find the directions of the extreme values of direct stress we differentiate (4.21) with respect to θ and put the result equal to zero, giving:

$$\tan 2\theta = \frac{2\tau_{xy}}{\sigma_{xx} - \sigma_{yy}} \quad (4.23)$$

Equation (4.23) yields two roots θ_1 and θ_2 in the range 0 to 180°, such that $|\theta_1 - \theta_2| = 90°$. Substituting these back into (4.21) gives the largest resolved direct stress, σ_1, and the smallest, σ_2:

$$\sigma_1, \sigma_2 = \frac{\sigma_{xx} + \sigma_{yy}}{2} \pm \left[\frac{(\sigma_{xx} - \sigma_{yy})^2}{4} + \tau_{xy}^2\right]^{1/2} \quad (4.24)$$

Furthermore, substituting either θ_1 or θ_2 into (4.22) gives $\tau = 0$. Therefore, for any plane stress system, there are two perpendicular planes on which the shear stress is zero; on one the direct stress is a maximum and on the other it takes its minimum value. These planes are called *principal planes* and the stresses acting on them are *principal stresses*.

We also need to find the maximum shear stress. We can show by differentiating (4.22) that this occurs on a plane inclined at 45° to the principal planes. Substituting this angle back into (4.22) gives:

$$\tau_{max} = \left[\frac{(\sigma_{xx} - \sigma_{yy})^2}{4} + \tau_{xy}^2\right]^{1/2} = \frac{\sigma_1 - \sigma_2}{2} \quad (4.25)$$

EXAMPLE 4.5

Figure 4.13 shows a small element of a thin plate, lying in the x–y plane. The x-faces are subjected to a tensile stress of 50 MPa and the y-faces to a compressive stress of 30 MPa. In addition, there is a shear stress of 30 MPa on all the faces. Find the largest and smallest direct stresses, σ_1 and σ_2, and their orientation θ to the x and y axes.

The results required can be obtained directly from the formulae derived above. The stresses are given by equation (4.24):

$$\sigma_1 = \frac{50 + (-30)}{2} + \left[\frac{[50 - (-30)]^2}{4} + 30^2\right]^{1/2} = 10 + 50 = 60 \text{ MPa}$$

$$\sigma_2 = 10 - 50 = -40 \text{ MPa}$$

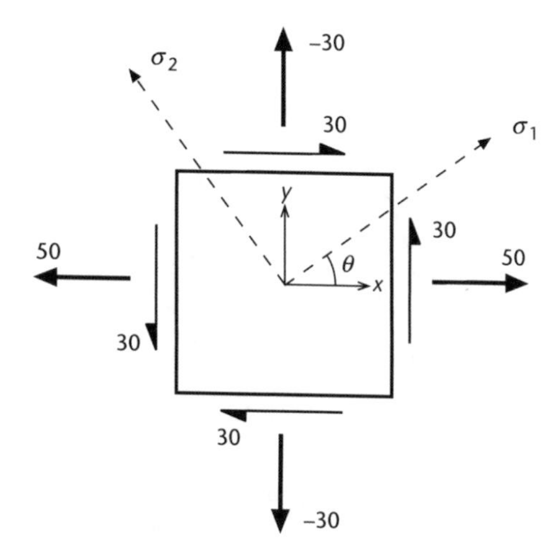

Figure 4.13

and the direction of the stresses is given by (4.23):

$$\theta = \frac{1}{2}\tan^{-1}\left[\frac{2 \times 30}{50 - (-30)}\right] = 18.4^\circ$$

So the system of x and y direction stresses given can be thought of as directly equivalent to two direct stresses of 60 MPa and −40 MPa acting in directions 18.4° anti-clockwise from the x and y axes, with no accompanying shear stress.

4.4.2 Mohr's circle for stress

The arithmetic involved in determining the principal stresses and their directions using the above equations can get rather tedious. However, the procedure can be greatly simplified by the use of a graphical construction known as Mohr's circle. If we take the square of equations (4.21) and (4.22) and add them together then, with a little rearrangement we get:

$$\left(\sigma - \frac{\sigma_{xx} + \sigma_{yy}}{2}\right)^2 + \tau^2 = \left(\frac{\sigma_{xx} - \sigma_{yy}}{2}\right)^2 + \tau_{xy}^2 \qquad (4.26)$$

If we plot equation (4.26) with shear on the y-axis and direct stress on the x-axis (Figure 4.14), then we get a circle with centre $(c, 0)$ and radius r, where:

$$c = \frac{\sigma_{xx} + \sigma_{yy}}{2} \qquad r = \left[\frac{(\sigma_{xx} - \sigma_{yy})^2}{4} + \tau_{xy}^2\right]^{1/2} \qquad (4.27)$$

The significance of this will become clear if we consider the meaning of a few points on the circle. The coordinates of point A are $(\sigma, \tau) = (c - r, 0)$, while those of point B are $(c + r, 0)$. Substituting for c and r and comparing with (4.24), we find that $c - r = \sigma_2$ and $c + r = \sigma_1$. Thus the points A and B represent the state of stress on the two principal planes. The coordinates of C are (c, r). Comparing our expression for r with (4.25) we can see that C gives us the maximum shear stress.

In fact, every point on the circumference of the circle represents the state of

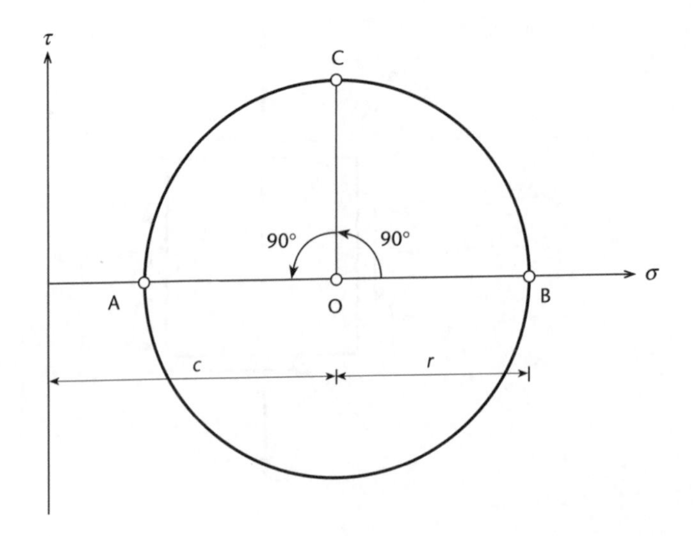

Figure 4.14
Mohr's circle for stress.

stress at a different orientation in the x–y plane. How are these orientations related to the position on the circle? We have seen that in the physical system the two principal stresses are separated by an angle of 90° and that the maximum shear stress acts on a plane at 45° to the principal planes. On the Mohr's circle the radii OA and OB are 180° apart and the radius OC is at 90° to these two. Thus the angle of rotation between sets of stresses in the Mohr's circle is *twice* the corresponding angle of rotation in the physical system.

Although the analysis from which it is derived is quite complex, the construction of Mohr's circle in practice is very simple. We start with a system of stresses on a small element in the x–y plane (Figure 4.15(a)). These stresses have all been drawn as positive. However, when plotting the Mohr's circle, we adopt a slightly different sign convention for the shear stresses. A shear stress is plotted as positive on the

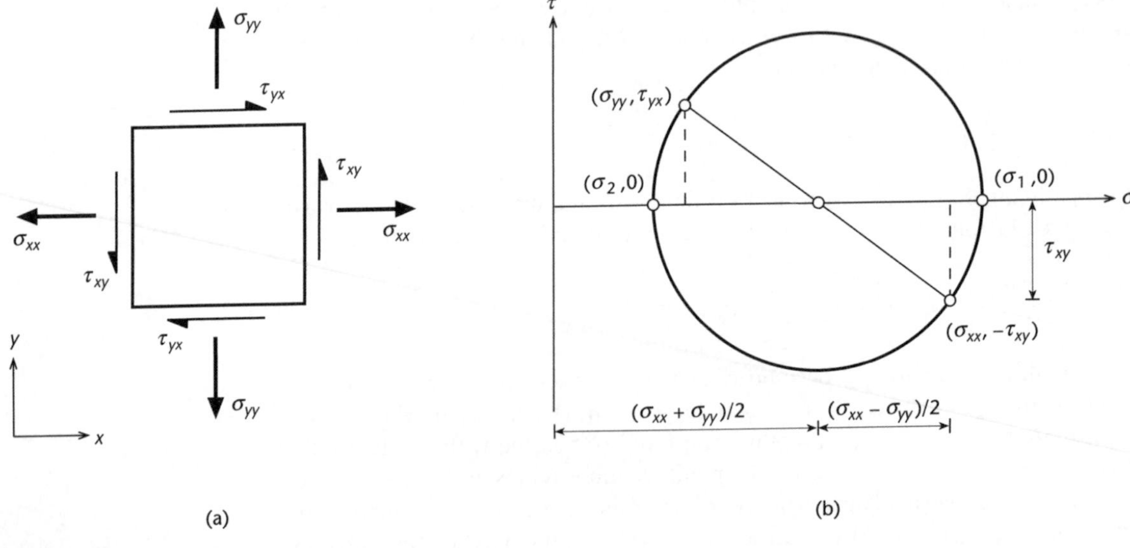

Figure 4.15
Construction of Mohr's circle.

Mohr's circle if it causes a *clockwise* moment about the centre of the element. For our element τ_{xy} therefore plots as negative and τ_{yx} as positive.

We thus have the coordinates of two points on a graph of τ against σ; the stresses on the x faces give the point $(\sigma_{xx}, -\tau_{xy})$ and those on the y-faces give (σ_{yy}, τ_{yx}). Since the x- and y-faces are 90° apart in the physical plane, their stresses will plot 180° apart on Mohr's circle. They therefore lie at opposite ends of a diameter, and where this crosses the σ-axis is the centre of the circle. We can now draw in the circle (Figure 4.15(b)). The σ coordinate of the centre, c, is always the mean of any two perpendicular direct stresses. The radius r can then be worked out from the geometry of the circle. Knowing c and r, it is a simple matter to determine the principal stresses, the maximum shear stress and their orientations.

EXAMPLE 4.6

Returning to the stress system of Example 4.5, determine:

(i) the stresses on an element rotated 45° anti-clockwise from the element shown;
(ii) the principal stresses and their directions;
(iii) the maximum shear stress and its direction.

The original stress system is reproduced in Figure 4.16(a). We first draw Mohr's circle (Figure 4.16(b)). The shear stresses on the x-faces give an anti-clockwise moment about the centre of the element and therefore plot as negative. So the stresses on the x-faces give us the point A = (50, −30) and those on the y-faces give us B = (−30, 30). These points lie at the ends of a diameter of the circle, so the circle can now be drawn and the centre coordinate c and radius r found:

- $c = (-30 + 50)/2 = 10$;
- the length OC = 50 − 10 = 40 and AC = 30 so, by Pythagoras: $r = \sqrt{(40^2 + 30^2)} = 50$;
- the angle $2\theta = \tan^{-1}(30/40) = 36.8°$.

(i) To find the stresses at 45° to those given, we rotate AB through 2 × 45 = 90° on the Mohr's circle. The required stresses are then given by the points D and E.

At D: $\sigma = 10 + 50\cos(90° - 36.8°) = 40$ MPa
$\tau = 50\sin(90° - 36.8°) = 40$ MPa

At E: $\sigma = 10 - 50\cos(90° - 36.8°) = -20$ MPa
$\tau = -50\sin(90° - 36.8°) = -40$ MPa

Figure 4.16(c) shows the stresses acting on the faces of a small element oriented at 45° to the element in (a). Note the directions of the shear stresses; those associated with the 40 MPa direct stress form a clockwise couple because they plot as positive on the Mohr's circle.

(ii) The principal stresses are simply:

$\sigma_1 = 10 + 50 = 60$ MPa $\qquad \sigma_2 = 10 - 50 = -40$ MPa

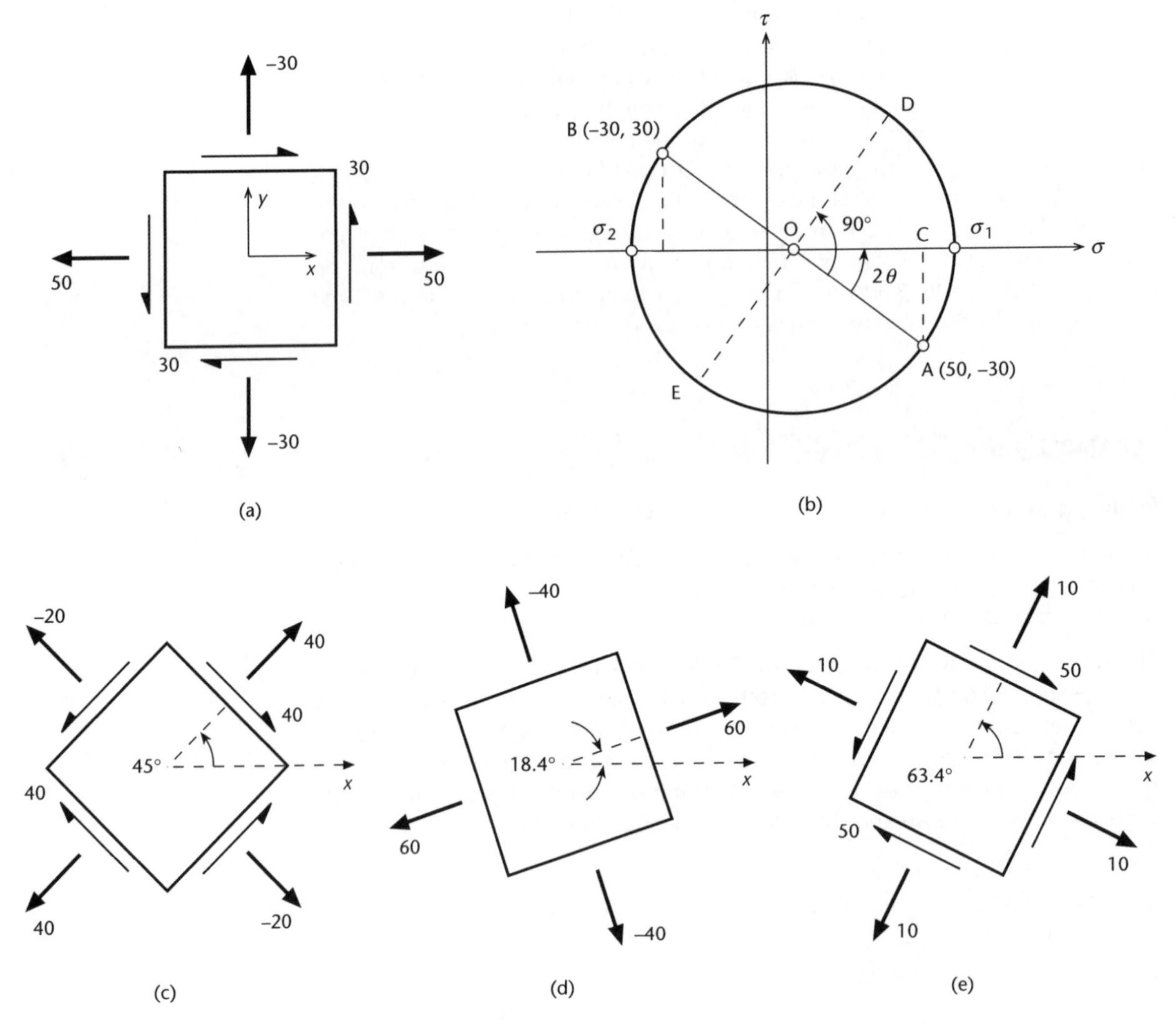

Figure 4.16

and σ_1 makes an angle of $36.8/2 = 18.4°$ with the *x*-axis. These results are in agreement with Example 4.5. The stresses on the principal planes are shown in Figure 4.16(d).

(iii) The maximum shear stress is equal to the radius of the circle: $\tau_{max} = 50$ MPa. On the Mohr's circle τ_{max} plots as positive at an angle of $90 + 36.8 = 126.8°$ from σ_{xx}. So, in the physical system, the planes on which the shear stresses act in a clockwise sense make an angle of $126.8/2 = 63.4°$ with the *x*-axis. The stresses are shown in Figure 4.16(e). Note that the four perpendicular faces on which the shear stress is a maximum must all be subjected to the same direct stress.

It should be remembered that the diagrams (a), (c), (d) and (e) all represent the same stress system. The only difference between them is that the stresses have been resolved into components in different directions.

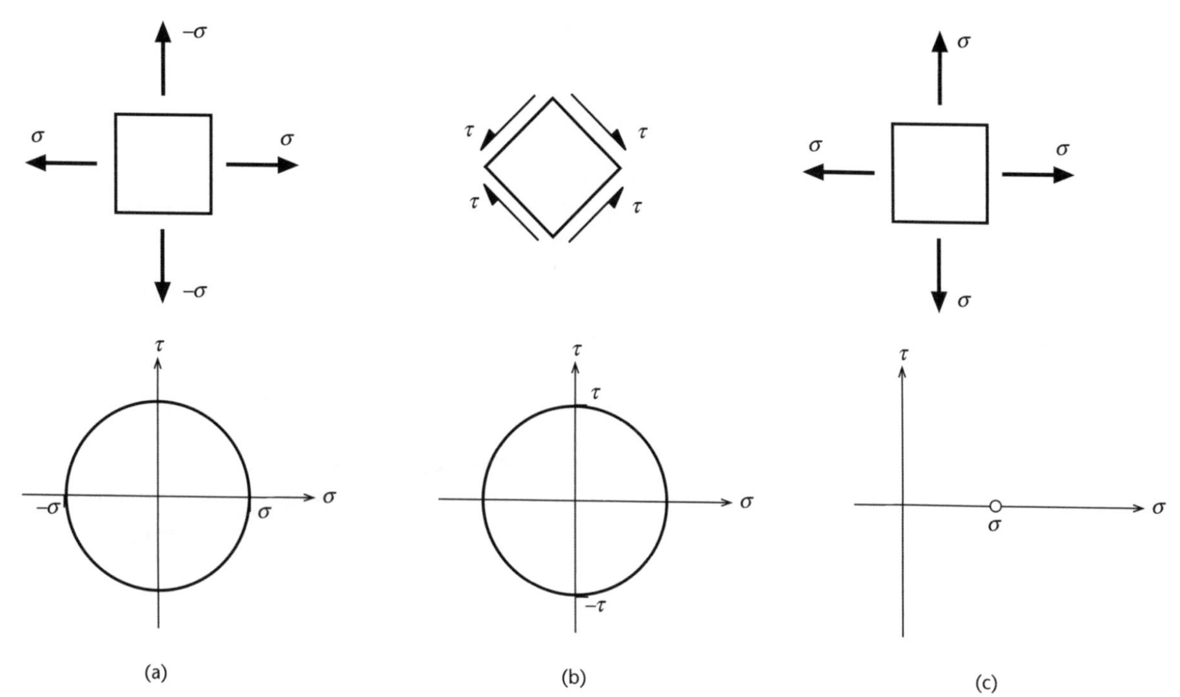

Figure 4.17
Mohr's circles for simple stress systems.

Figure 4.17 shows the Mohr's circles for some commonly encountered stress systems. In (a) the stresses on the element are principal stresses, since there are no shear stresses. Since the two stresses are equal in magnitude but opposite in sign, the Mohr's circle is centred on the origin. The element in (b) is loaded in pure shear, and again results in a circle centred on the origin. Since the two Mohr's circles are identical, it follows that (a) and (b) are two representations of the same stress system. We have now proved what we stated intuitively at the start of this section – that a set of shear stresses can be thought of as equivalent to direct tensile and compressive stresses of equal magnitude, acting at 45° to the shear stresses. This can be extremely useful in helping to visualise the effect of a shear loading.

In (c) there are two identical principal stresses. Since these lie at opposite ends of a diameter, the Mohr's circle must have zero radius. In this system, therefore, the stresses are the same in any direction!

4.4.3 Mohr's circle for strain

Two-dimensional systems of strain can be resolved to find components in various directions in much the same way as stresses. The principal strains are given by:

$$\varepsilon_1, \varepsilon_2 = \frac{\varepsilon_{xx} + \varepsilon_{yy}}{2} \pm \left[\frac{(\varepsilon_{xx} - \varepsilon_{yy})^2}{4} + \frac{\gamma_{xy}^2}{4} \right]^{1/2} \tag{4.28}$$

and the maximum shear strain by:

$$\frac{\gamma_{max}}{2} = \left[\frac{(\varepsilon_{xx} - \varepsilon_{yy})^2}{4} + \frac{\gamma_{xy}^2}{4}\right]^{1/2} \tag{4.29}$$

These equations are very similar in form to (4.24) and (4.25). Each direct stress term has simply been replaced by the corresponding direct strain, and each shear stress term has been replaced by *half* the shear strain.

Mohr's circle for strain can be deduced from the stress version in exactly the same way. We obtain a circle by plotting $\gamma/2$ against ε. In all other respects the procedure is identical to that outlined in the preceding section.

If we have already done a Mohr's circle analysis of the stresses, there is little point in repeating the exercise in terms of strain. The principal directions will, of course, be unchanged, and the desired strain values can be deduced from the stresses using the expanded form of Hooke's law. Where Mohr's circle for strain comes into its own is in the interpretation of experimental strain readings. When performing structural experiments we find it is very difficult to measure internal stresses and shear strains within members. It is, however, very easy to measure direct strains using electrical resistance strain gauges. If we measure the direct strains at a point in three different directions, we can use Mohr's circle to analyse the strain system and then convert the results to stresses using Hooke's law.

EXAMPLE 4.7

The strain gauge rosette shown in Figure 4.18(a) is glued to the surface of a thin steel plate and measures direct strains in the x and y directions, and at 45° to the x and y axes. The gauge readings under a particular loading are:

$$\varepsilon_A = -100\mu\varepsilon, \qquad \varepsilon_B = 250\mu\varepsilon, \qquad \varepsilon_C = 400\mu\varepsilon$$

Note: Because values of strain are generally small, a common unit is the *microstrain* ($\mu\varepsilon$). $1\mu\varepsilon = 1 \times 10^{-6}$.

Figure 4.18

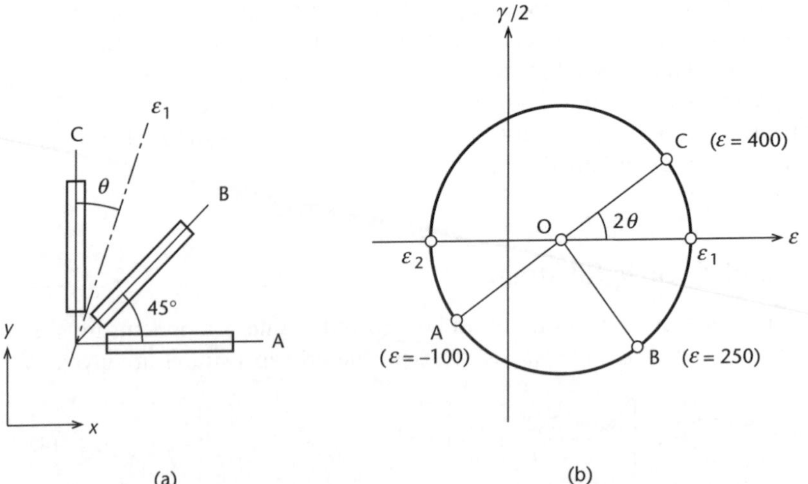

Determine the principal strains and their directions, and the corresponding principal stresses.

We start by sketching the Mohr's strain circle (Figure 4.18(b). Unless A and C are the principal directions, ε_1 will be larger than 400 and ε_2 smaller than -100. This allows us to get the dimensions of the circle roughly right.

Next, we mark the approximate positions of A, B and C on the circle. We know the values of direct strain in these directions, but not of the corresponding shear strains, so at this stage we cannot locate A, B and C exactly. However, the largest principal strain, ε_1, must occur in a direction quite close to the largest measured strain. We therefore assume that its direction is θ clockwise from ε_C; if this assumption is wrong it will quickly become obvious as we plot the diagram. The radius OC is then 2θ anti-clockwise from ε_1. OB is $2 \times 45 = 90°$ clockwise from OC and OA is a further 90° clockwise from OB.

Figure 4.18(b) contains three unknowns – the centre coordinate c, radius r and angle θ. These can be found using the three values of direct strain at A, B and C.

A: $\quad -100 = c - r\cos 2\theta$

B: $\quad 250 = c + r\sin 2\theta$

C: $\quad 400 = c + r\cos 2\theta$

Solving simultaneously: $\quad c = 150, r = 269, 2\theta = 21.8°$

So the principal strains are:

$$\varepsilon_1 = 150 + 269 = 419\mu\varepsilon \text{ at } 10.9° \text{clockwise from gauge C}$$

$$\varepsilon_2 = 150 - 269 = -119\mu\varepsilon \text{ at } 10.9° \text{clockwise from gauge A}$$

The principal stresses can be found using Hooke's law. With stresses of σ_1 and σ_2 in the plane of the plate and zero stress normal to the plate, equation (4.13) gives:

$$\varepsilon_1 = \frac{1}{E}(\sigma_1 - \nu\sigma_2) \qquad \varepsilon_2 = \frac{1}{E}(\sigma_2 - \nu\sigma_1)$$

These can be rearranged to give:

$$\sigma_1 = \frac{E}{1-\nu^2}(\varepsilon_1 + \nu\varepsilon_2) \qquad \sigma_2 = \frac{E}{1-\nu^2}(\varepsilon_2 + \nu\varepsilon_1)$$

From Table 4.1, for steel $E = 205$ GPa and $\nu = 0.3$, so:

$$\sigma_1 = 86.3 \text{ MPa}, \sigma_2 = 1.5 \text{ MPa}$$

The strains given are therefore caused by a state of stress very close to uniaxial tension.

4.4.4 Qualitative analysis of plane stress systems

To get a better feel for what is going on, and as a check on the mathematical analysis, it is instructive to perform some qualitative analysis of plane stress systems. In doing this, we will find it helpful to make use of the equivalence between a set of shear stresses and 45° tensile and compressive stresses (Figure 4.17).

EXAMPLE 4.8

For the element shown in Figure 4.19(a), make a rough estimate of the directions of the principal stresses and sketch the deformed shape.

Figure 4.19

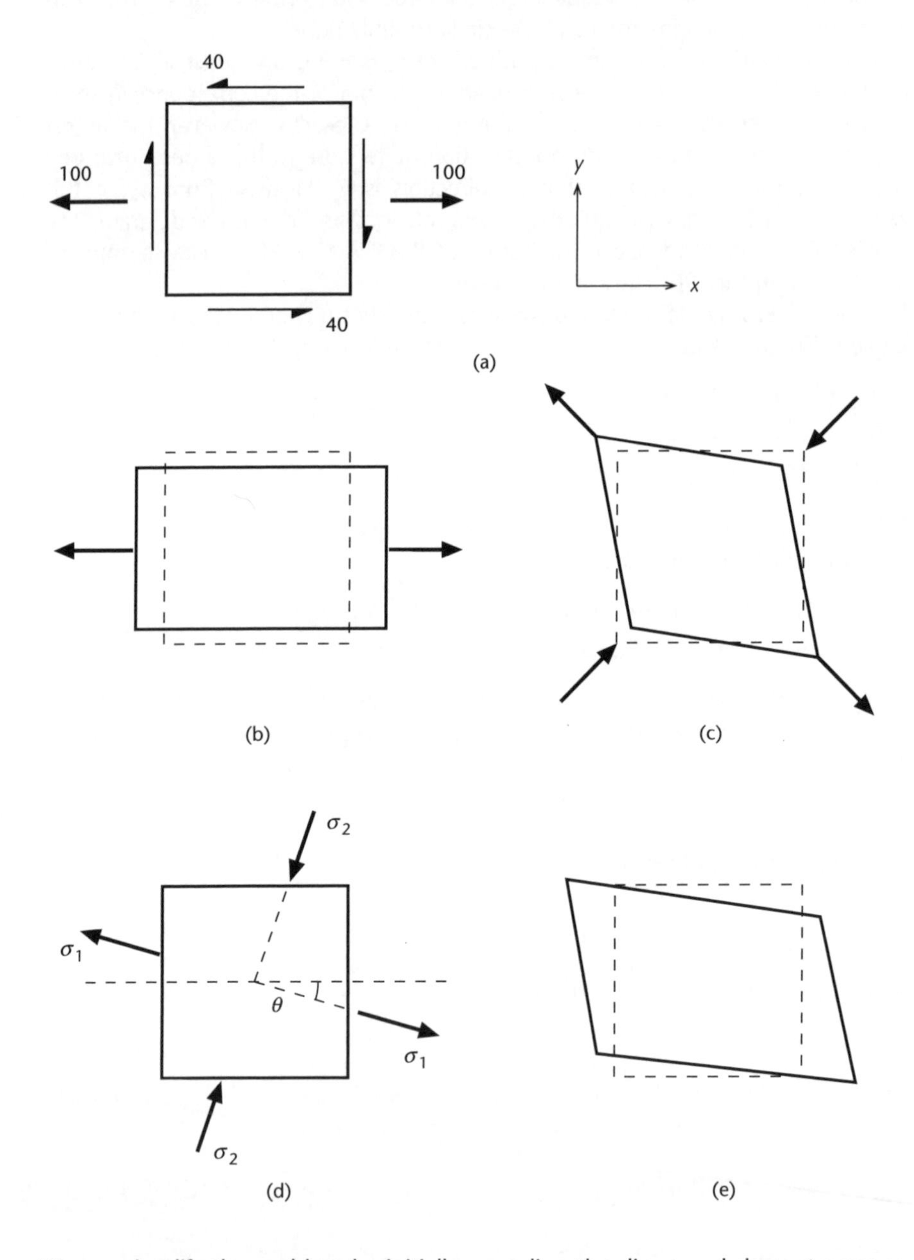

We can simplify the problem by initially regarding the direct and shear stresses as separate stress systems. The direct stresses and corresponding deformations are shown in Figure 4.19(b). There is a tensile principal stress in the x direction, zero stress in the y direction and as a result the element extends in the x direction and narrows slightly due to the Poisson's ratio effect. We have already seen that a system of shear stresses can be thought of as equivalent to a tensile principal stress along one diagonal and a compression along the other, distorting the square element into a rhombus (Figure 4.19(c)).

Combining the two sets of principal stresses in (b) and (c) gives the overall principal stresses shown in (d). The largest tensile stress σ_1 acts at an angle θ clockwise from the x-axis. θ must be between 0 and 45° – think of σ_1 as the vector sum of the tensile stresses in (b) and (c). Given the relative magnitudes of the applied direct and shear stresses, θ is likely to be less than 22.5°. (In fact, an exact calculation gives $\theta = 19.3°$.) σ_2 acts perpendicular to σ_1 and must be compressive.

Finally, the deformed shape of the element is found by combining the deformations in (b) and (c) and is shown in Figure 4.19(e).

4.4.5 Relations between elastic constants

In section 4.3 we introduced three constants governing the stress–strain behaviour of an elastic material: Young's modulus E, the shear modulus G and Poisson's ratio ν. In fact, only two constants (usually E and ν) are needed to define the material behaviour, since the third can be shown to be dependent on the other two.

Consider a small 2D element loaded in pure shear (Figure 4.20(a)). The shear stresses τ cause corresponding shear strains $\gamma = \tau/G$. As we have already seen, Mohr's circle for stress is centred on the origin and has radius τ (Figure 4.20(b)). The principal stresses are therefore:

$$\sigma_1 = \tau, \qquad \sigma_2 = -\tau$$

Similarly, the Mohr's circle for strain has radius $\gamma/2$, Figure 4.20(c), so that the principal strains are:

$$\varepsilon_1 = \gamma/2 = \tau/2G, \quad \varepsilon_2 = -\gamma/2 = -\tau/2G$$

Applying Hooke's law (equation (4.13)), in the direction of the largest principal stress:

$$\varepsilon_1 = \frac{1}{E}(\sigma_1 - \nu\sigma_2)$$

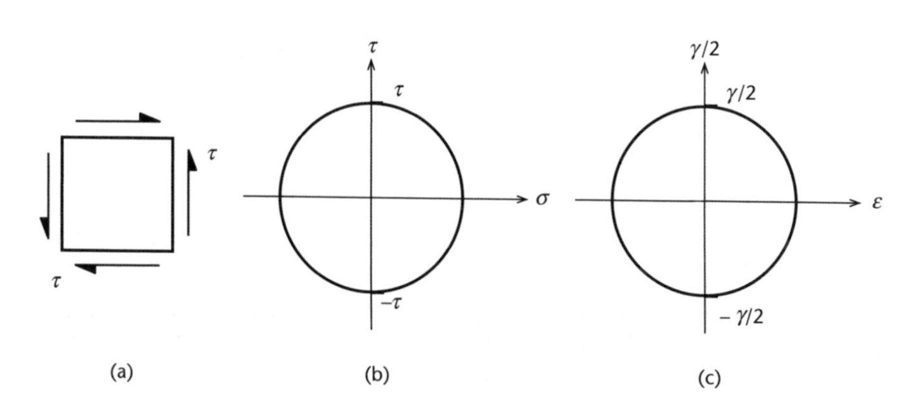

Figure 4.20
Derivation of expression for shear modulus.

Substituting from above:

$$\frac{\tau}{2G} = \frac{1}{E}\left(\tau - \nu(-\tau)\right)$$

and rearranging gives the expression for G:

$$G = \frac{E}{2(1+\nu)} \tag{4.30}$$

We shall also introduce another important elastic constant, the *bulk modulus K*. This is used in three-dimensional stress analysis to relate the average stress or pressure on a body to its volumetric strain $\mathrm{d}V/V$:

$$\sigma_{\text{ave}} = K\frac{\mathrm{d}V}{V} \tag{4.31}$$

Again we can show using Hooke's law that K is a function of E and ν. Imagine a small 3D element with dimensions $\mathrm{d}x \times \mathrm{d}y \times \mathrm{d}z$, subjected to stresses σ_{xx}, σ_{yy} and σ_{zz}. If the resulting strains are ε_{xx}, ε_{yy}, ε_{zz}, then the volumetric strain is the change in volume of the element divided by its original volume:

$$\frac{\mathrm{d}V}{V} = \frac{\mathrm{d}x(1+\varepsilon_{xx})\cdot\mathrm{d}y(1+\varepsilon_{yy})\cdot\mathrm{d}z(1+\varepsilon_{zz}) - \mathrm{d}x\cdot\mathrm{d}y\cdot\mathrm{d}z}{\mathrm{d}x\cdot\mathrm{d}y\cdot\mathrm{d}z}$$

If the strains are all small, then products of strain terms can be neglected and this simplifies to:

$$\frac{\mathrm{d}V}{V} = \varepsilon_{xx} + \varepsilon_{yy} + \varepsilon_{zz}$$

Hooke's law gives:

$$E\varepsilon_{xx} = \sigma_{xx} - \nu\sigma_{yy} - \nu\sigma_{zz}$$
$$E\varepsilon_{yy} = \sigma_{yy} - \nu\sigma_{zz} - \nu\sigma_{xx}$$
$$E\varepsilon_{zz} = \sigma_{zz} - \nu\sigma_{xx} - \nu\sigma_{yy}$$

Adding:

$$E(\varepsilon_{xx} + \varepsilon_{yy} + \varepsilon_{zz}) = (\sigma_{xx} + \sigma_{yy} + \sigma_{zz})(1 - 2\nu)$$

Noting that the average stress is $\sigma_{\text{ave}} = (\sigma_{xx} + \sigma_{zz} + \sigma_{zz})/3$ and substituting from above:

$$E\frac{\mathrm{d}V}{V} = 3\sigma_{\text{ave}}(1 - 2\nu)$$

and, comparing with (4.31), the bulk modulus is:

$$K = \frac{E}{3(1-2\nu)} \tag{4.32}$$

It can be seen that K becomes infinite if Poisson's ratio equals 0.5, implying that any set of stresses, however large, will cause no volume change. Many rubbers have values of ν very close to 0.5.

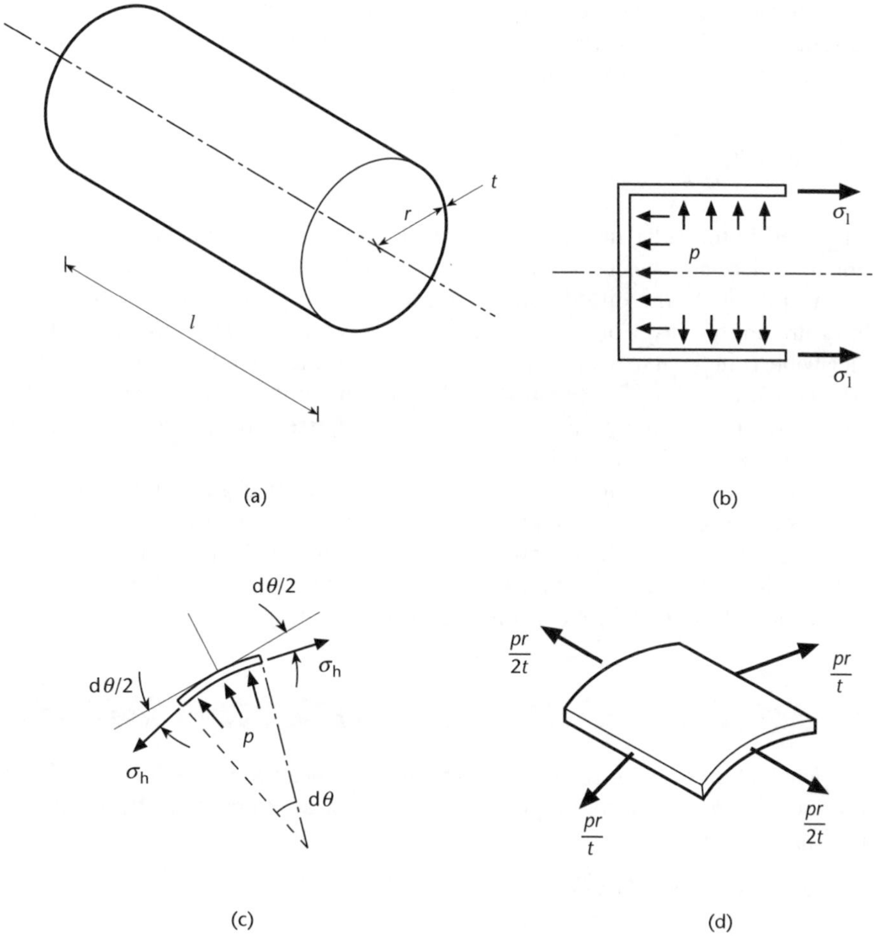

Figure 4.21
Stresses in a cylindrical pressure vessel.

(a) (b)

(c) (d)

4.4.6 Pressure vessels

One very important application of plane stress theory is the analysis of thin-walled pressure vessels. If a vessel contains a fluid under a pressure p, the fluid will apply this pressure uniformly and at right angles to every surface with which it comes into contact. The resulting loads will be carried almost entirely in the form of stresses in the plane of the vessel wall.

Consider, for example, the cylindrical pressure vessel with flat, closed ends in Figure 4.21(a). Instead of x, y and z components, it is more convenient to resolve the stresses into a *radial* component σ_r, a tangential or *hoop* stress σ_h and a longitudinal stress σ_l. Because of the axial symmetry of the problem, these stresses will take constant values around the circumference. As the vessel is thin-walled, we may also assume that σ_l and σ_h do not vary over the wall thickness.

The longitudinal stress can be found by considering equilibrium of a section formed by making an imaginary cut at right angles to the cylinder axis (Figure 4.21(b)). Resolving along the cylinder axis:

$$p \cdot \pi r^2 = \sigma_l \cdot 2\pi rt \qquad \text{hence } \sigma_l = \frac{pr}{2t} \tag{4.33}$$

Next consider the equilibrium of a section of the vessel wall formed by rotating through a small angle $\mathrm{d}\theta$ (Figure 4.21(c)). If the element has unit length going into the page then, resolving radially:

$$p \cdot r\mathrm{d}\theta = 2\theta_{\mathrm{h}} t \frac{\mathrm{d}\theta}{2} \qquad \text{hence} \quad \sigma_{\mathrm{h}} = \frac{pr}{t} \tag{4.34}$$

The stress in the radial direction can be deduced from the boundary conditions. At the inner face σ_{r} provides a normal reaction to the internal pressure and so takes the value $-p$. At the outside face it must reduce to zero as there is no applied pressure on this face. For a thin-walled vessel r is much larger than t, so that σ_{r} is negligible compared to σ_{l} and σ_{h}. The stresses on a small element of the vessel wall are therefore as shown in Figure 4.21(d). It can be shown by consideration of the axial symmetry of the problem that the shear stresses on this element must be zero, so that σ_{l} and σ_{h} are principal stresses.

If the vessel is not thin-walled then σ_{r} ceases to be negligible and none of the stresses can be assumed constant over the wall thickness. The stress system is then genuinely three-dimensional and the above analysis is no longer valid. The reader is referred to specialist solid mechanics texts for the solution of this problem (see the reading list in section 4.6 at the end of this chapter).

EXAMPLE 4.9

A cylindrical pressure vessel with closed ends has mean radius 600 mm and wall thickness 20 mm. Find the longitudinal and hoop stresses generated by an internal pressure of 25 MPa.

Equations (4.33) and (4.34) give:

$$\sigma_{\mathrm{l}} = \frac{25 \times 600}{2 \times 20} = 375 \text{ MPa} \qquad \sigma_{\mathrm{h}} = \frac{25 \times 600}{20} = 750 \text{ MPa}$$

Clearly very large stresses can be developed by quite modest internal pressures.

4.4.7 Plane strain problems

Many three-dimensional stress systems can be greatly simplified if the strains in one or more directions can be shown to be zero or negligibly small. This is best illustrated by means of an example.

EXAMPLE 4.10

A bridge bearing consists of a layer of rubber, 20 mm thick and 250 mm square in plan, completely enclosed by steel plates, as shown in Figure 4.22(a). Determine its stiffness under vertical loading. Take E for the rubber to be 10 MPa and ν to be 0.49.

Figure 4.22

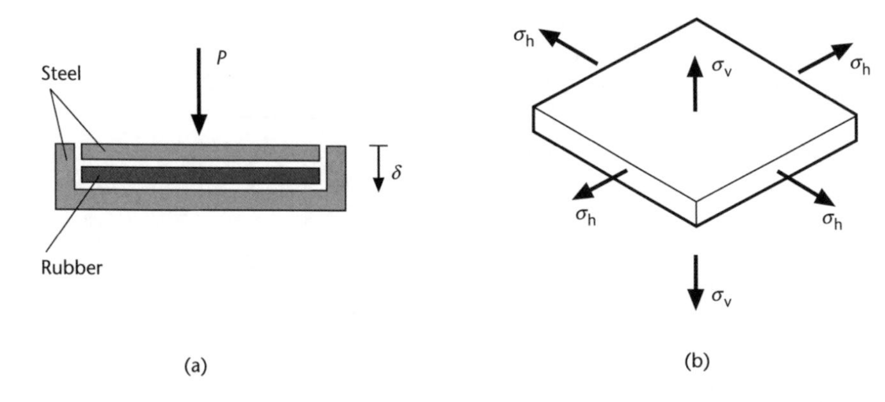

Since E for the steel is 20,500 times that for the rubber, the steel plates may be regarded as completely rigid. The rubber is entirely contained by the rigid steel casing and so can only deform vertically, that is, the horizontal strains are zero. The stresses exerted on the rubber are shown in Figure 4.22(b) – following convention these are drawn as tensile, though it is obvious that they will in fact be compressive. It is also obvious that the two horizontal stresses must be equal.

Applying Hooke's law (equation (4.13)), in either horizontal direction:

$$0 = \frac{\sigma_h - \nu(\sigma_v + \sigma_h)}{E} \qquad \text{hence } \sigma_h = \frac{\nu\sigma_v}{1-\nu}$$

And in the vertical direction:

$$\varepsilon_v = \frac{\sigma_v - \nu(\sigma_h + \sigma_h)}{E} = \frac{\sigma_v\left[1 - 2\nu^2/(1-\nu)\right]}{E}$$

Now if a vertical force P produces a displacement δ then $\sigma_v = P/A$ and $\varepsilon_v = \delta/l$, so:

$$\frac{\delta}{l} = \frac{P\left[1 - 2\nu^2/(1-\nu)\right]}{AE}$$

The vertical stiffness is the force divided by the corresponding displacement, so rearranging the above:

$$k_v = \frac{P}{\delta} = \frac{AE}{l[1 - 2\nu^2/(1-\nu)]}$$

The required values are $A = 62,500$ mm^2, $l = 20$ mm, $E = 10$ MPa, $\nu = 0.49$, giving:

$$k_v = 534,815 \text{ N/mm} = 535 \text{ kN/mm}$$

Clearly, the bearing is very stiff even though it is made from an apparently flexible material (that is, one with a low Young's modulus). In this case the stiffness comes from the combination of two effects:

- the confinement provided by the steel, preventing the rubber from expanding laterally under the load; and
- the high Poisson's ratio of the rubber, which in turn gives it a high value of bulk modulus (equation 4.32), so that it is highly resistant to volume change.

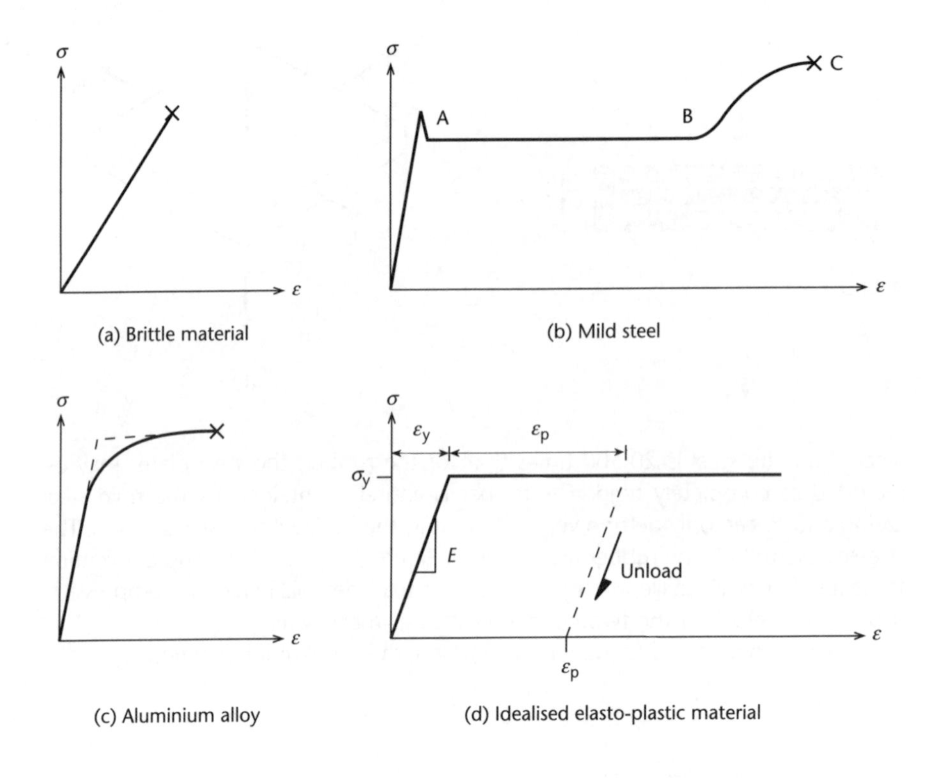

Figure 4.23 *Real and idealised stress–strain curves.*

4.5 Elementary plasticity

4.5.1 Yielding

Thus far we have confined ourselves to materials that are linear and elastic. While many materials initially behave in this way, as the load is increased their behaviour changes. Some *brittle* materials suddenly fracture when the stress reaches a critical level. However, most materials used in engineering undergo the phenomenon of *yielding* prior to fracture, and are termed *ductile*. Engineers generally avoid brittle materials, as they can break suddenly, whereas ductile materials first undergo large deformations, giving some warning of impending failure.

When a material yields, slip occurs between adjacent planes of atoms, and large deformations are sustained with little or no increase in load. These deformations are irrecoverable, so that if the load is removed the member does not return to its original dimensions. Yielding behaviour can be best characterised by looking at the stress–strain curve for a sample of material tested in tension. Some typical curves are shown in Figure 4.23.

(a) shows the behaviour of a brittle material such as glass. The curve is linear right up to the fracture point, indicated by the cross.

(b) is a typical stress–strain curve for mild steel (the cheapest and most common type of steel). The behaviour is initially linear elastic, until the yield point A is reached. At this point the stress drops slightly and there is then a long plateau AB, where the strain increases substantially with no increase in stress. This is known as the *plastic* range. The strain at B is typically twelve times that at A. Beyond B the specimen undergoes *work hardening*, causing an increase in stress before failure occurs at C.

(c) is the stress–strain curve for a material that does not have a sharply defined yield point, such as an aluminium alloy or a high-strength steel.

Calculations performed using very accurately defined material characteristics can become extremely complex. The accuracy cannot usually be justified given the level of uncertainty over other aspects of the analysis. We shall therefore make use of the idealised stress–strain curve shown in Figure 4.23(d). The material depicted is termed *perfectly elasto-plastic* since it switches sharply from elastic to plastic behaviour and has no complicating characteristics such as work hardening.

Consider this curve in a little more detail. The stress–strain relationship is initially linear, with gradient E. Yielding occurs at the *yield stress* σ_Y, at which point the *yield strain* is $\varepsilon_Y = \sigma_Y/E$. Beyond this point the strain increases while the stress remains unchanged. In the elastic region, each value of stress corresponded to a particular value of strain. In the plastic region, however, this is no longer true. If we know that the stress is σ_Y all that we can say about the strain is that it is greater than or equal to ε_Y. On the other hand, if we know the value of strain, then this *does* enable us to determine the value of stress, regardless of whether we are in the elastic or plastic region.

The strain in the plastic region can be thought of as the sum of an elastic component ε_Y and a *plastic strain* ε_P. If the specimen is loaded into the plastic region and then unloaded, the stress returns to zero along a line parallel to the initial elastic curve (shown dashed). Thus the elastic strain is recovered but the plastic strain remains as a permanent deformation.

The key material property governing yielding behaviour is the yield stress σ_Y. Unlike the elastic properties, the value of σ_Y is quite sensitive to the manufacturing process, and so may differ somewhat between batches of the same material. For mild steel a value of 275 MPa is commonly used, while for higher strength steels the yield strength may be as high as 1600 MPa. For aluminium alloys the yield stress is in normally the range 125 to 450 MPa.

In general, we aim to design structures so as to make sure they do not yield, but sometimes limited yielding is allowed under extreme loads. An understanding of the behaviour of structures loaded beyond their yield point is therefore vital for an engineer.

4.5.2 One-dimensional plasticity problems

We have seen that elastic structures can be analysed by making use of three basic principles: equilibrium of forces, compatibility of displacements, and Hooke's law, which relates internal forces to deformations. For perfectly elasto-plastic structures the principles of equilibrium and compatibility remain valid, but Hooke's law cannot be applied to elements which have yielded.

In most instances we are interested in the behaviour of structures in which only some of the members have yielded. In these cases we can deduce the deformations of the yielded members from those of adjoining members which are still elastic. If all the members within a structure have yielded, then we cannot proceed unless we are told the magnitude of the applied displacements.

EXAMPLE 4.11

A simple composite structure consists of two bars connected end to end and fixed between rigid endplates (Figure 4.24(a)). Both bars have the same cross-sectional area A, Young's modulus E and yield stress σ_y, but AB has length l while BC has length $2l$. A load P is applied at the interface B, producing a corresponding deflection δ. Describe the behaviour of the structure as P is increased.

Figure 4.24

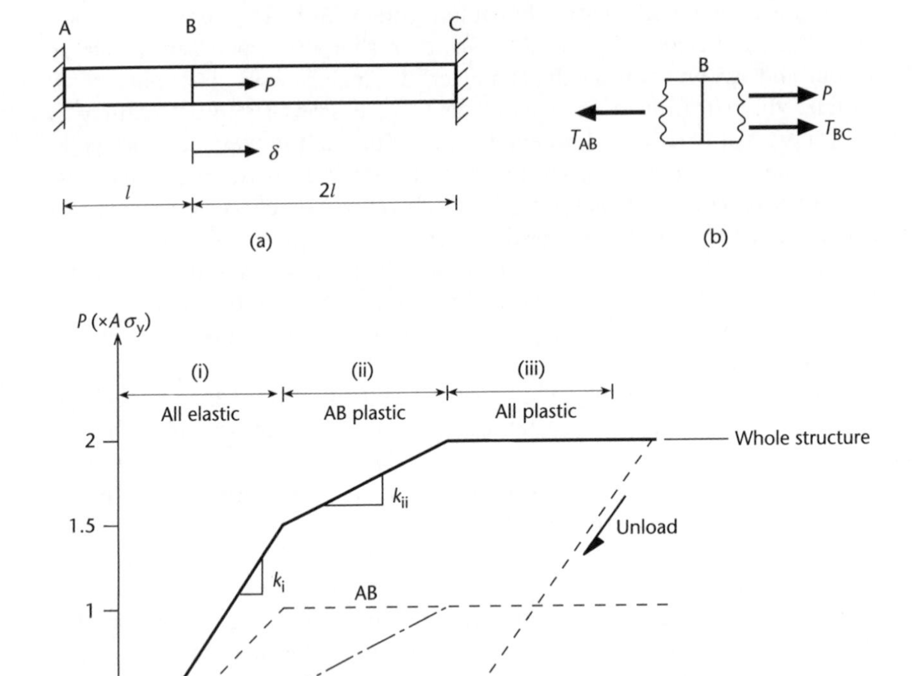

The analysis must be split into several stages. Initially, both bars behave elastically. At some point one bar yields then, as more load is applied the second bar reaches its yield point. At this stage the structure cannot sustain any more load. Consider each of the phases in turn:

(i) Small P – elastic behaviour

Equilibrium of B: the free body diagram is drawn in Figure 4.24(b), in which the internal bar forces are both assumed tensile:

$$T_{AB} - T_{BC} = P$$

Compatibility of displacements: a movement δ of B implies an extension of AB and a shortening of BC, so $e_{AB} = \delta$, $e_{BC} = -\delta$.

Hooke's law: from equation (4.15) $T = AEe/l$, so:

$$T_{AB} = \frac{AE\delta}{l} \qquad T_{BC} = -\frac{AE\delta}{2l} = -\frac{T_{AB}}{2}$$

Substituting into the equilibrium equation gives $T_{AB} = 2P/3$, $T_{BC} = -P/3$.

Thus, while the structure is elastic, the forces are distributed between the two bars in proportion to their stiffnesses (AB is twice as stiff because it is half as long as BC).

(ii) Intermediate values of P – partially plastic

We can see from (i) that bar AB carries the higher load, and therefore the higher stress (since both bars have the same cross-sectional area). So, as the applied load is increased, AB will be the first to reach yield. This will happen when $T_{AB} = A\sigma_y$, so:

$$T_{AB} = 2P/3 = A\sigma_y \qquad \rightarrow P = 3A\sigma_y/2$$

The corresponding value of δ can be obtained from Hooke's law for AB when just on the point of yielding:

$$\delta = \frac{(A\sigma_y)l}{AE} = \frac{\sigma_y l}{E}$$

When P is lower than $3A\sigma_y/2$, the entire structure is elastic and solution (i) applies. Above this value, the equilibrium and compatibility conditions still hold, but Hooke's law can no longer be applied to AB. Substituting for T_{AB} in the equilibrium equation, we get:

$$T_{BC} = A\sigma_y - P$$

(iii) Large values of P – fully plastic

As the load is increased above the first yield value, all the additional load is carried by BC until it too reaches yield. This occurs when $T_{BC} = -A\sigma_y$, so:

$$T_{BC} = -A\sigma_y = A\sigma_y - P \rightarrow P = 2A\sigma_y$$

The corresponding value of δ can be obtained from Hooke's law for BC when just on the point of yielding:

$$-\delta = \frac{(-A\sigma_y)2l}{AE} \qquad \rightarrow \delta = \frac{2\sigma_y l}{E}$$

Beyond this value, the displacement increases with no increase in load.

We can conveniently summarise the behaviour of the structure by plotting a graph showing how the load P varies as a function of displacement δ (Figure 4.24(c)). Initially, the entire structure is linear, with overall stiffness:

$$k_i = \frac{P}{\delta} = \frac{3A\sigma_y/2}{\sigma_y l/E} = \frac{3AE}{2l}$$

Beyond $\delta = \sigma_y l/E$ bar AB has yielded and makes no further contribution to the stiffness of the structure, which therefore drops to:

$$k_{ii} = \frac{2A\sigma_y - 3A\sigma_y/2}{2\sigma_y l/E - \sigma_y l/E} = \frac{AE}{2l}$$

Beyond $\delta = \sigma_y l/E$ bar AB has yielded and the structure has no further load resistance:

$$k_{iii} = 0$$

If the load were now removed, the elastic parts of the displacement would be recovered, that is, P would return to zero along a path parallel to the original elastic curve, with gradient k_i. Because of the plasticity, there would be a permanent *residual* displacement, whose value can easily be calculated from the graph.

4.6 Further reading

There are many specialist solid mechanics texts which provide a fuller treatment of the subject matter of Chapters 4–6 than is possible here. Two of the best are:

- P. P. Benham, R. J. Crawford and C. G. Armstrong, *Mechanics of Engineering Materials*, 2nd edn, Longman, 1996.
- J. M. Gere and S. P. Timoshenko, *Mechanics of Materials*, 4th edn, PWS, 1997.

For a more detailed explanation of material properties and how they relate to microstructures, manufacturing techniques, heat treatments etc., refer to specialist materials science textbooks such as:

- M. F. Ashby and D. R. H. Jones, *Engineering Materials 1: An Introduction to their Properties and Applications*, 2nd edn, Butterworth-Heinemann, 1996.
- M. F. Ashby and D. R. H. Jones, *Engineering Materials 2: An Introduction to Microstructures, Processing and Design*, 2nd edn, Butterworth-Heinemann, 1998.
- J. E. Gordon, *The New Science of Strong Materials, or Why You Don't Fall Through the Floor*, 2nd edn, Penguin, 1991.

4.7 Problems

Where appropriate, answers are given at the end of the book. The questions marked with an asterisk are a little more challenging.

4.1. A cube of rubber of side 50 mm is loaded in pure shear by a set of complementary shear forces of magnitude 10 kN acting in the x–y plane. Determine the shear stress and the relative displacement between opposite faces caused by the loading. Take $E = 1$ GPa and $\nu = 0.49$.

4.2. An aluminium alloy cylinder of height 100 mm, outer diameter 40 mm and inner diameter 25 mm rests on a flat surface, with its axis vertical. A steel cylinder having the same height, outer diameter 25 mm and inner diameter 10 mm fits inside it. The cylinders are joined in such a way that their axial deformations are constrained to be equal. Determine the axial stiffness of each cylinder and the load required to cause the cylinders to reduce in length by 0.1 mm. Take material properties from Table 4.1 and ignore any possible interaction effects due to changes in the horizontal dimensions caused by the loading.

4.3. The two cylinders of Problem 4.2 are now subjected to a temperature rise of 15°C with no external loading. Find the stress in each cylinder and the extension of the structure. Take α to be $12 \times 10^{-6}/°C$ for steel and $24 \times 10^{-6}/°C$ for the aluminium alloy.

4.4.* A rope of length l has weight w per unit volume and elastic modulus E. It hangs vertically and carries a vertical load P at its bottom end. The cross-sectional area A of the rope varies such that the axial stress σ is constant throughout. Determine:

(a) the cross-sectional area at the bottom of the rope;
(b) the cross-sectional area at a distance y from the bottom;
(c) the total extension of the rope.

4.5. A small rectangular element in the x–y plane is subjected to direct and shear stresses (in MPa). Sketch the Mohr's circle and hence deduce the principal stresses and the angle between σ_1 and the x-axis when:

(a) $\sigma_{xx} = 10,\ \sigma_{yy} = -10,\ \tau_{xy} = 10$ (b) $\sigma_{xx} = 10,\ \sigma_{yy} = 10,\ \tau_{xy} = 5$
(c) $\sigma_{xx} = 10,\ \sigma_{yy} = 0,\ \tau_{xy} = 10$ (d) $\sigma_{xx} = 5,\ \sigma_{yy} = 10,\ \tau_{xy} = -4$

4.6. A 120° strain gauge rosette comprises three gauges A, B and C positioned so that B and C are respectively 120° and 240° anti-clockwise from A. If the strain readings are: $\varepsilon_A = 200\mu\varepsilon$, $\varepsilon_B = 120\mu\varepsilon$ and $\varepsilon_C = -40\mu\varepsilon$, find the principal strains and the maximum shear strain, and their orientations relative to gauge A.

4.7. A pressure vessel consists of a cylinder of length 2.5 m and mean radius 400 mm, with closed ends. The vessel is made from high strength steel with $\sigma_y = 500$ MPa and is required to contain fluid under a pressure of 20 MPa. If a factor of safety of 2.0 against yielding is required, what thickness of steel should be used for the vessel walls?

4.8. Figure 4.25 shows the cross-section of a large concrete gravity dam. Because its length in the z direction is large, the cross-section is effectively subjected to plane

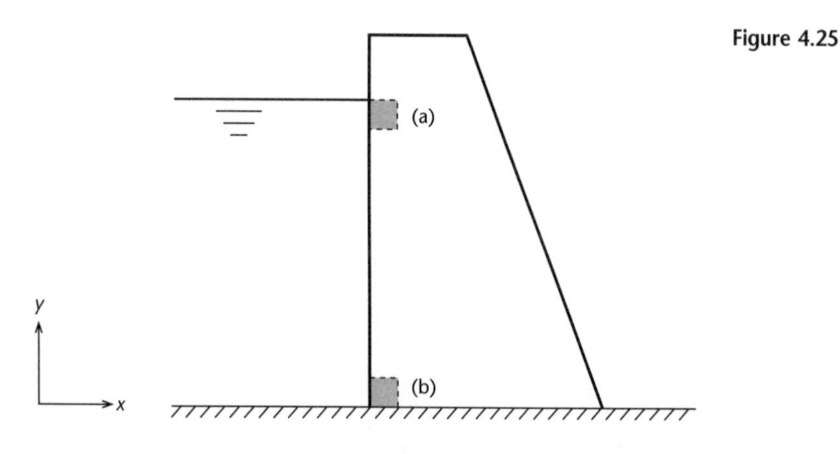

Figure 4.25

strain conditions ($\varepsilon_{zz} = 0$). The combined effects of water pressure and dam self-weight result in the following stresses on the small elements shown:

(a) at the water surface: $\sigma_{xx} = 0$, $\sigma_{yy} = -0.1$ MPa, $\tau_{xy} = 0$;
(b) at the base of the dam: $\sigma_{xx} = 4$ MPa, $\sigma_{yy} = -1$ MPa, $\tau_{xy} = 0$.

Taking the material properties for concrete to be $E = 25$ GPa and $\nu = 0.2$, determine the principal strains and the volumetric strain at each location.

4.9.* The two cylinders of Problem 4.2 are subjected to a steadily increasing compressive load. Taking the yield stresses to be 200 MPa for the aluminium alloy and 275 MPa for the steel:

(a) Determine the load at which the first of the cylinders just yields, and the corresponding change in length of the cylinders.
(b) Determine the load at which the second cylinder just yields, and the corresponding change in length.
(c) If the load is removed as soon as it reaches the value calculated in (b), what is the residual change in length?
(d) What is the residual stress in the outer cylinder? [*Hint: the residual strain can be easily deduced from (c).*]

CHAPTER 5

Bending of beams

CHAPTER OUTLINE

In practice, truly pin-jointed structures are rare, so that virtually all the elements in a frame structure are required to carry bending loads. An understanding of how beams carry loads in bending is therefore of fundamental importance.

In Chapters 2 and 3 we saw how to calculate bending moments in beams and frames. In this chapter we will explore how the bending moment at a certain location in a member is related to the distribution of internal stresses and strains. This is vital in design, since we generally wish to limit both the largest stress and the displacement to acceptable values.

Bending moments in a member are usually accompanied by shear forces and these in turn give rise to shear stresses within the section. The distribution of shear stresses in a flexural element can be quite complex, and so is dealt with separately in Chapter 6, after the rather simpler topic of shear stresses due to torsion.

On completion of this chapter the reader should understand:

- the relations between bending moment and axial stress;
- how to find second moments of area and section moduli for various cross-sections;
- how to calculate beam deflections by a variety of methods;
- how deflection calculations can be used to find reactions in redundant beams;
- how deflected shapes of beams relate to their bending moment distributions;
- the basic principles of yielding in flexure;
- how to calculate the plastic moment capacity of a section.

5.1 The basics of bending behaviour

Just as axial loads on bars give rise to internal forces and deformations, so too do transverse loads on beams. The pattern of internal stresses and strains now becomes rather more complex, however. For a uniform bar under axial load, we have seen that the direct stress is evenly distributed over the bar cross-section and takes a constant value over the length. The strain therefore also takes a constant value throughout the bar, which simply extends or shortens, with no curvature.

A loaded beam, on the other hand, deforms by bending into a curved shape, one edge of the beam shortening while the opposite one extends. This can easily be seen by, for example, taking a rubber eraser and drawing three vertical, parallel lines across one short face (Figure 5.1(a)). If you now bend the eraser between your fingers you can see the lines getting closer together towards one edge and further apart towards the other, as in Figure 5.1(b). Thus the stresses and strains are no longer constant over the cross-section. Instead, they vary from compressive at one edge (where the lines are closer together – the top edge in Figure 5.1) to tensile at the other.

The eraser can be used to illustrate some other important points about bending. As the bending moment is increased (simply push harder with your fingers) the eraser bends into a tighter arc. In fact, we will see that the curvature (that is, the second differential of the deflected shape) of a linear elastic member is directly proportional to the bending moment. Thus, if we know the bending moment distribution along a member we can find the deflections of various points by integration.

Now turn the eraser through 90° so that the longer sides are vertical and try to bend it again. You will find that the greater depth of section makes it much harder to bend – the flexural stiffness of the member varies depending on the axis about which the bending moment is applied. This characteristic can be described using a section property known as the *second moment of area.*

Before going any further, we must justify the fundamental assumption which underlies beam bending theory. Consider again the short length of beam with three straight, vertical lines drawn on one face (Figure 5.1(a)). If the beam is subjected to a uniform bending moment M, with no other applied loads, how will these lines move relative to one another?

- As already stated, the lines will get closer together at the top edge and further apart at the bottom.
- The deformation pattern must be symmetrical about the middle of the three lines.
- Since each beam cross-section is subjected to the same moment, their deformations must be the same.

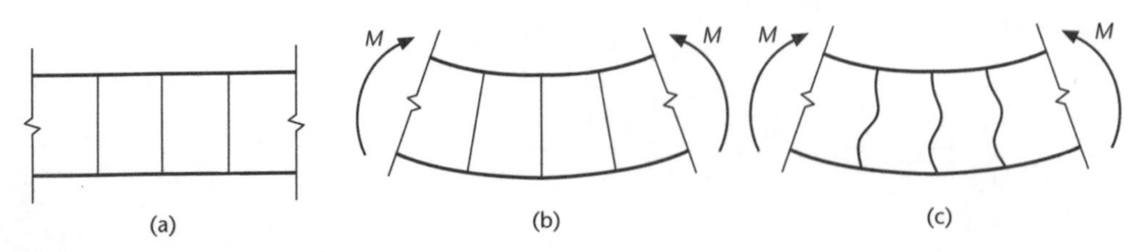

Figure 5.1
Pure bending of a beam segment.

It follows that the lines may rotate relative to one another, but they must remain straight (Figure 5.1(b)). (To convince yourself of this, assume for a moment that one of the lines becomes curved under the bending action. Since each line is subjected to the same bending moment, it follows that they must all deform into the same curved shape (Figure 5.1(c)). Clearly this violates the symmetry requirement and so is an impossibility.)

This finding, which is fundamental to elastic bending theory, is generally stated as follows: *when a body is subjected to pure bending, plane sections remain plane.*

5.2 Elastic bending stresses

5.2.1 Elementary bending theory

We can now go on to derive the basic equations used to analyse a member in bending. The following theory is valid for members subjected to bending about the y and/or z axes, and is valid so long as the member cross-section is symmetrical about at least one of those axes. If neither the y- nor the z-axis is an axis of symmetry, then *unsymmetrical bending* will occur, in which a moment applied about one axis will cause the beam to bend about both axes. This rather more complex case is dealt with elsewhere (see the Further Reading recommendations in section 4.6 at the end of Chapter 4).

Figure 5.2(a) shows an initially straight, uniform beam element which has been bent about the z-axis. So long as its length is arbitrarily small, we may assume that the element deforms into a circular arc. The top edge of the beam has shortened due to the curvature, while the bottom has extended. It follows that there must be a level in the beam at which no change in length has occurred. This is referred to as the *neutral axis*, and the distance from this line to the centre of the arc is the *radius of curvature*, R_y.

By definition, the longitudinal strain at the neutral axis is zero. At a distance y above the neutral axis the strain is given by the change in length of the line BB divided by its original length (which is the same as the length AA):

$$\varepsilon_{xx} = \frac{(R_y - y)\mathrm{d}\theta - R_y\mathrm{d}\theta}{R_y\mathrm{d}\theta} = -\frac{y}{R_y} \tag{5.1}$$

If the material is linear elastic with Young's modulus E, then the corresponding stress is:

$$\sigma_{xx} = -\frac{Ey}{R_y} \tag{5.2}$$

So the stress varies linearly with the distance y from the neutral axis, and the negative sign means that, for a beam that is sagging, the stress is compressive above the neutral axis and tensile below. If we draw the stresses acting on a short sagging element, we can see that they give rise to a couple, or bending moment, about the neutral axis (Figure 5.2(b)).

Figure 5.2(c) shows the cross-section of the beam, together with a plot of how the longitudinal stress varies with y. Consider the small horizontal strip shown shaded. From equation (5.2), the strip is subjected to a compressive force of magnitude $(Ey/R_y) \cdot \mathrm{d}A$. Since it is at a distance y from the neutral axis, this force

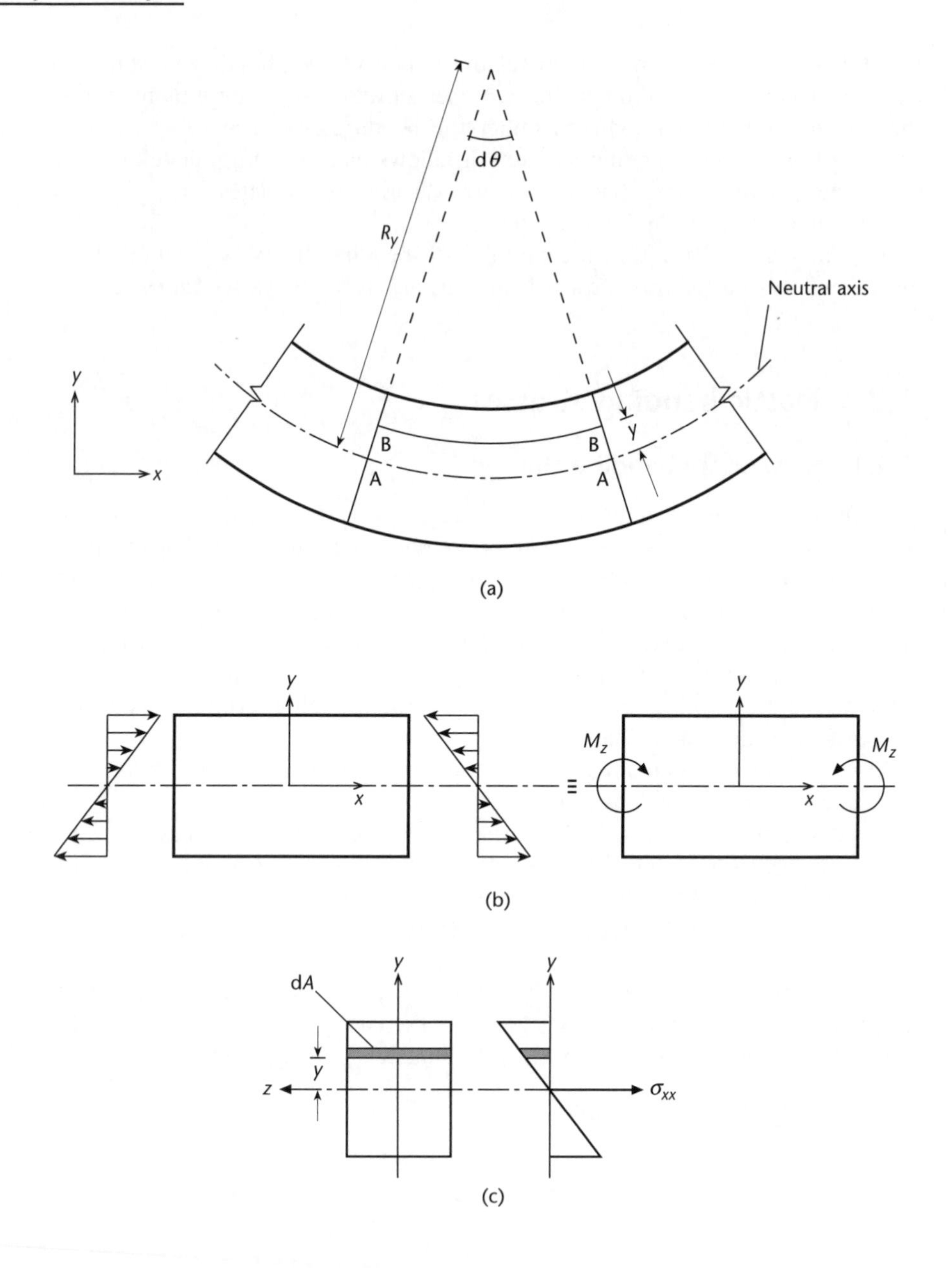

Figure 5.2
Elastic bending theory.

produces an anti-clockwise moment of $(Ey^2/R_y) \cdot \mathrm{d}A$. The bending moment M_z in the beam is simply the sum of the moments provided by the stresses on all such elements:

$$\frac{E}{R_y}\int_{y_1}^{y_2} y^2 \mathrm{d}A = M_z \tag{5.3}$$

where y_1 and y_2 are the y coordinates of the top and bottom edges of the section. Now the integral term is a function entirely of the cross-sectional geometry and is unrelated to the material characteristics and the applied loading. This is known as the *second moment of area* about the z-axis, denoted by I_{zz}:

$$I_{zz} = \int_{y_1}^{y_2} y^2 \mathrm{d}A \tag{5.4}$$

The second moment of area of a cross-section is a vital property in the flexural analysis of a member, and will be dealt with in some depth in the next section. Equations (5.2) and (5.3) are often combined and written as:

$$-\frac{\sigma_{xx}}{y} = \frac{M_z}{I_{zz}} = \frac{E}{R_y} \tag{5.5}$$

Similarly, if the bending moment were applied about the y-axis we could write:

$$\frac{\sigma_{xx}}{z} = \frac{M_y}{I_{yy}} = -\frac{E}{R_z} \tag{5.6}$$

Equations (5.5) and (5.6) are the governing equations for elastic bending, which can be used to determine stresses and deflections of beams. The negative signs in these equations are often omitted, since the sign of the stresses can normally be determined very easily by inspection.

In this chapter we shall restrict ourselves to uniaxial bending, that is sections subjected to M_y or M_z, but not both simultaneously. Before applying equations (5.5) and (5.6), we need to be able to work out the necessary section properties such as neutral axis position and second moment of area.

5.2.2 Bending properties of plane areas

The section properties which we generally need when analysing a member in bending are the neutral axis position $\bar{y}$, cross-sectional area A and second moment of area I_{zz}. The area is relatively easy to calculate, so we shall concentrate here on $\bar{y}$ and I_{zz}.

Returning to the cross-section of Figure 5.2(c), we have already seen that the force acting on the small shaded element is $-(Ey/R_y) \cdot \mathrm{d}A$. Now since the section is subjected only to a bending moment, we know that the resultant axial force on the section must be zero:

$$-\frac{E}{R_y}\int_{y_1}^{y_2} y\mathrm{d}A = 0$$

The non-zero term outside the integral can be cancelled leaving:

$$\int_{y_1}^{y_2} y\mathrm{d}A = 0 \tag{5.7}$$

This integral is known as the *first moment of area* of the section. Equation (5.7) only holds if the origin is at the centroid of the cross-section because, by definition, the first moment of area about the centroid must be zero. So for a linear elastic material the neutral axis *must* coincide with the centroidal axis. The problem of finding the neutral axis therefore reduces to that of finding the centroid. This is most easily done by taking moments of area about one edge of the section, as illustrated in the following example.

EXAMPLE 5.1

Find the position of the neutral axis of the cross-section shown in Figure 5.3 for bending about the y-axis and about the z-axis.

Figure 5.3

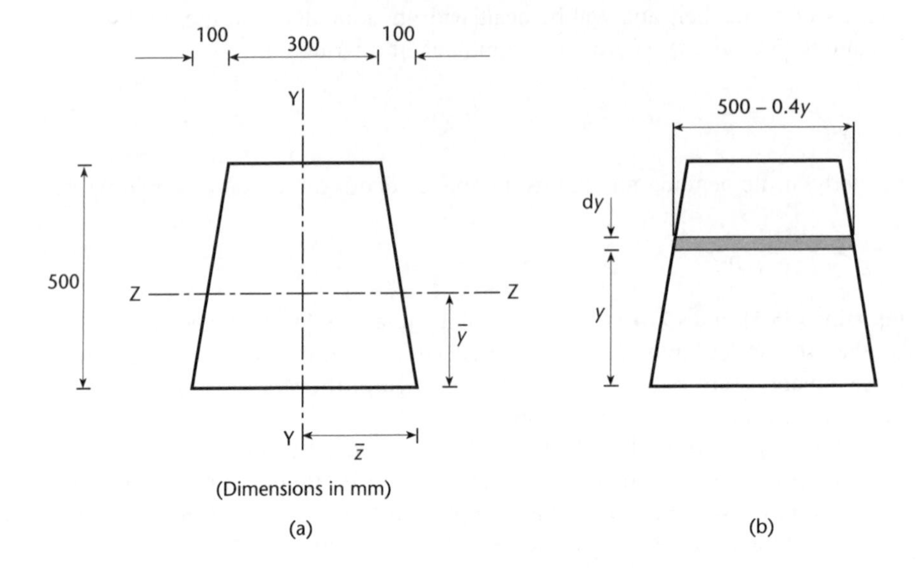

(Dimensions in mm)

(a) (b)

For a symmetrical section, it should be obvious that the centroid must lie on the axis of symmetry, so that the moments of the areas on either side of the axis are equal and opposite. Therefore, for bending about the y-axis, we can see by inspection that $\bar{z} = 250$ mm.

For bending about the z-axis, we divide the section into thin strips, as shown in Figure 5.3(b). The first moment of area of the strip about the bottom edge of the section is:

$$(500 - 0.4y) \cdot \mathrm{d}y \cdot y$$

and summing for all such strips gives the first moment of area of the whole cross-section about the bottom edge. Now the first moment of area is also given by the total area multiplied by the distance to the centroid, so

$$(500 \times 400)\bar{y} = \int_0^{500} (500 - 0.4y)y\mathrm{d}y$$

from which $\bar{y} = 229.2$ mm.

The other vital section property is the second moment of area. It is possible to calculate many different second moments of area for a section, as the value calculated depends on the axis about which moments are taken. In this chapter we will be concentrating on beams subjected to a bending moment about the z-axis, and so we require I_{zz}, the second moment of area about the neutral axis in the y-direction.

More generally, it is possible to define *major* and *minor* axes as those axes about which the second moment of area takes its largest and smallest values (these are the values usually quoted in tables of section properties). It can readily be shown that these two axes must always be perpendicular to each other and, for symmetrical sections, that they must lie on axes of symmetry. Of course, if one is designing a beam to carry moments about one or more axes, it makes sense to orient the cross-section so that the largest moment is imposed about the major axis. For the cross-section in Figure 5.3, *ZZ* is the major axis and YY the minor axis. The beam is therefore optimally positioned to carry a moment M_z which would be caused by vertical loading. In practice this orientation, that is, major axis horizontal and minor axis vertical, is extremely common.

The following example illustrates the calculation of the second moment of area by direct integration, using equation (5.4).

EXAMPLE 5.2

Find the second moment of area of the cross-section shown in Figure 5.3(a) for bending about the ZZ-axis.

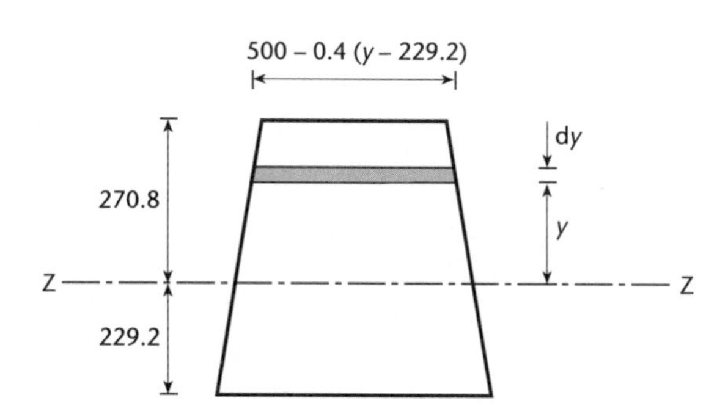

Figure 5.4

Consider the small strip shown in Figure 5.4, with y now measured from the ZZ-axis rather than the bottom of the section. The second moment of area of the strip about ZZ is

$$y^2 \mathrm{d}A = y^2 \cdot [500 - 0.4(y - 229.2)]\mathrm{d}y = y^2 \cdot (408.3 - 0.4y)\mathrm{d}y$$

and since *ZZ* is 229.2 mm from the bottom of the section, the limits of integration are from -229.2 to 270.8. Therefore:

$$I_{zz} = \int_{-229.2}^{270.8} (408.3y^2 - 0.4y^3)\mathrm{d}y = \left[136.1y^3 - 0.1y^4\right]_{-229.2}^{270.8} = 4.08 \times 10^9 \text{ mm}^4$$

In practice it is rarely necessary to perform calculations such as this. Most beams have cross-sections which can be divided into simple geometric shapes whose second moments of area are known, or can be looked up – some of the most useful

Table 5.1 Section properties of simple shapes.

	A	$\bar{y}$	I_{zz}	Z_e
Rectangle 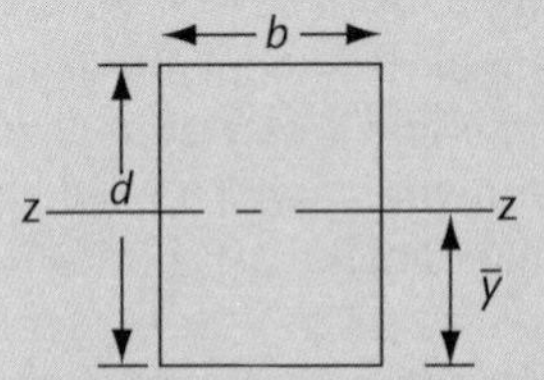	bd	$\frac{d}{2}$	$\frac{bd^3}{12}$	$\frac{bd^2}{6}$
Triangle 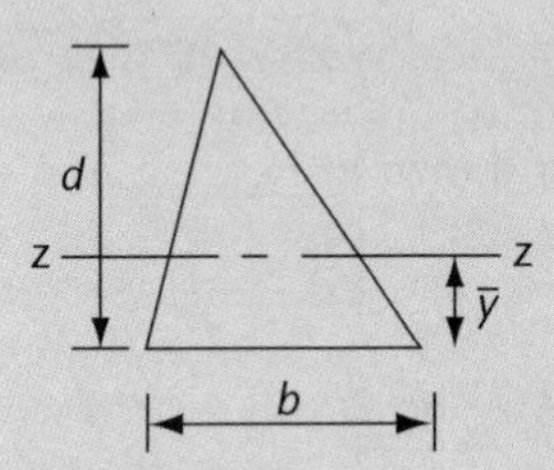	$\frac{bd}{2}$	$\frac{d}{3}$	$\frac{bd^3}{36}$	$\frac{bd^2}{24}$
Circle 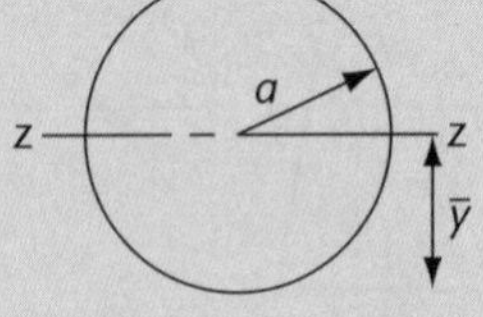	πa^2	a	$\frac{\pi a^4}{4}$	$\frac{\pi a^3}{4}$
Thin circle 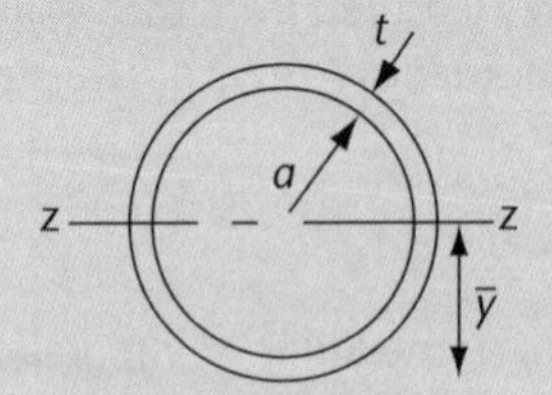	$2\pi at$	a	$\pi a^3 t$	$\pi a^2 t$

ones are given in Table 5.1. The second moment of area of the whole cross-section is then simply the sum of the I-values for the constituent parts, *so long as all the I-values relate to the same axis*. When using this approach a very useful aid is the parallel axis theorem. Suppose a cross-section has area A and second moment of area I_{AA} about an axis AA through its centroid (Figure 5.5). The I-value about a parallel axis BB a distance b away is given directly by the parallel axis theorem:

$$I_{BB} = I_{AA} + Ab^2 \tag{5.8}$$

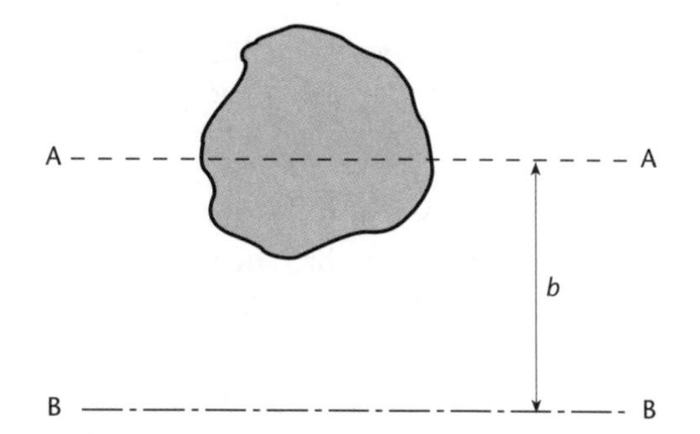

Figure 5.5
Parallel axis theorem.

EXAMPLE 5.3

Find the second moment of area of the I-section shown in Figure 5.6 for bending about the z-axis.

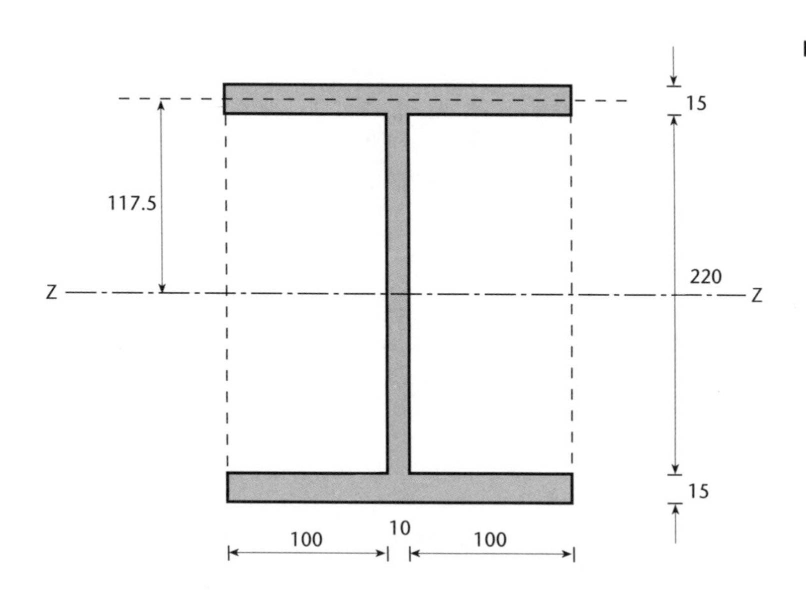

Figure 5.6

There are many ways in which this could be done; two will be illustrated here, both of which make use of the standard result for a rectangle with dimensions $b \times d$, $I = bd^3/12$. First we could calculate I for the large rectangle defined by the dashed lines and then subtract I for the unwanted area either side of the web. This approach has the advantage that all of the sub-elements considered have (by symmetry) the same neutral axis, and so no parallel axis terms are needed.

$$I_{zz} = \frac{210 \times 250^3}{12} - 2 \times \frac{100 \times 220^3}{12} = 95.97 \times 10^6 \text{ mm}^4$$

Alternatively, we could sum the I-values for the flanges and the web. With this approach, the flange calculation will require the use of the parallel axis theorem:

$$I_{zz} = \frac{10 \times 220^3}{12} + 2\left(\frac{210 \times 15^3}{12} + (210 \times 15) \times 117.5^2\right) = 95.97 \times 10^6 \text{ mm}^4$$

Here the first term represents the web and the terms in brackets the flanges. For each flange we first calculate I about its own centroidal axis then shift to the neutral axis of the section by adding a parallel axis term.

This expression gives a clue as to why the beams used in civil engineering are often I-shaped. For a given amount of material, we require a high second moment of area in order to give a high bending resistance. In this case easily the largest term in our expression for I_{zz} is the final, parallel axes term, and this is maximised by concentrating most of the material as far as possible from the neutral axis of the section.

One final section property that will be useful is the *elastic modulus* Z_e. From equation (5.5) we can see that the stress caused by a bending moment is a linear function of the distance y from the neutral axis:

$$\sigma_{xx} = \frac{M_z y}{I_{zz}}$$

When analysing a section we are usually interested in finding the largest stress, which will obviously occur at the furthest point from the neutral axis. It is convenient to introduce the elastic modulus

$$Z_e = \frac{I_{zz}}{y_{max}} \tag{5.9}$$

as this gives us a direct constant of proportionality between the applied bending moment and the maximum resulting stress, that is:

$$\sigma_{max} = \frac{M_z}{Z_e} \tag{5.10}$$

Note that we have been slightly loose with the sign convention here. Z_e is by convention taken as positive and the sign of the maximum stress deduced by inspection. Sometimes, for sections without a horizontal axis of symmetry, we define two elastic moduli, one based on the distance from the neutral axis to the bottom edge of the section and the other on the distance to the top edge. These two moduli allow us to determine the maximum tensile and compressive stresses caused by the bending (often called the *extreme fibre stresses*) rather than just the maximum absolute value.

When calculating bending stresses, we can use equation (5.5) when we wish to determine the distribution of stresses over the section, and (5.10) when we are interested solely in the peak stresses.

EXAMPLE 5.4

The beam whose cross-section is shown in Figure 5.6 is made from a material which can sustain a maximum stress (tensile or compressive) of 165 MPa. Find the largest bending moment that the beam can carry.

The second moment of area of the section has already been found in Example 5.3. The elastic modulus is found using equation (5.9):

$$Z_e = \frac{95.97 \times 10^6}{125} = 767.8 \times 10^3 \text{ mm}^3$$

and equation (5.10) then gives

$$M_z = 767.8 \times 10^3 \times 165 = 126.7 \text{ kNm}$$

Up till now, we have been very rigorous about the notation of bending stress calculations, indicating axes and directions by subscripts throughout. For the case of bending about a single axis, the subscripts can become rather cumbersome. From now on, therefore, we shall omit the subscripts when dealing with uniaxial behaviour, and include them only where they are necessary to prevent confusion.

5.2.3 Composite beams

Often beams are made of more than one material bonded together. Examples include reinforced concrete, where steel bars are embedded in the concrete, and timber beams strengthened by external steel plates. The different properties of the various materials make analysis more complicated; while the stress distribution in each material will be linear, there will be a discontinuity at the interface. However, we saw in section 5.1.1 that the strain can be expressed solely in terms of the geometry of the problem, without reference to the material properties. So the strain will vary linearly over a composite section, just as it would over a section with constant material properties.

In analysing composite beams, it is easiest to make use of the idea of the *equivalent section*, that is, the section made from a single material which would have the same bending properties as the composite section. Consider for example the beam shown in Figure 5.7(a), comprising a material A sandwiched between two layers of material B. The strain profile over the height of the section is linear, but since the two materials have different Young's moduli the stresses undergo a step change at the interface. In this example $E_B > E_A$. Now if the strain in material A at a distance y from the neutral axis is ε then the stress is $E_A\varepsilon$ and the force in an elemental strip of height dy is $E_A\varepsilon \cdot b\mathrm{d}y$. Suppose we wish to replace the actual cross-section by an equivalent section made entirely of material B. It is important

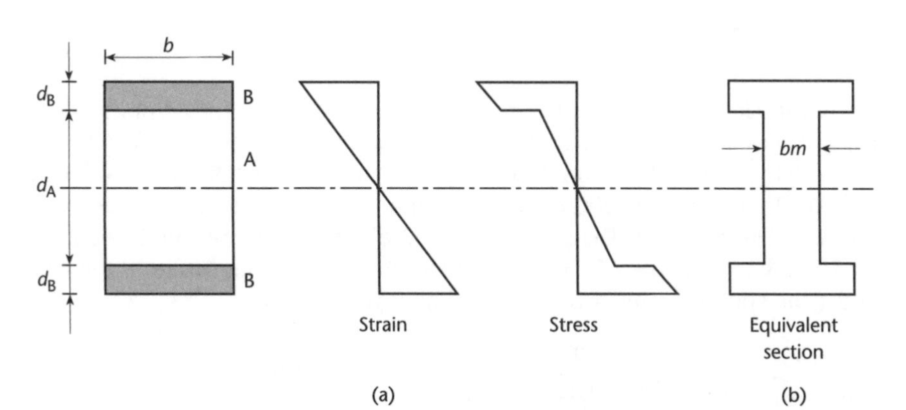

Figure 5.7
Equivalent section for a composite beam.

that the new section must transmit the same force at a given distance from the neutral axis, as otherwise the bending moment would be altered. This can be done by altering the breadth to some new value b' such that:

$$E_A \varepsilon b \mathrm{d}z = E_B \varepsilon b' \mathrm{d}z$$

$$b' = b\frac{E_A}{E_B} = bm$$

where m is known as the *modular ratio*. So, to produce an equivalent section made entirely of material B we simply have to factor the width of the central part by the modular ratio m, as shown in Figure 5.7(b).

Having derived the transformed section, the analysis is then relatively simple. The second moment of area can be calculated as before, and this can be used to relate an applied bending moment to the strain at any position in the cross-section, using the moment–curvature part of equation (5.5) together with equation (5.1). Stresses in the real section are then found by multiplying the strains by the appropriate Young's modulus.

EXAMPLE 5.5

Figure 5.8(a) shows a rectangular timber beam reinforced by steel plates on its top and bottom faces. The Young's moduli of the timber and steel are 10.25 GPa and 205 GPa respectively. Calculate the stress distribution in the beam under a bending moment of 10 kNm.

Figure 5.8

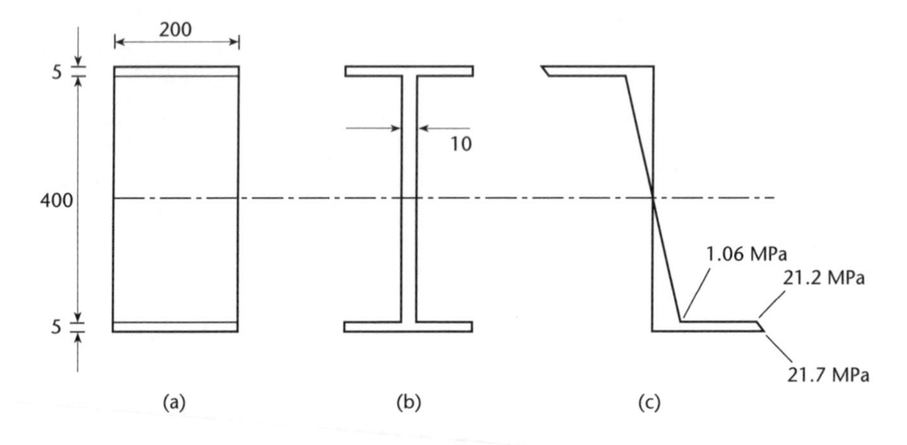

We will tackle this by finding the equivalent steel section for the composite beam (though we could just as well transform to an equivalent timber section). We do this by factoring the width of the timber part by the modular ratio $m = 10.25/205 = 0.05$. Therefore $b' = 200 \times 0.05 = 10$ mm and the equivalent section is as shown in Figure 5.8(b). The second moment of area for this section is found in a similar way to Example 5.3, and is 9.434×10^7 mm^4. The curvature of the beam under the applied moment is then given by equation (5.5) as

$$\frac{1}{R} = \frac{M}{EI} = \frac{10 \times 10^6}{(205 \times 10^3) \cdot (9.434 \times 10^7)} = 5.171 \times 10^{-7}\ \text{mm}^{-1}$$

and, from equation (5.1), the strain at a distance y from the neutral axis is simply obtained by multiplying the curvature by y. So, at the timber-steel interface $\varepsilon = 200 \times 5.171 \times 10^{-7} = 1.034 \times 10^{-4}$ and the stress in the timber at this point is $1.034 \times 10^{-4} \times 10.25 \times 10^3 = 1.06$ MPa.

The strain in the steel at this same level must be identical, since the two materials are bonded together, so its stress is $1.034 \times 10^{-4} \times 205 \times 10^3 = 21.2$ MPa, a factor of 20 greater.

The strain at the outer face of the steel plates is $\varepsilon = 205 \times 5.171 \times 10^{-7} = 1.060 \times 10^{-4}$ and the stress is again obtained by multiplying by E to give $1.06 \times 10^{-4} \times 205 \times 10^3 = 21.7$ MPa.

The stress distribution in the section is shown in Figure 5.7(c).

5.2.4 Strain energy

We showed in the previous chapter that the strain energy on a small element of volume $\mathrm{d}V$ under a uniaxial stress σ is given by equation (4.19) as

$$\mathrm{d}U = \frac{\sigma^2}{2E}\mathrm{d}V$$

For a beam subjected to a bending moment M about the major axis, the axial stress is $\sigma = My/I$. So for an elemental volume made up of a small element of area $\mathrm{d}A$ extending over a length of beam $\mathrm{d}x$ equation (4.19) becomes

$$\mathrm{d}U = \frac{M^2 y^2 \mathrm{d}A}{2EI^2}\mathrm{d}x$$

Integrating both over the cross-section of the beam and along its length, and noting that $\int y^2 \mathrm{d}A = I$, the total strain energy in the beam is

$$U = \int \frac{M^2}{2EI}\mathrm{d}x \tag{5.11}$$

5.3 Elastic deflections

While one aim of design is to ensure that stresses remain within acceptable limits, another is to ensure that the structure does not deform excessively. The calculation of deflections is therefore vital. Deflections in beams can be caused by both the bending moment and the shear force. However, if the beam is relatively long compared to its depth, as is usually the case, then the shear deflection will be small and can be neglected. This section is therefore concerned solely with bending deflections of beams; shear deflections are dealt with briefly in Chapter 6.

5.3.1 Direct integration method for simple beams

For a beam bent in the vertical plane, we have seen that the moment is related to the radius of curvature by equation (5.5):

$$\frac{M}{I} = \frac{E}{R}$$

Now the term $1/R$ is known simply as the *curvature*. If the beam deflects vertically by an amount v (measured positive in the positive y direction, that is, *upwards*), then it can be shown mathematically that the curvature and deflection are related by

$$\frac{1}{R} = \frac{\dfrac{d^2v}{dx^2}}{\left[1 + \left(\dfrac{dv}{dx}\right)^2\right]^{3/2}}$$

Since it is generally a design requirement that v is small compared to the overall member dimensions, it is normally safe to assume that $dv/dx \ll 1$, so that the denominator is very close to unity and this equation simplifies to:

$$\frac{1}{R} = \frac{d^2v}{dx^2}$$

If we now substitute for $1/R$ from equation (5.5) we get

$$\frac{d^2v}{dx^2} = \frac{M}{EI} \tag{5.12}$$

Knowing the bending moment distribution on a beam, equation (5.12) can be used to determine the deflected shape by integration. The term EI is known as the *flexural stiffness* or *flexural rigidity* of the section.

EXAMPLE 5.6

Calculate the slope and deflection at the tip of a cantilever of length l and flexural stiffness EI when it is subjected to a tip load P (Figure 5.9).

Figure 5.9

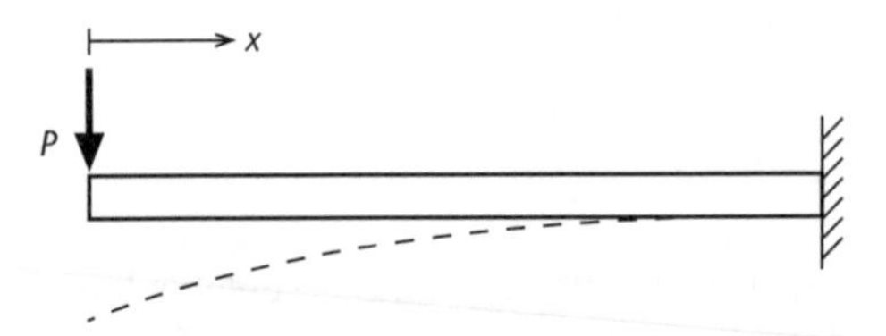

Taking x as the distance along the cantilever from the tip towards the fixed end, the bending moment at any point is simply $-Px$. Equation (5.12) therefore gives:

$$\frac{d^2v}{dx^2} = \frac{-Px}{EI}$$

Integrating:

$$\frac{dv}{dx} = -\frac{Px^2}{2EI} + A$$

where A is a constant of integration which must be found using the boundary conditions of the problem. Now at the fixed end we have $x = l$ and we know the slope is zero. Substituting these values into the above expression gives $A = Pl^2/2EI$. Integrating again:

$$v = -\frac{Px^3}{6EI} + \frac{Pl^2x}{2EI} + B$$

The second constant of integration B is found using the condition $v = 0$ at $x = l$, giving $B = -Pl^3/3EI$. Having obtained general expressions for the variations in slope and deflection along the beam, the values at the tip are found by putting $x = 0$:

$$\left.\frac{dv}{dx}\right|_{x=0} = \frac{Pl^2}{2EI}, \qquad v_{x=0} = -\frac{Pl^3}{3EI}$$

where the negative sign on the deflection indicates a downward movement.

5.3.2 Analysis of statically indeterminate beams

The direct integration method outlined above is extremely powerful. In some instances it can even be used to analyse statically indeterminate beams, as illustrated in the following example.

EXAMPLE 5.7

A fixed-ended beam of length l and flexural rigidity EI is subjected to a uniformly distributed load of intensity w per unit length, as shown in Figure 5.10. Find the midspan deflection.

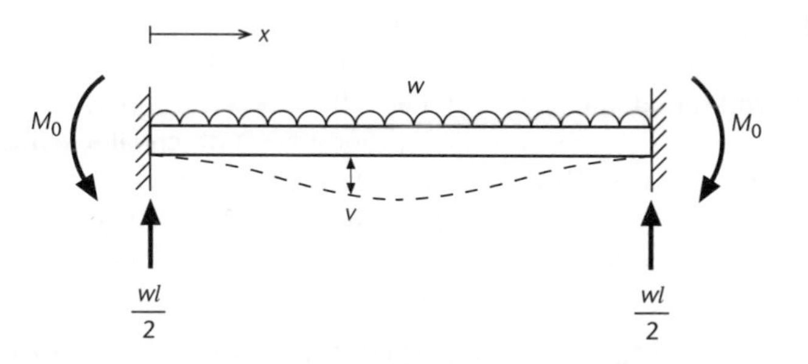

Figure 5.10

The beam is redundant, since there are six possible reactions – vertical, horizontal and rotational at each end. The horizontal reactions can be neglected, and by symmetry the vertical reactions are each $wl/2$. Let the moment reactions be M_0, acting in the directions shown. Formulating the bending moment expression and using equation (5.12):

$$M = -M_0 + \frac{wlx}{2} - \frac{wx^2}{2} = EI\frac{d^2v}{dx^2}$$

Integrating:

$$EI\frac{dv}{dx} = -M_0x + \frac{wlx^2}{4} - \frac{wx^3}{6} + A$$

Using the boundary condition that the slope is zero at $x = 0$ gives $A = 0$. But also, from symmetry, the slope is zero at $x = l/2$, which gives $M_0 = wl^2/12$. Integrating again:

$$EIv = \frac{wl^2x^2}{24} + \frac{wlx^3}{12} + \frac{wx^4}{24} + B$$

Since $v = 0$ at $x = 0$, the constant $B = 0$. Then, putting $x = l/2$ gives the midspan deflection:

$$v = -\frac{wl^4}{384EI}$$

This example introduces a very important principle, which we shall return to in later chapters; we were able to use our knowledge of the beams deformed shape to generate an extra equation that, together with the equilibrium equations, allowed us to determine the redundant reaction.

Having determined the reactions in this way, it is then a simple matter to calculate the shear force and bending moment distributions. Results for some of the more common types of redundant beam are shown in Figure 5.11. Diagrams (a) and (b) are for fixed-end beams, in which there is a moment reaction at each end. As we have seen in Example 5.7, these beams are highly redundant but, for the loadcases shown, can be greatly simplified by symmetry. It is interesting to compare these results with those for simply supported beams under similar loads (Figure 2.20). In each case the shear force diagram is identical and the bending moment diagram has simply had its axis shifted downwards by an amount equal to the moment reaction at each end.

Cases (c) and (d) are *propped cantilevers*, rigidly fixed at one end and with purely vertical support at the other. The lack of symmetry in the support conditions means that the vertical reactions are not equal even though the loading is symmetrical. This in turn means that the shear force diagrams are not anti-symmetric (as they are in (a) and (b)) and that, for the case of the uniformly distributed load, the peak bending moment occurs away from the centre.

5.3.3 Macaulay's notation for continuous beams

When we have a long beam with many different support reactions and/or applied loads, the direct integration approach can quickly become very long-winded. A neat way of keeping the algebra to a minimum is by the use of Macaulay's notation. Consider, for example the beam segment shown in Figure 5.12.

At a distance $x(< a)$ from the left-hand end the bending moment is simply Px. When $a < x < b$ the moment is $Px - W(x - a)$, and so on. Using Macaulay's notation, we can combine the bending moment expressions for the various parts into a single moment equation for the entire beam:

$$M = Px - W\{x - a\} + \frac{w\{x - b\}^2}{2}$$

Here the terms in curly brackets are known as *Macaulay terms* or *step functions*. A Macaulay term takes the general form $\{x - a\}^n$ where $n \geq 0$. If the term within the bracket is positive then the function takes the value $(x - a)^n$, otherwise its value is

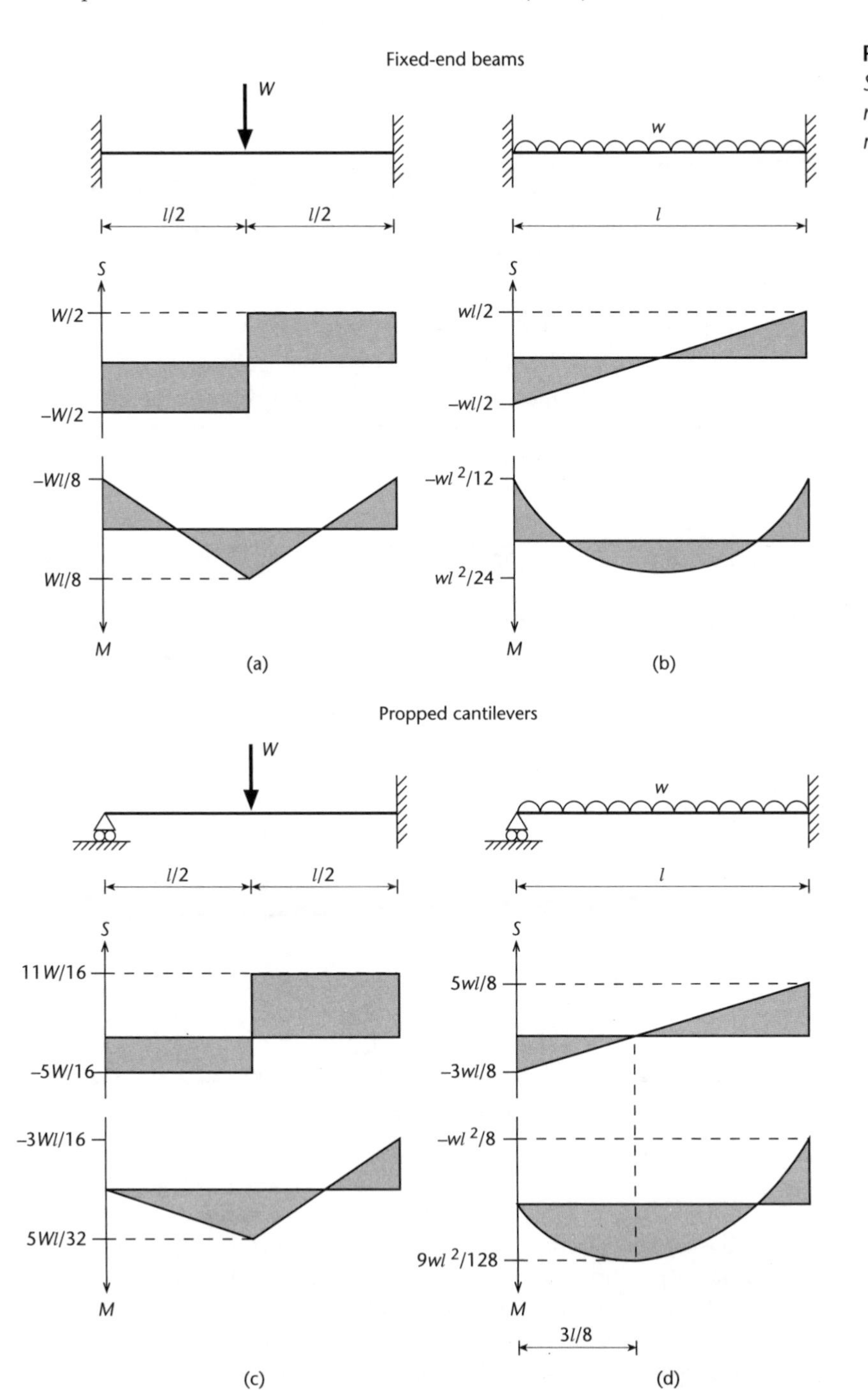

Figure 5.11
Shear force and bending moment diagrams for redundant beams.

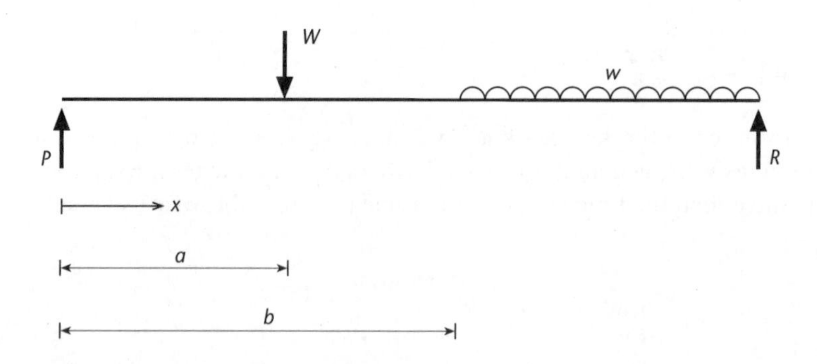

Figure 5.12
Loads on a beam segment.

zero. This allows us to switch terms on as we move along the beam and encounter additional loads. Another important property of a Macaulay term is that it integrates just like an ordinary bracketed term raised to a power, that is:

$$\int \{x-a\}^n \mathrm{d}x = \frac{\{x-a\}^{n+1}}{n+1} + C$$

So using Macaulay's notation, we can write down a single bending moment expression for a beam under any loading and can integrate that expression to give a single, continuous deflection equation.

EXAMPLE 5.8

Using Macaulay's notation, derive an expression for the deflected shape of the beam shown in Figure 5.13(a) and hence find the deflections of the points B and C.

Figure 5.13

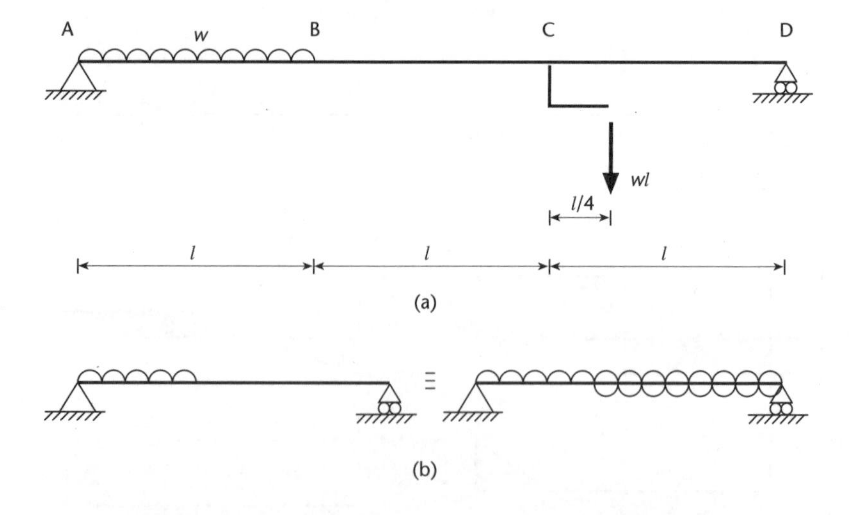

First, find the support reactions. By inspection, the horizontal reaction is zero, then resolving vertically and taking moments about any point gives $V_A = 13wl/12$, $V_D = 11wl/12$. We now write down the Macaulay bending moment expression for ABCD. Note that the short, right-angled element CE has the effect of imposing a vertical load wl and a moment $wl^2/4$ on the beam at C:

$$M = \frac{13}{12}wlx - \frac{wx^2}{2} + \frac{w\{x-l\}^2}{2} - wl\{x-2l\} + \frac{wl^2}{4}\{x-2l\}^0 = EI\frac{d^2v}{dx^2}$$

A couple of the terms in this equation need some explanation. First, it is not possible using Macaulay brackets to switch *off* the uniform load term ($-wx^2/2$) when we get beyond B. Instead, we must cancel it by switching *on* an opposing uniform load term ($w\{x-l\}^2/2$). We are thus representing the distributed loads on the beam as the sum of a downward and an upward load, as shown in Figure 5.13(b). Second, note that the Macaulay term $\{x-2l\}^0$ is required to switch on the concentrated moment term without applying any distance scaling to it. Integrating twice gives:

$$EIv = \frac{13}{72}wlx^3 - \frac{wx^4}{24} + \frac{w\{x-l\}^4}{24} - wl\frac{\{x-2l\}^3}{6} + \frac{wl^2}{8}\{x-2l\}^2 + Ax + B$$

The boundary conditions are $v = 0$ at $x = 0$ and $x = 3l$. Substituting these into the above gives $B = 0$ and $A = -17wl^3/24$. So the full deflected shape expression is:

$$v = \frac{w}{EI}\left(\frac{13lx^3}{72} - \frac{x^4}{24} + \frac{\{x-l\}^4}{24} - \frac{l\{x-2l\}^3}{6} + \frac{l^2\{x-2l\}^2}{8} - \frac{17l^3x}{24}\right)$$

and the deflections at B and C can now be found:

$$x = l \rightarrow v_B = -\frac{41wl^4}{72EI}, \qquad x = 2l \rightarrow v_C = -\frac{43wl^4}{72EI}$$

5.3.4 Moment–area methods

When we wish to determine the slope or deflection at just a few discrete points along a beam, a very direct solution can often be obtain by using the two moment–area theorems. These are easily derived from the moment–curvature relationships used above, as follows.

Figure 5.14 shows a segment of beam AB, together with a plot of its bending moment M divided by its flexural stiffness EI. From equation (5.12):

$$\frac{M}{EI} = \frac{d^2v}{dx^2}$$

Integrating between A and B:

$$\int_A^B \frac{M}{EI}dx = \int_A^B \frac{d^2v}{dx^2}dx = \left[\frac{dv}{dx}\right]_A^B = \theta_B - \theta_A \tag{5.13}$$

Equation (5.13) is the *first moment-area theorem,* and can be stated in words as: *the difference in slope between two points is equal to the area of the M/EI diagram between those points.*

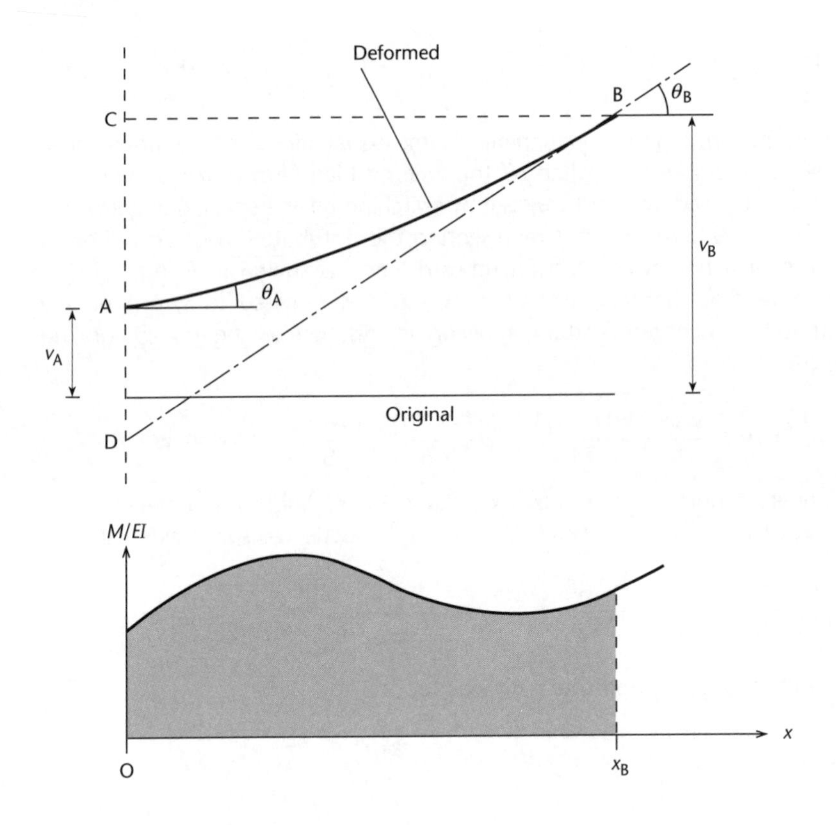

Figure 5.14
Moment–area theorems.

If we now take moments of the M/EI diagram about A we get:

$$\int_A^B \frac{Mx}{EI}\,dx = \int_A^B x\frac{d^2v}{dx^2}\,dx = \left[x\frac{dv}{dx}\right]_A^B - \int_A^B \frac{dv}{dx}\,dx = x_B\theta_B - v_B + v_A \tag{5.14}$$

since $x_A = 0$. The term $x_B\theta_B$ is equal to the distance CD in the diagram. Equation (5.14) is the *second moment–area theorem*, which states that *the moment about A of the M/EI diagram between A and B equals the deflection of A relative to a tangent drawn from B* (that is, the distance AD in Figure 5.14).

5.3.5 Use of standard solutions

The direct integration approach for determining deflections can be cumbersome, particularly as we are usually interested only in the peak value of the deflection in a beam. Often it is possible to come up with a very quick solution by the judicious combination of a few standard results. Table 5.2 lists the end slopes, end deflections and midspan deflections for cantilevers, simply supported beams and fixed-end beams under some simple loadings. Many of these come up so often in structural engineering that they are worth committing to memory.

Quite complicated loading configurations can be broken down and considered as the superposition of two or more of these cases, often with the help of arguments based on symmetry and anti-symmetry. The procedure is best illustrated by some examples.

Table 5.2 Slope and deflection formulae for beams.

	End slope	End deflection	Midspan deflection
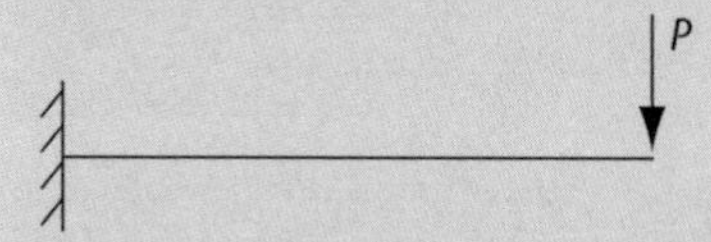 P	$-\dfrac{Pl^2}{2EI}$	$-\dfrac{Pl^3}{3EI}$	$-\dfrac{5Pl^3}{48EI}$
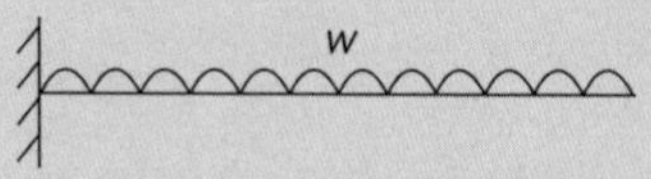 w	$-\dfrac{wl^3}{6EI}$	$-\dfrac{wl^4}{8EI}$	$-\dfrac{17wl^4}{384EI}$
 M	$\dfrac{Ml}{EI}$	$\dfrac{Ml^2}{2EI}$	$\dfrac{Ml^2}{8EI}$
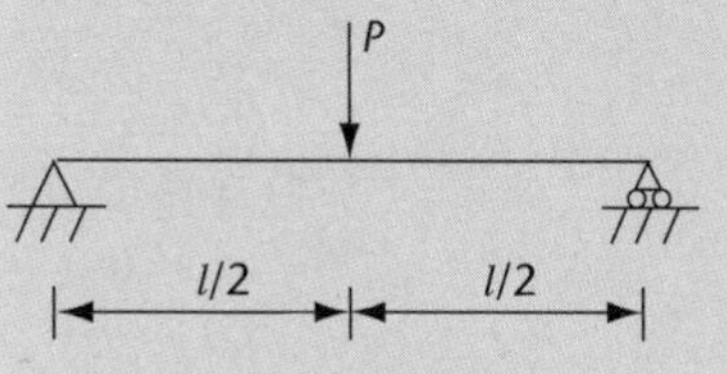 P; l/2, l/2	$\pm\dfrac{Pl^2}{16EI}$	0	$-\dfrac{Pl^3}{48EI}$
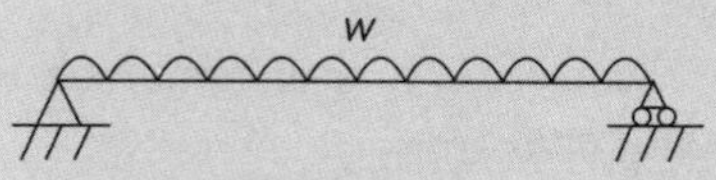 w	$\pm\dfrac{wl^3}{24EI}$	0	$-\dfrac{5wl^4}{384EI}$
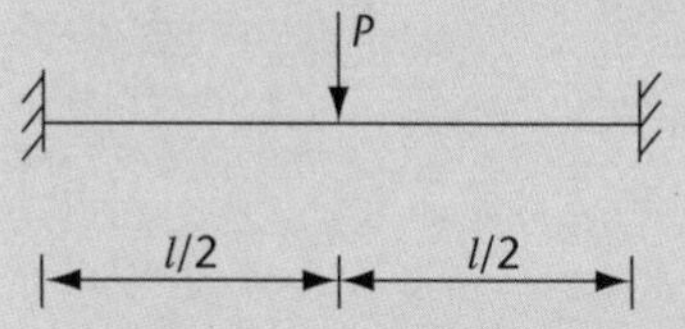 P; l/2, l/2	0	0	$-\dfrac{Pl^3}{192EI}$
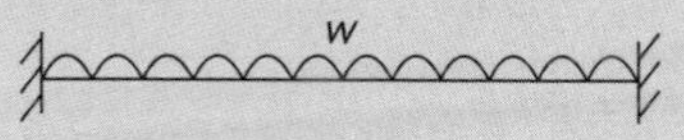 w	0	0	$-\dfrac{wl^4}{384EI}$

EXAMPLE 5.9:

Prove the result for the midspan deflection of a simply supported beam carrying a central point load, using only the result for a cantilever with a concentrated tip load.

If we sketch the deflected shape of the beam, Figure 5.15(a), we see from symmetry that the slope is zero at the centre. If we make an imaginary cut in the beam at the centre, there is a moment at the cut point preventing the two cut ends from rotating. The two halves thus behave as cantilevers fixed at the centre and

Figure 5.15

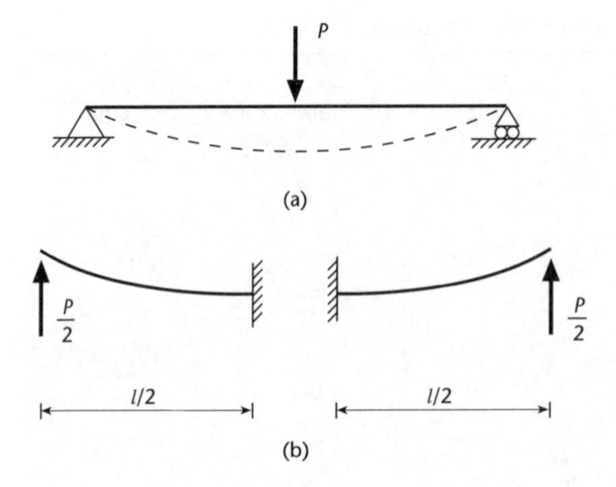

subjected to concentrated loads (the support reactions) at their tips (Figure 5.15 (b)). The deflection can therefore be found from the cantilever formula:

$$v_{\max} = -\frac{\frac{P}{2}\left(\frac{l}{2}\right)^3}{3EI} = -\frac{Pl^3}{48EI}$$

EXAMPLE 5.10

Find the tip deflection of a cantilever of length $2l$ subjected to a uniform load w over the half nearest the fixed end, with no loading elsewhere.

Figure 5.16

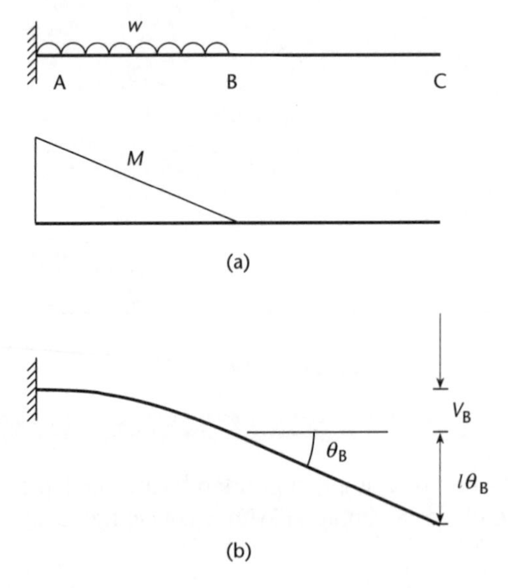

If we sketch the bending moment diagram for the cantilever (Figure 5.16(a)), it is obvious that the left-hand half behaves like a simple cantilever of length l, and that there is no moment in the right-hand half. It therefore follows that there will be no curvature in this portion, which will thus have a constant slope equal to the slope at B (Figure 5.16(b)). Assuming the slope is small compared to the overall dimensions, the deflection between B and C is simply $l\theta_B$; this is known as the *pointer effect*. The total vertical displacement of C is then $v_B + l\theta_B$. So, using the results in Table 5.2:

$$v_C = -\frac{wl^4}{8EI} - l\frac{wl^3}{6EI} = -\frac{7wl^4}{24EI}$$

As with direct integration, this approach can also be used to help find redundant support reactions, as in the following example.

EXAMPLE 5.11

A propped cantilever carries a concentrated moment M at the propped end. Calculate the vertical reaction at the prop.

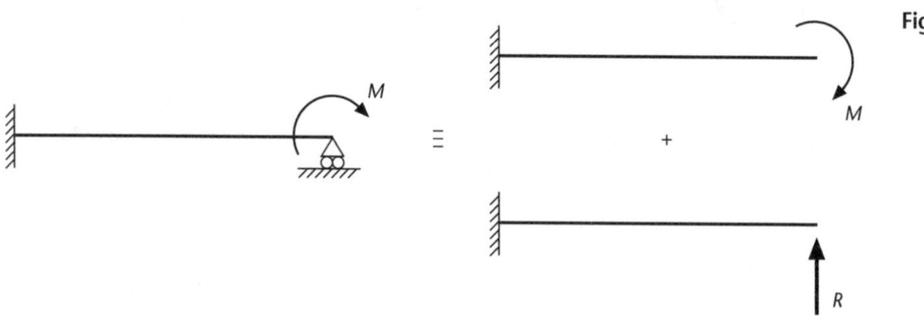

Figure 5.17

The force in the prop can be treated as a point load at the tip of a cantilever, as shown in Figure 5.17. Since we know that the deflection at the prop, due to the combined effects of the moment and the prop force, must be zero:

$$\frac{Ml^2}{2EI} - \frac{Rl^3}{3EI} = 0 \rightarrow R = \frac{3M}{2l}$$

5.3.6 Qualitative analysis

A qualitative idea of how beams deform is vital for a structural engineer. As in Chapter 2, this is most easily done by also considering the bending moment distribution, since the moment is proportional to the curvature. Superposition of simpler cases can again be used to good effect.

Consider, for example, the beam in Figure 5.18(a). The deflected shapes and bending moments due to the distributed load and the concentrated moment acting separately are easy to sketch, and are shown in (b) and (c). For each loadcase, the

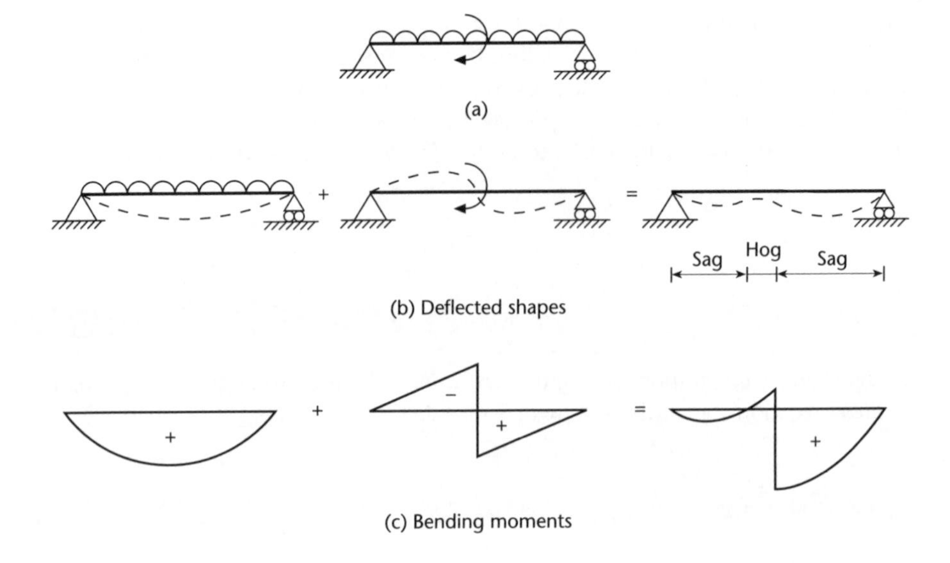

Figure 5.18
Superposition of bending moments and deflected shapes.

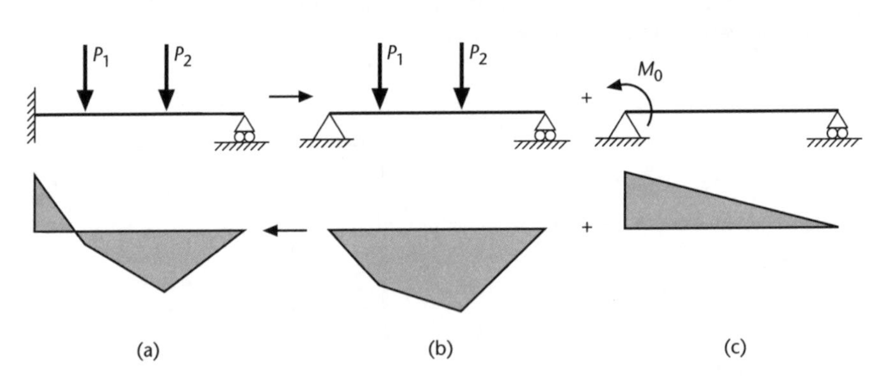

Figure 5.19
Bending moments in a propped cantilever.

largest moments occur near the centre of the beam, and so its curvature is greatest there. Near the supports the beam is almost straight, as the moments are small. The way the two effects are combined to give the overall moment and deflected shape depends to some extent on the relative magnitudes of the uniform and moment loads. It is obvious that the beam must be sagging (positive moment) over the right-hand half and close to the left-hand support. If the applied moment is large enough, then the hogging curvature it causes just to the left of centre will exceed the sagging due to the uniform load, and the beam will hog over a short length, as shown.

A qualitative understanding of bending moment distributions in redundant beams can also be obtained using superposition. Figure 5.19(a) shows a propped cantilever subjected to point loads. This differs from a simply supported beam only due to the existence of a moment reaction at the fixed end. If we ignore the moment reaction for the time being, then there is no difficulty in sketching the resulting bending moment diagram (Figure 5.19(b)). If we now consider the effect of the end moment in isolation, this produces the bending moment distribution in Figure 5.19(c). The total moments in the propped cantilever are then found by summing (b) and (c), and are plotted under the beam in (a). The end moment has the effect of reducing the moment by M_0 at the fixed end and by a proportionate amount at points along the beam.

5.4 Elasto-plastic bending

To conclude this chapter, we shall briefly look at the behaviour of beams when loaded beyond their yield points. The concept of yielding was introduced in Chapter 4, where it was applied to axially loaded bars. These are comparatively simple, since all points of a bar are subjected to the same stress, and therefore yield occurs simultaneously throughout the member. For a beam the situation is much more complex, because the stress varies over the cross-section and, usually, along the length. Consider the arbitrary section in Figure 5.20, subjected to a steadily increased moment M. Initially, the behaviour is elastic and the stress varies linearly from zero at the neutral axis to maximum values at the extreme fibres.

(a) As the moment is increased, the largest stress in the section becomes equal to the yield stress σ_y and yield occurs at that point (the bottom edge in this case). The moment at which this occurs is generally referred to as the *yield moment* M_y. Assuming perfectly elasto-plastic behaviour, the stress in the yielded material stays constant. However, material away from the yield point still has some reserve capacity, and so the section can still carry more moment.

(b) If the moment is increased further, then other highly stressed areas also reach σ_y and yielding gradually spreads through the section from the bottom fibre towards the centre.

(c) At some point yielding also occurs at the top edge – of course, if the section is symmetrical about the horizontal axis, then yielding occurs simultaneously at the top and bottom edges.

(d) Eventually the entire cross-section yields and it is now unable to sustain any further increase in moment, since no part of the cross-section can carry any

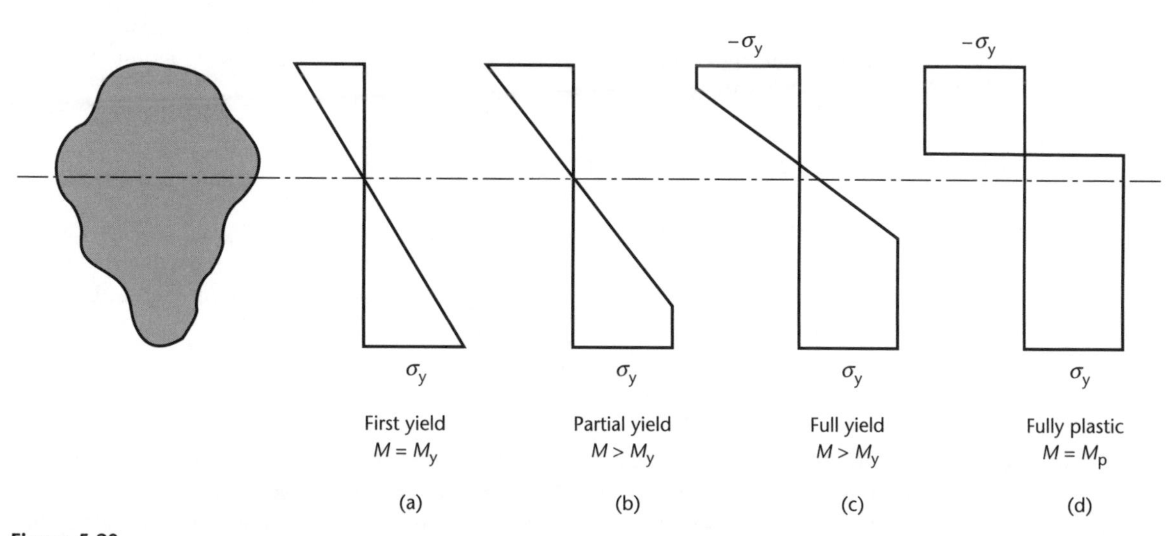

Figure 5.20
Stages in the flexural yield of a cross-section.

additional stress. The section is then described as *fully plastic* and the corresponding moment is known as the *plastic moment* M_p.

It can be seen that, where material has yielded, the stress distribution changes from linearly varying to constant. Thus, after partial yield there is a linear stress variation in the central, elastic region and constant stress in the outer, yielded parts. In (d) the stress consists of a uniform compressive stress of σ_y at all points above the neutral axis and a uniform tension of σ_y at all points below.

Note also that the neutral axis (that is, the level of zero stress) gradually moves as yielding increases. We found in section 5.1 that, while the section is elastic, the neutral axis passes through the centroid. Now the neutral axis position must satisfy the condition that the total axial force on the section is zero. Suppose that in (d) the neutral axis has moved so that there is an area A_1 of material above it and A_2 below. The total axial force is

$$P_{total} = A_1 \cdot (-\sigma_y) + A_2 \cdot \sigma_y = 0$$

and hence $A_1 = A_2$. Thus, when the section is fully plastic, the neutral axis simply divides it into two equal areas. If the section were symmetrical, yielding would, of course, occur symmetrically, with the result that the neutral axis would not move.

5.4.1 Plastic moment of a section

Once yielding has occurred, we can no longer analyse bending stresses using expressions involving I or Z_e, since they are based on the assumption that the whole section is linear elastic. Instead, we will find the moment on a section by taking moments of the stresses about the neutral axis.

Consider the partially yielded rectangular section in Figure 5.21(a). For the purposes of taking moments, the stress diagram can be divided into triangular (elastic) parts and rectangular (plastic) blocks and each replaced by a force acting at the centroid of the stress block. For example, the force in the elastic tension region is given by an average stress of $\sigma_y/2$ acting over an area ab, and its line of action is $2a/3$ below the neutral axis. Similarly, the force in the plastic tension region is given by a constant stress σ_y acting over an area cb, and is a distance $(a + c/2)$ from the neutral axis. Since the beam is symmetrical, the compressive terms have the same magnitudes and positions above the neutral axis. So, taking moments of the forces about the neutral axis:

$$M = 2\left[\frac{\sigma_y ab}{2} \cdot \frac{2a}{3} + \sigma_y cb \cdot \left(a + \frac{c}{2}\right)\right]$$

The partially yielded case is quite complicated. Much simpler is the fully yielded case (Figure 5.18(b)), where we simply have a uniform tension on one side of the neutral axis and a uniform compression on the other. Taking moments about the neutral axis gives the plastic moment for a rectangular section:

$$M_p = 2 \cdot \frac{\sigma_y bd}{2} \cdot \frac{d}{4} = \sigma_y \frac{bd^2}{4} \tag{5.15}$$

For elastic behaviour, Equation (5.10) defines the elastic modulus of a section Z_e as the constant of proportionality between the applied bending moment and the largest resulting stress. If a section is loaded just to the point of first yield, then equation (5.10) becomes:

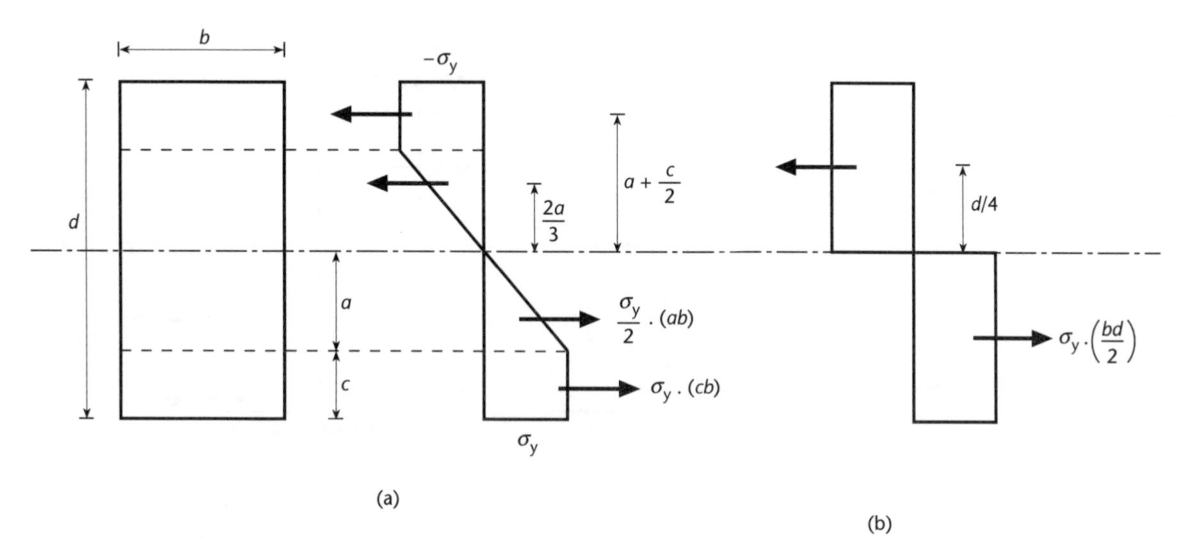

Figure 5.21
Stresses in a partially yielded rectangular section.

$$M_y = Z_e \sigma_y \tag{5.16}$$

Similarly, we can now define the *plastic modulus* as the constant of proportionality between the plastic moment and the yield stress:

$$M_p = Z_p \sigma_y \tag{5.17}$$

Comparing equations (5.15) and (5.17), we see that the plastic modulus of a rectangular section is $bd^2/4$; we saw earlier that the elastic modulus is $bd^2/6$. The ratio of M_p to M_y gives us a measure of how much additional moment a section can sustain between first yield and complete failure. From equations (5.16) and (5.17):

$$\alpha = \frac{M_p}{M_y} = \frac{Z_p}{Z_e} \tag{5.18}$$

The ratio is therefore dependent only on the dimensions of the section and is called the *shape factor*. For the rectangular section examined above $\alpha = 1.5$, meaning that the moment can be increased by 50% after first yield before complete collapse occurs. For sections more commonly used in structural engineering, such as I-sections, α is typically around 1.15.

EXAMPLE 5.12

In Example 5.4 we found the elastic modulus for the I-section shown in Figure 5.22 to be 767.8×10^3 mm^3. Find its plastic modulus and shape factor.

The plastic moment is found by taking moments of the forces about the neutral axis:

$$M_p = 2\left[(210 \times 15 \times \sigma_y) \times 117.5 + (110 \times 10 \times \sigma_y) \times 55\right] = 861.25 \times 10^3 \sigma_y$$

Figure 5.22

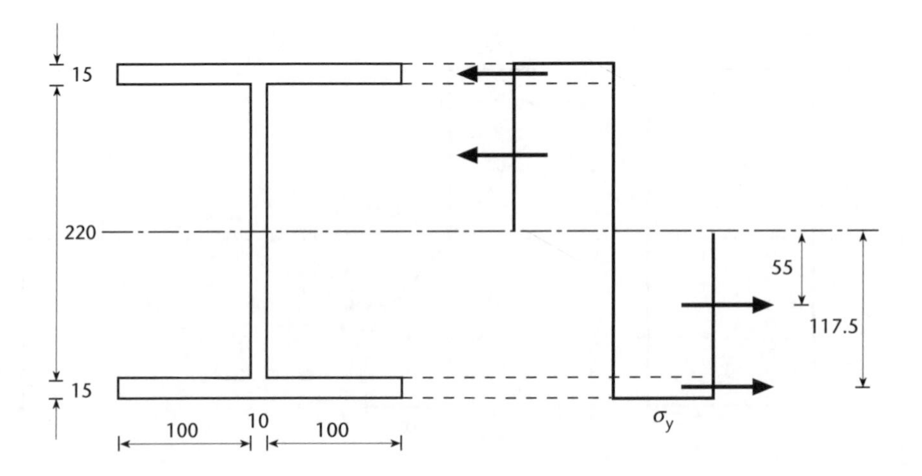

and from equation (5.17):

$$Z_p = \frac{M_p}{\sigma_y} = 861.25 \times 10^3 \text{ mm}^3$$

The shape factor is then

$$\alpha = \frac{861.25}{767.8} = 1.122$$

5.4.2 Residual deformations

As with axially loaded elements, if a beam is stressed beyond its yield point, plastic strains will be set up which cannot be recovered on unloading. The overall behaviour is most easily characterised by considering the moment–curvature relationship for a particular cross-section, as in the following example.

EXAMPLE 5.13

A rectangular solid steel section 25 mm wide and 40 mm deep is loaded by a moment M about its major axis. Taking $\sigma_y = 275$ MPa and $E = 205$ GPa, find:

(a) the moment and curvature at first yield;
(b) the moment and curvature when 50% of the cross-section has yielded;
(c) the residual curvature if the beam is unloaded from the point calculated in (b).

(a) The second moment of area of the section is $I = 25 \times 40^3/12 = 133.3 \times 10^3$ mm^4. Up to first yield the section obeys the elastic formulae, equation (5.5):

$$\frac{M}{I} = \frac{\sigma}{y} = \frac{E}{R}$$

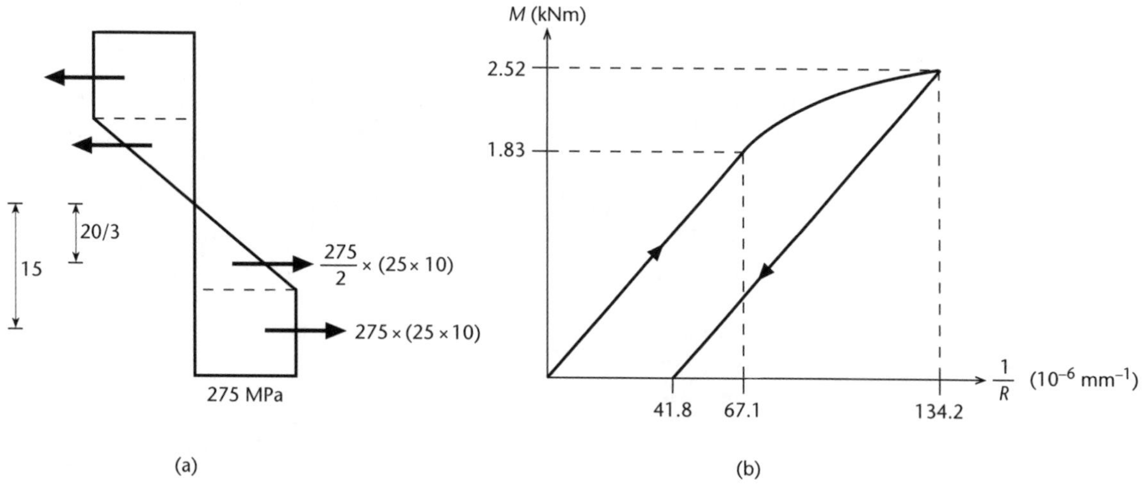

Figure 5.23

from which:

$$M = \frac{\sigma_y I}{y} = \frac{275 \times 133.3 \times 10^3}{20} = 1.83 \text{ kNm},$$

$$\frac{1}{R} = \frac{\sigma_y}{Ey} = \frac{275}{205 \times 10^3 \times 20} = 67.1 \times 10^{-6} \text{ mm}^{-1}$$

(b) After yield the above moment expression is no longer valid. When 50% of the section has yielded the stresses are as shown in Figure 5.23(a) and taking moments about the neutral axis gives

$$M = 2\left[(275 \times 25 \times 10) \times 15 + \left(\frac{275}{2} \times 25 \times 10\right) \times \frac{20}{3}\right] = 2.52 \text{ kNm}$$

The stress–curvature relationship used in (a) is still valid for that part of the section which remains elastic, that is, for values of y up to 10 mm. Therefore:

$$\frac{1}{R} = \frac{\sigma_y}{Ey} = \frac{275}{205 \times 10^3 \times 10} = 134.2 \times 10^{-6} \text{ mm}^{-1}$$

(c) If we plot M against $1/R$ (Figure 5.23(b)), the curve is linear up to point (a), then gradually reduces in slope as yielding spreads through the beam – the behaviour between (a) and (b) is a combination of elastic and plastic deformation. If we now unload from (b) there is an elastic recovery, so that we return to the $M = 0$ axis along a path parallel to the original elastic loading curve. The residual curvature can then be found easily from the geometry of the plot and is given by:

$$\frac{1}{R} = 41.8 \times 10^{-6} \text{ mm}^{-1}$$

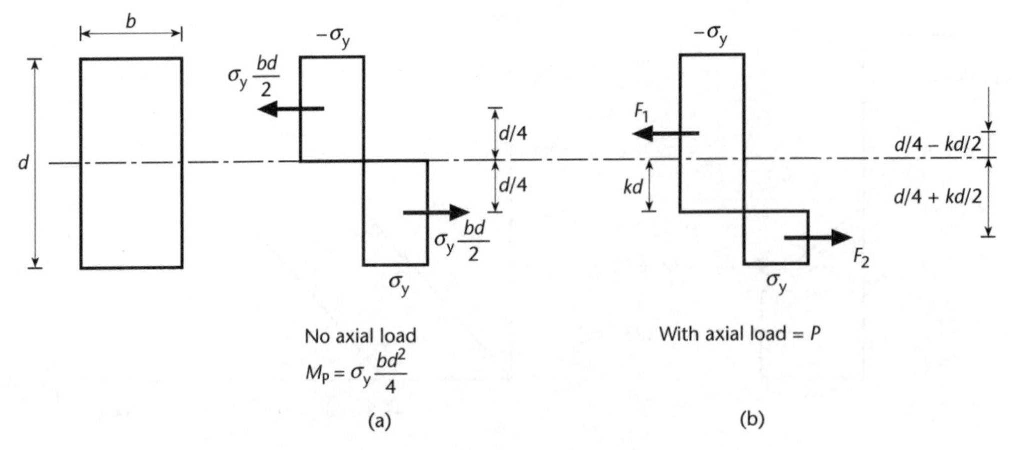

Figure 5.24
Effect of axial load on plastic moment capacity.

5.4.3 Influence of axial load

Whereas beams carry loads perpendicular to their axes, causing bending, columns generally sustain both axial and bending loads. It should be obvious that the plastic moment capacity of a section will be reduced by the presence of an axial load. For simplicity, we shall again examine the case of a simple rectangular section (Figure 5.24). With the member subjected only to a bending moment, as in Figure 5.24(a), equal areas of the cross-section yield in tension and compression and taking moments of the stresses about the neutral axis gives a plastic moment capacity of $\sigma_y bd^2/4$.

If instead the section is first subjected to an axial force P, and then a bending moment M is applied such that the section becomes fully plastic, then the axial compression has the effect of shifting the neutral axis down by some distance kd (Figure 5.24(b)). This means that a greater area of the section is in compression than in tension and that the resultant tensile and compressive forces are no longer at equal distances from the centroidal axis.

We can use equilibrium between the axial stresses and the applied axial force to find how far the neutral axis has been shifted:

$$P = F_1 - F_2 = \sigma_y b\left(\frac{d}{2} + kd\right) - \sigma_y b\left(\frac{d}{2} - kd\right) \qquad \text{hence} \quad k = \frac{P}{2\sigma_y bd}$$

Then, taking moments about the centroidal axis gives the plastic moment capacity of the section:

$$\begin{aligned} M_p &= F_1\left(\frac{d}{4} - \frac{kd}{2}\right) + F_2\left(\frac{d}{4} + \frac{kd}{2}\right) \\ &= \sigma_y bd^2\left(\frac{1}{2} + k\right)\left(\frac{1}{4} - \frac{k}{2}\right) + \sigma_y bd^2\left(\frac{1}{2} - k\right)\left(\frac{1}{4} + \frac{k}{2}\right) \\ &= M_{p0}\left(1 - 4k^2\right) \end{aligned}$$

where $M_{p0} = \sigma_y bd^2/4$ is the plastic moment capacity in the absence of any axial load. Thus the reduction in the plastic moment is proportional to the square of the axial load.

5.5 Problems

Where appropriate, answers are given at the end of the book. The questions marked with an asterisk are a little more challenging.

5.1. Show by direct integration that the second moment of area of a solid circular section of radius a about a centroidal axis is $\pi a^4/4$.

5.2. Calculate the properties of the channel section shown in Figure 5.25 for bending about the horizontal centroidal axis.

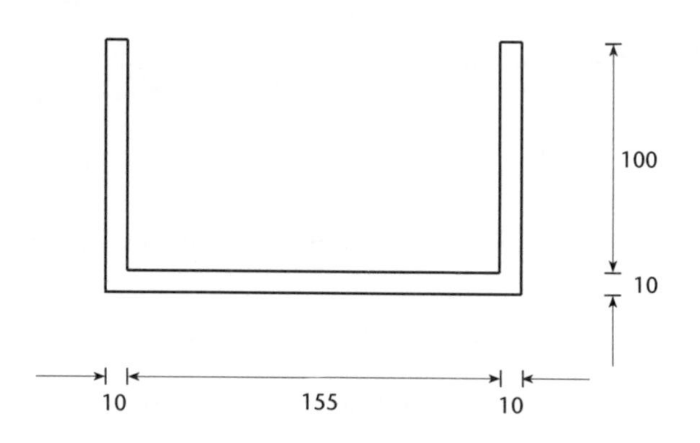

Figure 5.25

5.3. Find the maximum extreme fibre stresses when the section in Figure 5.25 is used as

(a) a simply supported beam of length 4.6 m carrying a uniform load of 9.5 kN/m;
(b) a cantilever of length 3.0 m carrying a tip load of 5.0 kN.

5.4. A rectangular timber beam 75 mm wide by 150 mm deep is strengthened by gluing a 75 mm wide by 5 mm thick steel plate to its bottom face. The timber has $E = 14$ GPa and $\sigma_y = 30$ MPa, while the steel has $E = 205$ GPa and $\sigma_y = 275$ MPa. If a bending moment is applied about the horizontal axis, determine which material reaches its yield stress first and at what moment.

5.5. Using direct integration, show that the midspan deflection of a simply supported beam under a uniform load is $5wl^4/384EI$.

5.6. Using either direct integration or superposition of standard solutions, find the reactions for the propped cantilever shown in Figure 5.26.

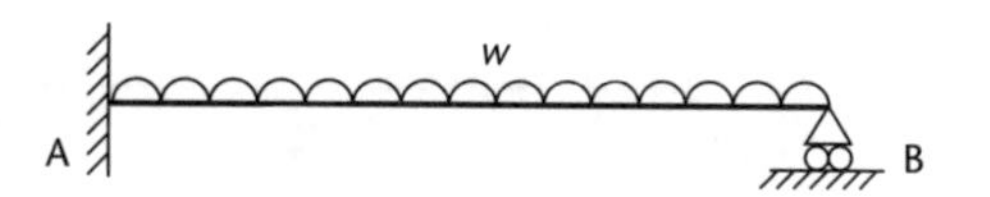

Figure 5.26

5.7. For the beam shown in Figure 5.27:

(a)* sketch the deflected shape under the loading shown;
(b) calculate the deflections at B and D and check they are in agreement with your sketch.

Figure 5.27

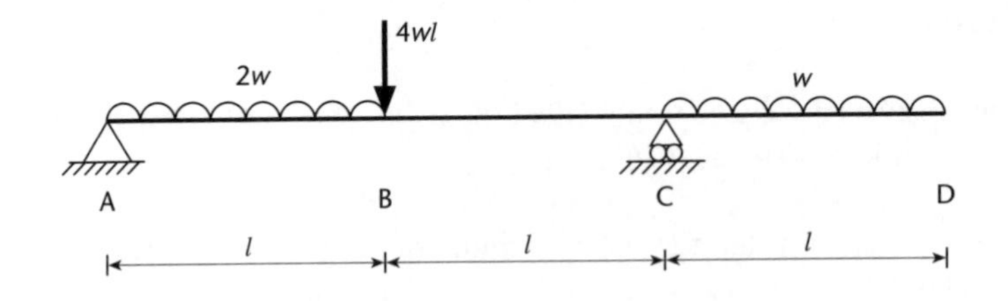

Figure 5.28

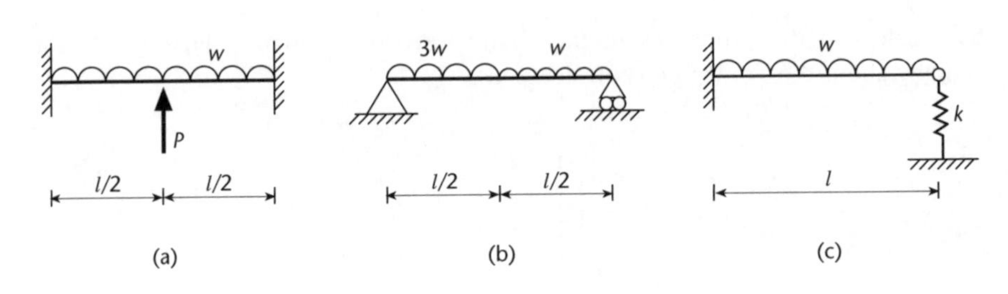

5.8. Solve the following using superposition of the standard solutions given in Table 5.2:

(a) For the beam in Figure 5.28(a), find the value of P required to halve the deflection due to w.

(b)* Find the midspan deflection of the beam in Figure 5.28(b). [*Hint: split into a symmetric and an anti-symmetric loadcase.*]

(c)* For the beam in Figure 5.28(c), find the deflection at the right-hand end. [*Hint: treat the spring force as an external load on the beam.*]

Figure 5.29

(a)

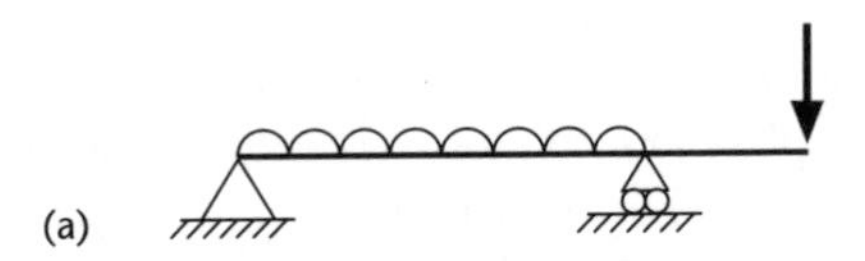

(b)

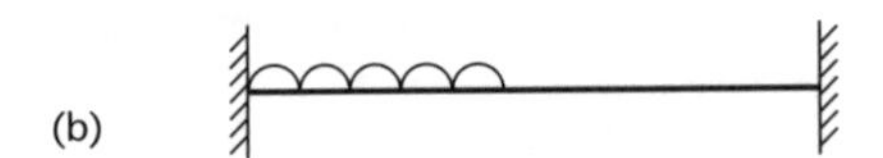

(c)

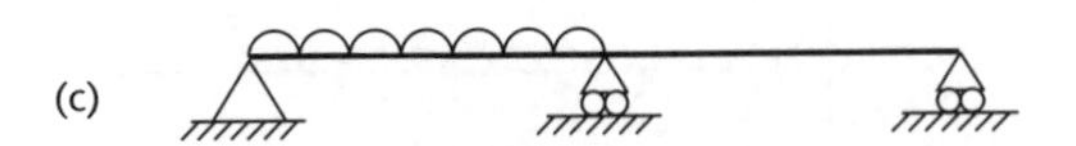

(d)

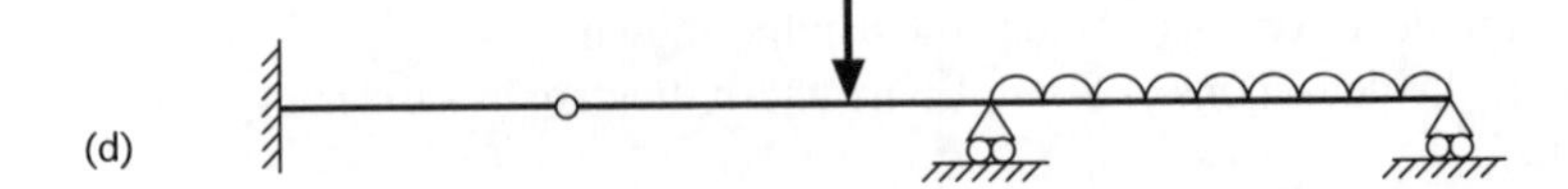

5.9.* Sketch the bending moment diagrams and deflected shapes for the beams shown in Figure 5.29, identifying any points of contraflexure.

5.10. A steel square hollow section has outer dimension 200 mm and thickness 10 mm. Taking $\sigma_y = 275$ MPa, find the bending moment at first yield and the fully plastic moment. What is the shape factor for the section?

CHAPTER 6

Torsion and shear

CHAPTER OUTLINE

In the previous chapter we went into some detail on how structural elements behave under the action of bending moments. These give rise to axial tensile and compressive stresses and to transverse deflections, and for many elements are the most significant form of loading.

However, there are other types of internal force whose effects must be checked; bending moments are generally accompanied by shear forces, and external loads may also give rise to torques (that is, moments about the longitudinal axis of the member). Although shear forces and torques are quite different types of load, they both cause shear stresses to be set up, and result in shear deformations. As with bending stresses and deflections, these must be calculated so that they can be limited to acceptable levels in design.

In this chapter we will look at the relationships between the external loads on a member and the shear stresses within it. We will first consider behaviour under torsion, for which the theory is relatively straightforward, before moving onto the rather more complex theory governing the shear in elements under bending loads. We will also look briefly at the behaviour of unsymmetrical sections in which the bending and torsional effects can interact. The chapter should enable the reader to:

- calculate elastic shear stress distributions and deformations caused by torques;
- analyse torsion problems involving plasticity;
- calculate stresses due to torsion combined with other types of load;
- determine shear stress distributions in sections subjected to bending loads;
- understand the concept of the shear centre;
- calculate shear deflections and appreciate in which instances they are likely to be significant.

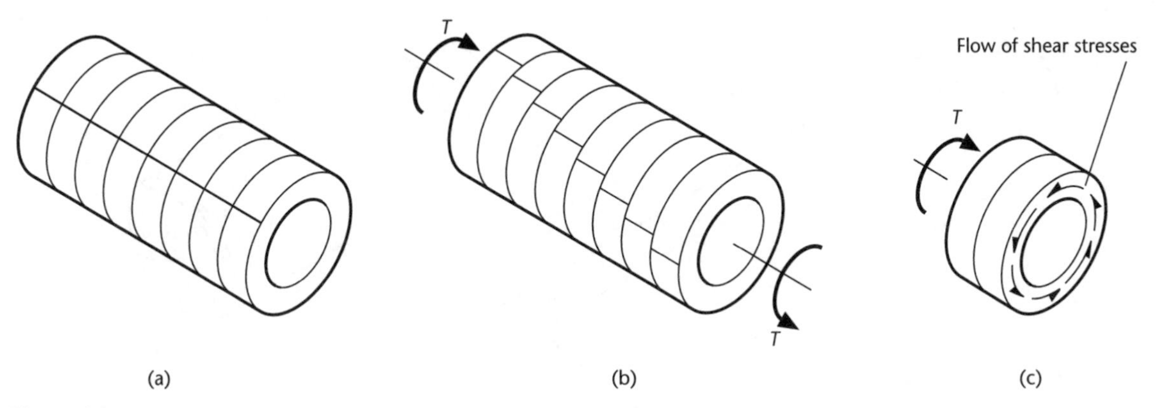

Figure 6.1
Torsional behaviour.

6.1 Torsion of circular sections

A torque is simply a moment applied about the axis of a member, causing it to twist. Imagine a hollow circular bar sliced into a series of thin, annular discs (Figure 6.1(a)), but with some resistance to slip between the discs. A straight line is drawn along one side of the bar. A torque is now applied at each end, which tends to cause the discs to rotate relative to each other, so that the initially straight line becomes stepped (Figure 6.1(b)). Also, shear stresses must be set up between the discs, tending to oppose the relative motion. If we consider equilibrium of just a few discs, we can see that the shear stresses tend to *flow* round the discs so as to balance the applied torque (Figure 6.1(c)). This concept of shear flow is a useful one, which we will return to later in the chapter.

It can be seen that the only deformation induced by the torque is a slip between the discs – the discs themselves remain undeformed and simply rotate as rigid bodies. It follows that torsion involves *only* shear stresses and deformations between the discs, with no direct stresses or strains.

Of course, to obtain a better approximation to reality, we must let the thickness of the discs tend towards zero. The shear deformation then becomes continuous rather than stepped.

6.1.1 Elastic torque–stress relationships

Consider first a thin-walled circular tube of length l, radius r and wall thickness $t(\ll r)$. If a constant torque is applied about the longitudinal axis the tube will twist as shown in Figure 6.2(a). This deformation involves only shearing of the tube; there are no longitudinal deformations and the cross-sectional dimensions do not change. If the material behaves elastically then the deformation will be uniformly distributed over the length. Thus the shear strain γ is constant over the length of the tube and is equal to the angle of rotation of the line AB. So the displacement of B to B′ can be expressed either as γl or as $r\phi$. Remembering the shear stress-strain relationship, we can write:

$$\tau = G\gamma = \frac{Gr\phi}{l} \tag{6.1}$$

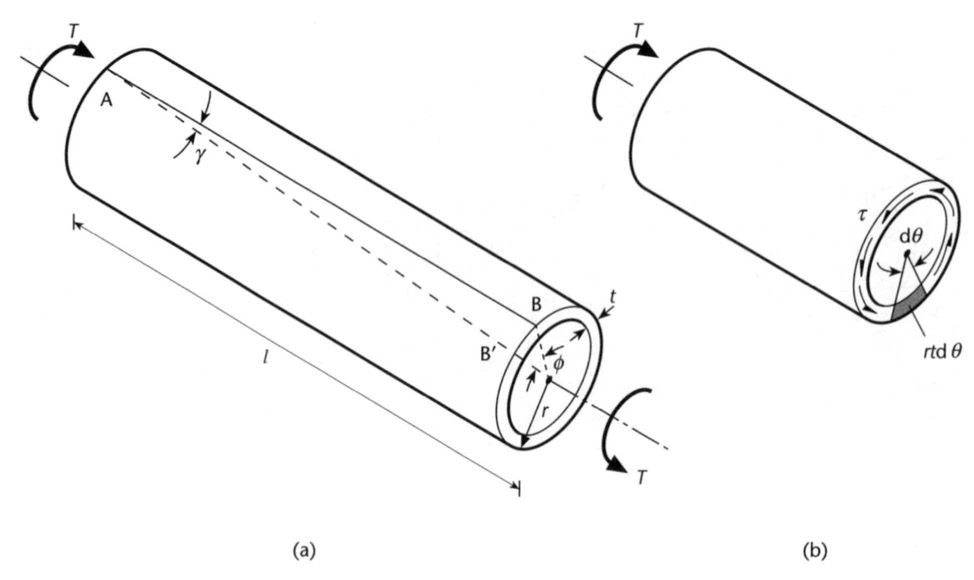

Figure 6.2
Elastic torsion of a thin circular tube.

The shear stress can be related to the torque by sectioning the tube at any point (Figure 6.2(b)). The shear force on a small element of the circumference formed by taking an angle $d\theta$ from the tube axis is $\tau \cdot trd\theta$, and taking moments of this force about the axis gives a resisting torque of $\tau \cdot trd\theta \cdot r$. For equilibrium, the total resisting torque obtained by integrating this expression around the circumference must balance the applied torque T. The symmetry of the problem means that τ is constant around the circumference, so the integration is trivial and we get:

$$T = 2\pi r^2 t\tau \tag{6.2}$$

Combining (6.1) and (6.2) gives:

$$\frac{\tau}{r} = \frac{T}{2\pi r^3 t} = \frac{G\phi}{l} \tag{6.3}$$

The quantity $2\pi r^3 t$ is the *polar second moment of area* of the tube, that is, the second moment of area of the cross-section about the tube's longitudinal axis, usually denoted by J. So (6.3) could be written:

$$\frac{\tau}{r} = \frac{T}{J} = \frac{G\phi}{l} \tag{6.4}$$

Equation (6.4) can be thought of as the torsional equivalent of the governing equation for elastic bending, Equation (5.5), since each term is directly analogous. For example, the twist per unit length ϕ/l is related to the torque T by the *torsional stiffness* GJ, just as in bending the curvature $1/R$ is related to the bending moment M by the flexural stiffness EI. Similarly, the shear stress is related to the torque in almost exactly the same way as a direct stress is related to a bending moment.

Note that the general definition of the polar second moment of area of a section is:

$$J = I_{xx} = \int r^2 \mathrm{d}A \tag{6.5}$$

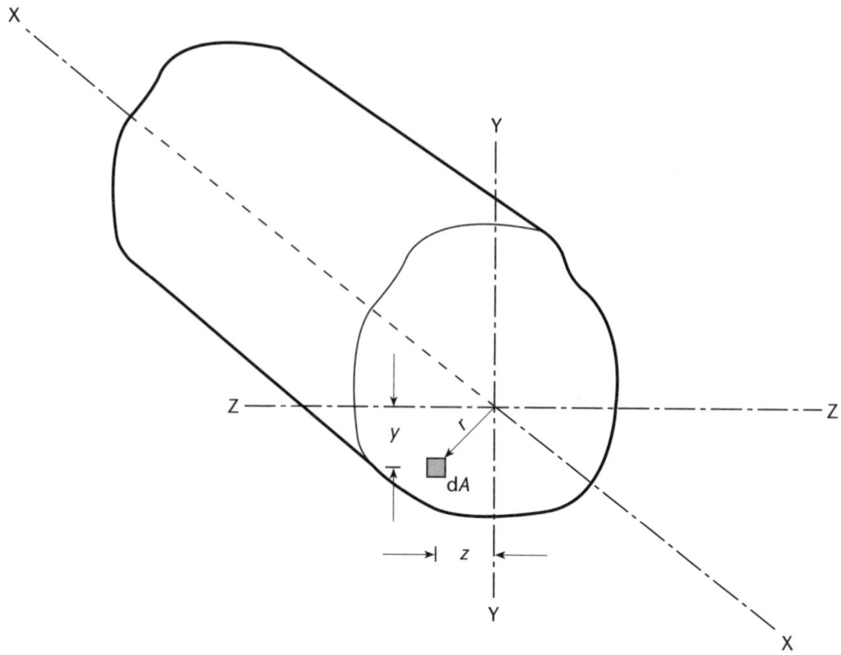

Figure 6.3
Perpendicular axes theorem.

Now from Figure 6.3 we can see that $r^2 = y^2 + z^2$, where y and z are the distances from the major and minor axes, ZZ and YY. So Equation (6.5) can be written:

$$J = \int z^2 \mathrm{d}A + \int y^2 \mathrm{d}A = I_{yy} + I_{zz} \tag{6.6}$$

Equation (6.6) is the *perpendicular axes theorem*. It states that the second moment of area about an axis perpendicular to the plane of the section is equal to the sum of the I-values about two perpendicular axes lying in the plane of the section, so long as all three axes pass through a common point. This often represents the easiest way of calculating J.

Now consider the case of a thick cylinder with inner and outer radii r_i and r_o, (Figure 6.4). The cylinder may be visualised as a series of concentric thin tubes of radius r and thickness dr, where $r_\mathrm{i} \leq r \leq r_\mathrm{o}$. In addition to the observations made above for a thin tube, we now note that considerations of symmetry mean that, as the cylinder deforms, radial lines in the cross-section remain straight. This means that each thin tube twists through the same angle of rotation ϕ, whatever its radius. Equation (6.1) can be applied to any tube, and hence we can see that the stress and strain vary linearly with radius, from a minimum at the inside edge to a maximum on the outside.

The resisting torque provided by any thin tube can be found from equation (6.3), simply replacing t by dr. Then integrating for all possible tubes between r_i and r_o gives the total torque carried by the cylinder:

$$T = \int_{r_\mathrm{i}}^{r_\mathrm{o}} \frac{G\phi}{l} 2\pi r^3 \mathrm{d}r = \frac{G\phi}{l} \cdot \frac{\pi(r_\mathrm{o}^4 - r_\mathrm{i}^4)}{2}$$

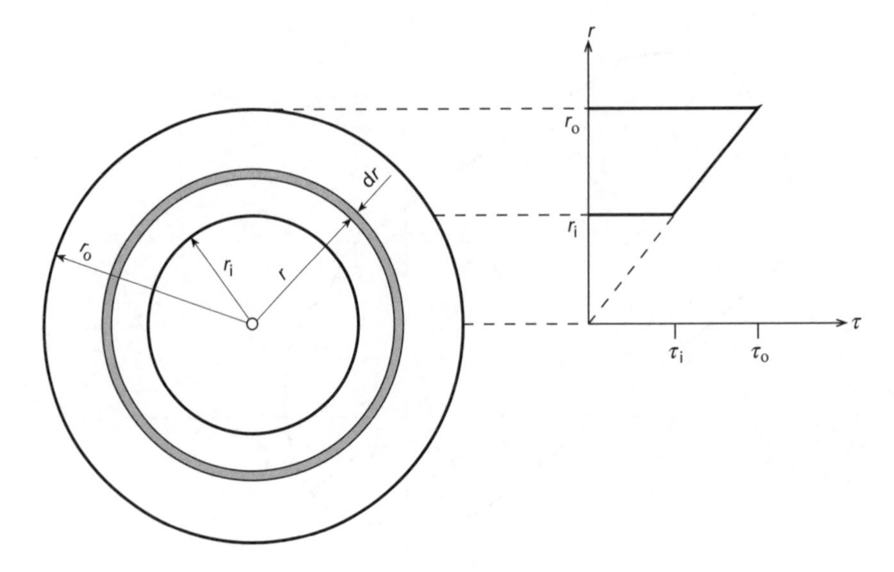

Figure 6.4
Torsional stresses in a thick cylinder.

This is, in fact, exactly the same result as equation (6.4) since it can easily be shown, either by direct integration or using the perpendicular axes theorem, that the polar second moment of area of a thick tube is

$$J = \frac{\pi(r_o^4 - r_i^4)}{2} \tag{6.7}$$

Thus equation (6.4) can be applied to a circular cylinder of any thickness, so long as the appropriate value of J is used. Of course, (6.4) and (6.7) can also be used for a solid cylinder, simply by setting r_i equal to zero.

EXAMPLE 6.1

A solid steel cylinder of length 1 m and diameter 80 mm is subjected to a torque of 5 kNm. Calculate the maximum shear stress and the total twist over the length of the cylinder. A second cylinder has identical exterior dimensions, but has a 40 mm diameter hole bored along its axis. Calculate its shear stress and twist when subjected to the same torque. Take G for steel to be 79 GPa.

For the solid cylinder, equation (6.7) gives $J = \pi \times 40^4/2 = 4.021 \times 10^6$ mm^4

The shear stress is a maximum on the outer edge, that is, at $r = r_o$. Using equation (6.4):

$$\tau_{max} = Tr_o/J = 5 \times 10^6 \times 40/4.021 \times 10^6 = 49.7 \text{ MPa}$$

The angle of twist can also be found from (6.4):

$$\phi = Tl/GJ = (5 \times 10^6 \times 1000)/(79 \times 10^3 \times 4.021 \times 10^6) = 0.0157 \text{ rads}$$

For the hollow cylinder the same approach can be used, except that the value of J must be modified to $J = \pi \times (40^4 - 20^4)/2 = 3.770 \times 10^6$ mm^4. Hence:

$$\tau_{max} = 5 \times 10^6 \times 40/3.77 \times 10^6 = 53.1 \text{ MPa}$$

$$\phi = (5 \times 10^6 \times 1000)/(79 \times 10^3 \times 3.77 \times 10^6) = 0.0168 \text{ rads}$$

Hollowing out the cylinder has increased both the shear stress and the twist by 7%. However, the mass has been reduced by 25%, so it is clear that the hollow cylinder is far more efficient. This is because the shear stress and strain under torsion vary linearly along the radius, so that parts of the tube near the centre contribute very little to the overall torsional capacity.

6.1.2 Torsion combined with other loads

In practice it is very common for torsional loads to be combined with other load types, particularly bending. For elastic behaviour, such combined load effects can be considered using the principle of superposition. That is, the stresses due to each load are calculated independently and then simply added to give the total effect. This is best illustrated by an example.

EXAMPLE 6.2

The hollow cylinder of Example 6.1 is mounted as a cantilever of length 0.6 m. A vertical concentrated load of 20 kN is applied at the free end, but displaced from the tube axis by 125 mm (Figure 6.5(a)). Calculate the largest principal stress and the maximum shear stress.

We first calculate the bending moment and torque distributions caused by the loading. The torque is caused by the offset of the load from the tube axis. It is therefore constant over the length and equal to:

$$T = 20 \times 0.125 = 2.5 \text{ kNm}$$

The bending moment varies linearly from zero at the tip to a maximum at the root:

$$M_{max} = 20 \times 0.6 = 12 \text{ kNm}$$

Considering these independently, the torque gives rise to a shear stress on the outer surface of the cylinder given by equation (6.4):

$$\tau = Tr/J = 2.5 \times 10^6 \times 40/3.77 \times 10^6 = 26.5 \text{ MPa}$$

For a circular section, the second moment of area about an axis in the plane of the section is simply half the polar second moment of area, so from equation (5.5) the longitudinal stress due to the bending moment is:

$$\sigma_{xx} = My/I = 12 \times 10^6 \times 40/1.885 \times 10^6 = 254.6 \text{ MPa}$$

We must now think about how these stresses are combined. Figure 6.5(b) shows the stresses acting on a small element on the top surface of the tube, at the root of the cantilever. This simple 2D stress system can be analysed using the Mohr's circle

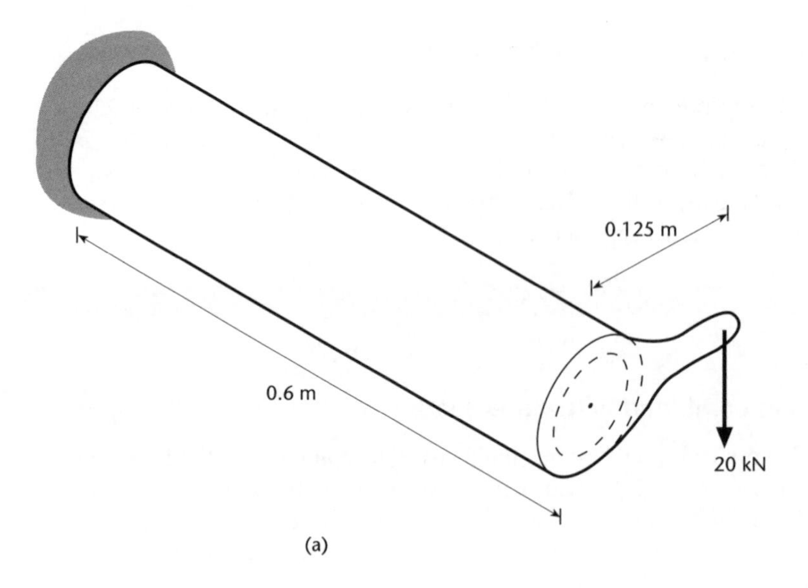

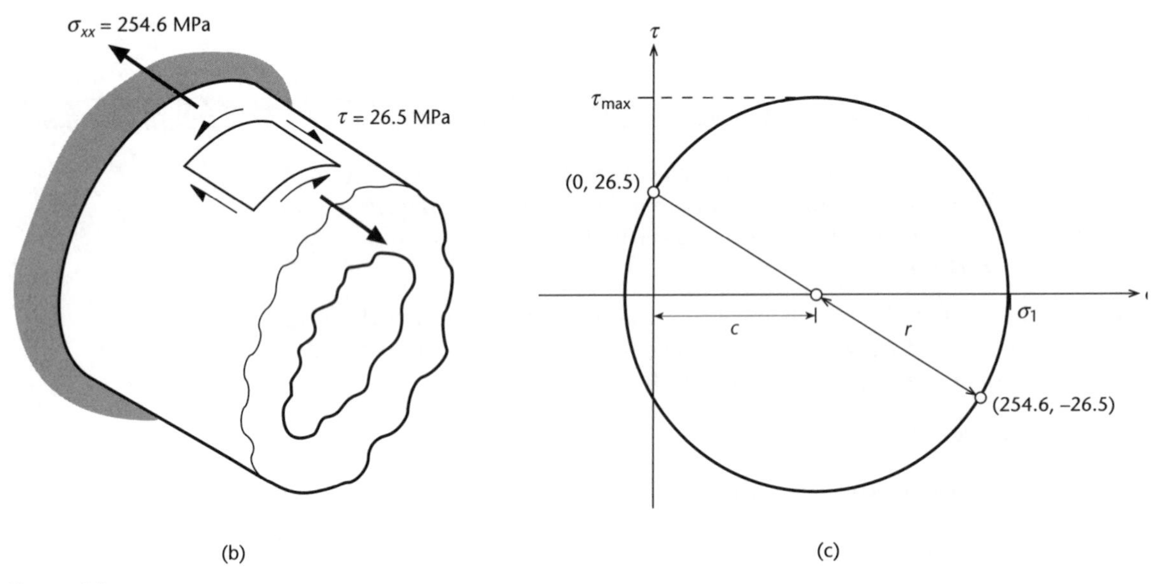

Figure 6.5

construction introduced in section 4.4. The Mohr's circle is shown in Figure 6.5(c), from which it can be seen that:

Centre $c = 254.6/2 = 127.3$ MPa, Radius $r = \sqrt{(127.3^2 + 26.5^2)} = 130.0$ MPa

Hence the maximum principal stress is $\sigma_1 = c + r = 257.3$ MPa

and the maximum shear stress $\tau_{max} = r = 130.0$ MPa

6.1.3 Yielding in torsion

Materials yield under shear stress in much the same way as under direct stress. In fact, yielding of ductile materials under any form of loading is always closely related to the shear stresses present. In this section we shall again restrict ourselves to the idealised case of perfectly elasto-plastic behaviour. That is, the material has a linear shear stress–strain curve with gradient G, up to a limiting shear stress τ_y beyond which the gradient is zero.

As the torque on a circular section is steadily increased, the behaviour remains linear until the maximum shear stress reaches τ_Y. Since the shear stress increases linearly with radial distance, the first yield must take place on the outside surface. Once yielded, this part of the tube can carry no additional stress, so any further torque applied to the tube must be carried by an increase in stress in the inner part of the tube, which is still elastic. As the torque continues to increase, plasticity spreads inwards from the outer surface of the tube until all of the material has yielded, and at this stage the tube can carry no additional torque.

As with elastic torsion, there is a close analogy with the behaviour in bending, as described in section 5.3, and much the same approach can be taken to analysing elasto-plastic torsion problems. When yielding has occurred we must evaluate the torque–shear stress relationship by taking moments about the member axis. However, the shear stress–twist part of equation (6.4) can still be used for those parts of the section that remain linear.

EXAMPLE 6.3

A circular bar of radius 25 mm is made of a material with $G = 50$ GPa and $\tau_y = 150$ MPa. Find the torque and the twist per unit length when (a) first yield occurs, (b) yielding has extended to a depth of 10 mm, (c) the load is then removed.

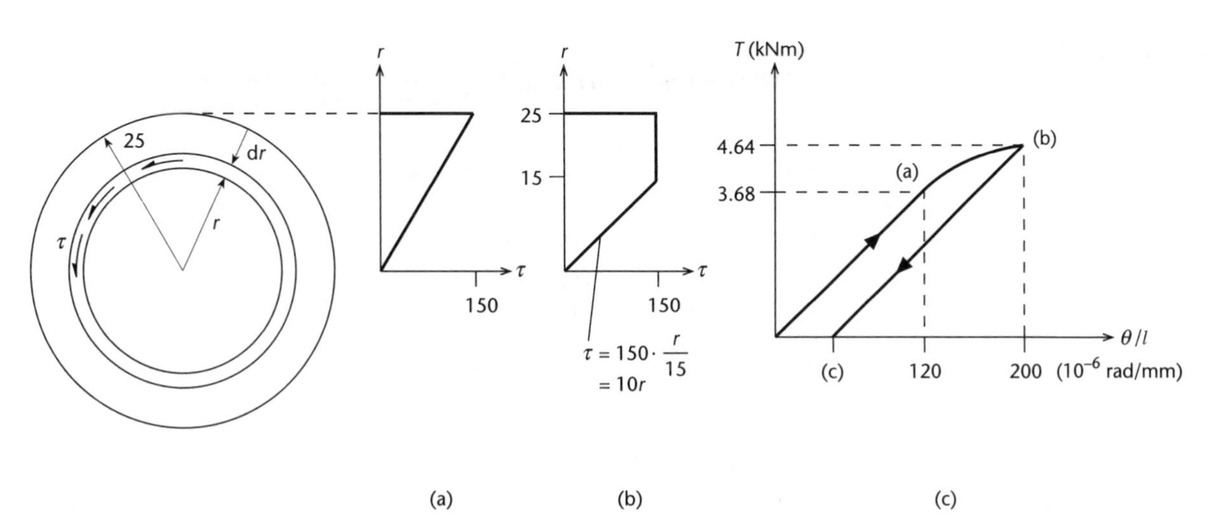

Figure 6.6

(a) First yield will occur on the outer face of the bar, where the stress is largest (Figure 6.6(a)), and up to this point the behaviour will be elastic. Equation (6.4) can therefore be used:

$$J = \frac{\pi \times 25^4}{2} = 613.6 \times 10^3 \text{ mm}^4$$

$$T_y = \frac{J\tau_y}{r} = \frac{613.6 \times 10^3 \times 150}{25} = 3.68 \times 10^6 \text{ Nmm} = 3.68 \text{ kNm}$$

$$\frac{\theta}{l} = \frac{T}{GJ} = \frac{3.68 \times 10^6}{5 \times 10^3 \times 613.6 \times 10^3} = 120 \times 10^{-6} \text{ rad/mm}$$

(b) The stress distribution is now as shown in Figure 6.6(b). If the stress on a thin annulus at radius r is τ then the force on the annulus is $\tau \cdot 2\pi r dr$ and the torque generated by this force about the axis is $\tau \cdot 2\pi r^2 dr$. So the total torque is found by integrating this expression over the radius of the bar, noting the discontinuity in the stress term at $r = 15$ mm:

$$T = \int_0^{15} 10r \cdot 2\pi r^2 dr + \int_{15}^{25} 150.2\pi r^2 dr = 4.64 \times 10^6 \text{ Nmm} = 4.64 \text{ kNm}$$

and we can still use the elastic shear stress–twist relationship up to $r = 15$ mm:

$$\frac{\theta}{l} = \frac{\tau}{Gr} = \frac{150}{50 \times 10^3 \times 15} = 200 \times 10^{-6} \text{ rad/mm}$$

(c) If the load is now removed $T = 0$ and the bar recovers elastically. If we plot T against θ/l throughout the loading cycle (Figure 6.6(c)), the unloading part is parallel to the original loading part. The residual twist can easily be found from the geometry of the plot and is

$$\frac{\theta}{l} = 49 \times 10^{-6} \text{ rad/mm}$$

6.2 Torsion of thin-walled non-circular sections

The theory developed above is applicable only to circular sections, which can conveniently be treated as series of thin rings centred on the axis of rotation. Circular sections are the most effective at carrying torsional loads, and so are used when torsion is the dominant load type, for example for shafts in rotating machinery. However, in many structures, torsion occurs in combination with other significant loads, and in such cases a circular section may not be optimal. An understanding of how non-circular sections carry torques is therefore important. We shall concentrate on thin-walled sections, for which the theory is comparatively straightforward.

6.2.1 Closed sections

Figure 6.7 shows a closed, non-circular section subjected to a torque. The cross-section is uniform over the length of the member. The thickness t may vary around

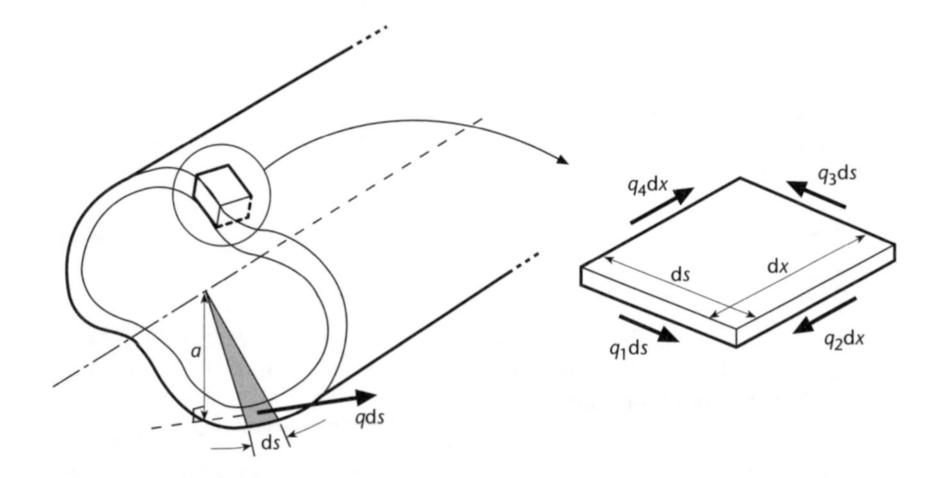

Figure 6.7
Torsion of a closed, thin-walled section of arbitrary shape.

the perimeter, so long as it is always much smaller than the overall cross-sectional dimensions. To account for the possibility of varying thickness we shall work in terms of the *shear flow*, defined as

$$q = \tau t \tag{6.8}$$

If we consider the small rectangular element, then for moment equilibrium about an axis perpendicular to the plane of the element we must have $q_1 = q_2 = q_3 = q_4$. From this we may infer that the shear stress is constant around the perimeter of the tube.

Now a small length of perimeter $\mathrm{d}s$ carries a force $q\mathrm{d}s$, and its torque about the tube axis is $\mathrm{d}T = q\mathrm{d}s \cdot a$, where a is the perpendicular distance from the tube axis to the line of action of the force. But the shaded area on the diagram is $\mathrm{d}A = ds \cdot a/2$, so $\mathrm{d}T = 2q\mathrm{d}A$, and integrating gives

$$T = 2qA_\mathrm{e} \tag{6.9}$$

where A_e is the area enclosed by the mean perimeter of the tube. This enables us to relate torque to shear stress, however we also require some way of assessing how much the tube deforms. This can be obtained by consideration of strain energy. We showed in equation (4.20) that the strain energy per unit volume for a member in pure shear is $\tau^2/2G$. Now if a length l of the section is subjected to a torque T, the work done by T will equal the stored strain energy integrated over the volume, that is

$$\frac{1}{2}T\phi = \int_V \frac{\mathrm{d}U}{\mathrm{d}V}\mathrm{d}V = \oint \frac{\tau^2}{2G} lt\mathrm{d}s$$

where $\oint$ means the integral around the whole of the perimeter. Then, substituting from equations (6.8) and (6.9) and rearranging gives:

$$\frac{G\phi}{l} = \frac{q}{2A_\mathrm{e}}\oint\frac{\mathrm{d}s}{t} = \frac{T}{4A_\mathrm{e}^2}\oint\frac{\mathrm{d}s}{t} \tag{6.10}$$

which allows us to determine the twist either in terms of the shear flow, or the torque. Earlier we showed that, for a circular section, the torque–twist relationship is given by equation (6.4) as

$$\frac{G\phi}{l} = \frac{T}{J}$$

Comparing this with equation (6.10), we see that exactly the same relationship applies to a closed, non-circular section, except that the polar second moment of area is replaced by

$$J = \frac{4A_e^2}{\oint \frac{ds}{t}} \tag{6.11}$$

The geometric parameter defined by equation (6.11) is generally termed the *torsion constant*, so as to distinguish it from the polar second moment of area, though the same symbol, J, is used for both.

EXAMPLE 6.4

The rectangular hollow section shown in Figure 6.8 is subjected to a torque of 20 kNm. Find the twist per unit length and the distribution of shear stresses. Take $G = 79$ GPa.

Figure 6.8

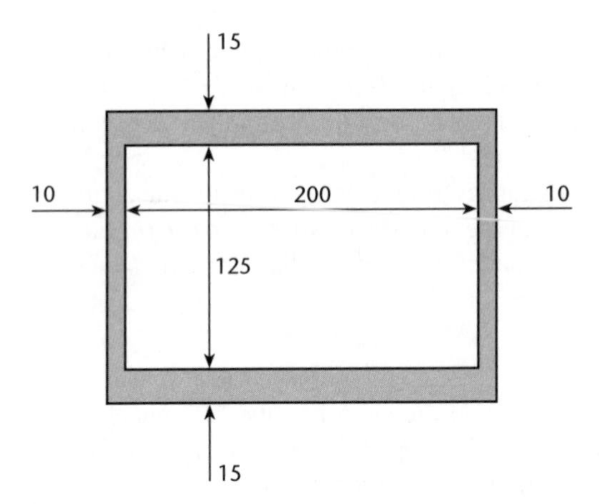

The enclosed area $A_e = 210 \times 140 = 29\,400$ mm^2, and

$$\oint \frac{ds}{t} = 2\left(\frac{210}{15} + \frac{140}{10}\right) = 56, \text{ so the torsion constant is:}$$

$$J = \frac{4 \times 29\,400^2}{56} = 61.74 \times 10^6 \text{ mm}^4$$

Then equation (6.10) gives:

$$\frac{\phi}{l} = \frac{T}{GJ} = \frac{20 \times 10^6}{79 \times 10^3 \times 61.74 \times 10^6} = 4.1 \times 10^{-6} \text{ rad/mm}$$

and (6.9) gives:

$$q = \frac{T}{2A_e} = \frac{20 \times 10^6}{2 \times 29\,400} = 340.1 \text{ N/mm}$$

The shear stresses in the various parts of the section are then found by dividing the shear flow by the relevant thickness:

$$\text{Flanges: } \tau = \frac{340.1}{15} = 22.7\ \text{MPa} \qquad \text{Webs: } \tau = \frac{340.1}{10} = 34.0\ \text{MPa}$$

6.2.2 Thin plates and open sections

A thin plate subjected to torsion can be analysed using the above theory by visualising it as a series of concentric, flat tubes. Figure 6.9 shows a plate of breadth b and thickness t, where $b \gg t$. A typical tube has its flanges at $\pm y$ from the neutral axis and has wall thickness dy. If the tube thickness is constant then its shear stress is also constant. To a first approximation the area enclosed by the tube is $A_e \approx 2by$ and $\oint \frac{ds}{t} \approx \frac{2b}{dy}$. The torsion constant is therefore given by

$$J = \oint_0^{t/2} \frac{4 \cdot (2by)^2}{2b} dy = \frac{1}{3}bt^3 \qquad (6.12)$$

The torsion constant for other thin-walled open sections can be approximated by breaking them down into rectangular elements and treating each as a thin plate. The overall torsion constant is then simply the sum of the $bt^3/3$ terms for each constituent rectangle.

It will be noted that, since t is small, the torsion constant for a thin plate is extremely low. Thin plates and open sections therefore have a very low stiffness under torsional loading compared to closed sections, and so the latter are nearly always preferred when significant torsion is present. The reason why open sections are so poor in torsion can be explained by considering the pattern of shear stresses within the cross-section. In a closed section, the continuity allows a one-directional shear flow around the section, setting up a substantial resisting torque (Figure 6.10 (a)). In a thin plate, on the other hand, the flow must occur within the very thin concentric hoops used in the above derivation, otherwise no resisting torque would be generated (Figure 6.10(b)). The small lever arm a means that for a given shear stress very little resisting torque is achieved. A similar principle applies to the T-section in Figure 6.10(c); again, in order to generate a resisting torque there must be a flow of shear stresses within the individual elements of the section, since no flow around the entire section is possible.

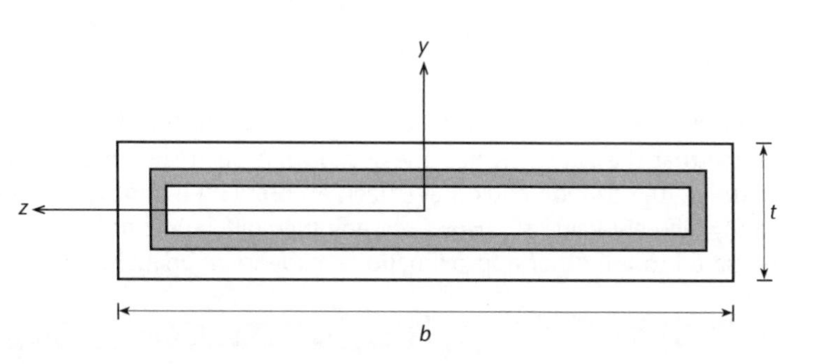

Figure 6.9
Torsion analysis of a thin plate.

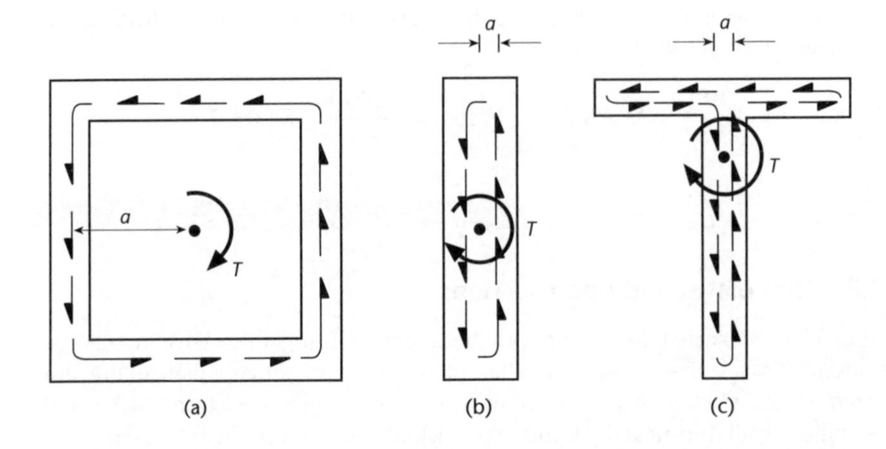

Figure 6.10
Shear flows in closed and open sections.

EXAMPLE 6.5

The cross-section shown in Figure 6.8 is modified by cutting a narrow slit along the centreline of the bottom flange. Calculate the reduction in the torsional stiffness.

Figure 6.11

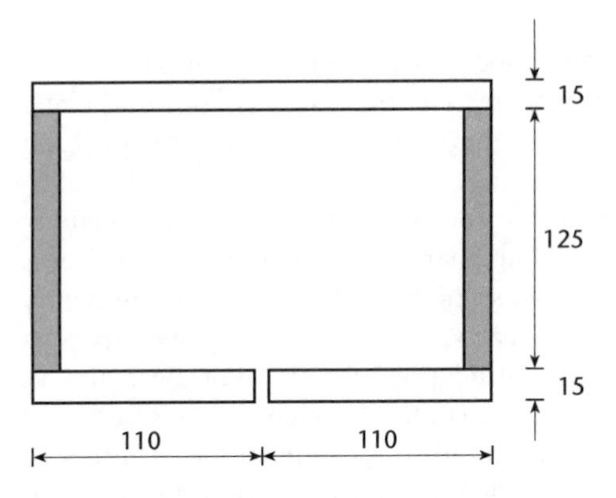

The torsional stiffness is GJ, but since G remains constant we need only consider the reduction in J. In Example 6.4 we found the torsion constant for the original section to be $J = 61.74 \times 10^6$ mm^4. The slit has the effect of breaking down the continuity of the section. J for the modified section can therefore be found by splitting it into a series of thin rectangles, as shown in Figure 6.11. Then, using equation (6.12):

$$J = \sum \frac{bt^3}{3} = 2 \times \frac{110 \times 15^3}{3} + 2 \times \frac{125 \times 10^3}{3} + \frac{220 \times 15^3}{3} = 0.578 \times 10^6 \text{ mm}^4$$

So the slit has reduced the torsional stiffness by a factor of 61.74/0.578 = 107.

Note that the way in which the section has been split into rectangles is rather arbitrary – the division could have been accomplished in other ways, resulting in a slightly different value of J. However, any such differences are likely to be very small compared to the other approximations made in an analysis.

6.3 Shear stresses in bending

6.3.1 Elastic shear stress distributions

Besides torsion, the other major causes of shear stresses in structural members are the shear forces that accompany the bending moments in flexural elements. Consider a beam element of length dx, subjected to a bending moment M about the z-axis at one end, increasing to $M + \mathrm{d}M$ at the other, (Figure 6.12(a)). Using simple bending theory, the stress at a distance y from the neutral axis is My/I at the left-hand end and $(M + \mathrm{d}M)y/I$ at the right-hand end.

Suppose we slice the beam along a horizontal plane BCDE. The difference in axial force between the two ends of the segment below the slice can be found by integrating the stresses over the end areas:

$$P_2 - P_1 = \int_A \frac{(M + \mathrm{d}M)y}{I} \mathrm{d}A - \int_A \frac{My}{I} \mathrm{d}A = \int_A \frac{\mathrm{d}M \cdot y}{I} \mathrm{d}A$$

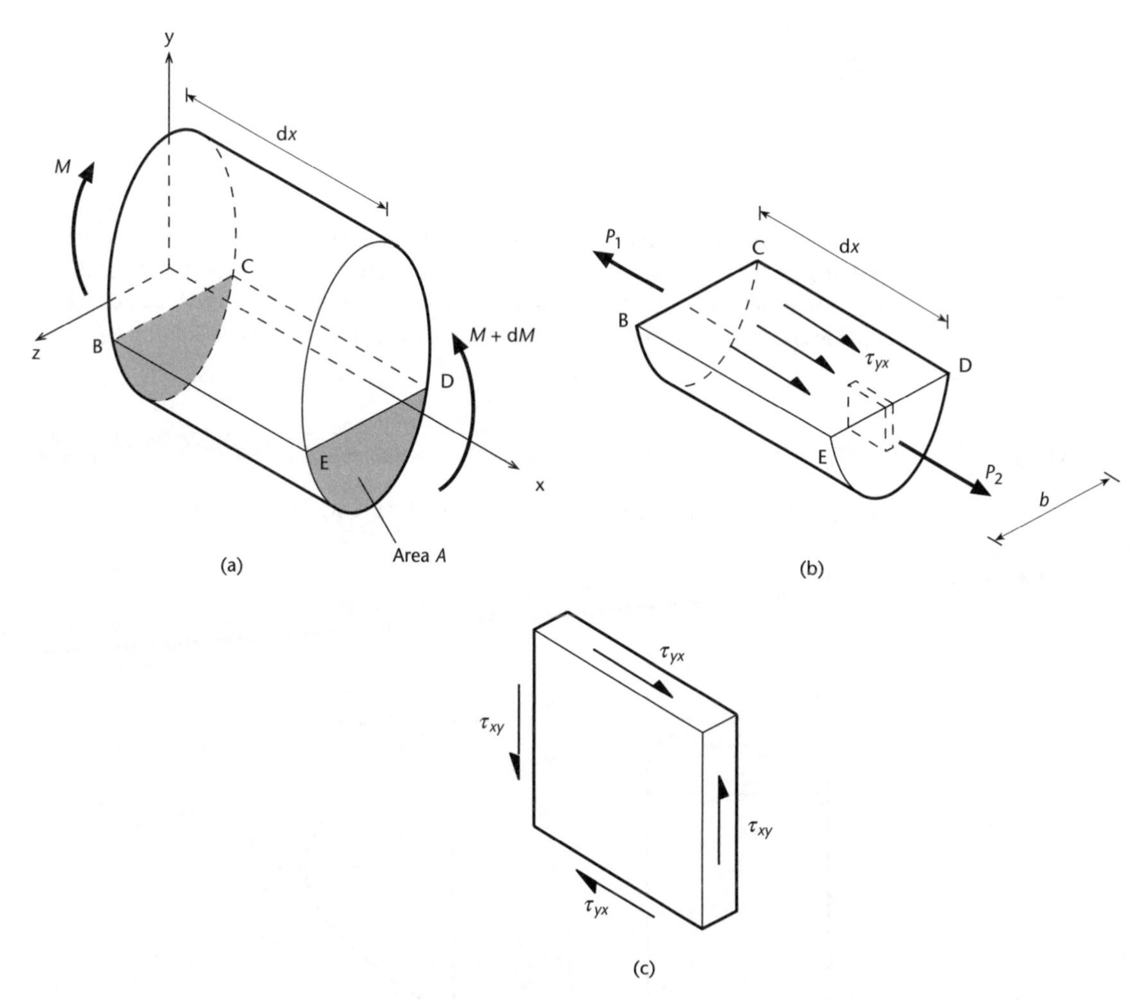

Figure 6.12
Shear stresses due to bending of a beam element.

Now the segment must be in equilibrium, and this can only be achieved if there is a shear stress τ_{yx} acting over the cut surface BCDE (Figure 6.12(b)), such that

$$\tau_{yx} b\mathrm{d}x + P_2 - P_1 = 0$$

Substituting for $(P_2 - P_1)$ from above:

$$\tau_{yx} = -\frac{\mathrm{d}M}{\mathrm{d}x} \frac{\int_A y\mathrm{d}A}{bI}$$

But from equation (2.9) we know that the shear force $S = -\mathrm{d}M/\mathrm{d}x$. Also, if we consider the stresses on a small rectangular element oriented along the beam (Figure 6.12(c)), then there must be a complementary shear stress τ_{xy} acting on the vertical face, and for equilibrium of the element we must have $\tau_{xy} = \tau_{yx}$. It therefore follows that the vertical shear stress at any point in the cross-section of our beam element is given by:

$$\tau_{xy} = \frac{S\int_A y\mathrm{d}A}{bI} \tag{6.13}$$

Equation (6.13) is often written as

$$\tau_{xy} = \frac{SA\bar{y}}{bI} \tag{6.14}$$

where A is the end area of the segment BCDE and $\bar{y}$ is the distance of the centroid of that area from the neutral axis of the section. This latter formulation is generally the simpler to use when the section is made up of simple geometric shapes whose areas and centroids can be determined by inspection, as in the following example.

EXAMPLE 6.6

Find the distribution of shear stresses in a solid rectangular cross-section of breadth *b* and depth *d* carrying a total shear force *S*.

Figure 6.13

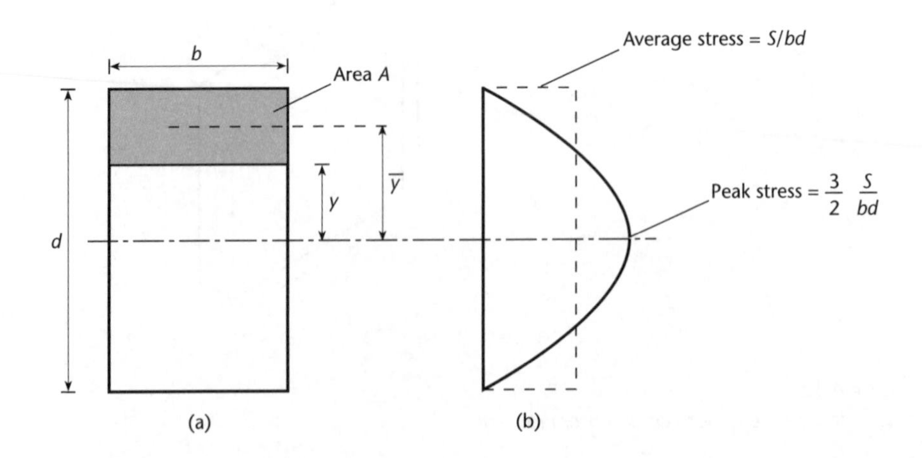

For a solid rectangle $I = bd^3/12$. To find the shear stress at a distance y from the neutral axis the section is divided up as shown in Figure 6.13(a). The shaded area is $A = b(d/2 - y)$ and the distance of its centroid from the neutral axis of the section is $\bar{y} = (d/2 + y)/2$, so equation (6.14) gives:

$$\tau = \frac{S \cdot b(d/2 - y) \cdot (d/2 + y)/2}{b \cdot bd^3/12} = \frac{6S}{bd^3}\left(\frac{d^2}{4} - y^2\right)$$

The shear stress thus varies parabolically with y from zero at $\pm d/2$ to a maximum value of $3S/2bd$ at the neutral axis ($y = 0$), as shown in Figure 6.13(b). So the maximum shear stress is 50% greater than the average shear stress over the section. Note that the vertical shear stress *must* be zero at the top and bottom edges of the section. If it were not then there would be complementary shear stresses acting along the top and bottom faces of the beam, which would violate the requirements of equilibrium.

We could check the above result by integrating the stress distribution over the cross-section since, for equilibrium, the sum of the vertical shear stresses must equal the applied shear force S.

There is one other point that should be noted before we move on. The simple bending theory we developed and used in Chapter 5 was based on the assumption that plane sections remain plane as the beam bends, so that the strain varies linearly over the cross-section. This is true when the bending moment is constant along the length of the beam, since the shear force is then zero ($S = -\mathrm{d}M/\mathrm{d}x$). If the bending moment varies, however, then there will be a non-zero shear force, and hence shear stresses and strains will be set up within the beam. Suppose, for example, the rectangular section of Example 6.6 were used as a cantilever carrying a point load at its tip (Figure 6.14). The beam is subjected to a constant shear force of magnitude P which gives rise to a parabolic shear stress distribution over the cross-section. A plane section therefore distorts as shown – at the top and bottom edges the shear stress is zero, so the section remains perpendicular to the top and bottom faces, while the greatest distortion occurs at the neutral axis, where the shear stress is largest. This phenomenon is known as *warping*.

Most beams used in structural engineering are quite slender (that is, their cross-sections are small compared to their lengths), with the result that warping effects are small and our simple bending theory is reasonably accurate. For very short, stocky sections, however, warping becomes significant and a more elaborate theory is required. This is quite a specialised topic which will not be covered here.

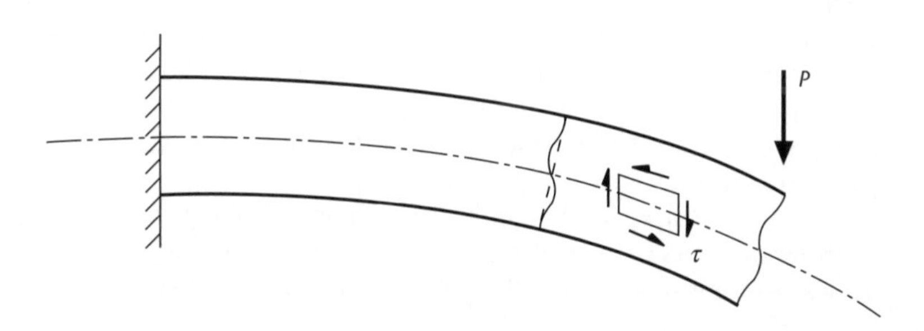

Figure 6.14
Warping.

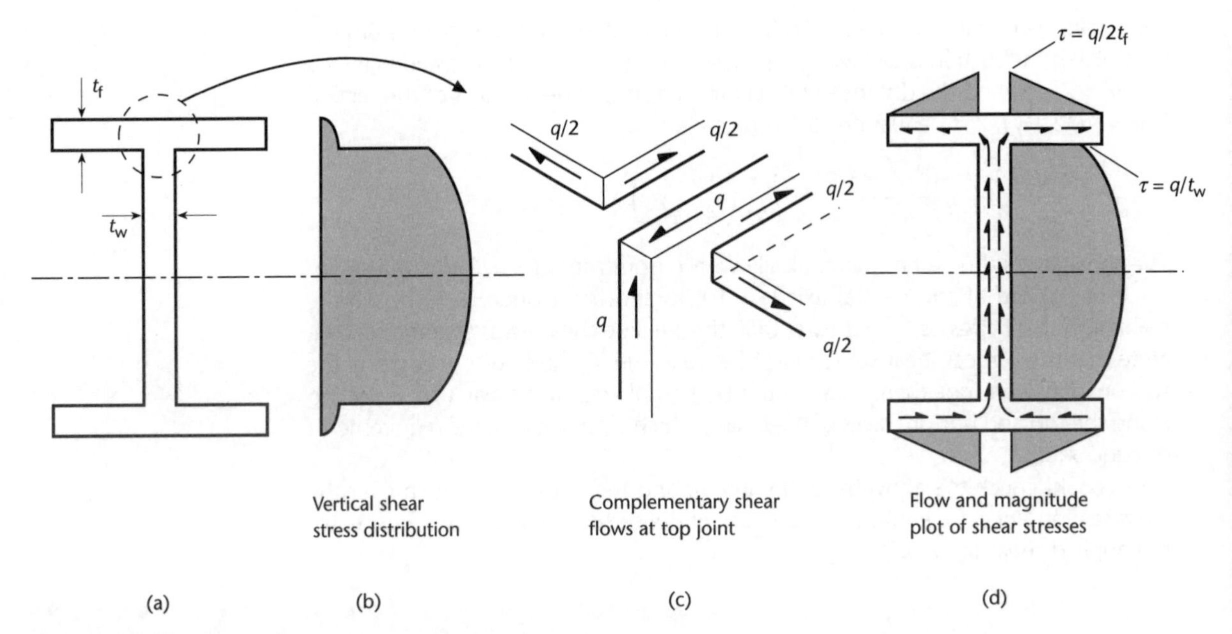

Figure 6.15
Shear stresses in an I-beam.

6.3.2 Application to typical structural sections

The use of sections such as solid rectangles is rare, since they are not structurally efficient. Much more common are thin-walled sections such as I- and T-beams, channels and box sections. Before proceeding with a mathematical analysis it is worth spending a little time thinking qualitatively about how thin-walled sections carry shear loads.

Consider, for example, the I-section shown in Figure 6.15(a). To sustain a vertical shear force we need to develop vertical shear stresses in the section. Applying equation (6.14) will give a parabolic shear stress distribution, as for the simple rectangle considered earlier, except that there will be a step change in magnitude going from the flange to the web, as the breadth b is suddenly increased (Figure 6.15(b)). In practice, the vertical shear stress in the flanges is very small and can be neglected. The web therefore provides virtually all of the vertical shear resistance.

Consider now the exploded view of the flange–web junction in Figure 6.15(c). The shear flow $q(= \tau \times \text{thickness})$ at the top of the web gives rise to a complementary shear flow along the length of the beam. For longitudinal equilibrium of the joint, this must be balanced by opposing shear flows on the adjoining faces of the flanges, and from symmetry these must each equal $q/2$. These in turn require complementary shear flows of $q/2$ running horizontally along the flanges. So at the joint there is a horizontal shear stress in each flange corresponding to half of the shear flow at the top of the web. This stress must reduce to zero at the ends of the flanges, since there can be no shear stress at a free end, and we will see later that the variation along the flange is linear. Figure 6.15 (d) shows a sketched plot of both the magnitudes and directions of the shear stresses, which can be seen to flow through the section. Note that the horizontal

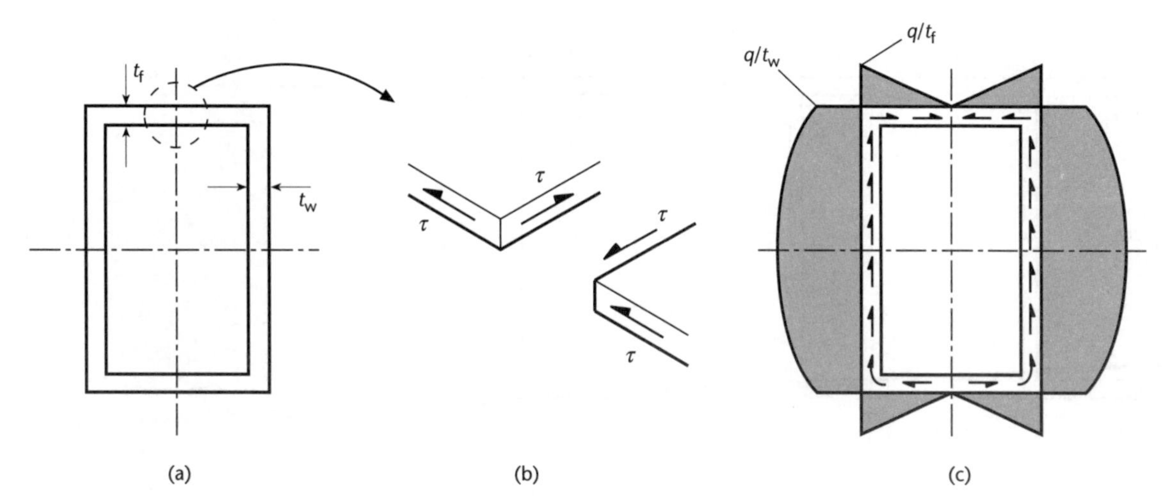

Figure 6.16
Shear stresses in a rectangular hollow section.

shear stresses in the flanges do not contribute directly to the vertical shear resistance, but are required to keep the section in overall equilibrium.

A similar line of argument can be used to sketch the shear stress distribution for the rectangular hollow section in Figure 6.16(a). Again, the bulk of the shear resistance is provided by vertical shear stresses in the two webs, distributed parabolically with a maximum at the neutral axis. Consideration of the complementary shear flows at the corners requires that the horizontal shear stresses at the ends of the flanges give the same shear flow q as the vertical shear stresses at the tops of the webs.

The stresses at the centres of the flanges can be determined by arguments based on symmetry. Figure 6.16(b) shows an isometric view of one flange, cut at the centre. Suppose that there is a non-zero stress τ just to the left of the centreline. This will result in a complementary shear stress on the perpendicular, longitudinal face, also of magnitude τ. For longitudinal equilibrium, this in turn requires the shear stresses shown on the right-hand side of the cut. These clearly violate the symmetry of the problem, and so the centre of the flange can only be in equilibrium if the shear stresses at that point are all zero. The flow of shear stresses and their magnitudes are therefore as shown in Figure 6.16(c).

The horizontal shear stresses in the flanges of thin-walled sections obey the same rules of equilibrium as the vertical stresses in the web – that is, they represent the difference between the integrated bending stresses at the two ends of a short element, as in Figure 6.11. Therefore shear stresses in thin-walled sections can be analysed using equation (6.13) or (6.14), in much the same way as solid sections.

EXAMPLE 6.7

Plot the shear stress distribution in the Universal Beam section shown in Figure 6.17 (a) when subjected to a shear force of 150 kN. The second moment of area about the major axis is $I = 102.1 \times 10^6$ mm^4.

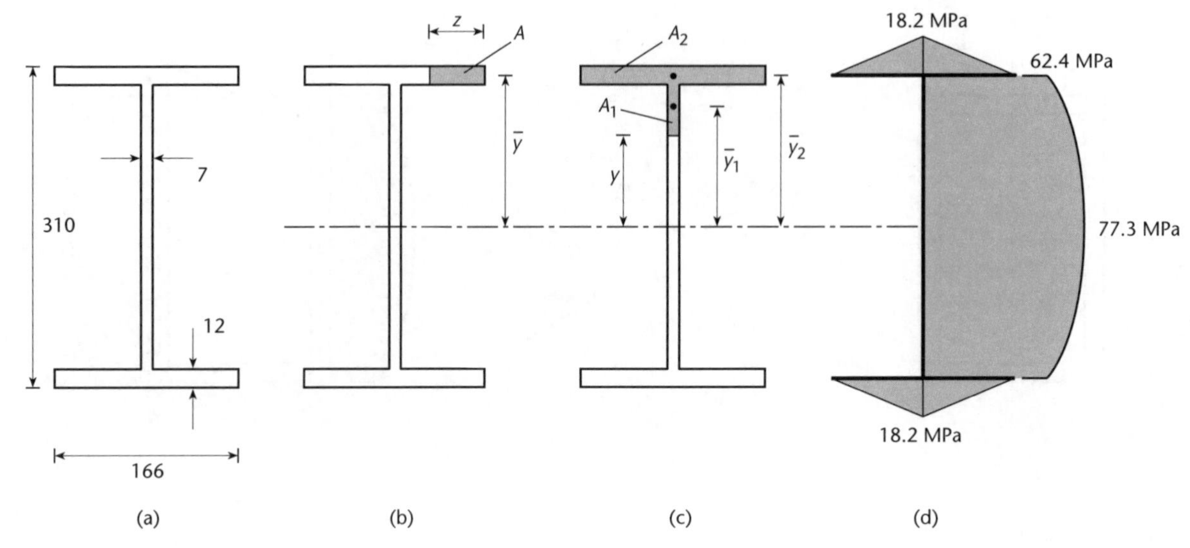

Figure 6.17

We will first find the horizontal shear stress in the flange – see Figure 6.17(b). It is easiest to work from the end of the flange, where we know the shear stress must be zero. If we take an arbitrary length z, the area of the shaded region is simply $A = 12z$ and the distance of its centroid from the neutral axis of the section is $\overline{y} = 149$ mm (that is, independent of z). Then, applying equation (6.14) gives:

$$\tau = \frac{150 \times 10^3 \times 12z \times 149}{12 \times 102.1 \times 10^6} = 0.2189z$$

So the shear force varies linearly along the flange from zero at the tip $(y = 0)$ to 18.2 MPa at the centre $(z = 83 \text{ mm})$.

To find the shear stress at some point in the web, we need to consider the shaded area shown in Figure 6.17(c), defined by the distance y from the neutral axis. The centroid of the shaded area is rather fiddly to find, but we can simplify the calculation by replacing $A\overline{y}$ by $(A_1\overline{y}_1 + A_2\overline{y}_2)$ where the subscripts 1 and 2 refer to the web and flange parts respectively. So:

$$A_1 = 7(143 - y), \qquad \overline{y}_1 = (143 + y)/2,$$
$$A_2 = 166 \times 12 = 1992 \text{ mm}^2, \qquad \overline{y}_2 = 149 \text{ mm}$$

and equation (6.14) gives:

$$\tau = \frac{150 \times 10^3 \times [7(143 - y) \times (143 + y)/2 + 1992 \times 149]}{7 \times 102.1 \times 10^6} = 77.3 - 7.346 \times 10^{-4}y^2$$

The stress varies parabolically over the web from a minimum of 62.4 MPa at the flanges $(y = 143 \text{ mm})$ to a maximum of 77.3 MPa at the centre $(y = 0)$. The distribution of shear stresses over the section is therefore as shown in Figure 6.17 (d).

As a check, the vertical shear flow at the top or bottom of the web is $q_{web} = 62.4 \times 7 = 436.8$ N/mm and the horizontal shear flow at the centre of the flanges is $q_{flange} = 18.2 \times 12 = 218.4$ N/mm $= q_{web}/2$, as required for equilibrium of the flange–web joints.

It is common practice in design to work out approximate shear stresses in I-sections by assuming that the web carries all the shear, and that the shear stress distribution over the web is uniform. In this case this approach would give a web shear stress of $150 \times 10^3/(310 \times 7) = 69.1$ MPa, whis is about 90% of the peak value calculated by the more accurate method.

6.3.3 Unsymmetrical sections and shear centre

So far we have looked only at shear distributions of sections with a vertical axis of symmetry. These behave in a comparatively simple way because the horizontal shear stresses in the flanges are in complete equilibrium. If the section does not have a vertical axis of symmetry then this will no longer be the case; the horizontal shear stresses will create a couple, causing the section to twist. The only way to prevent this twisting is to ensure that the external loading is applied not through the centroid of the section, but at some offset, so as to produce an opposing couple. This offset point is known as the *shear centre*.

EXAMPLE 6.8

Find the position of the shear centre for the channel section in Figure 6.18(a). I about the major axis is 16.91×10^6 mm^4.

Figure 6.18

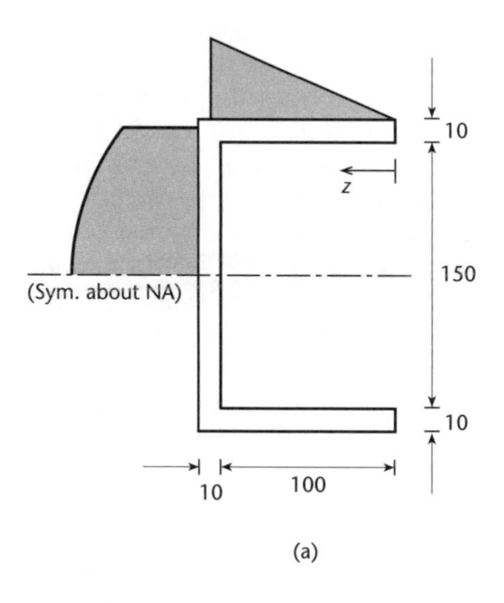

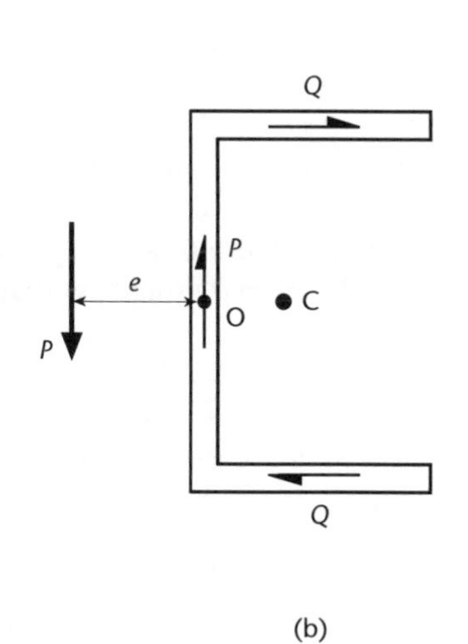

Suppose the section is used as a cantilever carrying a tip load P – the actual loading configuration does not affect the position of the shear centre, which is a geometric property, but choosing a simple loadcase makes the analysis easier to follow. We know that the vertical shear stresses in the web at any point along the beam must integrate to give a total upward shear force P, for equilibrium with the applied loading. For the flanges, the stress at a distance z from the end is

$$\tau = \frac{SA\bar{y}}{bI} = \frac{P \times 10z \times 80}{10 \times 16.91 \times 10^6} = 4.731 \times 10^{-6} Pz$$

and integrating gives the total flange force Q:

$$Q = \int_0^{105} \tau \cdot 10 \, \mathrm{d}z = \int_0^{105} 4.731 \times 10^{-5} Pz \, \mathrm{d}z = 0.2608P$$

Figure 6.18(b) shows the section with the shear forces in each part of the cross-section indicated and the external load applied through the shear centre, a distance e to the left of the point O, which is at the centre of the web. Taking moments about O:

$$P_e = 2Q \times 80 \rightarrow e = \frac{2 \times 0.2608P \times 80}{P} = 41.7 \text{ mm}$$

Thus, to prevent twisting of the section the load P must be offset by 76.4 mm from the centroid C of the section, since C is 34.7 mm to the right of O.

6.4 Shear deflections

A loaded beam will generally sustain deflections due to shear as well as due to bending. In most cases the shear deflections will be very small and can be neglected, but for very short, stocky elements they can be significant in comparison to the bending deflections.

We saw earlier (Figure 6.14) that the shear stresses set up in the beam cause shear strains and therefore warping of the cross-section. The stresses and strains vary from zero at the extreme fibres to maximum values at the neutral axis, so that a small rectangular element on the neutral axis deforms to a parallelogram, as shown in Figure 6.14. A small element of the beam remote from the neutral axis will be distorted much less than the one shown, causing the section to warp. We shall concentrate here on calculating the shear deflection at the neutral-axis level. The behaviour at other points is rather more complex due to the effects of warping, and will not be discussed further.

The slope of the top and bottom faces of a small element on the neutral axis is simply equal to its shear strain:

$$\frac{\mathrm{d}v}{\mathrm{d}x} = \gamma_{\mathrm{NA}} = \frac{\tau_{\mathrm{NA}}}{G} \tag{6.15}$$

where the subscript NA denotes the value at the neutral axis. Integrating equation (6.15) over a distance x from a point of zero deflection gives the shear deflection at x:

$$v_x = \int_0^x \frac{\tau_{\mathrm{NA}}}{G} \mathrm{d}x = \int_0^x \frac{KS}{AG} \mathrm{d}x \tag{6.16}$$

Here S/A is the average shear stress on the section and K is a constant to convert this average shear stress to the maximum value. K is a function of the section shape, and is equal to 1.5 for a simple rectangle. For a thin-walled section, the web carries virtually all the shear, so the average stress S/A should be calculated using the web area only, and K is then the ratio of the maximum to the average web shear stress. For an I-section K is generally in the range 1.0 to 1.1, depending on the exact section shape.

EXAMPLE 6.9

A cantilever is made from an aluminium alloy with Young's modulus $E = 80$ GPa and shear modulus $G = 30$ GPa. It has a rectangular cross-section with breadth 50 mm and depth 125 mm, and carries a uniformly distributed load of 10 kN/m. Calculate the shear deflection at the tip if the length of the cantilever is (a) 1000 mm, and (b) 250 mm. In each case, compare the result with the bending deflection.

First, calculate the section properties:

$$A = 50 \times 125 = 6250 \text{ mm}^2 \text{ and } I = \frac{50 \times 125^3}{12} = 8.138 \times 10^6 \text{ mm}^4$$

(a) The applied load is 10 kN/m = 10 N/mm and the total load on the 1000 mm beam is 10,000 N. So the shear force a distance x mm from the fixed end is $S = 10x - 10{,}000$ (N), and equation (6.16) gives:

$$v = \int_0^l \frac{KS}{AG} dx = \int_0^{1000} \frac{1.5 \times (10x - 10{,}000)}{6250 \times 30 \times 10^3} dx = -0.04 \text{ mm}$$

The bending deflection can be found from the formula in Table 5.2:

$$v = -\frac{wl^4}{8EI} = -\frac{10 \times 1000^4}{8 \times 80 \times 10^3 \times 8.138 \times 10^6} = -1.92 \text{ mm}$$

The ratio of the two deflections is $\frac{v_{shear}}{v_{bending}} = \frac{0.04}{1.92} = 2.1\%$. The shear deflection could therefore be neglected without significant loss of accuracy in this case.

(b) The shear force expression for the 250 mm beam is $S = 10x - 2500$ (N), and equation (6.16) gives:

$$v = \int_0^{250} \frac{1.5 \times (10x - 2500)}{6\,250 \times 30 \times 10^3} dx = -0.0025 \text{ mm}$$

and the bending deflection is:

$$v = -\frac{10 \times 250^4}{8 \times 80 \times 10^3 \times 8.138 \times 10^6} = -0.0075 \text{ mm}$$

The ratio of the two deflections is $\frac{v_{shear}}{v_{bending}} = \frac{0.0025}{0.0075} = 33.3\%$. So for the shorter beam the shear deflection is significant and cannot be neglected without considerable loss of accuracy.

6.5 Problems

Where appropriate, answers are given at the end of the book. The questions marked with an asterisk are a little more challenging.

6.1. A steel shaft is required to transmit a torque of 25 kNm. To ensure against yielding, the shear stress in the steel must be limited to 100 MPa. Find the radius of shaft required using

(a) a 10 mm thick circular hollow tube (*hint: use the equations for thin-walled sections*),
(b) a solid circular section.

Compare the weights of the two shafts.

6.2. Problem 6.1 is now modified by the addition of a second design criterion; in addition to the stress limit already given, the twist per unit length must not exceed 20×10^{-6} rad/mm. How does this affect the answers to 6.1? Take $G = 79$ GPa.

6.3. A solid circular shaft of radius 25 mm is loaded by a torque of 4 kNm and an axial tensile force of 130 kN. Find the largest direct stress in the shaft, and its direction.

6.4. A hollow steel shaft of length 1.0 m has inner and outer radii 40 mm and 60 mm respectively. It is loaded in torsion until the entire section becomes plastic. Taking $G = 79$ GPa and $\tau_y = 125$ MPa, find:

(a) the torque T and angle of twist ϕ at first yield;
(b) T and ϕ when the section has just become fully plastic;
(c)* T and ϕ when yielding has extended to a radius of 50 mm.

Hence sketch graphs showing how T varies with ϕ and how the radius to which yielding has occurred varies with ϕ.

6.5. A circular and a square thin-walled section are made of the same material, and both have constant thickness t and perimeter length p. Compare their torsional stiffnesses and the torques required to cause yielding.

6.6. Sketch the flow of shear stresses in the thin-walled sections shown in Figure 6.19 when subjected to a torque.

6.7. Calculate and sketch the shear stress distribution for the T-section shown in Figure 6.20 when subjected to a vertical shear force of 200 kN. By assuming that all the vertical shear resistance is provided by the web, find the ratio K between the maximum and the average web shear stress. (The neutral axis of the section is 166 mm from the bottom of the web and the second moment of area about that axis is $I = 18.05 \times 10^6$ mm^4.)

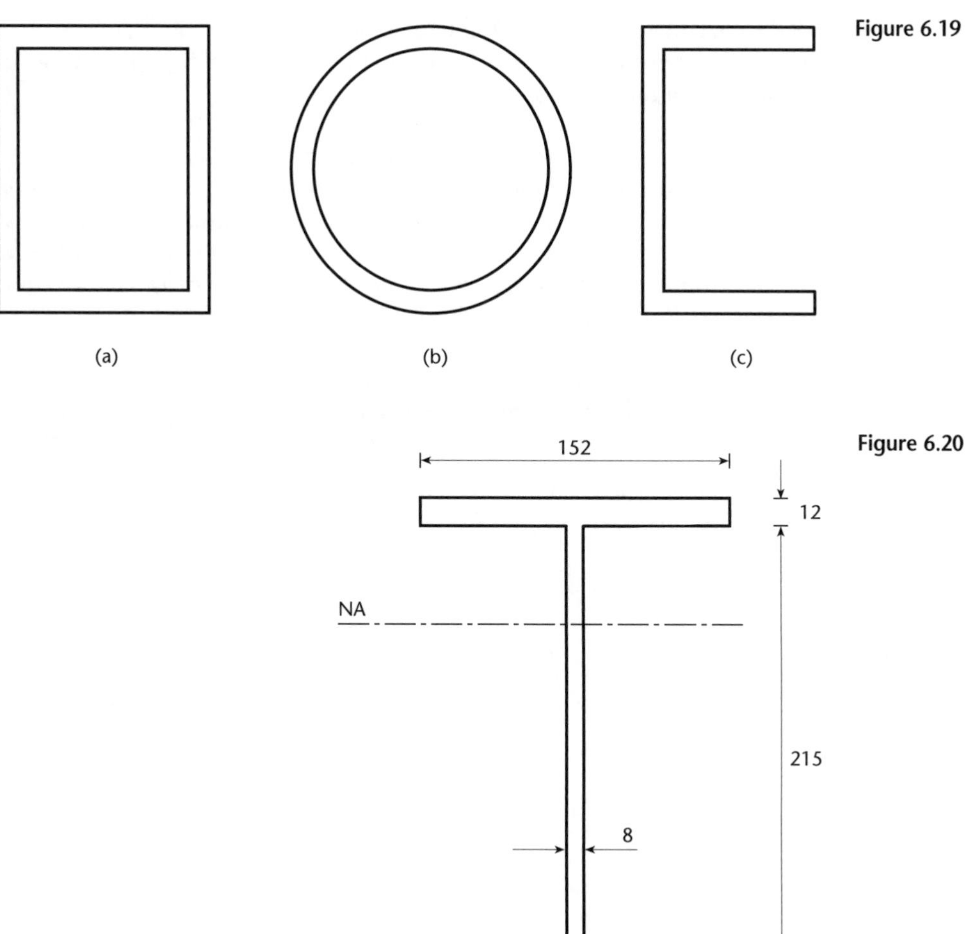

Figure 6.19

Figure 6.20

6.8.* Find the shear stress distribution for the thin-walled circular section in Figure 6.21 when loaded by a vertical shear force S. Check your answer by finding the integral of the vertical components of shear stress over the cross-section.

[*Hint: express* $\int_A y\mathrm{d}A$ *in terms of the angle* θ *shown.*]

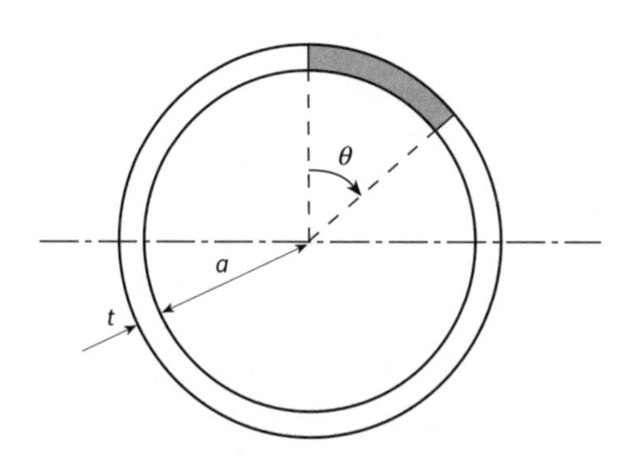

Figure 6.21

6.9. Sketch the shear stress distributions for the sections in Figure 6.19 when subjected to a vertical shear force. Show also the directions of the shear stresses in the various parts of the sections.

6.10.* A square tube has side dimension d and wall thickness $t(\ll d)$. A thin horizontal slit is cut along the tube at the mid-height of one of the vertical faces. Sketch the shear stress distribution when the section is loaded by a vertical shear force and find the distance e of the shear centre from the centroid of the section. [*Hint: the shear stress at the slit must be zero, and this destroys the symmetry of the problem – how does this affect the flow of shear stresses around the section?*]

6.11. The T-section of Figure 6.20 is made from steel with $E = 205$ GPa and $G = 79$ GPa. It is used as a cantilever of length 500 mm, loaded at its tip by a vertical point load of 50 kN and a torque of 0.5 kNm. Calculate the values at the tip of:

(a) the bending deflection;
(b) the shear deflection;
(c) the angle of twist.

CHAPTER 7

Virtual work and influence lines

CHAPTER OUTLINE

As has been stressed in earlier chapters, most structures are statically indeterminate, or redundant. A statically determinate structure requires the loss of only one member, or one reaction, to reduce it to a mechanism, whereas a structure with many redundancies usually requires the loss of several members before it can collapse. For this reason, redundant structures are generally considered safer, and so are more widely used. It is therefore vital that we have ways of analysing them.

Redundant structures cannot be analysed using simple statics alone, since insufficient equilibrium equations will be generated to allow all the reactions and internal forces to be determined. The additional equations required must be generated by relating the unknown forces to the displacements of the structure, using the stiffness properties of the members, and then imposing the conditions of displacement compatibility. This approach was briefly introduced in Chapters 4 and 5, and will be a recurring theme in Chapters 7–10.

One of the most powerful and elegant ways of combining force equilibrium and displacement compatibility into a single analysis is by the use of virtual work. This approach lends itself to very direct hand analysis of simple structures and is also useful for the derivation of computer methods for more complex structures. In this chapter we will first introduce the principle of virtual work in the context of statically determinate problems, before concentrating on its use for the solution of indeterminate structures. Lastly, we shall look briefly at another kind of structural problem that can be very neatly solved using virtual work: the derivation of influence lines for beams.

On completion of this chapter you should understand the principle of virtual work and how it relates to the principles of force equilibrium and displacement compatibility, and be able to use virtual work to:

- calculate reactions and deflections in statically determinate structures;
- determine bar forces and reactions in redundant, pin-jointed trusses;
- analyse redundant beams and moment frames;
- derive influence lines for beams.

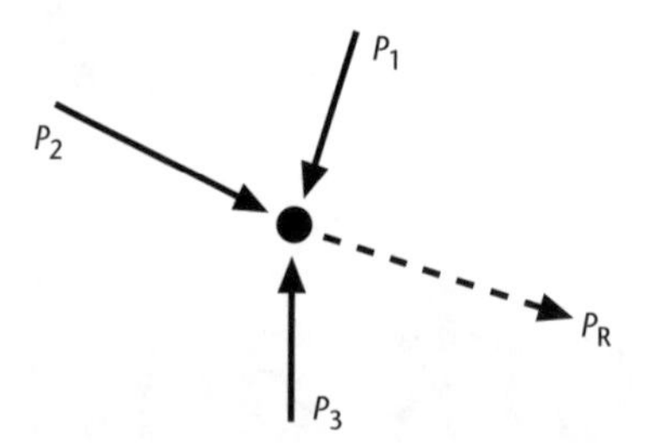

Figure 7.1
Forces acting on a particle.

7.1 The principle of virtual work

A force P undergoing a displacement δ in the direction of the force does work $P\delta$. Consider now a single particle under the actions of several forces P_i, with resultant P_R (Figure 7.1).

If a displacement δ_R is applied to the particle in the direction of P_R, then obviously work will be done by the forces. By resolving δ_R into components δ_i in the direction of each force, we can equate the work done by the individual forces to that done by the resultant:

$$\sum_i P_i\delta_i = P_R\delta_R$$

But if the forces on the particle are in equilibrium then $P_R = 0$, so:

$$\sum_i P_i\delta_i = 0 \tag{7.1}$$

This is the most basic form of the principle of virtual work, which can be expressed in words as:

If any set of compatible displacements is imposed on a body in equilibrium, then the virtual work done by the forces acting on the body is zero.

The work done is termed *virtual work* because the displacements used in calculating it need not be the ones that would in reality be caused by the applied forces. The only restrictions are that the forces must be in equilibrium and the displacements must form a compatible set. This means that we can pick any set of displacements we like (so long as they are compatible with each other) in order to give a virtual work equation from which the unknown forces can be easily calculated. The technique can also be applied the other way round; that is, we could just as easily apply a set of fictitious forces in order to calculate some unknown displacements.

On first encounter, virtual work can seem rather a strange concept – it may be difficult to accept that it is legitimate to use totally fictitious forces or displacements to solve a problem. However, as should become clear over the course of this chapter, the principle is really just a neat way of combining the concepts of equilibrium of forces and compatibility of displacements into a single statement.

Having established the principle with respect to a particle, we can extend it to a structure quite easily. The total virtual work done in a structure is simply the sum of the work done by the external and internal forces. However, care must be taken with the sign of the internal work term. Consider, for example, the simple bar in tension shown in Figure 7.2.

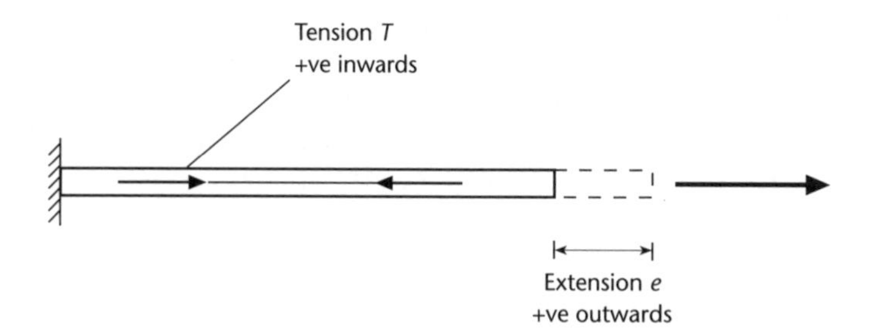

Figure 7.2
Positive forces and deformations for a bar in tension.

The internal force in the bar and its extension are both normally regarded as positive, but they act in opposite directions, that is, the force opposes the movement, so the work done is negative. The virtual work equation can therefore be written as:

$$\sum_i P_i\delta_i - \sum_j T_j e_j = 0$$

or

$$\sum_i P_i\delta_i = \sum_j T_j e_j \tag{7.2}$$

where the subscript i refers to the external forces and corresponding displacements imposed on the structure and j refers to summation over the various bars within the structure.

Equation (7.2) suggests an alternative way of stating the principle of virtual work; we can simply say that the virtual work done by the external forces must be equal to the work stored within the structure in the form of elastic energy.

It is worth re-emphasising that in equation (7.2) the bar tensions T_j must be in equilibrium with the external forces P_i, while the displacements δ_i and member extensions e_j must satisfy the displacement compatibility requirements. However, there is no need for the displacements and extensions to be the ones that would actually be caused by the external forces and bar tensions.

The above form of the virtual work equation is adequate for simple structures such as pin-jointed frames, where the internal forces do not vary within a given member. For more complex structures, where member forces are not constant, the right-hand summation must be replaced by an integral. This is often the case in problems involving flexure, which will be dealt with in section 7.4.

7.2 Application to statically determinate problems

In order to illustrate the use of virtual work, we shall look briefly at some simple statically determinate problems. While these could be solved by other methods, we shall see that virtual work often gives a more direct solution, since it enables us to write the governing equations in a more compact form.

7.2.1 Statically determinate reactions

In general, unknown forces are determined by applying a set of virtual displacements. To find a support reaction, the most direct solution is obtained by applying a unit virtual displacement at the support in question, with the displacements at all other supports kept at zero. Virtual displacements of other points on the structure are then found by assuming that there are no member deformations, so that the structure behaves as a mechanism. This means that the internal virtual work is zero and equation (7.2) becomes simply:

$$\sum_i P_i \delta_i = 0$$

in which P_i are the real external forces and δ_i are the corresponding virtual displacements. The equations thus derived are very closely related to the equilibrium equations for the structure, but for complex structures the virtual work approach generally gives a quicker solution.

EXAMPLE 7.1

Calculate the reaction at support B for the simply supported beam in Figure 7.3.

Figure 7.3

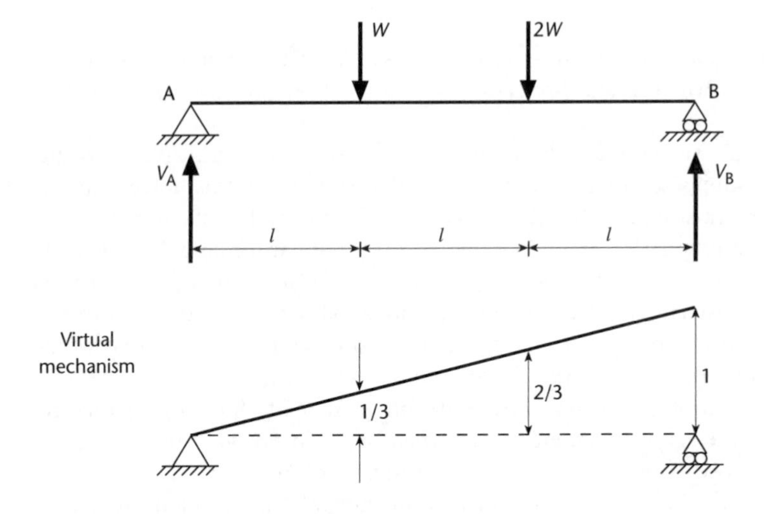

To solve this problem by virtual work, we create a virtual displacement set by applying a unit vertical deflection at B with beam AB remaining straight, giving the mechanism shown in Figure 7.3. Since the beam is undeformed, the internal virtual work is zero. Therefore the virtual work equation is simply:

$$W \cdot (-1/3) + 2W \cdot (-2/3) + V_B \cdot 1 = 0$$

$$\rightarrow V_B = 5W/3$$

Note that the virtual displacements are negative where they oppose the corresponding real forces. It can be seen that the virtual work equation is the same as would be obtained by taking moments about A, so that, for this particular case, the virtual work approach is no quicker than simple statics.

EXAMPLE 7.2

Calculate the reaction at the central support B for the two-span beam in Figure 7.4.

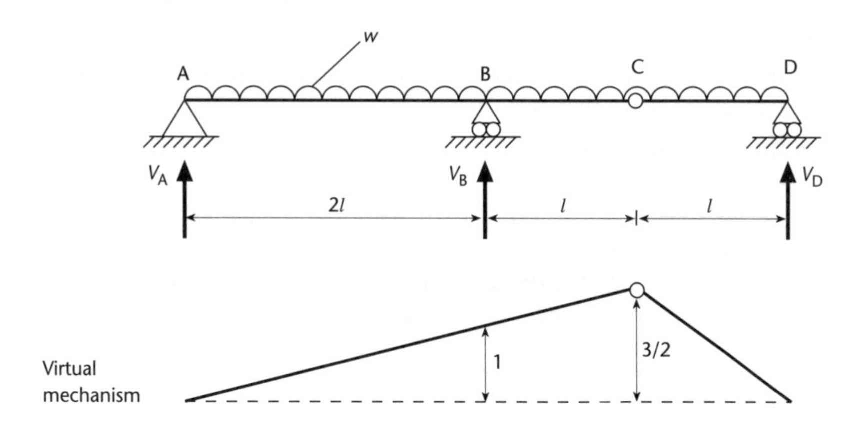

Figure 7.4

Again we draw a virtual mechanism by applying a unit displacement at B, with the beam remaining straight and in contact with the other supports. The difference in this case is that there is a pin in the beam at C, allowing the two beam segments on either side to rotate relative to one another. Because the structure remains an assemblage of straight elements, the vertical displacement of the centroid of the UDL on each segment is one half of the displacement of C. Thus the virtual work equation is:

$$V_B \cdot 1 + 4wl.(-3/4) = 0$$
$$\rightarrow V_B = 3wl$$

In this case, the virtual work equation is that which would be obtained by combining two equilibrium equations, so that the virtual work solution is more direct.

It can be seen from these examples that the virtual displacement set used need bear no relation to what is actually happening in the structure under the applied loads; the virtual displacements are chosen solely for mathematical convenience, so as to give the most direct solution possible.

7.2.2 Deflections of statically determinate trusses

The examples above show how a virtual work approach can be used in place of simple equilibrium equations. The virtual work method has not allowed us to calculate anything we could not have found by statics, just to perform the calculations a little more quickly. A more interesting and useful application is the calculation of deflections, which involves considerations of both equilibrium and compatibility.

Just as forces can be found by imposing a set of imaginary displacements on a structure, so displacements can be found by using a set of fictitious virtual forces.

As before, there is no need for the force set and the displacement set to bear any relationship to each other, but the virtual forces must be in equilibrium. In order to find the real displacements of a structure we must, of course, use real member extensions together with the virtual forces. The solution procedure therefore takes the following form:

(a) Use statics to find the real forces in the truss caused by the applied loads. Convert these to real extensions using the elastic properties of the members.

(b) Create a virtual force set by applying a unit load corresponding to the required displacement.

(c) Set up and solve the virtual work expression (equation (7.2)), in which the force terms P_i and T_j are virtual, and the displacement terms δ_i and e_j are real. Note that the virtual external force set P_i comprises only a single unit load, so that the left-hand side of this equation is simply equal to one times the required displacement.

EXAMPLE 7.3

Find the displacement of the point S for the frame shown in Figure 7.5(a), in which all members have Young's modulus E. The horizontals have cross-sectional area A and the diagonals $A\sqrt{2}$.

Figure 7.5

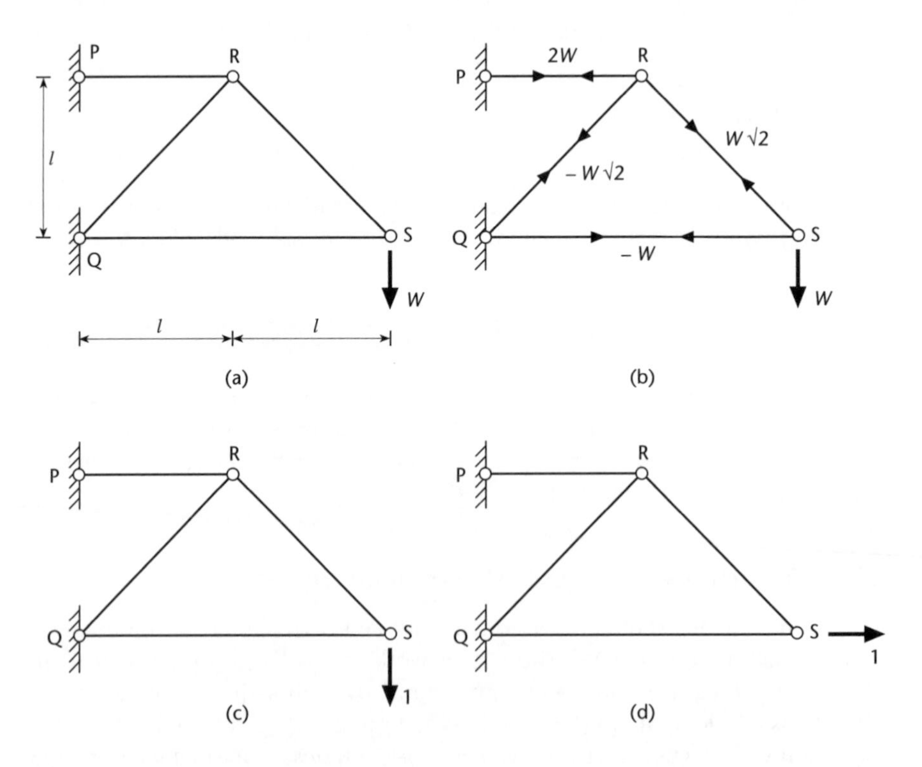

We first find the bar forces by resolving at joints around the frame; the results are shown in Figure 7.5(b). These forces can then be combined with the elastic properties of the members to give the bar extensions – this can be conveniently set out in tabular form, as shown below:

	PR	QS	QR	RS
Real force	$2W$	$-W$	$-W\sqrt{2}$	$W\sqrt{2}$
Length	l	$2l$	$l\sqrt{2}$	$l\sqrt{2}$
AE	AE	AE	$AE\sqrt{2}$	$AE\sqrt{2}$
Real extension ($= Tl/AE$)	$2Wl/AE$	$-2Wl/AE$	$-Wl\sqrt{2}/AE$	$Wl\sqrt{2}/AE$

We now need one set of virtual forces to find each displacement component. For the vertical displacement of S, we apply a unit vertical load at S, (Figure 7.5(c)). The resulting virtual forces are:

	PR	QS	QR	RS
Virtual force	2	-1	$-\sqrt{2}$	$\sqrt{2}$

Then, equating the external and internal virtual work:

$$1 \times \delta_V = (Wl/AE) \cdot [2 \times 2 + (-2) \times (-1) + (-\sqrt{2}) \times (-\sqrt{2}) + \sqrt{2} \times \sqrt{2}]$$
$$\rightarrow \delta_V = 10Wl/AE$$

Similarly, to find the horizontal displacement of S, we apply a unit horizontal load, (Figure 7.5(d)), giving:

	PR	QS	QR	RS
Virtual force	0	1	0	0

and the virtual work equation is then just:

$$1 \times \delta_H = (Wl/AE) \cdot [(-2) \times 1]$$
$$\rightarrow \delta_H = -2Wl/AE$$

7.3 Forces in redundant trusses

7.3.1 Single redundancy

We now come to the most useful application of virtual work – determining forces in redundant structures. In these cases, equilibrium considerations alone are not enough to solve for all the unknown forces. We must therefore also make a statement that the structure, when deformed by the external loads, actually fits together. Thus the problem becomes one of compatibility of the *real* displacements, and to solve the problem we must use a set of *virtual* forces. The easiest virtual force set to use is a unit force corresponding to the redundancy we are trying to find, with all other external forces (apart from the reactions) set to zero. In this way, the redundant force becomes the only unknown in the virtual work equation. The steps in the analysis procedure are:

(a) Let the redundant bar force be R.

(b) Use equilibrium methods to calculate the real forces in the remaining members in terms of R. Hence use the member stiffnesses to find the real extensions.

(c) Now create a virtual force set by letting the redundant force equal unity and using equilibrium methods to find the resulting member forces. These internal forces are often referred to as *self-equilibrating,* since they form an equilibrium set without any external loads.

(d) Form the virtual work equation using the virtual member forces from (c) and the real extensions from (b). Note that the external virtual work is *always* zero, since there are no external virtual forces – this is why the self-equilibrating force set is so useful. Solve for R.

(e) Substitute back into (b) to find the other internal forces. If displacements are now required, these can be found by applying virtual forces at the relevant points, as for statically determinate structures.

A very similar procedure can be used in the case of a redundant reaction. In this case, the virtual force set is found by replacing the redundant reaction by a unit load. The virtual force set will therefore include some external loads in the form of reactions. However, the external virtual work is still zero as these virtual reactions all correspond to zero real displacements.

EXAMPLE 7.4

Find the bar forces in the frame shown in Figure 7.6(a). All members have cross-sectional area 0.01 m^2 and Young's modulus 205 GPa.

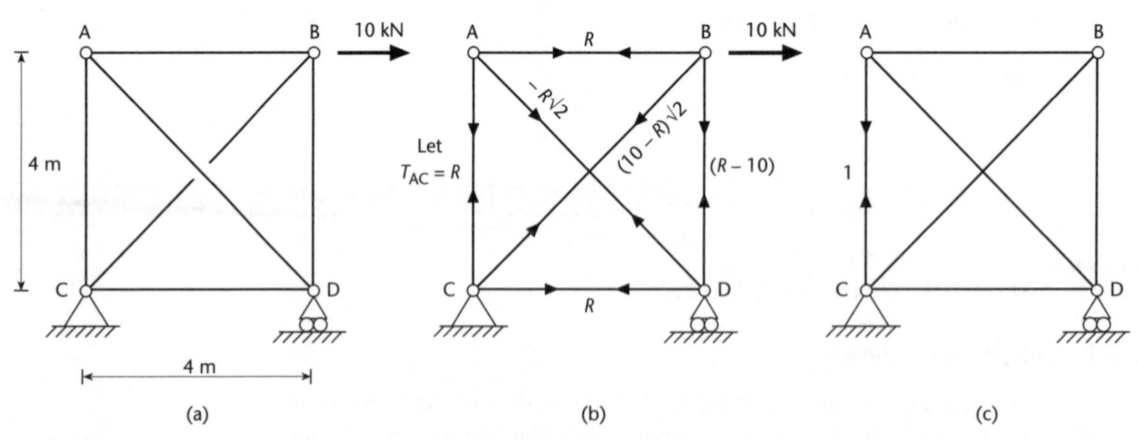

Figure 7.6

The frame has one redundant member. If we let the force in member AC be R, then by resolving at joints we can work out all the member forces in terms of R and the

10 kN load (Figure 7.6(b)). (Note that we could have selected any member to be the redundant one, but not a reaction.) From these, we can calculate the real extensions:

	AB	AC	BD	CD	AD	BC
Real force T(kN)	R	R	$R-10$	R	$-R\sqrt{2}$	$(10-R)\sqrt{2}$
Length l (m)	4	4	4	4	$4\sqrt{2}$	$4\sqrt{2}$
Real extension $e = Tl/AE(10^{-6}$ m)	$1.951R$	$1.951R$	$1.951(R-10)$	$1.951R$	$-3.902R$	$3.902(10-R)$

To find R, we replace it by a unit force and find the resulting virtual force set in the absence of any external loads (Figure 7.6(c)). The results are:

	AB	AC	BD	CD	AD	BC
Virtual force	1	1	1	1	$-\sqrt{2}$	$-\sqrt{2}$

Note that these forces can be obtained very easily from the forces in the table above, simply by setting R equal to unity and the external load to zero; this is much quicker than resolving around the frame again.

The external virtual work is zero because there are no external virtual forces, and the internal virtual work is found by multiplying the real extensions by the virtual forces. Equation (7.2) therefore gives:

$$[1.951R + 1.951R + 1.951(R-10) + 1.951R + 3.902\sqrt{2}R - 3.902\sqrt{2}(10-R)] \times 10^{-6} = 0$$

$$\rightarrow R = 3.96 \text{ kN}$$

7.3.2 Multiple redundancies

The general approach to structures with multiple redundancies simply involves repetition of the approach for a single redundancy, using a different self-equilibrating force set for each redundancy. Usually, this results in a set of simultaneous equations for the redundant forces. As before, the redundancy of the structure means that there are many possible virtual force sets which satisfy the equilibrium requirements, and since any of these is acceptable, it makes sense to pick one which makes the algebra as simple as possible.

EXAMPLE 7.5

Find the forces in the frame shown in Figure 7.7(a), which has two redundancies. All members have the same properties.

Figure 7.7

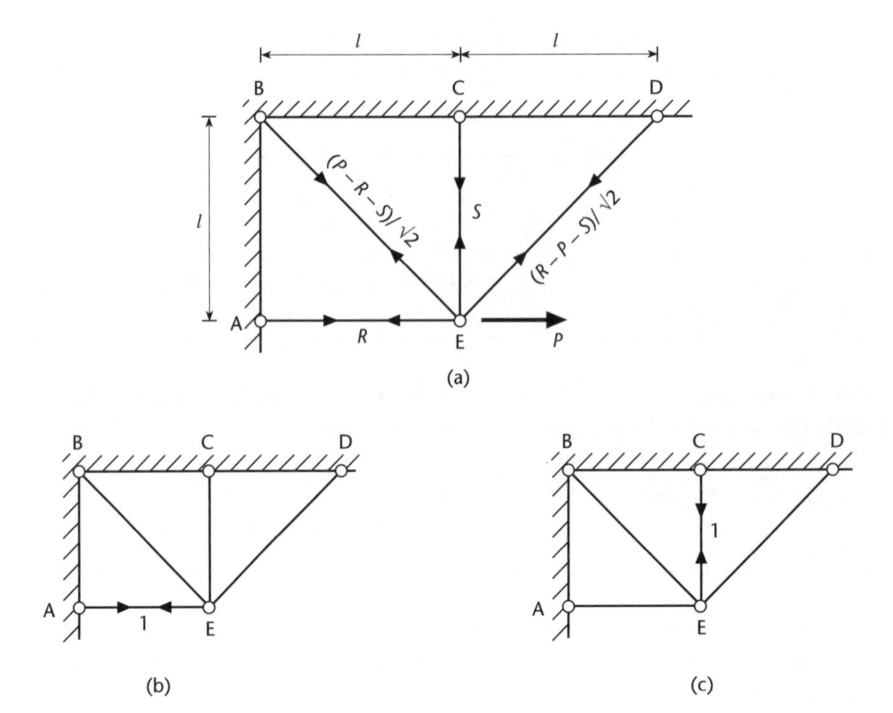

We commence by selecting any two redundant bar forces. In this case, we let the force in AE be R and that in CE be S. The remaining forces, and hence the real extensions, can then be determined in terms of these two unknowns.

	AE	BE	CE	DE
Real force	R	$(P-R-S)/\sqrt{2}$	S	$(R-P-S)/\sqrt{2}$
Length	l	$l\sqrt{2}$	l	$l\sqrt{2}$
Real extension	Rl/AE	$(P-R-S)l/AE$	Sl/AE	$(R-P-S)l/AE$

To solve for the two redundancies, we now need two virtual force sets. The first is obtained by setting the force in AE to unity (Figure 7.7(b)), and the second by setting the force in CE to unity (Figure 7.7(c)).

As in the previous example, there is no need to resolve around the frame each time to find the resulting member forces. In the first case we can simply take the force terms from the table above and put $R = 1$ and the other loads equal to zero, and in the second case we put $S = 1$ and set the other loads to zero.

	AE	BE	CE	DE
Virtual force set 1	1	$-1/\sqrt{2}$	0	$1/\sqrt{2}$
Virtual force set 2	0	$-1/\sqrt{2}$	1	$-1/\sqrt{2}$

Virtual work equations can now be written by taking the product of each virtual force set with the real extensions:

Set 1 gives: $[R - (P - R - S)/\sqrt{2} + (R - P - S)/\sqrt{2}] \cdot l/AE = 0$

Set 2 gives: $[-(P - R - S)/\sqrt{2} + S - (R - P - S)/\sqrt{2}] \cdot l/AE = 0$

Solving gives $R = P(2 - \sqrt{2})$ and $S = 0$. These values can then be substituted back into the real force expressions to give the other bar forces.

7.3.3 Temperature effects and lack of fit

In redundant structures, temperature changes and/or initial lack of fit of members can cause significant internal forces to be set up. This is not the case with statically determinate structures, which can move to accommodate these effects without any additional loads being generated.

In most traditional methods of analysing redundant structures, the forces due to temperature change and lack of fit can be quite complicated to work out. They can, however, be calculated very easily using virtual work. This is because the real displacement set we use in the virtual work equation does not have to be related to the virtual force set. It simply has to satisfy the condition of compatibility, that is, the deformed structure must fit together. Thus, when we calculate the set of real extensions, we must simply add up all the reasons why the various joints of the structure are not in their original positions:

(a) Internal forces cause member extensions, which can be found in the usual way.

(b) A temperature change of ΔT simply causes a change in length of $\alpha l \Delta T$, where α is the coefficient of linear thermal expansion for the bar material.

(c) Lack of fit is slightly more complicated, since it is not immediately obvious whether the lack of fit term should be added to or subtracted from the other deformation terms. Essentially, we are adding up all the reasons why the bar is longer than it ought to be. Thus if the bar is initially too long by an amount ε then we must add ε to the extension.

This last point can be clarified by considering a simple example. Suppose two joints A and B in a frame are initially 5.0 m apart, and that the bar intended to fit between them has been manufactured with a length of 5.005 m. The joints A and B are jacked apart by the additional 5 mm required so that the bar can be inserted, and the jack is then removed. Subsequently, the internal force in the bar AB causes it to shorten by 20 mm. Then the final distance between A and B is $5.005 - 0.02 = 4.985$ m. Thus, the total change in distance between A and B is -15 mm, which is made up of an axial deformation of -20 mm plus an initial lack of fit of 5 mm.

Having found the total real extension terms, these are then multiplied by a set of virtual forces derived by replacing the redundancy by a unit force, just as for an ordinary redundant frame.

EXAMPLE 7.6

The frame shown in Figure 7.8 carries no external loads, but the three exterior members undergo a temperature rise of 10°C, and the diagonal BD is initially 2 mm too short. Find the member forces. All members have $AE = 500$ MN and $\alpha = 12 \times 10^{-6}$/°C.

Figure 7.8

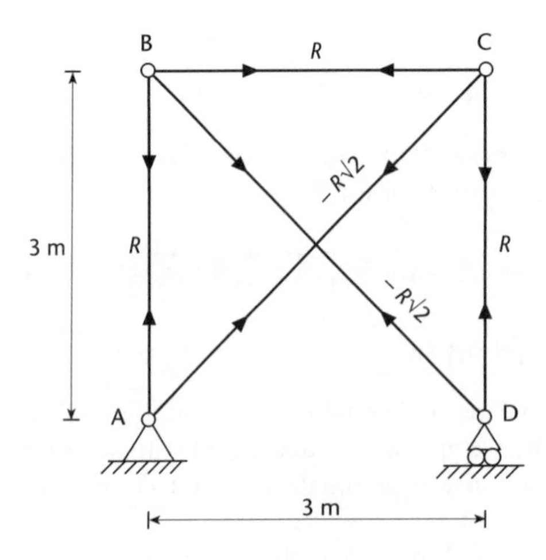

If we let the force in AB be R, then we can calculate the real member forces and extensions as usual. We also tabulate the lack of fit and temperature terms, noting that, since BD is initially too short, the lack of fit term is negative. Lastly, we find the virtual forces by replacing the redundant force R by unity.

	AB	BC	CD	AC	BD
Real force (kN)	R	R	R	$-R\sqrt{2}$	$-R\sqrt{2}$
Length (mm)	3000	3000	3000	$3000\sqrt{2}$	$3000\sqrt{2}$
Real extension (mm)	$0.006R$	$0.006R$	$0.006R$	$-0.012R$	$-0.012R$
Lack of fit (mm)	0	0	0	0	-2.0
Temp. effect (mm)	0.36	0.36	0.36	0	0
Virtual force (kN)	1	1	1	$-\sqrt{2}$	$-\sqrt{2}$

The virtual work equation is then formed using the virtual forces and the *total* real extensions, including the temperature and lack of fit terms:

$$3 \times (0.006R + 0.36) + \sqrt{2} \times 0.012R + \sqrt{2} \times (0.012R + 2.0) = 0$$
$$\rightarrow R = -75.2 \text{ kN}$$

If there were only a lack of fit term, and no temperature change, then we would get $R = -54.5$ kN. It is noticeable that even a very small lack of fit can generate quite large member forces.

7.4 Flexural problems

7.4.1 The virtual work equation for flexure

As well as being useful for pin-jointed structures, virtual work can also be applied to a wide range of flexural problems. A slightly different form of equation is now required. Whereas the tensile or compressive forces in trusses are constant within a given structural member, the bending moments in flexural elements often vary

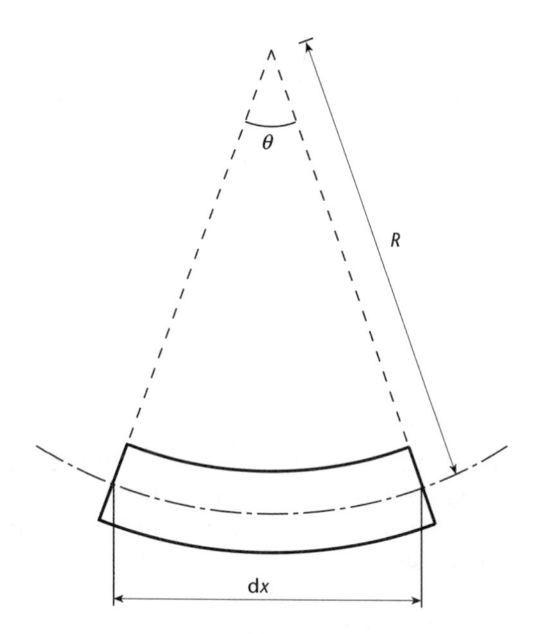

Figure 7.9
Relationship between rotation and curvature.

along the length of a member. Integration must therefore be used instead of discrete summation.

Just as the work done by a force is equal to the force multiplied by the corresponding displacement, so the work done by a moment is equal to the moment multiplied by the corresponding rotation. For a small element of beam of length dx bent at a radius of curvature R (Figure 7.9), the rotation is simply dx/R. The virtual work equation therefore has the form:

$$\sum_i P_i\delta_i + \sum_i Q_i\theta_i = \int M\left(\frac{1}{R}\right)\mathrm{d}x \tag{7.3}$$

where P_i and Q_i are the external forces and moments on the structure, which must be in equilibrium with the internal moments M, and δ_i and θ_i are the corresponding displacements and rotations, which must form a compatible set with the curvatures $1/R$.

The above form assumes, as is usually the case in flexural problems, that axial deformations are negligibly small compared to flexural ones. If not, then the member tensions multiplied by their extensions must be included in the internal work expression. Equation (7.3) then becomes:

$$\sum_i P_i\delta_i + \sum_i Q_i\theta_i = \int M\left(\frac{1}{R}\right)\mathrm{d}x + \sum_j T_j e_j \tag{7.4}$$

As in equation (7.2), the subscript i refers to summation over all the load points, while j refers to the members.

EXAMPLE 7.7

Use virtual work to calculate the tip deflection and rotation for the cantilever shown in Figure 7.10(a), which has constant flexural stiffness EI.

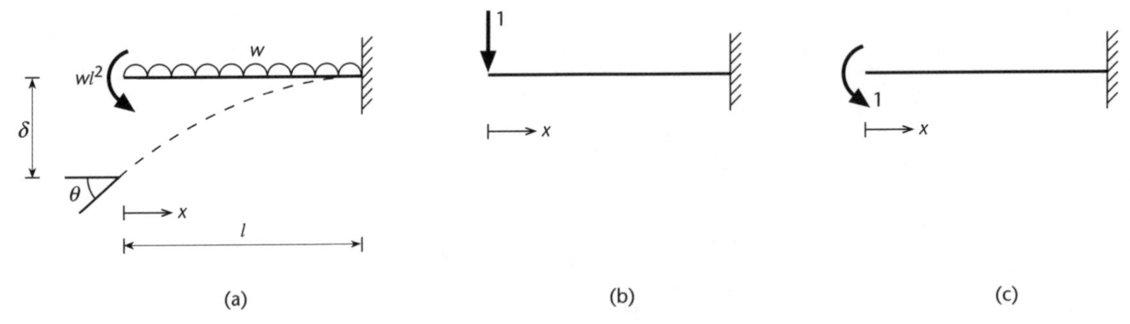

Figure 7.10

As with pin-jointed trusses, in order to find real displacements we must use real member deformations (in this case curvatures) together with appropriate sets of virtual forces. We start by finding the real bending moment in the beam and hence, from equation (5.5), the real curvature:

$$M = -wl^2 - wx^2/2$$
$$(1/R) = M/EI = -(wl^2 + wx^2/2)/EI$$

To find the tip displacement, we apply a unit virtual load, as shown in Figure 7.10 (b). The virtual bending moment is then simply $M = -x$, and the virtual work equation can now be written as:

$$1 \cdot \delta = \int_0^l x \cdot \left(\frac{wl^2 + wx^2/2}{EI}\right) \mathrm{d}x = \frac{5wl^4}{8EI}$$

Similarly, to find the tip rotation we apply a unit virtual moment (Figure 7.10(c)), giving $M = -1$ and hence:

$$1 \cdot \theta = \int_0^l 1 \cdot \left(\frac{wl^2 + wx^2/2}{EI}\right) \mathrm{d}x = \frac{7wl^3}{6EI}$$

Of course, these results could also have been found by integration of the moment–curvature relationship or by superposition of standard solutions, as described in Chapter 5.

7.4.2 Analysis of indeterminate beams and frames

Indeterminate beams and frames can be dealt with using virtual work in exactly the same way as redundant trusses. Each redundancy in turn is replaced by a unit force in order to calculate a set of virtual forces. These are then combined with the real curvatures in order to solve the problem.

EXAMPLE 7.8

Calculate the support reactions for the frame shown in Figure 7.11(a), in which both members have the same properties.

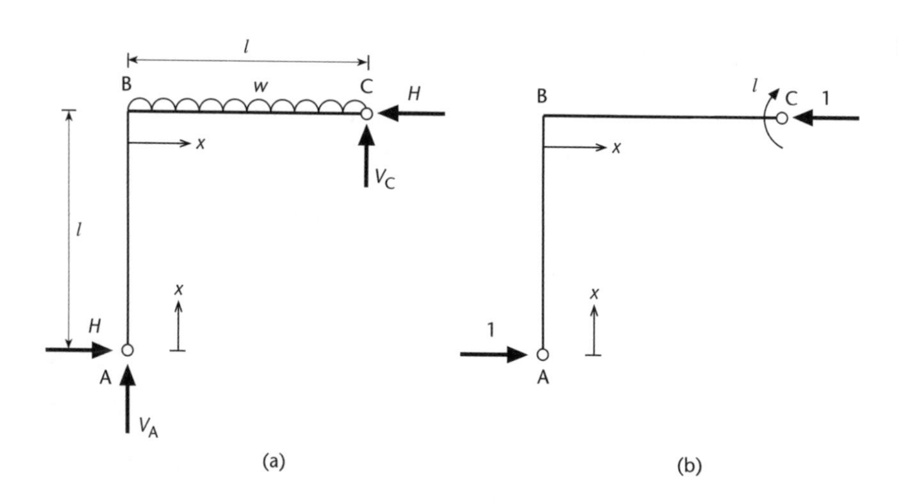

Figure 7.11

Resolving vertically and taking moments about one of the pinned ends gives:

$$V_A = wl/2 + H \quad \text{and} \quad V_C = wl/2 - H$$

Bending moment expressions for the structure can now be formulated in terms of the unknown horizontal reaction H. For AB, the variable distance x is measured from A towards B, while for BC it is measured from B towards C.

For AB: $M = -Hx$

For BC: $M = -Hl + (wl/2 + H)x - wx^2/2$

The real curvatures are equal to these moments divided by EI. To find the horizontal reactions, we use a virtual loadcase obtained by putting unit forces in their place (Figure 7.11(b)). Note that, for equilibrium, an external moment of magnitude l is also required. The virtual moment expressions are then:

For AB: $M = -x$

For BC: $M = -l$

The virtual work equation can now be written as:

$$0 = \int_0^l x \cdot \frac{Hx}{EI} \mathrm{d}x + \int_0^l l \cdot \left(\frac{Hl - (wl/2 + H)x + wx^2/2}{EI} \right) \mathrm{d}x$$

where the first integral relates to member AB and the second to BC. This equation can be solved to give $H = 2wl/5$. Hence $V_A = 9wl/10$ and $V_C = wl/10$.

7.4.3 Curved elements

In Chapter 3 we looked at the analysis of some simple curved members such as three-pinned arches, parabolic arches and suspension cables. All of these can be analysed using simple statics; the three-pinned arch is statically determinate, while parabolic members, so long as they are subjected to a uniform load per unit span, carry only axial forces.

In practice, many other forms of curved element may be encountered, and these can be rather harder to analyse. Virtual work provides one of the easiest ways of performing calculations for curved members, for which most other approaches quickly get very confusing. The approach is exactly as for other flexural members; first, expressions are derived for the real curvatures, then appropriate virtual force sets are applied so as to determine the required displacements or redundant forces.

EXAMPLE 7.9

Calculate the support reactions for the two-pinned semi-circular arch shown in Figure 7.12(a).

Figure 7.12

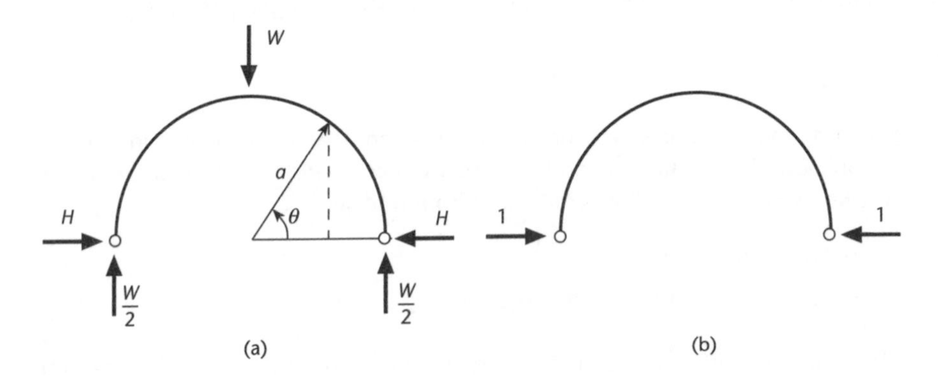

Using symmetry, it is obvious that the vertical reactions are each equal to $W/2$. We define the position on the arch in terms of the angle θ made by the radius, as shown in Figure 7.12(a). The real bending moment can then be written as

$$M = Wa(1 - \cos\theta)/2 - Ha\sin\theta$$

and the real curvature is obtained by dividing this expression by EI. To find the horizontal reactions, we create a virtual force set consisting of unit loads in place of H, (Figure 7.12(b)), with no other external loads. The resulting virtual moment expression is

$$M = -a\sin\theta$$

Noting that a small increment of length along the arch is $ds = a \cdot d\theta$, we can then form the virtual work equation using real curvatures and virtual moments:

$$2\int_0^{\pi/2} -a\sin\theta\cdot\left(\frac{Wa}{2EI}(1-\cos\theta)-\frac{Ha}{EI}\sin\theta\right)\cdot a d\theta = 0$$

Integrating and rearranging gives $H = W/\pi$.

7.5 Betti's reciprocal theorem

This simple and extremely useful theorem can be derived very easily using virtual work. Consider the structure shown in Figure 7.13, which is subjected to two different sets of loads, each generating corresponding internal forces and deformations. We can apply equation (7.4) using forces from (a) and deformations from (b):

$$\sum P_A\delta_B = \int M_A\left(\frac{1}{R}\right)_B dx + \sum T_A e_B$$

If we express the member deformations as functions of the internal forces and member properties, this equation becomes:

$$\sum P_A\delta_B = \int M_A\left(\frac{M_B}{EI}\right)dx + \sum T_A\frac{T_B l}{AE} \tag{7.5}$$

We can repeat this process, but this time using forces from (b) and deformations from (a):

$$\sum P_B\delta_A = \int M_B\left(\frac{1}{R}\right)_A dx + \sum T_B e_A = \int M_B\left(\frac{M_A}{EI}\right)dx + \sum T_B\frac{T_A l}{AE} \tag{7.6}$$

The right-hand sides of equations (7.5) and (7.6) are identical so it follows that:

$$\sum P_A\delta_B = \sum P_B\delta_A \tag{7.7}$$

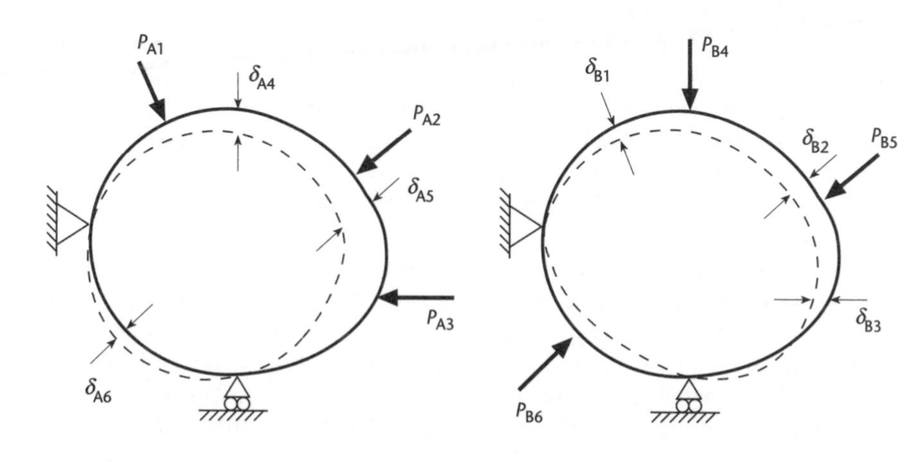

Figure 7.13
Betti's reciprocal theorem.

Equation (7.7) is Betti's reciprocal theorem, which states that, for a linear elastic structure subjected to two sets of loads A and B, *load set A multiplied by the corresponding displacements caused by load set B must always equal load set B multiplied by the corresponding displacements due to load set A.*

7.6 Influence lines

Many structures, particularly bridges, are subjected to moving loads, which cause the forces in the structure to vary with the position of the load. This variation can be conveniently displayed and analysed using *influence lines*. An influence line is a graph showing the variation of a parameter at a chosen point as a unit load traverses the structure, with the value of the parameter plotted against the position of the unit load. The parameter could be, for instance, shear force or bending moment. It is vital to distinguish at the outset between, say, a bending moment diagram for a beam and an influence line for bending moment:

- a bending moment diagram shows how the bending moment varies *along the beam* for a particular, *fixed* set of applied loads;
- an influence line for bending moment shows how the bending moment *at a particular point* varies as a unit load *moves across* the beam.

With the advent of modern computer methods, the use of influence lines in design has declined sharply. However, they remain useful for gaining insights into the effects of simple moving load systems.

7.6.1 Determination of influence lines for statically determinate beams

For very simple structures, influence lines can be determined using elementary statics. For example, to find the influence line for the shear force at the midpoint of a cantilever we simply have to analyse the loadcase in Figure 7.14(a), in which the cantilever carries a unit load at a distance x from the tip. To find the shear force at the midpoint A, we cut the beam at that point and consider equilibrium of the left-hand portion. Two cases must be considered – when $x < l/2$ the free body diagram is shown in Figure 7.14(b), from which $S = 1$. When $x > l/2$ the unit load has moved to the right of A and so $S = 0$. Hence the influence line is as shown in Figure 7.14(c).

For most structures the use of statics is rather cumbersome, and influence lines can be derived much more efficiently using either a virtual work approach, as in the example that follows, or Betti's reciprocal theorem. Using these methods, it is possible to show that the influence line for a given force can be represented by the deflected shape of the structure when a unit displacement is applied corresponding to that force. This is known as the *Muller–Breslau principle*.

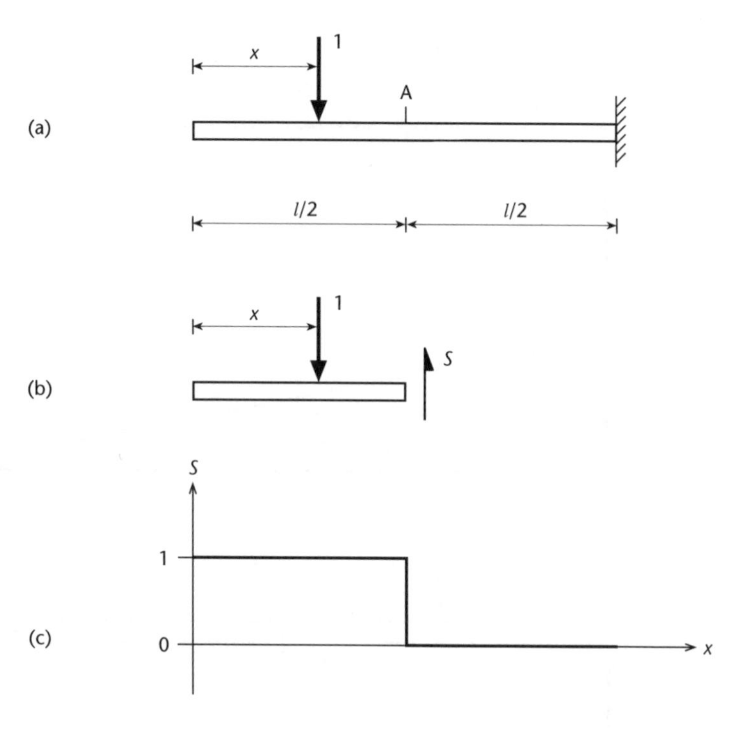

Figure 7.14
Influence line for shear force at the midpoint of a cantilever.

EXAMPLE 7.10

For the simply supported beam in Figure 7.15(a), draw the influence lines for the vertical reaction at A and the shear force and bending moment at C.

Reaction

To find the influence line for the vertical reaction at A, a unit virtual displacement is applied at A, with no displacement at the other support (Figure 7.15(b)). The external virtual work is found using the real forces on the beam in (a) multiplied by the corresponding virtual displacements in (b). Since the beam is assumed to deflect as a rigid body, the internal virtual work is zero, so equation (7.2) gives:

$$V_A \times 1 - 1 \times \delta = 0, \quad \text{or} \quad V_A = \delta$$

Thus the simple deflected shape drawn in (b) is the influence line for the vertical reaction at A.

Shear force

If we cut the beam at C, the shear force S acts in opposite directions on the two cut faces (Figure 7.15(c)). The displacement corresponding to this loading is a vertical sliding between the cut faces. We therefore create a virtual displacement set by imposing a unit vertical displacement between the parts of the beam either side of C, while keeping the two portions of beam parallel (Figure 7.15(d)). Putting the sum of the internal and external virtual work equal to zero:

$$S \times 1 + 1 \times (-\delta) = 0, \quad \text{or} \quad S = \delta$$

Thus the displaced shape in (d) is the influence line for the shear force at C. In this process, it is essential that the virtual displaced shape consists of parallel beam

Figure 7.15

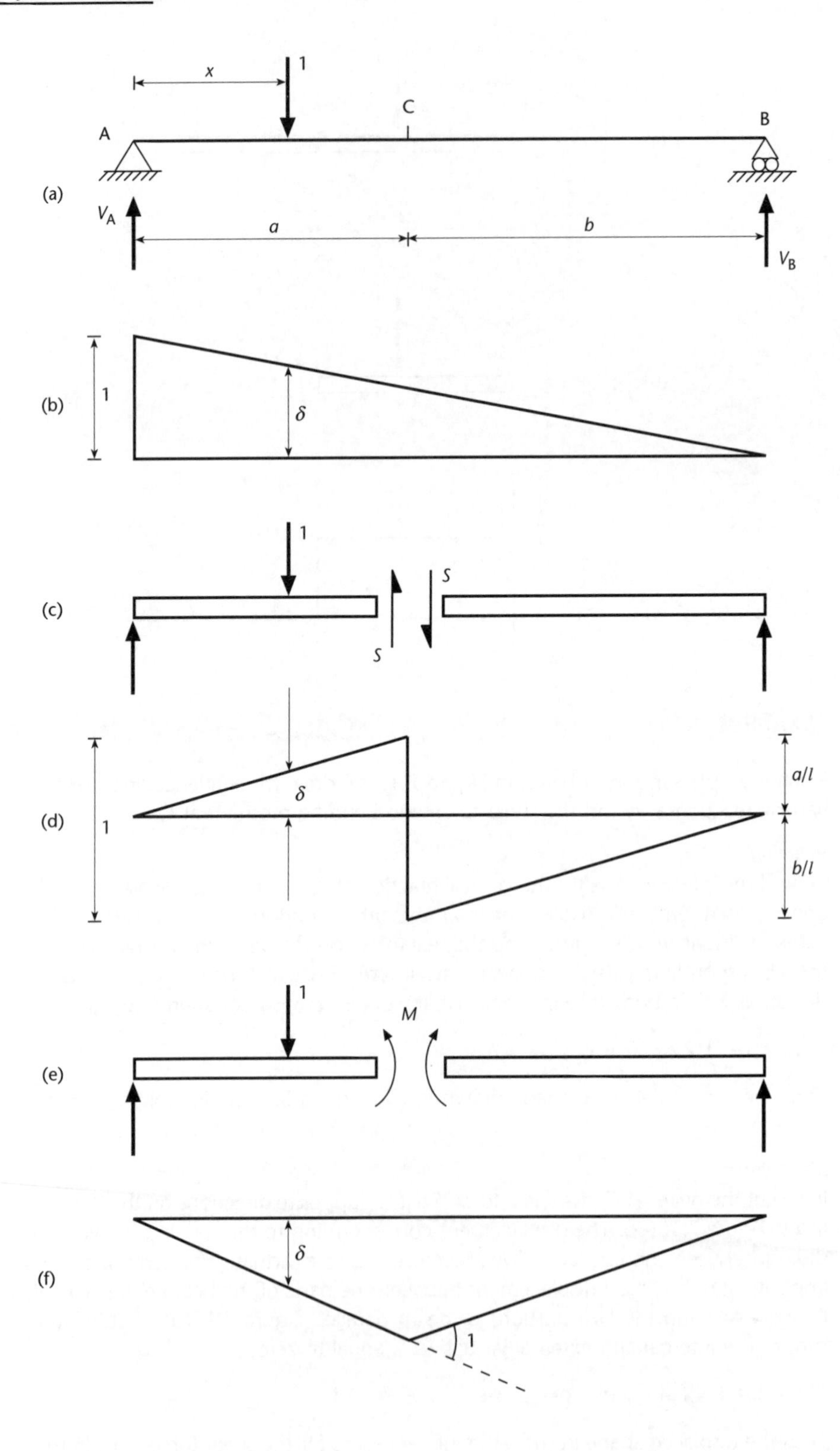

portions. Otherwise an additional term would be introduced into the virtual work equation, involving the bending moment at C and the relative rotation between the two parts of the beam.

Bending moment

For the bending moment M at C, a similar approach is adopted. If we cut the beam at C, M acts in opposite directions on the two cut faces (Figure 7.15(e)). So the required virtual displacement set is obtained by cutting the beam at C and applying a unit rotation between the two beam segments (Figure 7.15(f)). The virtual work equation is then:

$$1 \times \delta + M \times (-1) = 0, \quad \text{or} \quad M = \delta$$

and the influence line for bending moment has therefore been constructed in (f). As with bending moment diagrams, this has been drawn with sagging moments plotted downwards.

7.6.2 Influence lines for redundant beams

Influence lines for beams with redundant reactions can be determined using either virtual work or Betti's reciprocal theorem, in much the same way as for statically determinate beams. The latter approach is illustrated below.

EXAMPLE 7.11

For the two-span beam in Figure 7.16(a), calculate the influence lines for (a) the vertical reaction at A, and (b) the bending moment at B.

(a) To find the vertical reaction at A, create a second loadcase as shown in Figure 7.16(b). The support at A is removed and replaced by an upwards force P. The beam is restrained by the supports at B and C and so deflects as shown. Applying the reciprocal theorem (equation (7.7)), to loadcases (a) and (b):

$$1 \cdot (-v_x) + V_A \cdot v_A = 0, \quad \text{so} \quad V_A = v_x / v_A$$

The influence line is therefore given by the deflection curve in Figure 7.16(b), normalised so as to give a value of unity at A. The shape of this curve can be found by direct integration of the moment–curvature relationship, as described in Chapter 5:

$$\frac{d^2v}{dx^2} = \frac{M}{EI} = \frac{Px - 2P\{x - l\}}{EI}$$

Integrating twice and using the boundary conditions $v = 0$ at $x = l$ and $x = 2l$ gives:

$$v = \frac{Pl^3}{6EI}\left(\alpha^3 - 2\{\alpha - 1\}^3 - 5\alpha + 4\right), \quad \text{where} \quad \alpha = \frac{x}{l}$$

Figure 7.16

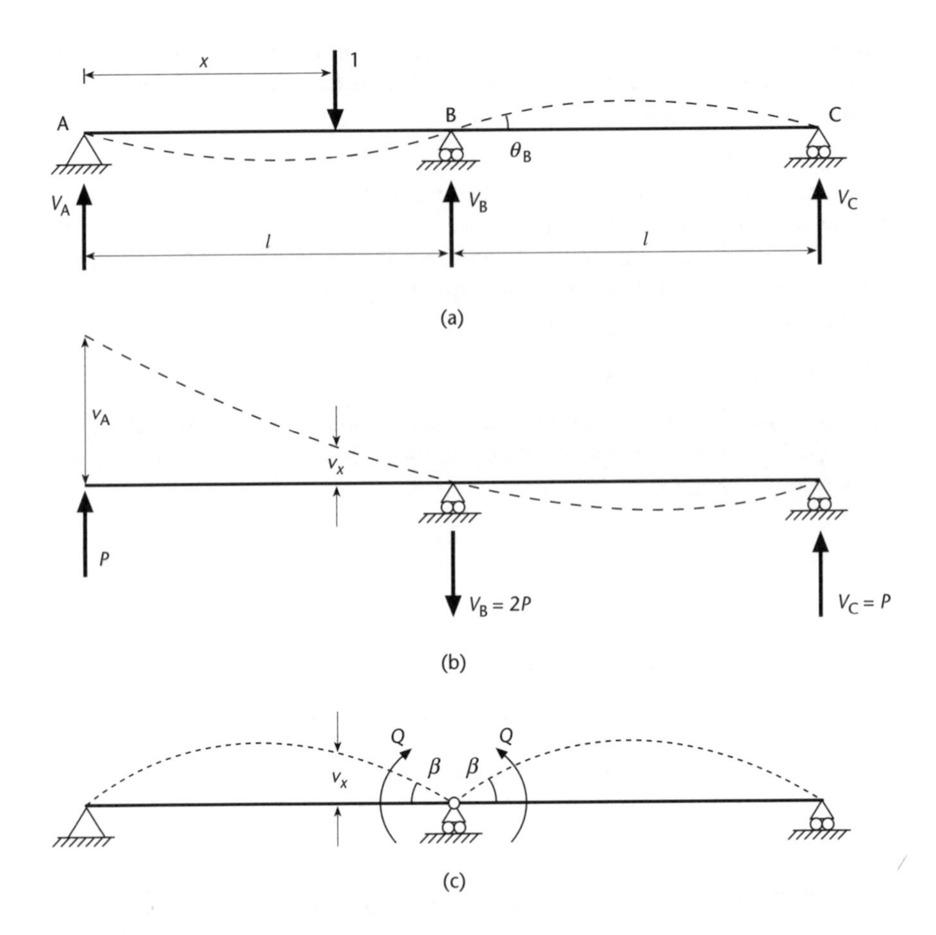

So, normalising to a value of unity at A, the influence line for V_A is given by:

$$V_A = \frac{\alpha^3 - 2\{\alpha - 1\}^3 - 5\alpha + 4}{4}$$

(b) For the bending moment, we use the loadcase shown in Figure 7.16(c). A pin is inserted between the two sides of the beam meeting at B, and moments Q are applied on either side, causing rotations β. Now in the original beam there is a bending moment M_B and a rotation θ_B at B. So, applying the reciprocal theorem to loadcases (a) and (c) gives:

$$1 \cdot (-v_x) + M_B \cdot (-\beta) + M_B \cdot (-\beta) = Q \cdot (-\theta_B) + Q \cdot \theta_B, \quad \text{so } M_B = -v_x/2\beta$$

The influence line is therefore given by the deflection curve in (c), normalised so as to give a unit change in slope at B. Again, v_x and β are found by setting up and integrating the moment–curvature expression. The results for span AB are:

$$\beta = \frac{Ql}{3EI}, \qquad v_x = \frac{Ql^2}{6EI}(\alpha - \alpha^3), \quad \text{where } \alpha = \frac{x}{l}$$

and the deflected shape for BC can be deduced from symmetry. The influence line for M_B is then:

$$M_B = -\frac{v_x}{2\beta} = \frac{l^2(\alpha^3 - \alpha)/6}{2 \cdot l/3} = \frac{l}{4}(\alpha^3 - \alpha)$$

Note that M_B is negative for $0 < \alpha < 1$, implying that the beam is always hogging at B, whatever the position of the load. We can show by differentiation that the moment reaches a peak value of $-l/6\sqrt{3} = -0.096l$ when the unit load is at $\alpha = 1/\sqrt{3} = 0.577$.

7.6.3 Applications of influence lines

For linear elastic structures, influence lines can be used together with the principle of superposition to give the value of a function when any loading system is applied to the structure. In this section, we will consider the application of influence lines to very simple load systems.

Consider the simply supported beam in Figure 7.17(a). Two equal loads of magnitude W travel across the beam a fixed distance a apart – these might be, for instance, the axle loads imposed by a single, very large vehicle. The influence line for the midspan bending moment is shown in (b) and is denoted by m. The influence line corresponds to a unit load crossing the beam, so if a load W crosses the beam, the bending moment will be $M = Wm$. The scaled influence lines for the

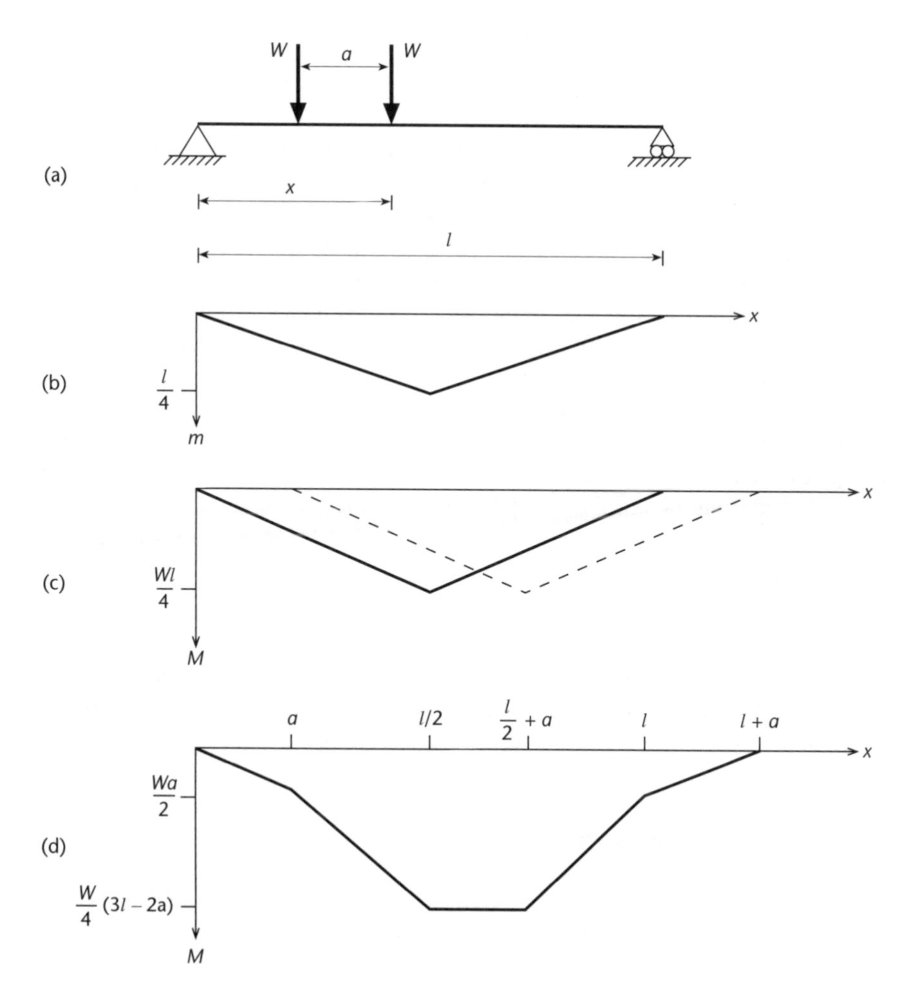

Figure 7.17
Bending moments due to two moving loads.

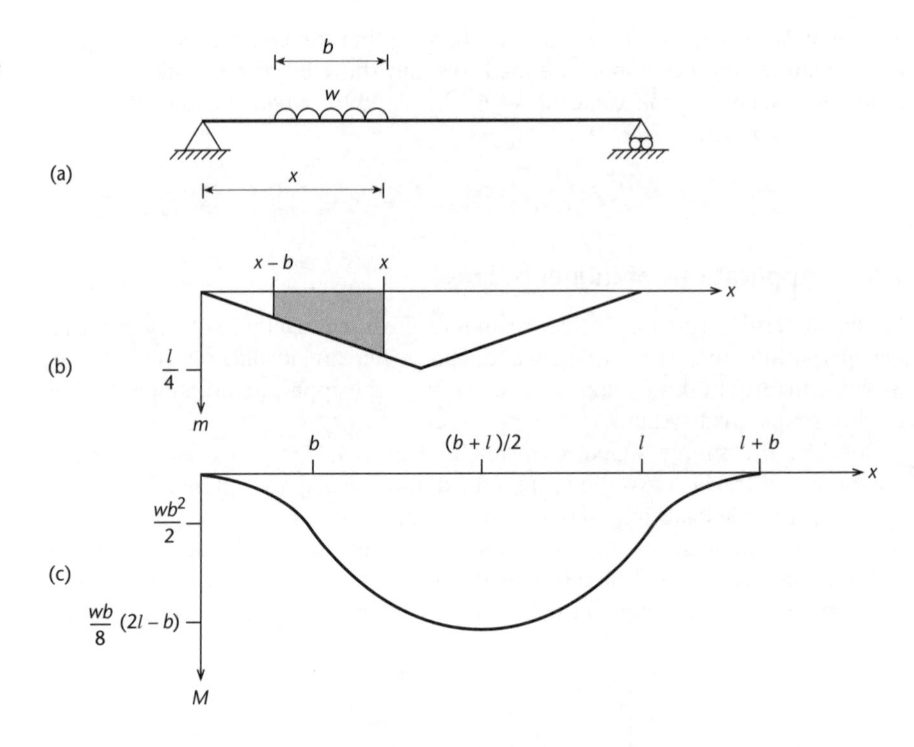

Figure 7.18
Bending moments due to a moving patch load.

two point loads are shown in (c). The influence line for the rear load (the dashed line) must be shifted to the right by the distance a, so that the rear load has no effect on the beam until $x > a$ and remains on the beam when $l < x < l + a$. In (d), the two lines in (c) have been added to give a plot of how the total midspan bending moment varies with x. It is now possible to see at a glance the value of the maximum midspan moment, and the load positions which cause it.

Figure 7.18(a) shows a uniform patch load of intensity w and length b crossing the same beam; the influence line for the midspan moment is again drawn underneath in (b). Now the load acting on a small element dx of the beam is $w \cdot \mathrm{d}x$. As $\mathrm{d}x \to 0$ the load can be considered as acting at a point and this small portion of load gives rise to a moment $\mathrm{d}M = wm \cdot \mathrm{d}x$. The total moment due to the patch load can be found by integrating this expression over the whole length of the patch:

$$M = w \int_{x-b}^{x} m\mathrm{d}x$$

Of course, the integral is just the area of the influence line diagram underneath the patch load, shown shaded. The variation of the midspan moment with the position of the patch load is therefore as shown in Figure 7.18(c).

7.7 Further reading

In spite of its importance as an analysis technique for indeterminate structures, there are comparatively few good books devoted to virtual work and its applications. Two of the better ones are:

- G. A. O. Davies, *Virtual Work in Structural Analysis*, Wiley, 1982.
- F. Thompson and G. G. Haywood, *Structural Analysis Using Virtual Work*, Chapman & Hall, 1986.

7.8 Problems

Where appropriate, answers are given at the end of the book. The questions marked with an asterisk are a little more challenging.

7.1. For each of the beams shown in Figure 7.19, use an appropriate virtual mechanism to determine the reaction at C.

Figure 7.19

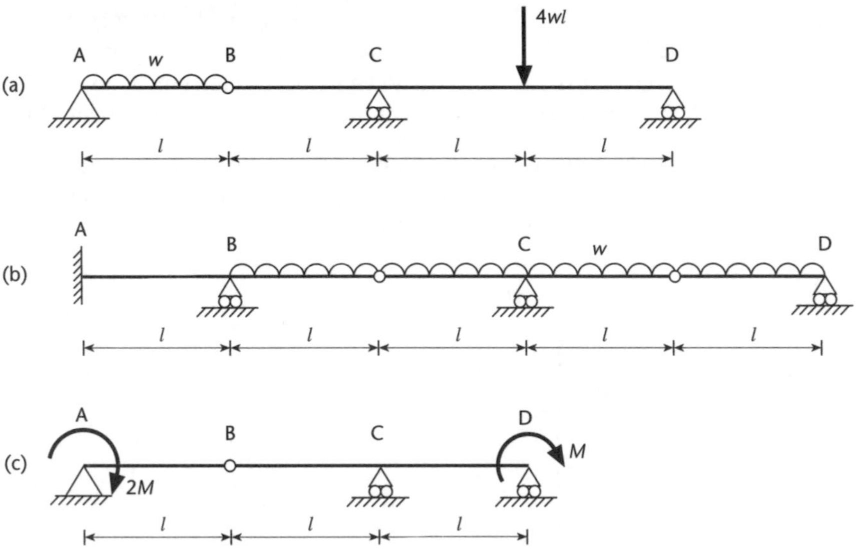

7.2. Find the displacement of joint B of the pin-jointed truss in Figure 7.20. All members have length 4.0 m, cross-sectional area 750 mm^2 and Young's modulus 205 GPa.

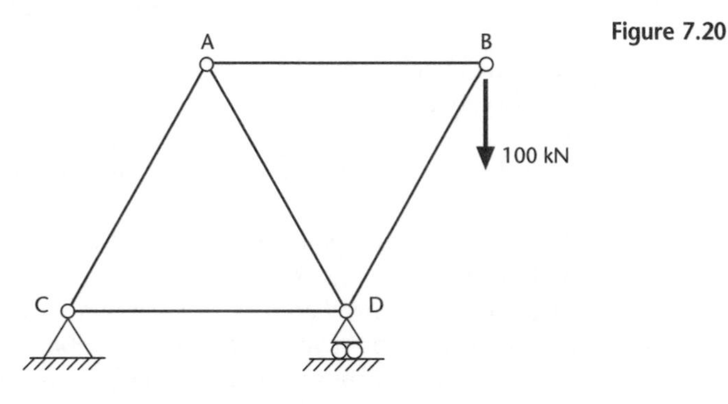

Figure 7.20

7.3. Find the vertical displacement of point D of the Warren truss in Figure 7.21, in which all members have length l, cross-sectional area A and Young's modulus E.

Figure 7.21

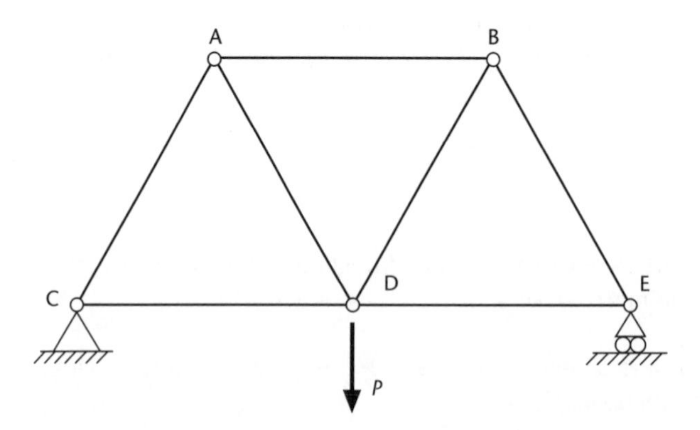

7.4. (a) The horizontal and vertical members of the pin-jointed truss in Figure 7.22 have length 4.0m; the inclined members are at 45° to the horizontal. All members have cross-sectional area 500 mm^2 and Young's modulus 205 GPa. Find the vertical displacement of B and the horizontal displacement of D.

(b)* An extra roller support is now introduced at B. Find the vertical reaction at this support and the new horizontal displacement of D. [*Hint: most of the results you need have already been calculated in part (a).*]

Figure 7.22

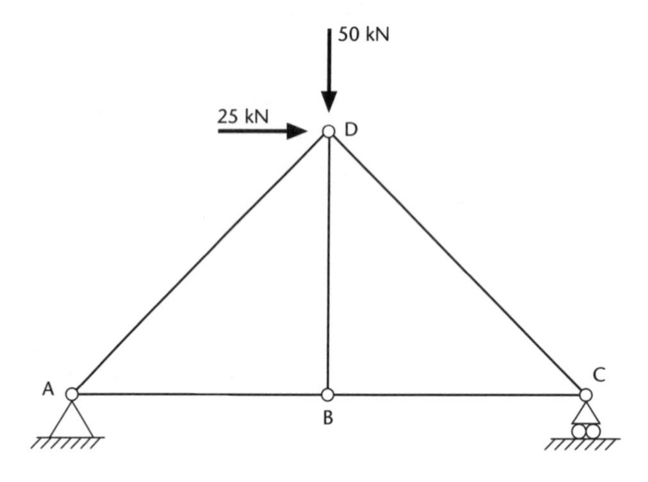

7.5. The members of the pin-jointed frame shown in Figure 7.23 all have the same cross-sectional area A and Young's modulus E. The lengths AD and CD are each l and ACD is an equilateral triangle. Find the forces in the members when the load P is applied if member BD is too short by a length ε before assembly.

7.6. The plane pin-jointed roof truss shown in Figure 7.24 is pinned at its ends to walls which may be regarded as rigid. The horizontal and vertical members are all of length 3.0 m. All members have cross-sectional area 450 mm^2, Young's modulus 205 GPa and coefficient of linear thermal expansion 12×10^{-6}/°C. Snow loading on the roof imposes the loads shown, and also lowers the temperature of members AB, BC, CD and DE by 20°C. Show that the horizontal reaction at the walls is 50 kN and determine the vertical deflection of G, assuming the truss remains elastic. [*Hint: use symmetry to reduce the problem size.*]

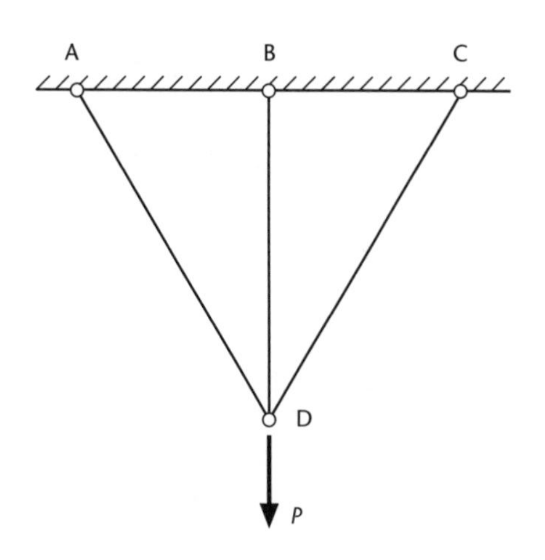

Figure 7.23

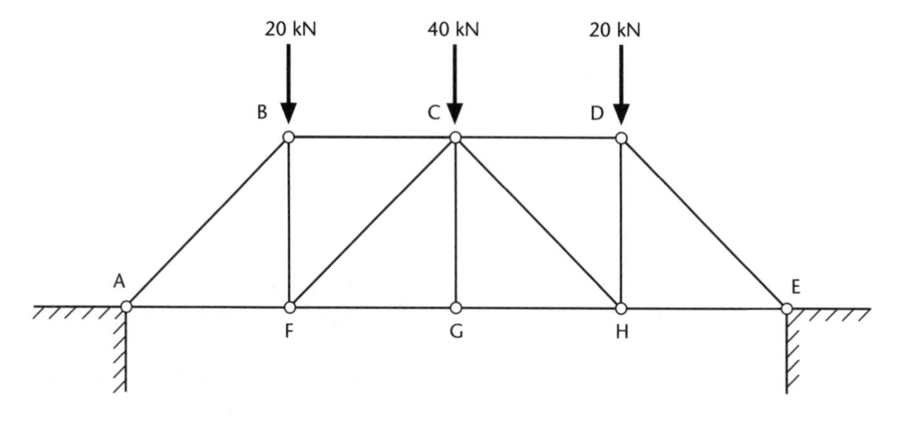

Figure 7.24

7.7. The beam shown in Figure 7.25 has flexural stiffness EI. Calculate the support reactions.

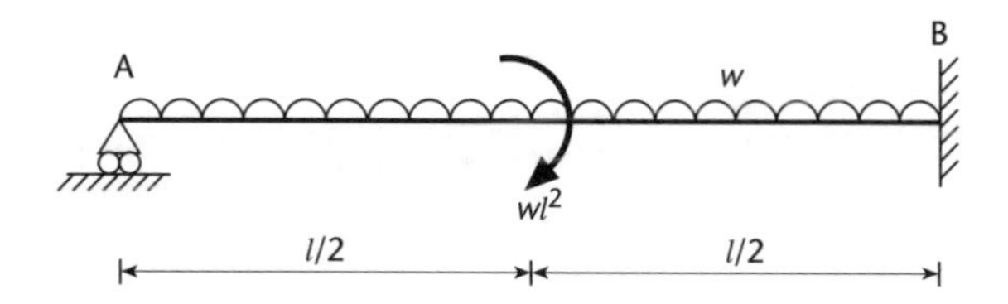

Figure 7.25

7.8. A semi-circular three-pinned arch of radius a and constant flexural stiffness EI has pinned supports and a third pin at the crown. Find the deflection of the central pin joint when a vertical point load W is applied there.

7.9.* The proving ring shown in Figure 7.26 has radius a and flexural stiffness EI. It can be used to measure tensile or compressive loads; under the action of tensile loads the faces of the cut separate, while under compressive loads only axial forces can be transmitted across the cut. The faces of the cut just touch under zero load. For the compressive case, find the normal force at the cut and derive an expression for the stiffness of the ring. Hence show that the ring is approximately 7.2 times as stiff in

Figure 7.26

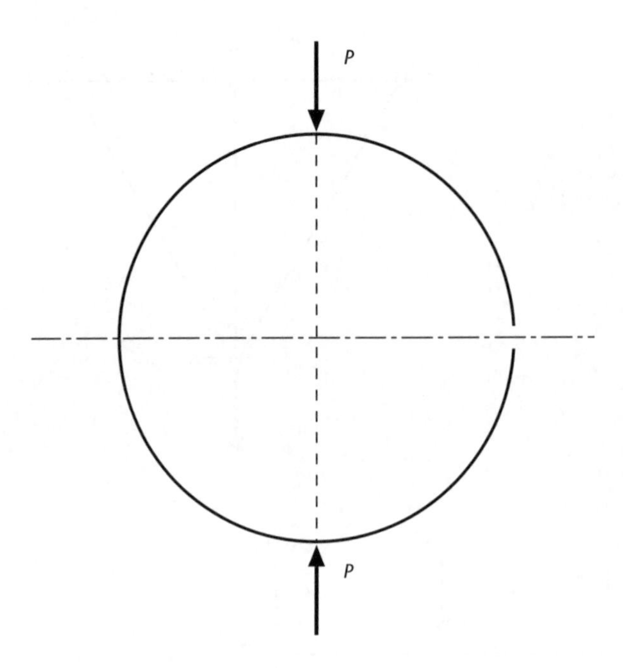

compression as in tension. [*Hint: the solution to the tensile case can be deduced very easily from the compressive case.*]

7.10. Draw the influence lines for the end reactions and for the shear force and bending moment at the quarter-span point of a simply supported beam of length l.

7.11.* A uniform, continuous beam ABC of length 20 m rests on simple supports at its ends A and C and at the point B, 8 m from A. Draw the influence lines for the vertical reaction and the bending moment at B. Determine the maximum vertical reaction and bending moment at this point when two loads of 10 kN and 5 kN spaced 5 m apart traverse the beam.

CHAPTER 8

Moment distribution

CHAPTER OUTLINE

Moment distribution is a powerful iterative method of determining bending moments in statically indeterminate structures such as continuous beams and rigid-jointed frames, and was for many years one of the most popular ways of analysing such structures. In recent years the use of the computerised methods described in Chapters 9 and 10 has become extremely widespread, and the importance of moment distribution has waned correspondingly. However, it remains useful as a quick hand calculation method for simple structures, and can sometimes be helpful in developing a feel for the behaviour of redundant structures.

This chapter should enable the reader to:

- understand the basic principles of moment distribution;
- calculate rotational stiffnesses, carry-over factors and distribution factors for flexural members;
- use moment distribution to determine bending moments in redundant beams;
- apply moment distribution to redundant plane frames;
- plot bending moment diagrams from the results of a moment distribution analysis;
- develop a feel for the way bending moments are distributed within a structure according to the relative stiffnesses of the various members.

8.1 Outline of method

Moment distribution is an example of a *displacement* or *stiffness method*, in which the stiffness properties of the members play a central role in the calculation of the internal forces. It is suitable for structures whose deformations are dominated by bending, so that shear and axial deformations may be neglected. The general approach is as follows.

Consider the structure in Figure 8.1(a). Imagine that we have available a set of clamps that enable us to prevent rotation of any joint within the structure. We first apply the clamps to all the joints and then apply the external loads (Figure 8.1(b)). With no end rotations allowed, each member behaves as a fixed-ended beam. Moments are therefore developed at the ends of any member which is subjected to external loads along its length. These are known as *fixed-end moments*, denoted by M^F. In our structure, there will be fixed-end moments at the ends of AB due to the load P, but not at the ends of BC, since it carries no external loads.

Suppose joint B is now unclamped. We can see that the moments acting at B are not in equilibrium – there is a clockwise fixed-end moment at the end of AB but none in BC. The lack of equilibrium implies that the structure has been forced into

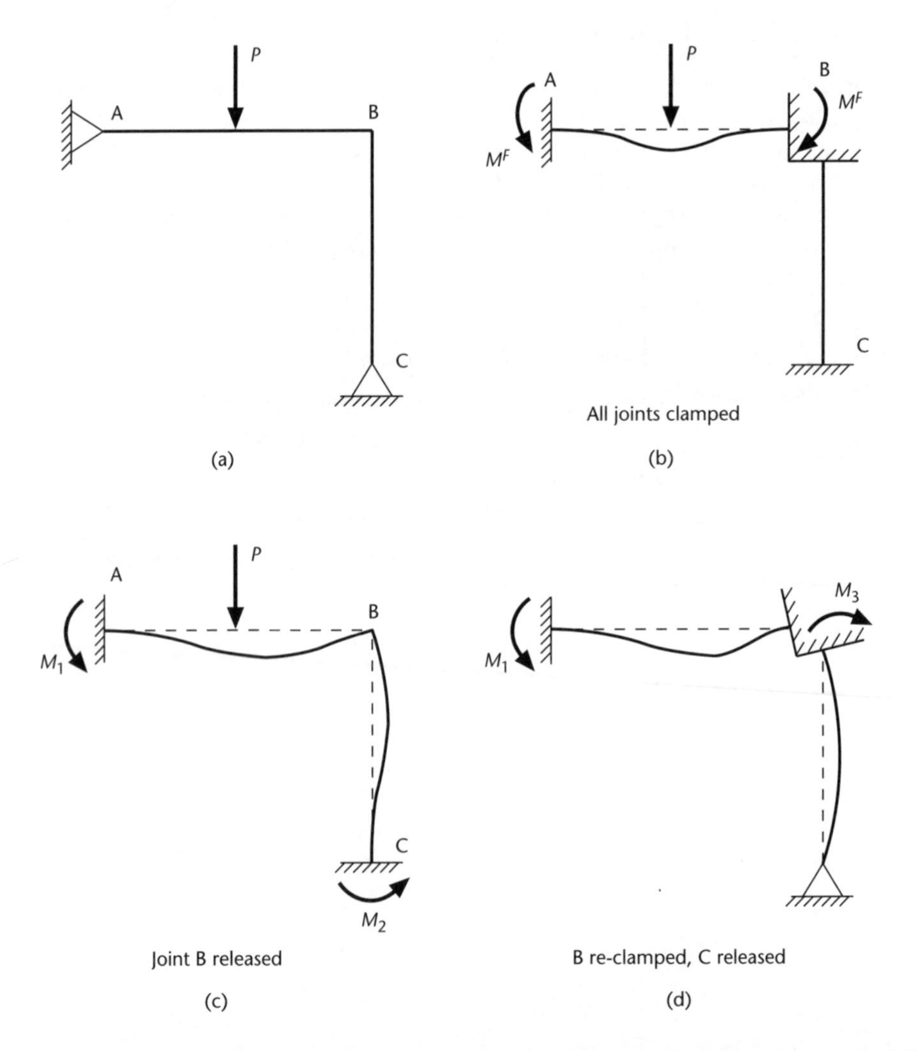

Figure 8.1
Moment distribution.

a deflected shape that it cannot sustain once the clamps are removed. The joint therefore rotates anti-clockwise until it reaches an equilibrium position (Figure 8.1 (c)). Note that both AB and BC must rotate by the same amount at B, since they are rigidly joined and the angle between them cannot change.

Now this movement of the structure does not only affect the moments at B. It also induces additional moments at A and C, so that the total moment at A now takes some value M_1 and the moment at C takes a value M_2. In effect, the out-of-balance moment at B has been redistributed to other parts of the structure. However, the redistribution has not yet resulted in a satisfactory set of bending moments throughout the structure – if either A or C is now released there will again be an out-of-balance moment.

The joint B is now clamped in its new position and another joint, say C, is released. Again, the member rotates at C until the joint is in equilibrium – since C is a pin joint the member must rotate until the moment has dropped from M_2 to zero (Figure 8.1(d)). As before, the rotation also causes a moment (M_3) to be developed at the other end of member BC, and this will affect the equilibrium of joint B.

This process of releasing joints and then re-clamping them in their new positions is repeated at each joint in the structure, and for several cycles around the structure. As this is done, the out-of-balance moments at the joints tend to reduce, until eventually they change by a negligible amount. When this stage is reached then it follows that an equilibrium solution has been achieved and all the clamps can therefore be removed without any further deformation occurring.

Moment distribution provides a simple mathematical way of performing this process. Fixed-end moments are calculated, and then out-of-balance moments at joints are redistributed repeatedly until equilibrium is achieved. In order to do this, we need a way of calculating the fixed-end moments, and of determining how the out-of-balance moment at a joint is distributed between the members meeting at that joint. These will be discussed in the following sections.

8.1.1 Sign convention

For moment distribution we must use a different sign convention from that adopted up to this point. All quantities will be considered positive if they are *anti-clockwise*, whether they are moments, rotations, or the deflection of one end of a member relative to another. For bending moments, this differs from our normal convention, where we took sagging moments as positive and hogging as negative. Consider, for example, the beam segment in Figure 8.2, which is bent under the action of end moments M_{AB} and M_{BA}. The bending moments are both sagging, and therefore would normally be considered positive, but in our new convention M_{AB} is negative and M_{BA} positive. The end rotations are both positive and so is the relative displacement δ between the two ends, because it corresponds to an anti-clockwise rotation of the line AB.

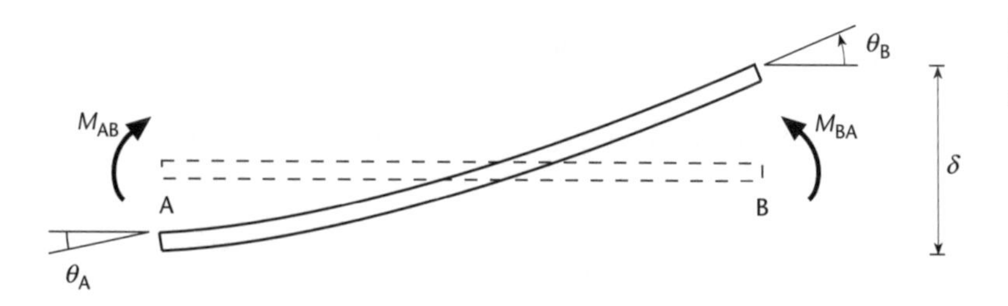

Figure 8.2
Sign convention for moment distribution.

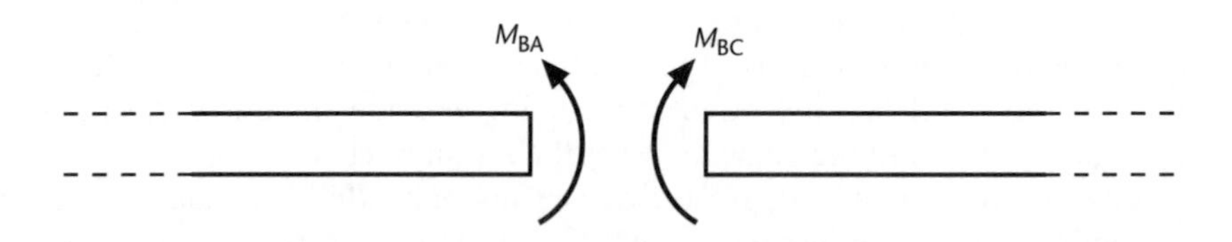

Figure 8.3
Moment equilibrium at a point.

The notation used for the moments is also worth mentioning. We use M_{AB} to denote the moment in member AB at joint A, and M_{BA} to denote the moment in AB at joint B. This helps to minimise confusion when there are several members meeting at a joint.

The new sign convention has important implications for the way that we express moment equilibrium at a point within the structure. Suppose we cut a member AC at a point B – the bending moments on the two cut faces are as shown in Figure 8.3. Under our old sign convention we would simply have said that M_{BA} and M_{BC} are both sagging moments of equal magnitude. However, with anti-clockwise taken as positive the two moments must have the same magnitude but opposite sign, so that the equilibrium of point B can be written as:

$$M_{BA} + M_{BC} = 0 \tag{8.1}$$

8.1.2 Fixed-end moments

These are the moments that are initially developed when all the joints are clamped. It is important to remember that they are the moments *imposed on the member by the clamp,* and not vice versa. A fixed-ended beam is, of course, redundant, but it is a relatively simple matter to determine the end moments using the moment–curvature relations, as in Chapter 5. Consider, for example, the member in Figure 8.4, carrying a point load at a distance a from one end. The horizontal reactions have no influence on the bending behaviour and so can be neglected. The other reactions are as shown, and we wish to find M^F_{AB} and M^F_{BA}. Note that all moments are drawn as positive, that is anti-clockwise. The moment – curvature equation is:

$$M = -M^F_{AB} + V_A x - P\{x - a\} = EI\frac{d^2v}{dx^2}$$

Integrating gives:

$$EI\frac{dv}{dx} = -M^F_{AB}x + V_A\frac{x^2}{2} - P\frac{\{x-a\}^2}{2} + A$$

where A is a constant of integration. Then, using the boundary conditions that the slope is zero at each end gives $A = 0$ and $V_A = Pb^2/l^2 + 2M^F_{AB}/l$ and integrating again we get:

$$EIv = -M^F_{AB}\frac{x^2}{2} + \left(\frac{Pb^2}{l^2} + \frac{2M^F_{AB}}{l}\right)\frac{x^3}{6} - P\frac{\{x-a\}^3}{6} + B$$

Using the boundary condition that $v = 0$ at $x = 0$ gives $B = 0$, and then putting $v = 0$ at $x = l$ gives:

$$M^F_{AB} = +\frac{Pab^2}{l^2} \tag{8.2}$$

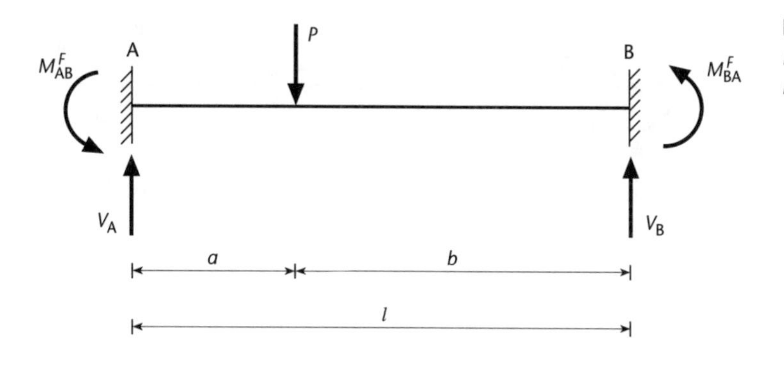

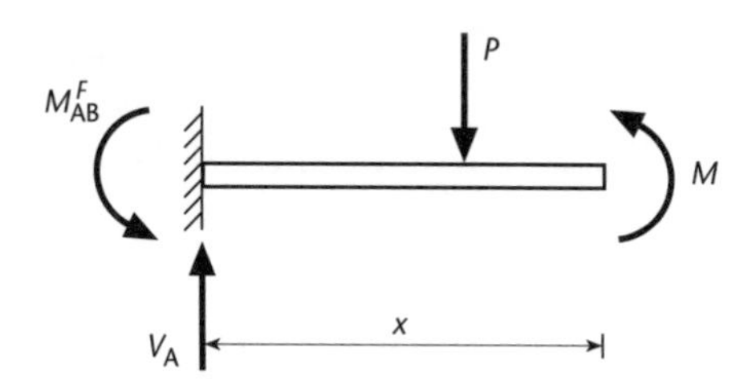

Figure 8.4
Determination of fixed-end moments.

By a similar procedure:

$$M_{AB}^F = -\frac{Pa^2b}{l^2} \tag{8.3}$$

The signs of these terms indicate that the fixed-end moment acts anti-clockwise at the left-hand end and clockwise at the right-hand end.

Some other useful results will now be stated. For a central point load P we simply put $a = b = l/2$ in equations (8.2) and (8.3) giving:

$$M_{AB}^F = +\frac{Pl}{8}, \qquad M_{BA}^F = -\frac{Pl}{8} \tag{8.4}$$

For, the case of a UDL of intensity w, an analysis similar to that presented above gives:

$$M_{AB}^F = +\frac{wl^2}{12}, \qquad M_{BA}^F = -\frac{wl^2}{12} \tag{8.5}$$

8.1.3 Stiffness, carry-over and distribution factors

Every time we release a joint during moment distribution, the out-of-balance moment must be distributed between the members meeting at the joint – this is done using *distribution factors*. In addition, a moment applied at the released end of a member will cause a moment to be set up at the fixed end – the ratio of these two moments is known as the *carry-over factor*. Both of these factors are closely related to the rotational stiffness of the members.

Consider a typical member AB. During moment distribution one end is held fixed and the other is allowed to rotate, giving rise to the forces and deformations shown in Figure 8.5(a). We define the rotational stiffness s_{AB} as the moment required at end A to produce a unit rotation there, and the carry-over factor as the

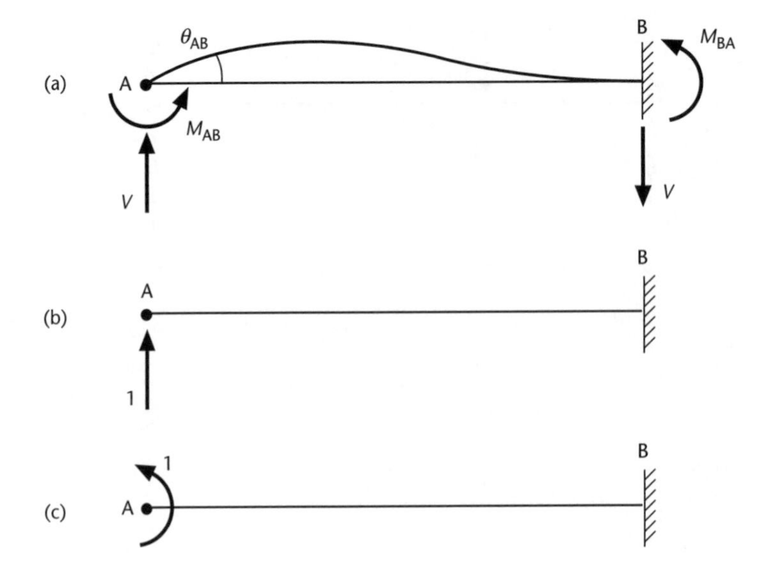

Figure 8.5
Moments and deformations of a pinned-fixed beam.

ratio of the moment developed at the fixed end to that at the pinned end, therefore:

$$s_{AB} = \frac{M_{AB}}{\theta_{AB}}, \qquad c_{AB} = \frac{M_{BA}}{M_{AB}} \tag{8.6}$$

For a uniform member $s_{AB} = s_{BA}$ and $c_{AB} = c_{BA}$. Now the anti-clockwise bending moment at a distance x from A is:

$$M = Vx - M_{AB} = EI\frac{d^2v}{dx^2} \tag{8.7}$$

The unknown force V can be found using virtual work. We use the virtual loadcase in Figure 8.5(b), for which the moment at x from A is $M = x$. Then, applying equation (7.3) using the real deformations together with the virtual forces and moments:

$$\int_0^l x \cdot \frac{(Vx - M_{AB})}{EI} dx = 0 \quad \rightarrow V = \frac{3M_{AB}}{2l}$$

Substituting back into equation (8.7) and putting $x = l$ gives $M_{BA} = M_{AB}/2$. The rotation θ_{AB} can now be found using the virtual loadcase in Figure 8.5(c), for which the bending moment expression is simply $M = -1$. The virtual work equation is then:

$$\int_0^l -1 \cdot \frac{(Vx - M_{AB})}{EI} dx = 1 \cdot \theta_{AB} \quad \rightarrow \theta_{AB} = \frac{M_{AB}l}{4EI}$$

So substituting these results into (8.6), the rotational stiffness and carry-over factor are:

$$s_{AB} = \frac{4EI}{l}, \quad c_{AB} = \frac{1}{2} \tag{8.8}$$

Now consider how the moment at a joint is distributed between the members meeting there. Suppose joint A is released and an out-of-balance moment M is applied. If A is a rigid joint then each member meeting there must rotate through the same angle θ at end A. The moment thus generated in a typical member AB is $s_{AB}\theta$ and the total moment is

$$M = s_{AB}\theta + s_{AC}\theta + s_{AD}\theta + \dots$$

The distribution factor d_{AB} for a typical member AB is defined as the proportion of the total moment carried by AB:

$$d_{AB} = \frac{M_{AB}}{M} = \frac{s_{AB}\theta}{s_{AB}\theta + s_{AC}\theta + s_{AD}\theta\ + \dots} = \frac{s_{AB}}{\sum s} \tag{8.9}$$

Thus the moment is shared between the members in proportion to their rotational stiffnesses.

8.2 Application to statically indeterminate structures

8.2.1 Continuous beams

We have now introduced all the elements needed for a moment distribution analysis, and are ready to apply the method to some indeterminate structures. We shall begin with the relatively simple case of a continuous beam.

EXAMPLE 8.1

Plot the bending moment diagram for the beam ABC in Figure 8.6(a), which has constant flexural stiffness EI.

We first imagine joint B clamped and determine the fixed-end moments. AB has no loads along its length, therefore $M^F_{AB} = M^F_{BA} = 0$. For BC, equations (8.2) and (8.3) give

$$M^F_{BC} = \frac{50 \times 4 \times 6^2}{10^2} = 72 \text{ kNm}, \qquad M^F_{CB} = \frac{-50 \times 4^2 \times 6}{10^2} = -48 \text{ kNm}$$

Next, find the distribution factors. For joint B, equation (8.9) gives

$$d_{BA} = \frac{4EI/5}{4EI/5 + 4EI/10} = \frac{2}{3}, \quad d_{BC} = \frac{4EI/10}{4EI/5 + 4EI/10} = \frac{1}{3}$$

At a fixed end, such as A, the fixed support can be thought of as an infinitely stiff beam, so that the distribution factors are 1 for the support and 0 for AB. The same applies at C. The moment distribution calculations are set out in tabular form below. We first enter the distribution factors and then the fixed-end moments (FEMs).

We now release joint B. There is an out-of balance moment of 72 kNm, so we must add -72 kNm in order to put the joint into equilibrium. This additional moment is apportioned according to the distribution factors at the joint: $(2/3) \times -72 = -48$ kNm is added to M_{BA} and $(1/3) \times -72 = -24$ kNm is added to M_{BC}. These figures are entered and a line drawn under them to indicate that

Figure 8.6

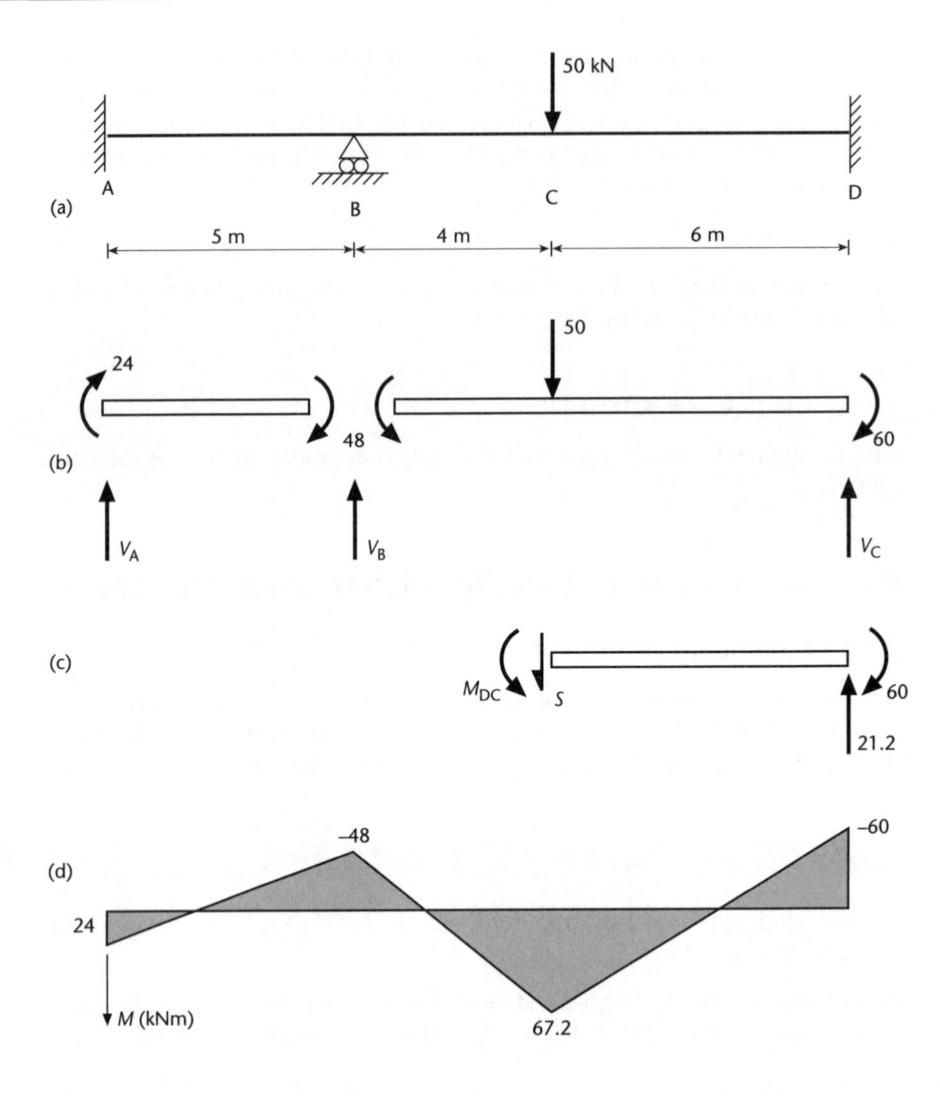

these are balancing moments, putting the joint into equilibrium. The distributed moments in turn generate carry-over moments at the far ends of the members. Since the carry-over factors are all equal to 1/2, this means that we must add $1/2 \times -48 = -24$ kNm to M_{AB} and $\frac{1}{2} \times -24 = -12$ kNm to M_{CB}.

Joint	A	B		C
	M_{AB}	M_{BA}	M_{BC}	M_{CB}
d	0	2/3	1/3	0
FEM			72	−48
Release B		−48	−24	
Carry-over	−24			−12
Totals	−24	−48	48	−60

The distribution is now complete for this very simple case, since there was only one clamped joint (B) to release, and this is now in equilibrium. The moments generated are simply summed to give the total moments at each location. Note that the total moments either side of B *must* sum to zero, otherwise the joint is not in equilibrium.

While the moment distribution phase of the analysis was quite short in this case, there is still some work to be done before we can plot the bending moment diagram. We need to convert the moments from the anti-clockwise positive system back into sagging/hogging, and we need to work out the bending moments at points in the beam away from the joints. Figure 8.6(b) shows the beam split at B, with the moments that we have just calculated drawn on in the appropriate directions, and the external loads and reactions also shown. Now that we have determined the end moments, the reactions can be found by simple statics. For instance, taking moments about B for BC gives:

$$48 - 60 - (50 \times 4) + 10V_C = 0 \quad \rightarrow V_C = 21.2 \text{ kN}$$

We can now find the moment at D by considering the segment DC (Figure 8.6(c)). Taking moments about D gives:

$$M_{DC} = -21.2 \times 6 + 60 = -67.2 \text{ kNm}$$

We can now draw the bending moment diagram (Figure 8.6(d)), in which sagging is plotted below the axis and hogging above. We do this by referring to diagrams (b) and (c) and remembering that:

- a sagging moment in a horizontal beam acts *either* clockwise on a left-hand face *or* anti-clockwise on a right-hand face (and vice-versa for hogging);
- with only point loads present, the bending moment diagram consists of straight line segments between the load and support points.

EXAMPLE 8.2

Repeat the moment distribution part of Example 8.1, but with the fixed joint at C replaced by a pin.

The FEMs and distribution factors are the same as before *except* for the distribution factor at the pinned end C. We must now have $d_{CB} = 1$, since there is zero rotational stiffness to the right of C. The moment distribution calculations are shown below. For compactness, a slightly different layout is now adopted; when we release a joint, both the balancing moments and the resulting carry-over moments are written on the same line. As before, the balancing moments are indicated by underlining.

After entering the distribution factors and FEMs, we first release joint C. There is an out-of-balance moment of −48 kNm, so we must apply a correction of 48 kNm, all of which is added to M_{CB}. A carry-over moment of $(1/2) \times 48 = 24$ kNm is added to M_{BC}. C is now in equilibrium, so we draw a horizontal line under the 48, clamp the joint and move on to B.

At B there is now an out-of-balance moment of $72 + 24 = 96$ kNm, so we must add -96 kNm to restore equilibrium. We add $(2/3) \times -96 = -64$ kNm to M_{BA} and $(1/3) \times -96 = -32$ kNm to M_{BC}. This in turn generates carry-over moments of $(1/2) \times -64 = -32$ kNm at M_{AB} and $(1/2) \times -32 = -16$ kNm at M_{CB}.

B is now in equilibrium but we have disturbed the equilibrium at C. We must therefore apply another correcting moment, this time 16 kNm, at C. The process contiues until there is little change in the results and we then sum the columns to give the total moments. Having calculated the joint moments, the reactions and the bending moment diagram, if required, can be found using simple statics in the same way as Example 8.1.

Joint	A	B		C
	M_{AB}	M_{BA}	M_{BC}	M_{CB}
d	0	2/3	1/3	1
FEM			72.0	−48.0
Release C			24.0	48.0
Release B	−32.0	−64	−32.0	−16.0
Release C			8.0	16.0
Release B	−2.6	−5.3	−2.7	−1.4
Release C			0.7	1.4
Release B	−0.2	−0.4	−0.2	−0.1
Release C				0.1
Totals	−34.8	−69.7	69.8	0

The pinned end causes the solution to Example 8.2 to converge very slowly. The convergence can be greatly improved by using a modified stiffness for a pin-ended member. Consider a member BC, where B is a normal joint and C is a pinned support. To satisfy the zero moment condition at C we must perform the following operations:

Clamp C, impose a rotation θ at B:	$M_{BC} = 4EI\theta/l \rightarrow M_{CB} = 2EI\theta/l$
Clamp B, apply balancing moment at C:	$M_{BC} = -EI\theta/l \leftarrow M_{CB} = -2EI\theta/l$
Add:	$M_{BC} = 3EI\theta/l \qquad M_{CB} = 0$

We can see from the last line that the effect of these operations is simply to reduce the member stiffness to $3EI/l$. So where we have a pin-ended member, we can set the moment at the pin to zero in a single step by using this modified stiffness *and* setting the carry-over factor from B to C equal to zero.

EXAMPLE 8.3

Repeat Example 8.2 using the modified stiffness approach.

The FEMs are unchanged. The member stiffnesses are now $s_{AB} = 4EI/5$ and $s_{BC} = 3EI/10$, so the distribution factors are:

$$d_{BA} = \frac{4EI/5}{4EI/5 + 3EI/10} = \frac{8}{11}, \qquad d_{BC} = \frac{3EI/10}{4EI/5 + 3EI/10} = \frac{3}{11}$$

The moment distribution is tabulated below. We start by releasing C and applying a balancing moment of 48 kNm, which results in a carry-over moment of 24 kNm to B. C is now in equilibrium and need not be visited again, since there is no carry-over from B to C. We now balance joint B, which causes a carry-over to A but *not* to C. The same result has now been achieved as in Example 8.2, but with considerably less arithmetic.

Joint	A	B		C
	M_{AB}	M_{BA}	M_{BC}	M_{CB}
d	0	8/11	3/11	1
FEM			72.0	−48.0
Release C			24.0	48.0
Release B	−34.9	−69.8	−26.2	
Totals	−34.9	−69.8	69.8	0

8.2.2 Beams with support movements

The cases considered so far have involved members whose ends may rotate, but not deflect. We shall now go on to look at the case where there is a relative vertical movement between the two ends of a beam, as might occur in a building suffering differential settlements. Consider the fixed-ended beam undergoing a displacement δ at the right-hand support (Figure 8.7(a)). The deflected shape is anti-symmetric, so that there is a point of contraflexure at the centre. There is therefore a shear force but no bending moment at that point, so that splitting the beam at the centre gives the free body diagrams in Figure 8.7(b). Each half of the beam thus behaves as a cantilever of length $l/2$, with the tip force related to the deflection by one of the standard results given in Table 5.2:

$$\frac{\delta}{2} = \frac{P(l/2)^3}{3EI} \rightarrow P = \frac{12EI\delta}{l^3} \tag{8.10}$$

and the end moments are then:

$$M^F_{AB} = M^F_{BA} = -\frac{Pl}{2} = -\frac{6EI\delta}{l^2} \tag{8.11}$$

Figure 8.7
Fixed-ended beam with a support deflection.

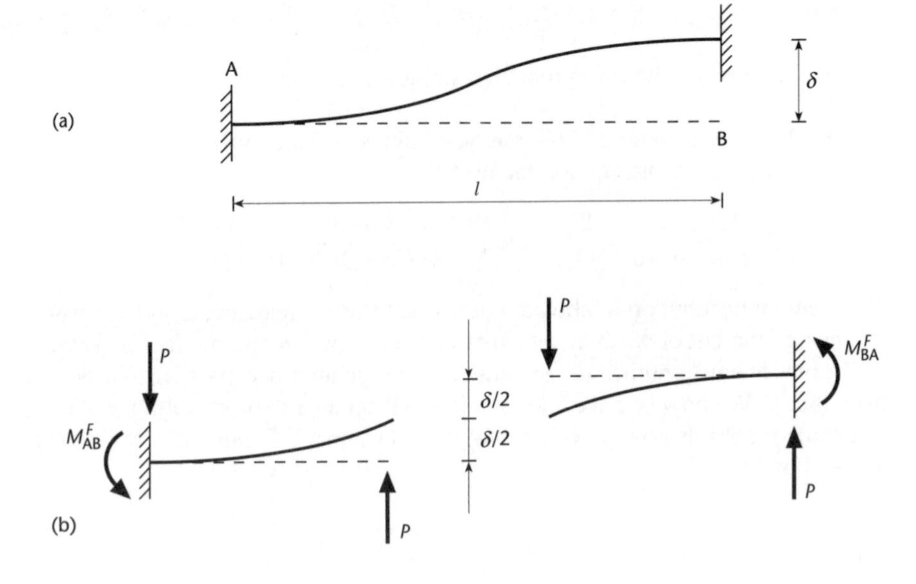

We can now include support movements in a moment distribution analysis simply by introducing the corresponding fixed-end moment terms using equation (8.11).

EXAMPLE 8.4

The three-span beam in Figure 8.8(a) has constant flexural stiffness $EI = 25 \times 10^3$ kNm^2. It carries no external loads, but differential ground movements cause

Figure 8.8

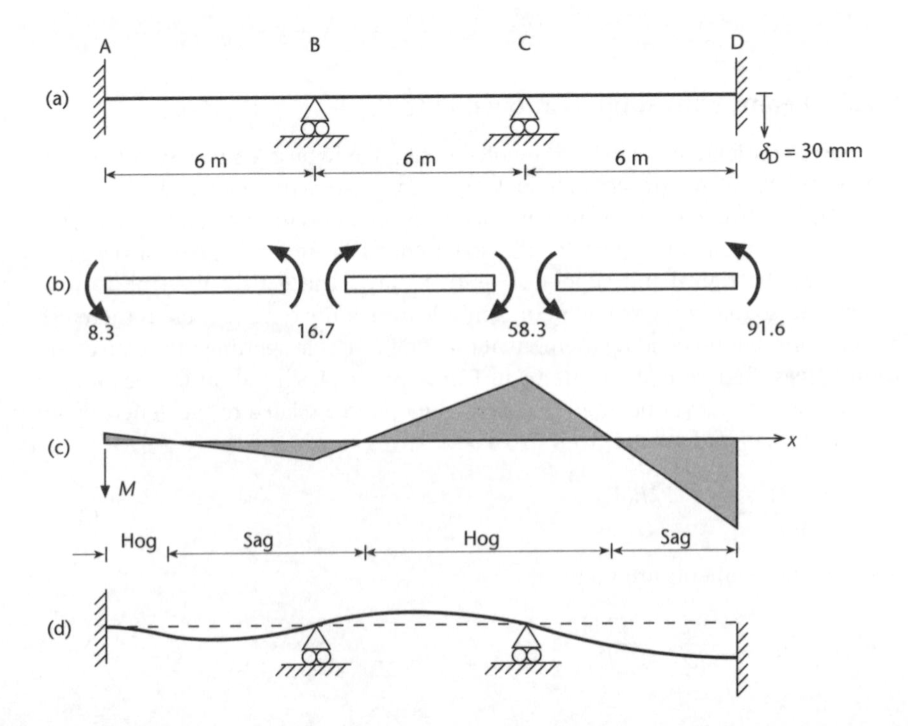

support D to sink by 30 mm relative to A, B and C. Calculate the resulting joint moments.

There are no fixed-end moments in AB or BC, and the FEMs in CD are found from equation (8.11):

$$M^F_{CD} = M^F_{DC} = -\frac{6 \times 25 \times 10^3 \times (-0.03)}{6^2} = 125 \text{ kNm}$$

Since all spans have the same length and properties, the distribution factors at B and C are all equal to 1/2. The moment distribution can now proceed as usual, balancing first joint C, then joint B, and repeating until equilibrium is achieved – the calculations are shown below.

Joint	A		B		C		D
	M_{AB}	M_{BA}	M_{BC}	M_{CB}	M_{CD}	M_{DC}	
d	0	1/2	1/2	1/2	1/2	0	
FEM					125.0	125.0	
Release C			−31.3	−62.5	−62.5	−31.3	
Release B	7.8	15.7	15.7	7.8			
Release C			−2.0	−3.9	−3.9	−2.0	
Release B	0.5	1.0	1.0	0.5			
Release C			−0.1	−0.2	−0.2	−0.1	
Totals	8.3	16.7	−16.7	−58.3	58.3	91.6	

Although not requested in the question, it is always a good idea to check that these results make sense by sketching the bending moment diagram and deflected shape. We first split the beam at the joints and mark on the calculated bending moments, with positive moments drawn anti-clockwise (Figure 8.8(b)). This enables us to see which moments are sagging (clockwise on a left-hand end or anti-clockwise on a right-hand end) and which hogging. We can now plot the bending moment diagram in (c) and use this to sketch the deflected shape in (d), in which the pattern of sagging and hogging curvatures matches the bending moment diagram.

8.2.3 Rigid-jointed frames

Rigid-jointed frames can be analysed in much the same way as continuous beams. The first example will deal with a simple plane structure in which the joints can rotate, but not deflect.

EXAMPLE 8.5

Find the joint moments for the frame in Figure 8.9(a). All members have the same EI.

Figure 8.9

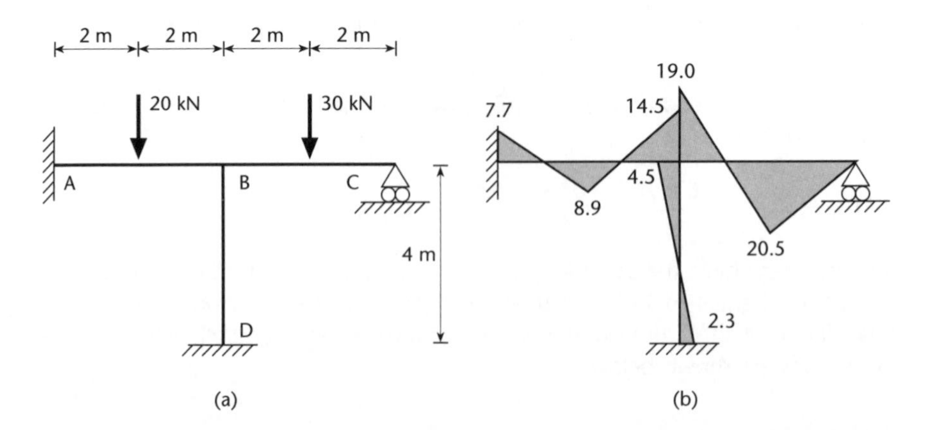

In reality, of course, joint B would move slightly due to the axial deformations of the members, but the resulting deflections are very small compared to those due to bending. From equation (8.4) the FEMs are:

$$M^F_{AB} = 20 \times 4/8 = 10 \text{ kNm}, \qquad M^F_{BA} = -10 \text{ kNm},$$
$$M^F_{BC} = 30 \times 4/8 = 15 \text{ kNm}, \qquad M^F_{CB} = -15 \text{ kNm}$$

The stiffnesses of AB and BD are $s_{AB} = s_{BD} = 4EI/4 = EI$. Because BC has a pinned end we use a modified stiffness $s_{BC} = 3EI/4$ and allow no carry-over from B to C. The distribution factors at B are then:

$$d_{BA} = d_{BD} = \frac{1}{1+1+3/4} = \frac{4}{11}, \quad d_{BC} = \frac{3/4}{1+1+3/4} = \frac{3}{11}$$

The moment distribution can now be performed. Because we have used the modified stiffness approach the results converge very quickly – we first apply a balancing moment of 15 kNm at C. There is then an out-of-balance moment of $-10 + 15 + 7.5 = 12.5$ kNm at B, so a balancing moment of -12.5 kNm is distributed between the three members meeting at B. The joint is now in equilibrium, since it is subjected to clockwise moments of 14.5 kNm and 4.5 kNm, and an anti-clockwise moment of 19.0 kNm. From the results, the bending moment diagram can be plotted (Figure 8.9(b)). This follows the usual convention that the lines are plotted on the tension sides of the members.

Joint	A	B			C	D
	M_{AB}	M_{BA}	M_{BC}	M_{BD}	M_{CB}	M_{DB}
d	0	4/11	3/11	4/11	1	0
FEM	10.0	−10.0	15.0		−15.0	
Release C			7.5		15.0	
Release B	−2.3	−4.5	−3.5	−4.5		−2.3
Totals	7.7	−14.5	19.0	−4.5	0	−2.3

The frame in Example 8.5 is quite unusual – most frames are fixed at one level only (the ground) and will sway under horizontal loads such as wind. In such cases we must make use of the results for a beam with an end displacement (equations (8.10) and (8.11)), resulting in a rather more complex solution procedure.

EXAMPLE 8.6

Find the bending moments at the joints and the lateral deflection of the beam for the portal frame in Figure 8.10(a). For AB and CD, $EI = 15 \times 10^3$ kNm^2 and for BC, $EI = 45 \times 10^3$ kNm^2.

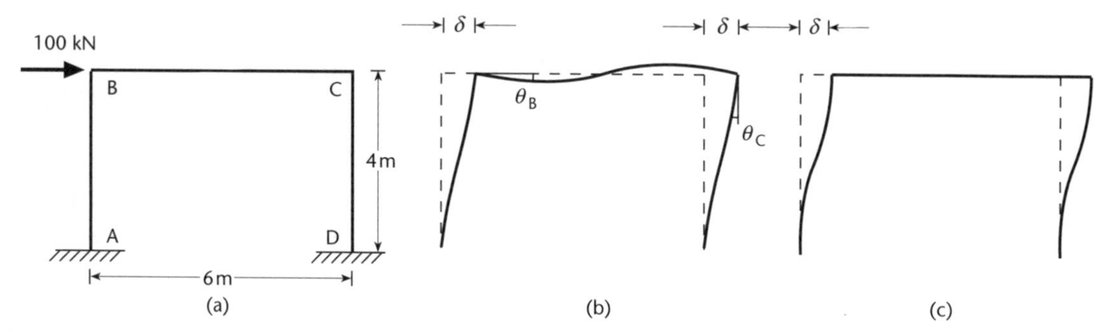

Figure 8.10

Before performing the moment distribution, we need to think carefully about how the structure deforms. The lateral load will cause the joints B and C both to deflect horizontally *and* to rotate (Figure 8.10(b)). The horizontal deflections of B and C can be assumed to be the same, because the axial shortening of BC is negligibly small compared to the bending of the columns.

For the purposes of analysis, we can treat the overall deformation as a two-stage process by separating out the lateral sway and the joint rotations. Thus we first apply a lateral sway to BC with the joints clamped to prevent rotation (Figure 8.10 (c)). Each column deforms in the same way as the beam with a support movement in Figure 8.9. The lateral sway and the corresponding fixed-end moments can therefore be calculated using equations (8.10) and (8.11). Since the columns have the same properties and undergo identical deformations, the 100 kN load must be divided equally between them, so:

$$\delta = \frac{Pl^3}{12EI} = \frac{50 \times 4^3}{12 \times 15 \times 10^3} = 0.0178 \text{ m}$$

and

$$M^F_{AB} = M^F_{BA} = M^F_{CD} = M^F_{DC} = \frac{Pl}{2} = \frac{50 \times 4}{2} = 100 \text{ kNm}$$

Having imposed the lateral sway, we now allow the joints to rotate and distribute the out-of-balance moments in the usual way. The member stiffnesses are:

$$s_{AB} = s_{CD} = \frac{4 \times 15 \times 10^3}{4} = 15 \times 10^3 \text{ kN/m},$$

$$s_{BC} = \frac{4 \times 45 \times 10^3}{6} = 30 \times 10^3 \text{ kN/m}$$

The distribution factors are:

$$d_{BA} = d_{CD} = \frac{15}{15+30} = \frac{1}{3}, \quad d_{BC} = d_{CB} = \frac{30}{15+30} = \frac{2}{3}$$

The moment distribution can now be carried out by repeatedly releasing B and C until equilibrium is achieved – the calculations are shown below.

Joint	A	B		C		D
	M_{AB}	M_{BA}	M_{BC}	M_{CB}	M_{CD}	M_{DC}
d	0	1/3	2/3	2/3	1/3	0
FEM	100.0	100.0			100.0	100.0
Release B	−16.7	−33.3	−66.7	−33.3		
Release C			−22.2	−44.5	−22.2	−11.1
Release B	3.7	7.4	14.8	7.4		
Release C			−2.5	−4.9	−2.5	−1.3
Release B	0.4	0.8	1.7	0.8		
Release C			−0.2	−0.5	−0.3	−0.2
Totals	87.4	74.9	−75.1	−75.0	75.0	87.4

8.3 Problems

Where appropriate, answers are given at the end of the book. The questions marked with an asterisk are a little more challenging.

8.1. A propped cantilever AB of length 8 m is rigidly fixed at end A and pinned at end B. Find the moment at A and the vertical reactions at A and B when a vertical point load of 20 kN is applied 3 m from A. Hence draw the bending moment diagram.

8.2. A continuous beam ABCD has three equal spans of length 5 m. The beam is rigidly fixed at its ends A and D and rests on roller supports at B and C. Calculate the moments at the joints when a UDL of 10 kN/m is applied to spans AB and BC. *EI* is constant throughout the beam.

8.3. For the beam in Problem 8.2, draw the bending moment diagram clearly indicating all peak values and where they occur. Check that this diagram corresponds to a credible deflected shape.

8.4.* For the beam in Problem 8.2, taking $EI = 20 \times 10^3$ kNm^2, calculate the vertical movement of support D which would be required in order to reduce the moment there to zero. Sketch the new bending moment diagram and deflected shape. [*Hint: the simplest approach is to analyse the beam using a guessed value of support movement then scale the results so that, when added to those in 8.2, the moment at D is zero.*]

8.5. Find the joint moments for the frame shown in Figure 8.11, in which the verticals have flexural stiffness EI and the horizontals $2EI$.

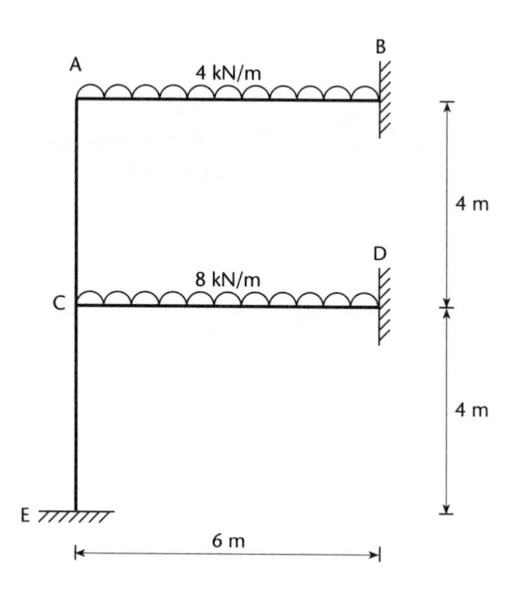

Figure 8.11

8.6. Find the bending moments at the joints of the frame in Figure 8.12, in which all members have the same properties. Find also the horizontal reactions at A, C and D.

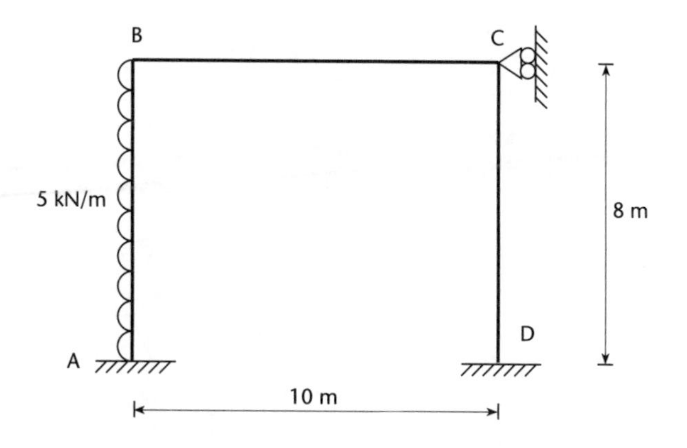

Figure 8.12

8.7.* For the frame in Figure 8.12, the roller support at C is now removed, so that side-sway occurs due to the horizontal load. Calculate the new bending moments at the joints. [*Hint: using the principle of superposition, this problem can be treated as the sum of two loadcases. The case analysed in Problem 8.6 is one of them – what is the other?*]

Figure 8.13

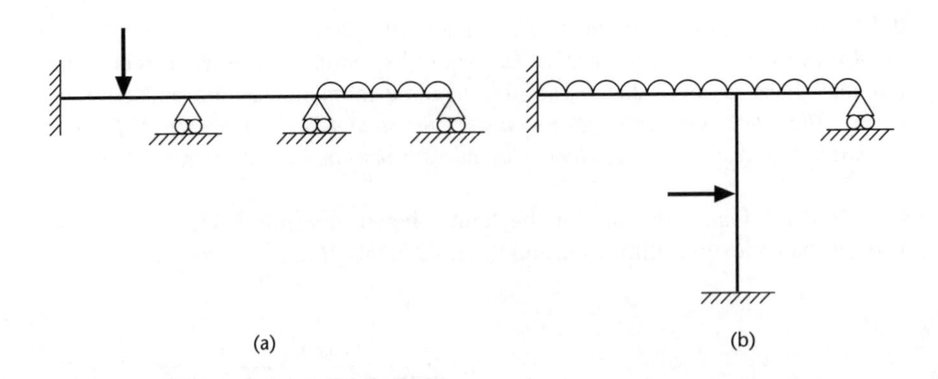

8.8.* Without calculation, sketch the bending moment diagrams and deflected shapes for the structures in Figure 8.13.

CHAPTER 9

The stiffness matrix method

CHAPTER OUTLINE

Nowadays most structural analysis is performed computationally using the methods that will be introduced in the next two chapters – the stiffness matrix method for frame structures made up of bars, beams and columns, and the finite element method for continuum structures such as plates and solid bodies.

As an introduction to stiffness matrices, we shall first look briefly at a simple hand analysis procedure for redundant, rigid-jointed frames known as the slope–deflection method. Like moment distribution, this is a stiffness method of analysis, in which consideration of the way the structure deforms under load plays a vital role in the solution procedure. Indeed, we now use the joint displacements as the *primary variables* in our analysis; that is, we *first solve* for the joint displacements and *then* find the internal forces by back-substitution. This is the reverse of all the approaches we have studied so far.

Even for quite simple structures, the slope–deflection method tends to require the solution of several simultaneous equations. This can be a tedious and error-prone procedure to do by hand, so it makes sense to write the equations in matrix form and solve them computationally. In the stiffness matrix method we take this idea a step further, by developing a procedure whereby *both* the setting up of the problem *and* its solution are carried out computationally. The stiffness matrix method is an extremely powerful way of analysing redundant frames, though it can also be applied to statically determinate structures.

On completion of this chapter the reader should be able to:

- make reasonable predictions of deflected shapes of rigid-jointed frames, so as to establish which displacements must be considered in a slope–deflection analysis;
- analyse simple rigid-jointed frames by the slope–deflection method;
- derive stiffness matrices for bars and beams;
- understand how the global stiffness matrix for a structure is assembled from the local matrices for the constituent elements;
- plot bending moment diagrams and deflected shapes from the output of a slope–deflection or stiffness matrix analysis;
- appreciate the underlying method, limitations and potential pitfalls of stiffness matrix computer programs.

9.1 The slope–deflection method

Like moment distribution, the slope–deflection approach is suitable for statically indeterminate, rigid-jointed frames in which axial and shear deformations may be considered negligible compared to those due to bending. The basic approach is as follows:

(a) from a consideration of the structure and its loading, identify the unknown joint displacements that will be the primary analysis variables;

(b) using the stiffness properties of the members, formulate equations relating the applied loads and joint displacements to the moments at the member ends;

(c) solve for the displacements using conditions of moment equilibrium at the joints;

(d) back-substitute into the stiffness equations formulated in (b) to find the unknown end moments;

(e) reactions, shear forces etc. can now be found using equilibrium methods, if required.

This is in contrast to the equilibrium methods introduced in earlier chapters, where we first used simple statics to find the member forces, then found displacements (if required) using the force–extension and moment–curvature relationships.

9.1.1 Deflected shapes of rigid-jointed frames

In all structures problems, it is useful to have a rough idea of the expected structural behaviour before starting the analysis. This allows the most suitable analysis approach to be chosen and enables you to check that the answers make sense. For stiffness methods such as slope–deflection this step becomes *essential*, as the analysis cannot proceed correctly unless it is set up in terms of the right displacement variables.

For reasonably simple structures, a qualitative assessment can be made by thinking carefully about the relationships between displacements, bending moments and reactions. In particular:

- Look for points of zero displacement and points of equal displacement (treating axial deformations as negligible).
- Look for points of contraflexure.
- Remember that the angle between members meeting at a rigid joint does not change – thus if a rigid joint rotates, then the ends of all members meeting at that joint undergo the same rotation.
- Think carefully about the directions of reactions.
- Use superposition to simplify complicated load cases or support conditions.

A very common type of rigid-jointed structure is the portal frame. In Figures 9.1 and 9.2 approximate deflected shapes and bending moment diagrams are shown for flat-roofed portal frames subjected to simple loadcases; Figure 9.1 shows frames with pinned feet and Figure 9.2 frames with fixed feet. Many other results can be derived in approximate form using superposition of these results.

Consider first the frame with pinned feet under a central point load (Figure 9.1 (a)). Both the deflected shape and the bending moment diagram are symmetrical. The downwards deflection of the cap beam under the load causes rotations at the

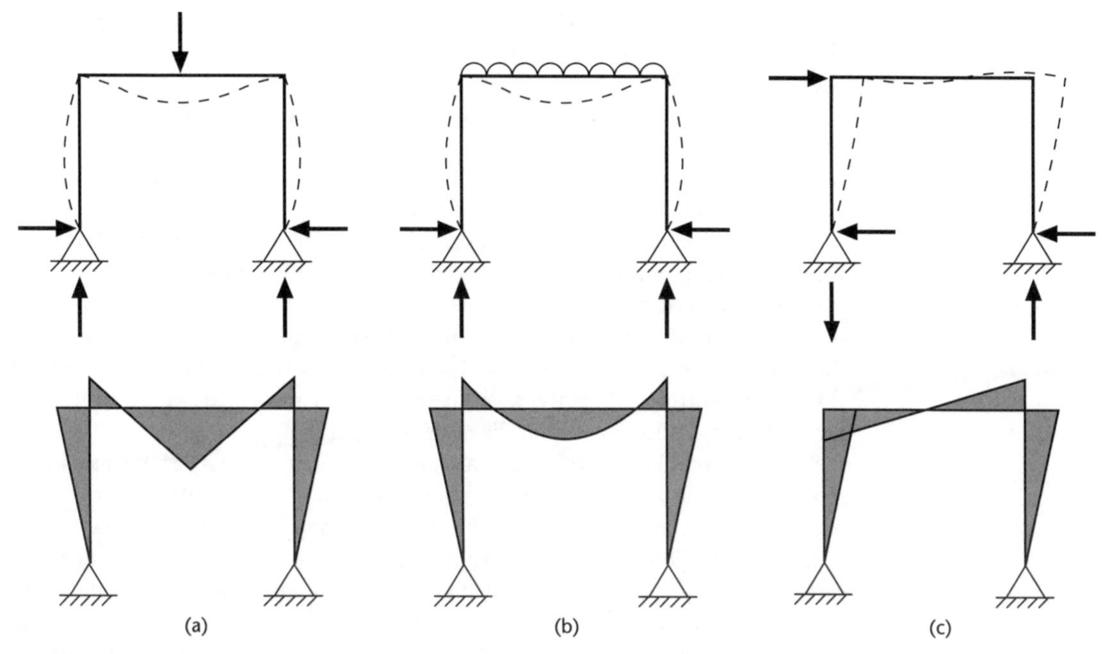

Figure 9.1
Deflected shapes and bending moment diagrams for portal frames with pinned feet.

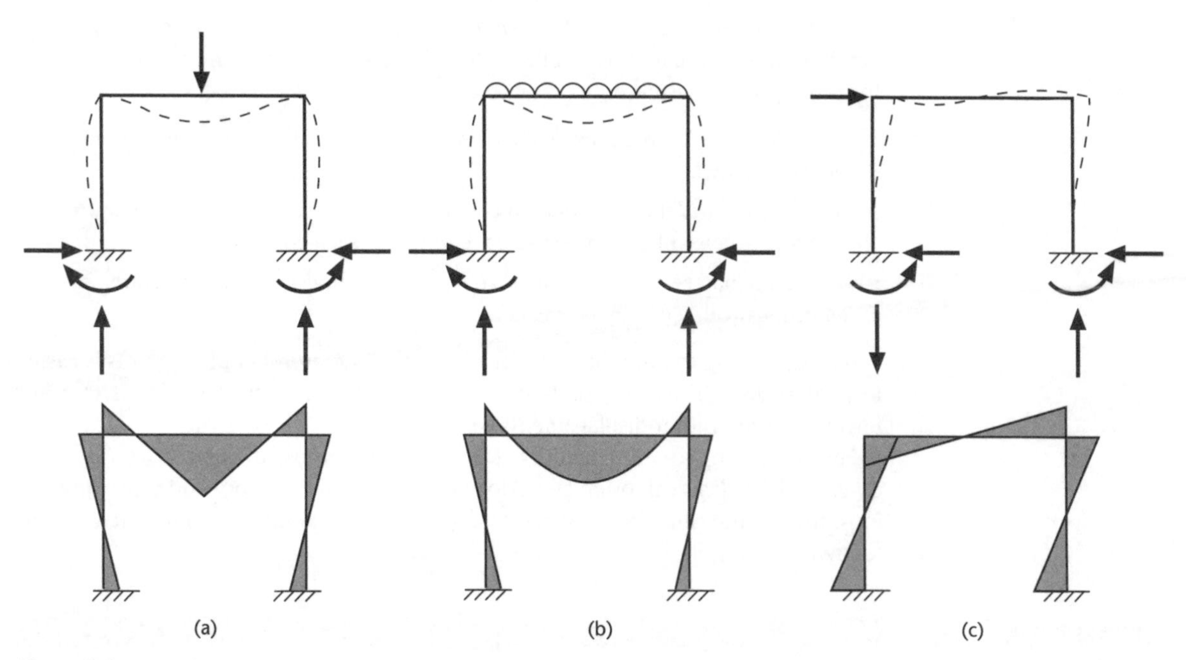

Figure 9.2
Deflected shapes and bending moment diagrams for portal frames with fixed feet.

beam–column joints; if there were no horizontal restraint at the column feet, then these rotations would cause the columns to splay outwards. It follows that the horizontal reactions at the base must act *inwards*, to prevent the splaying effect.

The bending moment diagram can now be sketched. Since only point loads are present, it consists entirely of straight lines. As usual, the lines are plotted on the tension faces of members and the moments in members meeting at a joint are equal. From the bending moment diagram, it is then obvious that there are points of contraflexure symmetrically positioned in the cap beam – the central part of the beam is sagging, but short lengths at either end are hogging.

If we were performing a slope–deflection analysis on this frame, we would need to include *four* displacement variables: the rotations at the two feet and at each end of the cap beam. This could, of course, be reduced to two unknowns using symmetry.

Case (b), with a uniform vertical load, is similar to (a) except that the bending moment distribution in the cap beam is now parabolic.

The horizontal load in (c) causes an anti-symmetric deflected shape and bending moment diagram, so there must be a point of contraflexure at the centre of the cap beam. It can be tempting to draw the deflected shape assuming no rotation at the beam–column joints, but this would only be correct if the cap beam was completely rigid. Note that there will now be equal and opposite vertical reactions, upwards at the right-hand support, downwards at the left. These create a couple, opposing the tendency of the structure to overturn under the horizontal load. An analysis of this frame would require *five* displacement variables: the four joint rotations and the horizontal deflection of the cap beam – since we are ignoring axial deformations, all parts of the cap beam are assumed to have the same horizontal deflection. We could reduce the number of unknown rotations to three using the anti-symmetry of the problem.

Turning to the frames with fixed feet (Figure 9.2), the deflected shapes and bending moment diagrams can be sketched in a similar way. The main differences from the pinned case are:

- Since there are no rotations at the feet, the number of unknown displacements is reduced by two.
- In order to satisfy the zero rotation condition at the feet a reversal of curvature, and therefore a point of contraflexure, is required in each column.
- The additional restraint increases the overall stiffness of the structure, so that deflections will generally be smaller.

It is interesting to note that a structure with a large number of restraints is easier to analyse by stiffness methods than one with fewer restraints, since there are fewer unknown joint displacements to calculate.

For more complicated structures or loadcases than those in Figures 9.1 and 9.2, a qualitative analysis can often be performed using superposition, and sometimes by introducing and subsequently removing artificial restraints, as illustrated in the following example.

EXAMPLE 9.1

Sketch the deflected shape and bending moment diagram for a rectangular portal frame with pinned feet, subjected to an off-centre vertical point load.

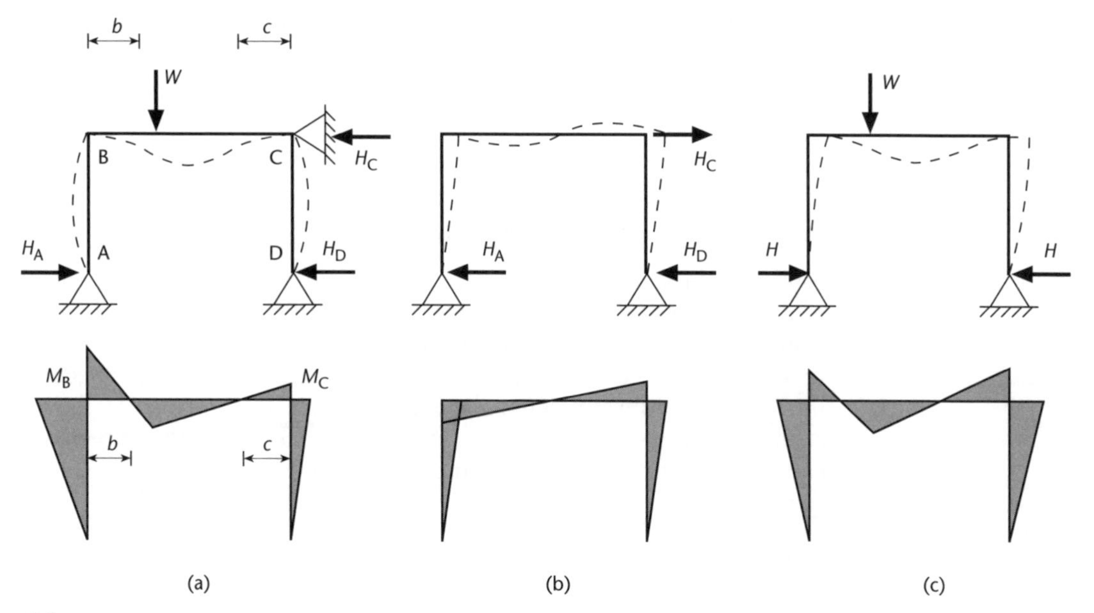

Figure 9.3

This loadcase is similar to that in Figure 9.1(a), but the off-centre load means that the frame is likely to sway and we can no longer make use of symmetry. Suppose we first introduce an artificial horizontal restraint at the cap beam level. The deflected shape and bending moment diagram will be as in Figure 9.3(a). Note that the maximum vertical deflection of the beam is to the left of centre (though not directly under the load). This in turn means that the points of contraflexure in the beam are not symmetrically positioned – the distance b on the diagram is less than c. Therefore, $M_B > M_C$. If we now consider the columns, the corner moment is simply equal to the horizontal reaction at the base multiplied by the height, so it follows that $H_A > H_D$. Hence, for horizontal equilibrium, the artificial restraint at C must be imposing a force to the left.

We now consider a second loadcase, consisting solely of the force H_C but this time acting to the right. This is identical to one of the cases considered above and gives the deflected shape and bending moment diagram in Figure 9.3(b). If we now add the two cases (a) and (b) then the artificial forces H_C cancel and we get the results for the unrestrained frame (Figure 9.3(c)). We see that a vertical force displaced to the *left* causes the frame to sway to the *right*. Note also that, for horizontal equilibrium, we must have $H_A = H_D$, which in turn means that $M_B = M_C$.

The displacements that would have to be included in a slope–deflection analysis are the rotations of joints A, B, C and D, and the horizontal displacement of BC.

9.1.2 Sign convention and notation

For slope–deflection analysis we shall use the same sign convention and notation as for moment distribution. Thus moments, rotations and relative displacements will all be considered positive if they act in an anti-clockwise sense. With this convention, equilibrium requires that the moments acting at a point sum to zero. Moments are denoted by M_{AB} for the moment at end A of member AB, M_{BA} for the moment at end B of member AB, and so on. Refer to section 8.1.1 for a fuller explanation.

9.1.3 The slope–deflection equations

Having established which unknown displacements we wish to find, the next step is to relate them to the bending moments within the members. Consider a typical plane frame element AB (Figure 9.4(a)). The element sustains applied loads along its length, and bending moments (M_{AB} and M_{BA}) and shear forces are generated at the member ends. The deformation of the member can be defined by the rotations at the two ends (θ_A and θ_B) and the relative deflection between them (δ). This configuration can be analysed as the superposition of two separate loadcases:

- First apply the external loads with the joints rigidly clamped, so that no rotation or displacement is allowed (Figure 9.4(b)). This gives rise to *fixed-end moments* M^F_{AB} and M^F_{BA}.
- Now remove the clamps and the external loads and impose the rotations θ_A and θ_B and the relative deflection δ (Figure 9.4(c)). This requires a set of displacement-related shear forces (S^D_{AB} and S^D_{BA}) and bending moments (M^D_{AB} and M^D_{BA}) to be set up at the member ends.

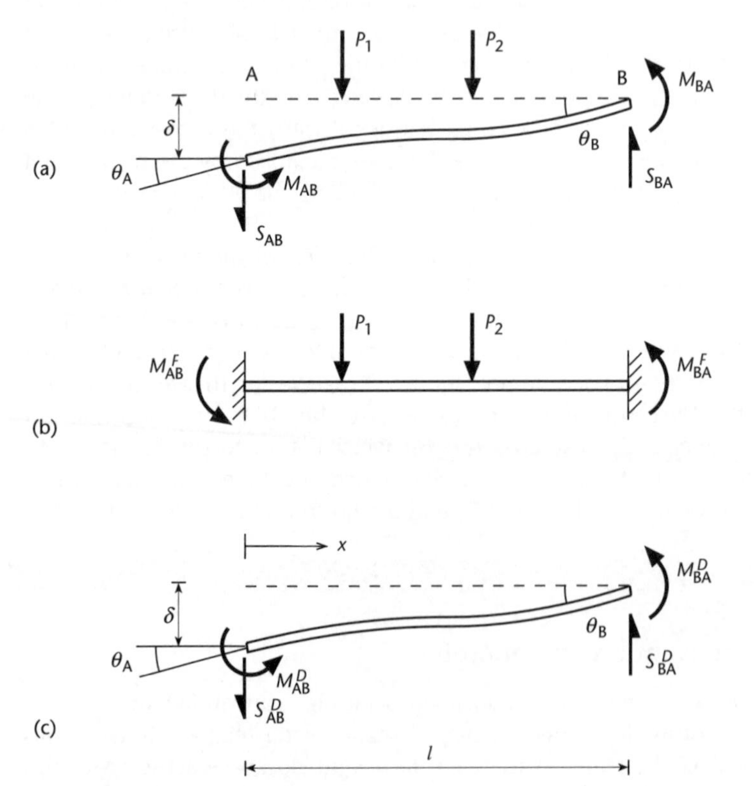

Figure 9.4
Fixed-end and displacement-related moments in a plane frame element.

The total moments in the frame element in Figure 9.4(a) are then simply the sums of the fixed-end and displacement-related terms, that is:

$$M_{AB} = M^F_{AB} + M^D_{AB}, \quad M_{BA} = M^F_{BA} + M^D_{BA} \tag{9.1}$$

The procedures for calculating the fixed-end moments and formulae for common loadcases have already been presented in section 8.1.2. We shall now show how the displacement-related moments can be found using the *slope–deflection equations*.

Returning to Figure 9.4(c), the moment at a distance x from A can be found in the usual way and related to the curvature:

$$M = -M^D_{AB} - S^D_{AB}x = EI\frac{d^2v}{dx^2}$$

The boundary conditions are:

$$\left.\frac{dv}{dx}\right|_{x=0} = \theta_A, \quad v_{x=0} = 0, \quad \left.\frac{dv}{dx}\right|_{x=l} = \theta_B, \quad v_{x=l} = \delta$$

Integrating twice and using the boundary conditions at $x = 0$ gives:

$$EI\frac{dv}{dx} = -M^D_{AB}x - S^D_{AB}\frac{x^2}{2} + EI\theta_A$$

$$EIv = -M^D_{AB}\frac{x^2}{2} - S^D_{AB}\frac{x^3}{6} + EI\theta_A x$$

Substituting in the boundary conditions at $x = l$, these become:

$$EI\theta_B = -M^D_{AB}l - S^D_{AB}\frac{l^2}{2} + EI\theta_A$$

$$EI\delta = -M^D_{AB}\frac{l^2}{2} - S^D_{AB}\frac{l^3}{6} + EI\theta_A l$$

Finally, eliminating S^D_{AB} from these two equations gives:

$$M^D_{AB} = \frac{2EI}{l}\left(2\theta_A + \theta_B - \frac{3\delta}{l}\right) \tag{9.2}$$

And by a similar process we can show that:

$$M^D_{BA} = \frac{2EI}{l}\left(2\theta_B + \theta_A - \frac{3\delta}{l}\right) \tag{9.3}$$

Equations (9.2) and (9.3) are the slope–deflection equations. They use stiffness coefficients to relate the moments at the ends of a member to the corresponding rotations and to the relative deflection between the member ends.

9.1.4 Application to indeterminate structures

To solve a problem by the slope–deflection method, we first write down expressions for the moments at the two ends of each member in the frame – each moment will be the sum of a fixed-end moment and a slope–deflection term. We then impose the equilibrium conditions; the sum of the moments at a joint must be zero, and the moment at any pinned end must be zero. This leads to a set of simultaneous equations for the rotations and deflections, which can be solved.

EXAMPLE 9.2

Find the moments at A and B for the frame in Figure 9.5. Both members have flexural stiffness $EI = 420 \times 10^3$ kNm^2.

Figure 9.5

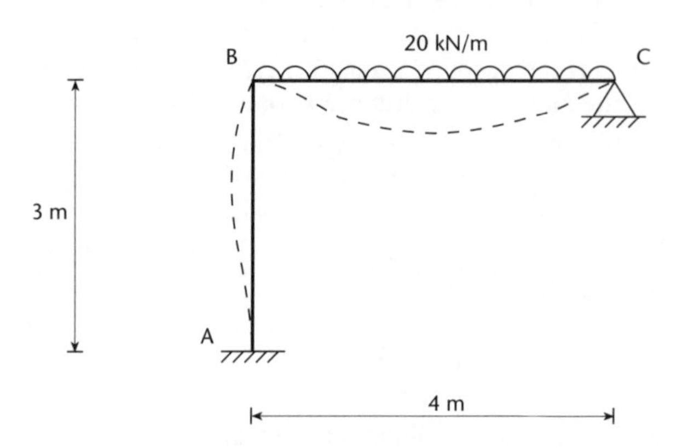

The deformed shape is shown dashed on the figure. In reality B would deflect slightly due to changes in length of the members, but these are treated as negligible in a slope–deflection analysis. We therefore have only two displacement unknowns: the rotations θ_B and θ_C.

First calculate the fixed-end moments. AB carries no external loads and therefore sustains no fixed-end moments. For BC, equation (8.5) gives:

$$M^F_{BC} = \frac{20 \times 4^2}{12} = 26.67 \text{ kNm}, \quad M^F_{CB} = -26.67 \text{ kNm}$$

Now apply the slope–deflection equations, (9.2) and (9.3), to AB and BC. For AB we have $\theta_A = 0$, and for both members the relative deflection between the ends is $\delta = 0$, so we get:

$$M^D_{AB} = \frac{2 \times 420 \times 10^3}{3}(\theta_B) = 280 \times 10^3 \theta_B$$
$$M^D_{BA} = 280 \times 10^3 (2\theta_B) = 560 \times 10^3 \theta_B$$
$$M^D_{BC} = \frac{2 \times 420 \times 10^3}{4}(2\theta_B + \theta_C) = 210 \times 10^3 (2\theta_B + \theta_C)$$
$$M^D_{CB} = 210 \times 10^3 (2\theta_C + \theta_B)$$

The total moments are found by summing the fixed-end and slope–deflection terms. We now apply the equilibrium conditions. At B, the moments must sum to zero:

$$M_{BA} + M_{BC} = 0 \quad \rightarrow 560 \times 10^3 \theta_B + 210 \times 10^3 (2\theta_B + \theta_C) + 26.67 = 0$$
$$\rightarrow 980\theta_B + 210\theta_C = -0.02667$$

Also, the moment in the pinned support at C must be zero:

$$M_{CB} = 0 \quad \rightarrow 210 \times 10^3 (2\theta_C + \theta_B) - 26.67 = 0$$
$$\rightarrow 210\theta_B + 420\theta_C = 0.02667$$

Solving these simultaneous equations: $\theta_B = -45.7 \times 10^{-6}$ rads, $\theta_C = 86.4 \times 10^{-6}$ rads. The signs indicate a clockwise rotation at B and anti-clockwise at C, which agrees with our diagram. We find the bending moments by substituting these values back into the moment expressions derived earlier:

$$M_{AB} = 280 \times 10^3 \times -45.7 \times 10^{-6} = -12.8 \text{ kNm}$$

$$M_{BA} = 560 \times 10^3 \times -45.7 \times 10^{-6} = -25.6 \text{ kNm}$$

$$M_{AB} = 210 \times 10^3 \times (2 \times -45.7 \times +86.4) \times 10^{-6} + 26.67 = 25.6 \text{ kNm}$$

$$M_{CB} = 210 \times 10^3 \times (2 \times 86.4 - 45.7) \times 10^{-6} - 26.67 = 0$$

Sometimes there will not be enough moment equilibrium equations to allow the joint displacements to be determined. In such cases it is often possible to generate an extra equation by consideration of the horizontal shear in the structure. Figure 9.6 shows a portal frame under a horizontal load. If we look at the end forces and moments on the vertical members, then taking moments about B for AB and about C for CD we get:

$$H_A h = M_{AB} + M_{BA}, \qquad H_D h = M_{CD} + M_{DC}$$

But for equilibrium of the whole frame, the horizontal reactions must equal the applied horizontal load, so:

$$P = H_A + H_D = \frac{(M_{AB} + M_{BA})}{h} + \frac{(M_{CD} + M_{DC})}{h} \tag{9.4}$$

Equation (9.4) is known as a *shear equation*. It is independent of the moment equilibrium equations at the joints, and so provides an extra piece of information that can be used to solve the problem.

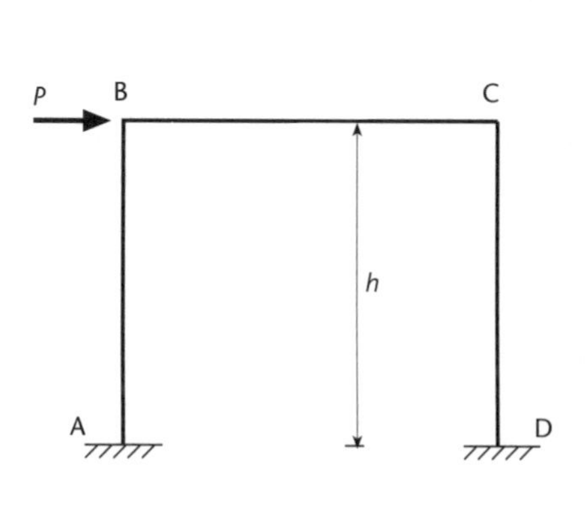

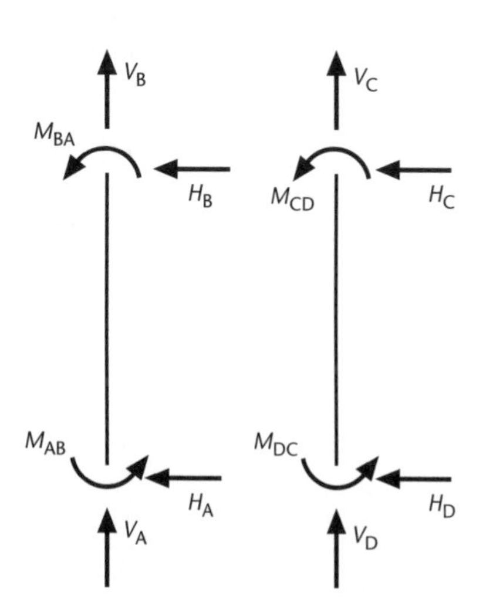

Figure 9.6
Derivation of the shear equation.

EXAMPLE 9.3

Draw the bending moment diagram for the frame in Figure 9.7(a). All members have the same properties.

Figure 9.7

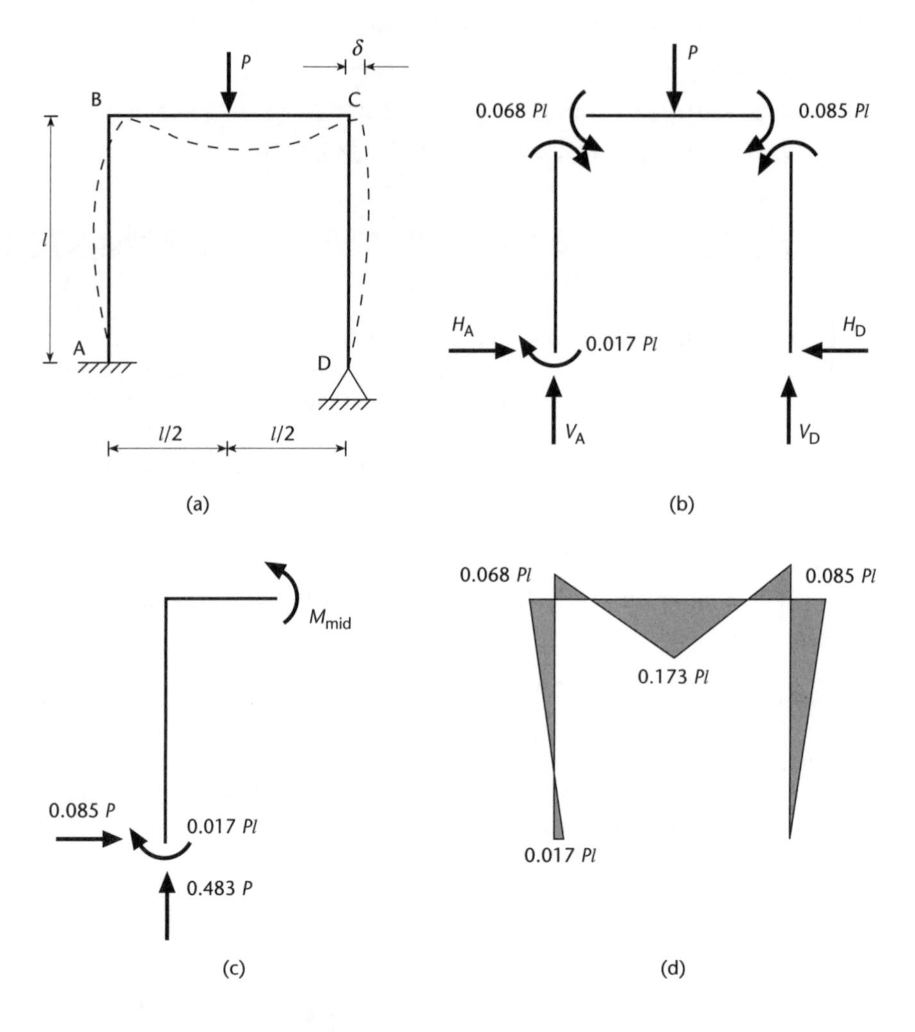

The deflected shape is shown dashed. Since the supports are unsymmetrical, there is likely to be some sway, and there are therefore four unknown displacements: θ_B, θ_C, θ_D and the horizontal sway δ. To solve, we first write out the moment equations, comprising a slope–deflection term and, in member BC, a fixed-end moment:

$$M_{AB} = \frac{2EI}{l}\left(\theta_B - \frac{3\delta}{l}\right)$$

$$M_{BA} = \frac{2EI}{l}\left(2\theta_B - \frac{3\delta}{l}\right)$$

$$M_{BC} = \frac{2EI}{l}(2\theta_B + \theta_C) + \frac{Pl}{8}$$

$$M_{CB} = \frac{2EI}{l}(2\theta_C + \theta_B) - \frac{Pl}{8}$$

$$M_{CD} = \frac{2EI}{l}\left(2\theta_C + \theta_D - \frac{3\delta}{l}\right)$$

$$M_{DC} = \frac{2EI}{l}\left(2\theta_D + \theta_C - \frac{3\delta}{l}\right)$$

The equations of moment equilibrium at B, C and D are:

$$M_{BA} + M_{BC} = 0 \rightarrow \frac{2EI}{l}\left(4\theta_B + \theta_C - \frac{3\delta}{l}\right) + \frac{Pl}{8} = 0$$

$$M_{CB} + M_{CD} = 0 \rightarrow \frac{2EI}{l}\left(\theta_B + 4\theta_C + \theta_D - \frac{3\delta}{l}\right) - \frac{Pl}{8} = 0$$

$$M_{DC} = 0 \rightarrow \frac{2EI}{l}\left(2\theta_D + \theta_C - \frac{3\delta}{l}\right) = 0$$

We can also formulate a shear equation; since there are no horizontal loads on the structure the horizontal reactions must sum to zero, so:

$$\frac{M_{AB} + M_{BA}}{l} + \frac{M_{CD} + M_{DC}}{l} = 0 \rightarrow \frac{2EI}{l}\left(3\theta_B + 3\theta_C + 3\theta_D - \frac{12\delta}{l}\right) = 0$$

We now have four equations for the four displacement unknowns, which can be solved to give:

$$\theta_B = -0.0256\frac{Pl^2}{EI}, \quad \theta_C = 0.0227\frac{Pl^2}{EI}, \quad \theta_D = -0.0199\frac{Pl^2}{EI}, \quad \delta = -0.00568\frac{Pl^3}{EI}$$

We can check the signs of these terms by comparing with our sketched deflected shape: as expected, θ_B and θ_D are clockwise, while θ_C is anti-clockwise. The negative value of δ implies a clockwise movement of A about B and of C about D, again in agreement with our diagram. We can now substitute back into our moment expressions:

$$M_{AB} = -0.017Pl, \qquad M_{BA} = -M_{BC} = -0.068Pl,$$

$$M_{CB} = -M_{CD} = -0.085Pl, \qquad M_{DC} = 0$$

The slope–deflection analysis is now complete, but a little more work is needed before we can plot the bending moment diagram. Figure 9.7(b) shows the structure split at the joints, with the moments drawn on in the appropriate directions. The support reactions are also shown. Taking moments about B for AB gives $H_A = 0.085P$, and taking moments about D for the whole frame gives $V_A = 0.483P$. To find the midspan moment in BC we then consider equilibrium of the section shown in Figure 9.7(c) giving $M_{mid} = 0.173Pl$. The bending moment diagram can now be plotted (Figure 9.7(d)), with the lines drawn on the tension sides of the members.

9.2 Stiffness matrices for bars and trusses

In the examples of section 9.1, we obtained a set of equilibrium equations involving the joint displacements and the fixed-end moments, which are functions of the external loading. Separating out the load and displacement terms, we could write the simultaneous equations as a matrix equation of the form:

{External loads} = [Stiffness coefficients] × {Joint displacements}

where the curly brackets denote a column vector (that is, an $n \times 1$ matrix, where n is the number of degrees of freedom) and the square brackets denote a square ($n \times n$) matrix. Solution then requires the inversion of the *stiffness matrix* to give the unknown displacements, which can be done very efficiently by computer. This approach is known as the stiffness matrix method, and can be summarised as follows:

(a) Formulate the stiffness matrix for each member, relating forces and displacements at the member ends. This is often known as the *local* stiffness matrix.

(b) Combine the local matrices using conditions of equilibrium and compatibility at the joints to give a *global* stiffness matrix for the entire structure.

(c) Invert the stiffness matrix and hence determine the unknown displacements.

(d) Substitute back into the *local* matrix equations (a) to find the member forces.

The approach has obvious similarities to the slope–deflection method, but there are also some important differences:

- To make it suitable for computer implementation, we take a very systematic approach to setting up the matrix equations, eliminating the use of the irritatingly non-systematic shear equation, for example.
- We no longer treat axial deformations as negligible. For a problem dominated by bending, this will result in an increase in the number of equations to be solved, while giving little improvement in accuracy. However, it has the advantage of making the same automated approach applicable to a very wide variety of problems.

To introduce the stiffness matrix approach, we will first look at the very simple case of bars carrying only axial loads, before moving on to the more general case of members carrying both axial and bending loads.

9.2.1 Sign convention and notation for computer applications

Figure 9.8(a) shows the global axis system for a plane frame. As previously, a right-handed set is used, with x horizontal and y vertical, positive upwards. A positive rotation acts from the positive x-axis towards the positive y-axis, that is, *anti-clockwise*.

In addition to the global system, we often need to specify local axes for the members within a frame. We define the local x-axis to be the longitudinal axis of the member, regardless of its orientation. As with the moment distribution and slope–deflection approaches, an anti-clockwise moment or rotation is taken as positive, regardless of the face on which it acts. Thus all the quantities shown in Figure 9.8(b) are positive.

The notation we will use in the remainder of this chapter is as follows. External forces are denoted by F, moments by M; the corresponding deflections are denoted by u and rotations by θ. The subscripts x, y and z denote local axis directions, X, Y and Z global axis directions. The subscripts 1 and 2 refer to the two ends of a member, the positive local x-axis running from end 1 towards end 2. The vector of

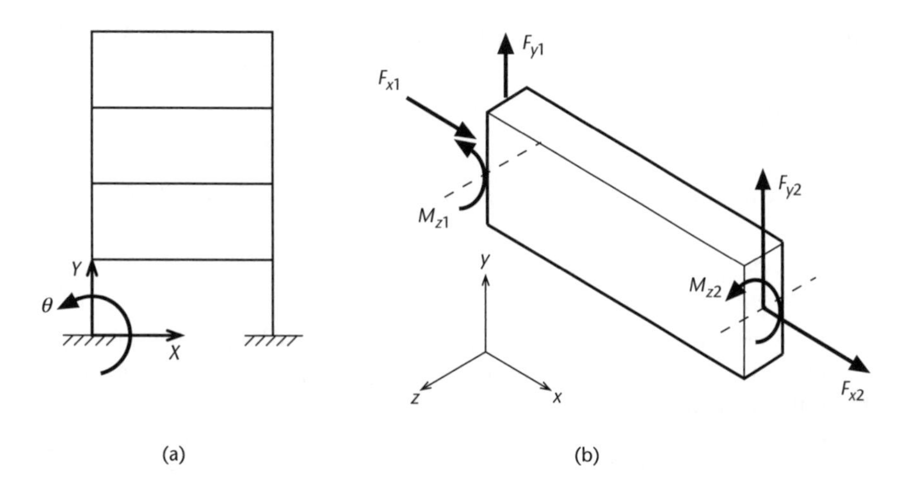

Figure 9.8
Axis systems for computer applications.

forces F and displacements M is written as **f** in local coordinates or **F** in global coordinates. Similarly the vector of deflections and rotations is **u** in local or **U** in global coordinates, and the stiffness matrix relating the two is **k** (local) or **K** (global).

A full explanation of the sign conventions used in this book is given in Appendix A.

9.2.2 Local stiffness matrix for an axially loaded element

The simplest element for which a stiffness matrix can be derived is a bar subjected to external loads along its axis, and undergoing corresponding axial displacements (Figure 9.9(a)). Each possible displacement in a problem is referred to as a *degree of freedom*, and the dimension of the stiffness matrix must equal the number of degrees of freedom. In this case, only two displacements are permitted and we therefore wish to find a 2 × 2 matrix.

The simplest way to find a stiffness matrix is to calculate the forces generated by each displacement in turn, with all other displacements set to zero. The resulting stiffness equations can then be combined using superposition. First, consider the

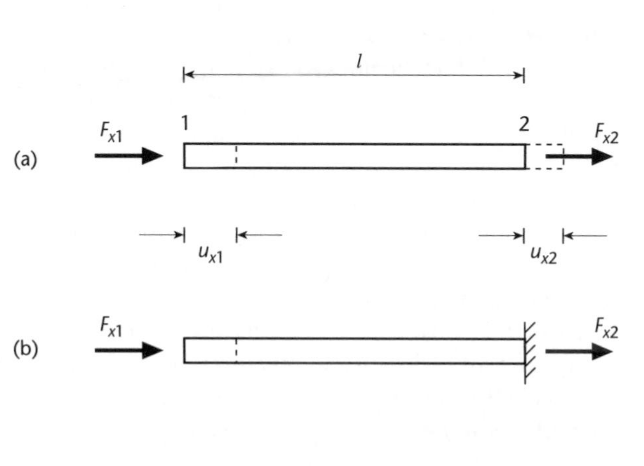

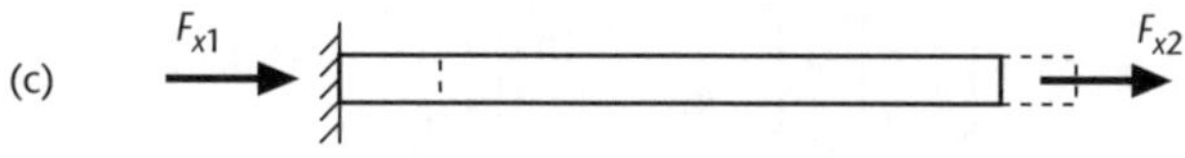

Figure 9.9
Forces and displacements in an axially loaded bar.

effect of applying a displacement u_{x1} with u_{x2} set to zero (Figure 9.9(b)). Applying the usual axial force–extension relationships, the end forces are:

$$F_{x1} = \frac{AE}{l}u_{x1}, \qquad F_{x2} = -\frac{AE}{l}u_{x1}$$

Next apply a displacement u_{x2} with u_{x1} set to zero (Figure 9.9(c)). The end forces are:

$$F_{x1} = -\frac{AE}{l}u_{x2}, \qquad F_{x2} = \frac{AE}{l}u_{x2}$$

Using the principle of superposition, the forces acting when both displacements occur simultaneously can be found by simply summing these force terms. Writing the result in matrix form:

$$\begin{bmatrix} F_{x1} \\ F_{x2} \end{bmatrix} = \begin{bmatrix} k & -k \\ -k & k \end{bmatrix} \begin{bmatrix} u_{x1} \\ u_{x2} \end{bmatrix} \quad \text{or} \quad \mathbf{f} = \mathbf{ku} \tag{9.5}$$

where $k = AE/l$ and $\mathbf{k}$ is the stiffness matrix for the bar, relating the end forces and displacements. For example, if u_{x1} and u_{x2} are equal then the end forces are both zero; if a displacement of +1 unit is applied at end 1 and −1 unit at end 2 then $F_{x1} = 2k$ and $F_{x2} = -2k$.

Consider the meaning of the terms in the stiffness matrix. Each leading diagonal term represents the force at a point and in a certain direction corresponding to a unit displacement at the same point and in the same direction. The off-diagonal stiffnesses are cross-coupling terms, representing the force at one point due to a unit deflection at the other. We can show quite easily using the reciprocal theorem (see Chapter 7) that for a linear elastic structure these *must* be equal. That is, the force at end 1 due a deflection at end 2 is equal to the force at end 2 due to the same deflection at end 1. The stiffness matrix is therefore always symmetric, and this has important implications for the way that it is stored in a computer – we can roughly halve the required storage by saving only the upper or lower triangle.

9.2.3 Global stiffness matrix for a simple two-bar structure

We would not, of course, bother with a matrix approach for a single-bar structure – the method is useful for analysing frameworks of bars or beams. In such cases, it would be extremely tedious to derive the stiffness matrix from first principles each time. Instead, we can write down the local stiffness matrix for each element straight away, using the standard result of equation (9.5), then derive the global matrix by applying the conditions of force equilibrium and displacement compatibility at the joints.

As an illustration of the approach, consider the two-bar structure shown in Figure 9.10. This is an exceptionally simple case since both bars have the same orientation and carry only axial loads, so that the problem is one-dimensional, and the local and global axis systems have the same orientation.

There are three degrees of freedom (the axial displacements at A, B and C) and we therefore wish to derive a 3 × 3 stiffness matrix. We can immediately write down the local matrix equations, from equation (9.5):

$$\text{For AB:} \begin{bmatrix} F_{x1} \\ F_{x2} \end{bmatrix} = \begin{bmatrix} k_1 & -k_1 \\ -k_1 & k_1 \end{bmatrix} \begin{bmatrix} u_{x1} \\ u_{x2} \end{bmatrix}, \quad \text{and for BC:} \begin{bmatrix} F_{x1} \\ F_{x2} \end{bmatrix} = \begin{bmatrix} k_2 & -k_2 \\ -k_2 & k_2 \end{bmatrix} \begin{bmatrix} u_{x1} \\ u_{x2} \end{bmatrix}$$

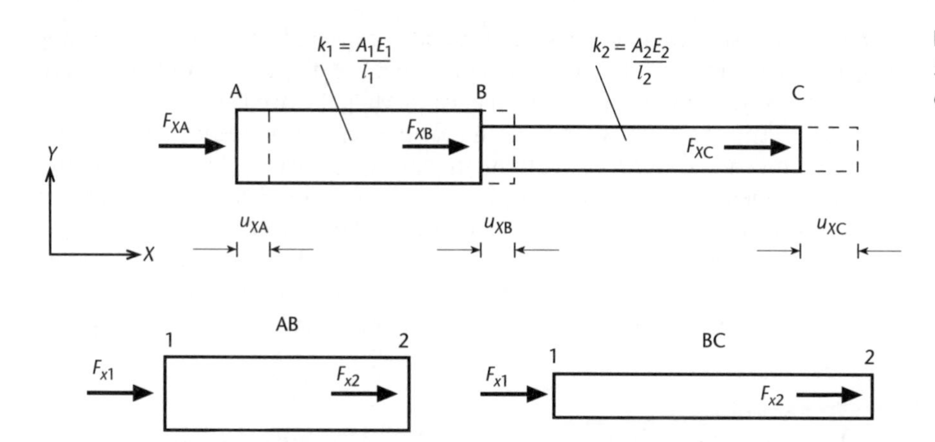

Figure 9.10
Structure comprising two concentric bars.

where $k_1 = A_1E_1/l_1$ and $k_2 = A_2E_2/l_2$. Now for displacement compatibility, the deflections of the joints A, B and C must correspond to the end displacements of the two members, that is:

$$u_{XA} = u_{x1}^{AB}$$
$$u_{XB} = u_{x2}^{AB} = u_{x1}^{BC}$$
$$u_{XC} = u_{x2}^{BC}$$

We can use these conditions to write our local matrix equations in terms of the global displacement variables u_{XA}, u_{XB} and u_{XC}:

$$\text{AB:}\begin{bmatrix} F_{x1} \\ F_{x2} \end{bmatrix} = \begin{bmatrix} k_1 & -k_1 \\ -k_1 & k_1 \end{bmatrix}\begin{bmatrix} u_{XA} \\ u_{XB} \end{bmatrix}, \qquad \text{BC:}\begin{bmatrix} F_{x1} \\ F_{x2} \end{bmatrix} = \begin{bmatrix} k_2 & -k_2 \\ -k_2 & k_2 \end{bmatrix}\begin{bmatrix} u_{XB} \\ u_{XC} \end{bmatrix}$$

Now consider equilibrium of the forces at the joints:

$$\begin{bmatrix} F_{XA} \\ F_{XB} \\ F_{XC} \end{bmatrix} = \begin{bmatrix} F_{x1} \\ F_{x2} \\ 0 \end{bmatrix}_{AB} + \begin{bmatrix} 0 \\ F_{x1} \\ F_{x2} \end{bmatrix}_{BC}$$

We can now obtain the global stiffness matrix equation by substituting the local matrix equations into these equilibrium equations. In doing this it is convenient first to inflate the local equations to 3 × 3 form by adding zeros in the appropriate positions, as shown:

$$\begin{bmatrix} F_{XA} \\ F_{XB} \\ F_{XC} \end{bmatrix} = \begin{bmatrix} k_1 & -k_1 & 0 \\ -k_1 & k_1 & 0 \\ 0 & 0 & 0 \end{bmatrix}\begin{bmatrix} u_{XA} \\ u_{XB} \\ u_{XC} \end{bmatrix} + \begin{bmatrix} 0 & 0 & 0 \\ 0 & k_2 & -k_2 \\ 0 & -k_2 & k_2 \end{bmatrix}\begin{bmatrix} u_{XA} \\ u_{XB} \\ u_{XC} \end{bmatrix}$$
$$= \begin{bmatrix} k_1 & -k_1 & 0 \\ -k_1 & k_1 + k_2 & -k_2 \\ 0 & -k_2 & k_2 \end{bmatrix}\begin{bmatrix} u_{XA} \\ u_{XB} \\ u_{XC} \end{bmatrix} \tag{9.6}$$

Thus, once we have managed to get all our local stiffness matrix equations expressed in terms of the same set of displacement variables, we simply add the corresponding terms to give the global stiffness matrix.

In our 3 × 3 stiffness matrix, some zeros have appeared away from the leading diagonal. This is because the movement of A does not directly affect the force at C (and vice-versa), as there is not a member connecting the two joints. In general, an off-diagonal stiffness term will be zero unless the degrees of freedom corresponding to the row and column of the term are directly connected. In most structures only a few members meet at each joint. This means that stiffness matrices for structures with many degrees of freedom are usually *banded* in form, that is, all their non-zero terms are contained in quite a narrow band around the leading diagonal. This is important to the programmer, since it is possible to come up with very efficient numerical solutions of banded matrix equations.

9.2.4 Boundary conditions

Once we have set up a stiffness matrix equation, we must invert the matrix in order to solve for the unknown displacements. If, however, we attempt to invert any of the stiffness matrices derived so far, we will find that they are singular. This is because they represent structures that are free from any restraints, as though they were floating in space. In reality, all structures are subject to some boundary conditions, and these usually take the form (at least for the purposes of an idealised analysis) of setting one or more displacements to zero. We can use the zero-displacement boundary conditions both to eliminate the singularity and to reduce the overall problem size.

Consider the two-bar structure from the previous section, whose stiffness matrix is given in Equation (9.6). Suppose we fix joint A and apply external loads to joints B and C (Figure 9.11). At B and C we have known forces and unknown displacements, which we wish to find, while at A we have a known (zero) displacement and an unknown reaction force R. (For this simple structure we could, of course, find R by resolution, but in more complex structures this may be difficult or impossible.)

Equation (9.6) therefore becomes:

$$\begin{bmatrix} R \\ F_{XB} \\ F_{XC} \end{bmatrix} = \begin{bmatrix} k_1 & -k_1 & 0 \\ -k_1 & k_1 + k_2 & -k_2 \\ 0 & -k_2 & k_2 \end{bmatrix} \begin{bmatrix} 0 \\ u_{XB} \\ u_{XC} \end{bmatrix}$$

Terms in the first column of the stiffness matrix are always multiplied by zero, and so serve no purpose. The remaining terms in the first row relate unknown displacements (u_{xB}, u_{xC}) to an unknown force R. These are therefore also of no use. So we can simply remove the first row and column of the matrix leaving us with a non-singular, 2 × 2 problem:

$$\begin{bmatrix} F_{XB} \\ F_{XC} \end{bmatrix} = \begin{bmatrix} k_1 + k_2 & -k_2 \\ -k_2 & k_2 \end{bmatrix} \begin{bmatrix} u_{XB} \\ u_{XC} \end{bmatrix}$$

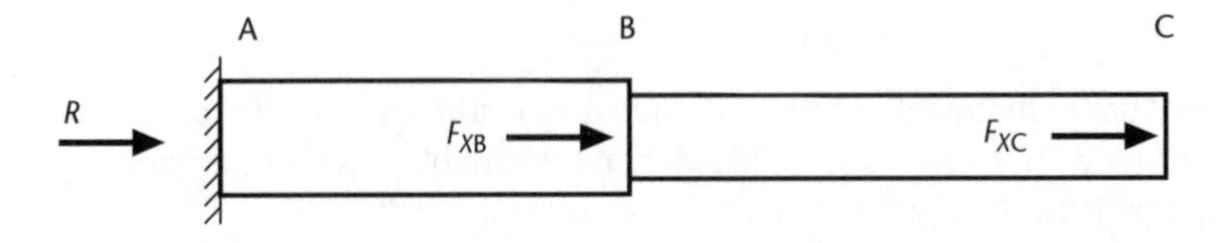

Figure 9.11
Imposition of zero-displacement boundary condition.

which can be inverted to give expressions for the unknown displacements:

$$\begin{bmatrix} u_{XB} \\ u_{XC} \end{bmatrix} = \frac{1}{k_1 k_2} \begin{bmatrix} k_2 & k_2 \\ k_2 & k_1 + k_2 \end{bmatrix} \begin{bmatrix} F_{XB} \\ F_{XC} \end{bmatrix}$$

In general, a stiffness matrix equation can be modified by deleting all the rows and columns corresponding to zero displacements without any useful information being lost.

9.2.5 Global stiffness matrices for plane trusses

Lastly in this section, we shall assemble the global stiffness matrix for a slightly more realistic structure, a plane truss, involving members with varying orientations and support conditions. The member orientations make this slightly more complicated than previously. In generating the element stiffness matrices, it is convenient to use the standard result given in equation (9.5), in which the x-axis runs along the member, but before combining the stiffness matrices we must ensure they are all in the same axis system. This can be done using a transformation matrix **T**.

Consider the element shown in Figure 9.12; the angle from the global X to the local x-axis is ϕ (positive anti-clockwise). The bar sustains only x-direction forces and deformations in the local axis system (a), but when these are resolved into the global axis system (b) there are components in both the X and Y directions. By resolving force components along the bar, we can express the local forces in terms of the global:

$$\begin{bmatrix} F_{x1} \\ F_{x2} \end{bmatrix} = \begin{bmatrix} \cos\phi & \sin\phi & 0 & 0 \\ 0 & 0 & \cos\phi & \sin\phi \end{bmatrix} \begin{bmatrix} F_{X1} \\ F_{Y1} \\ F_{X2} \\ F_{Y2} \end{bmatrix} \quad \text{or} \quad \mathbf{f} = \mathbf{TF} \tag{9.7}$$

where **T** is the *transformation matrix*. Similarly, resolving displacement components gives the corresponding relationship $\mathbf{u} = \mathbf{TU}$ where **T** is the same transformation

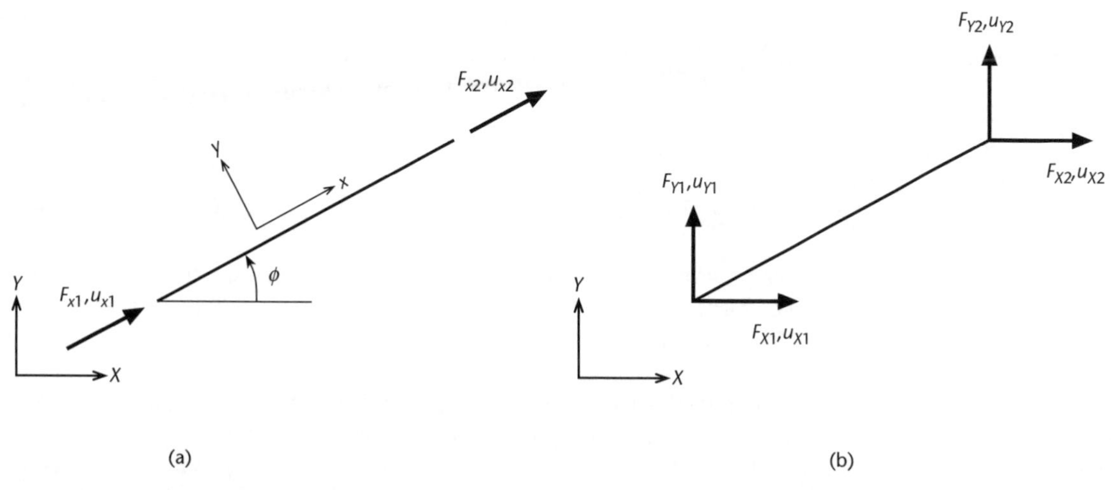

Figure 9.12
Axis transformation for an axially loaded bar element.

matrix as in equation (9.7). From equation (9.5) our stiffness matrix equation in local coordinates is

$$\mathbf{f} = \mathbf{ku}$$

Writing this in terms of the global forces and displacements:

$$(\mathbf{TF}) = \mathbf{k}(\mathbf{TU})$$

Now the transformation matrix $\mathbf{T}$ has the special property that the product of $\mathbf{T}$ with its transpose is a unit matrix, that is $\mathbf{T}^{\mathrm{T}}\mathbf{T} = \mathbf{I}$. So if we pre-multiply the stiffness equation by $\mathbf{T}^{\mathrm{T}}$ we get:

$$\mathbf{F} = (\mathbf{T}^{\mathrm{T}}\mathbf{kT})\mathbf{U}$$

The term in brackets relates the end forces and displacements in *global* coordinates and is therefore the stiffness matrix of the element in global coordinates, that is:

$$\mathbf{K} = \mathbf{T}^{\mathrm{T}}\mathbf{kT} \tag{9.8}$$

So to convert elements with varying orientations into a common axis system, we first determine the transformation matrix between the local and global systems and then apply equation (9.8). Applying the transformation matrix of equation (9.7) to the local stiffness matrix for a bar element (equation (9.5)) gives:

$$\mathbf{K} = k\left[\begin{array}{cc|cc} c^2 & sc & -c^2 & -sc \\ sc & s^2 & -sc & -s^2 \\ \hline -c^2 & -sc & c^2 & sc \\ -sc & -s^2 & sc & s^2 \end{array}\right] = \left[\begin{array}{c|c} \mathbf{K}_{11} & \mathbf{K}_{12} \\ \hline \mathbf{K}_{21} & \mathbf{K}_{22} \end{array}\right] \tag{9.9}$$

where $s = \sin\phi$ and $c = \cos\phi$. Equation (9.9) is the stiffness matrix for a bar element in global coordinates, with the four rows and columns corresponding to the degrees of freedom u_{X1}, u_{Y1}, u_{X2} and u_{Y2} respectively. The way the matrix is set out deserves some comment. It has been divided by *partition lines* into four 2×2 sub-matrices; $\mathbf{K}_{11}$ relates the forces at end 1 to the displacements at end 1, $\mathbf{K}_{12}$ relates the forces at end 1 to the displacements at end 2, and so on. This is a widely used shorthand in stiffness matrix equations. Because stiffness matrices are always symmetric, it follows that $\mathbf{K}_{11}$ and $\mathbf{K}_{22}$ must themselves be symmetric, and that $\mathbf{K}_{21} = \mathbf{K}_{12}^{\mathrm{T}}$.

EXAMPLE 9.4

Determine the stiffness matrix equation for the pin-jointed structure in Figure 9.13. For the diagonals $AE = 8\sqrt{2} \times 10^3$ kN and for the other members $AE = 40 \times 10^3$ kN.

First, we identify the active degrees of freedom in the problem. In this case there are five: u_{XB}, u_{YB}, u_{XC}, u_{YC} and u_{YD}. We must therefore derive a 5×5 stiffness matrix.

Next, the stiffness matrices for the individual bars, in global coordinates, are determined. We could do this by writing down the matrices in local coordinates and then applying a transformation, but it is quicker to work directly from equation

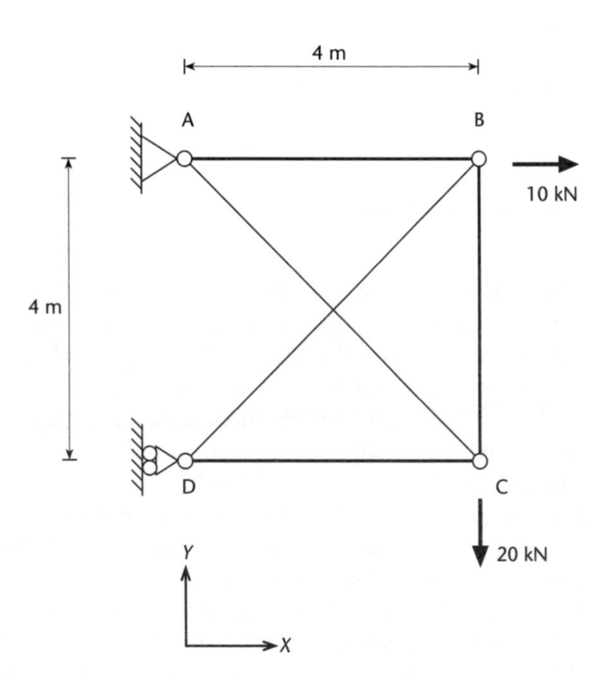

Figure 9.13

(9.9). Starting with member AB. The angle ϕ is measured *anti-clockwise from* the global X-axis *to* the member axis so, for AB, $\phi = 0$, giving $s = 0$, $c = 1$. We can therefore write out the stiffness matrix – the corresponding degree of freedom is entered next to each row so that there is no ambiguity about the ordering of the elements:

$$\mathbf{K}_{AB} = k_1 \left[\begin{array}{cc|cc} 1 & 0 & -1 & 0 \\ 0 & 0 & 0 & 0 \\ \hline -1 & 0 & 1 & 0 \\ 0 & 0 & 0 & 0 \end{array}\right] \begin{array}{l} u_{XA} \\ u_{YA} \\ u_{XB} \\ u_{YB} \end{array}$$

where $k_1 = AE/l = 40 \times 10^3/4 = 10 \times 10^3$ kN/m. But since $u_{XA} = u_{YA} = 0$, we can delete the first two rows and columns, leaving:

$$\mathbf{K}_{AB} = k_1 \begin{bmatrix} 1 & 0 \\ 0 & 0 \end{bmatrix} \begin{array}{l} u_{XB} \\ u_{YB} \end{array}$$

The matrices for the other members are found in the same way:

$$\text{AC} : \phi = 315^\circ, \quad u_{XA} = u_{YA} = 0, \text{ hence: } \mathbf{K}_{AC} = k_2 \begin{bmatrix} 0.5 & -0.5 \\ -0.5 & 0.5 \end{bmatrix} \begin{array}{l} u_{XC} \\ u_{YC} \end{array}$$

$$\text{BC} : \phi = 270^\circ, \text{ hence: } \quad \mathbf{K}_{BC} = k_1 \left[\begin{array}{cc|cc} 0 & 0 & 0 & 0 \\ 0 & 1 & 0 & -1 \\ \hline 0 & 0 & 0 & 0 \\ 0 & -1 & 0 & 1 \end{array}\right] \begin{array}{l} u_{XB} \\ u_{YB} \\ u_{XC} \\ u_{YC} \end{array}$$

BD : $\phi = 225°, u_{XD} = 0$, hence: $$\mathbf{K}_{BD} = k_2 \begin{bmatrix} 0.5 & 0.5 & -0.5 \\ 0.5 & 0.5 & -0.5 \\ -0.5 & -0.5 & 0.5 \end{bmatrix} \begin{matrix} u_{XB} \\ u_{YB} \\ u_{YD} \end{matrix}$$

CD : $\phi = 180°, u_{XD} = 0$, hence: $$\mathbf{K}_{CD} = k_1 \begin{bmatrix} 1 & 0 & 0 \\ 0 & 0 & 0 \\ 0 & 0 & 0 \end{bmatrix} \begin{matrix} u_{XC} \\ u_{YC} \\ u_{YD} \end{matrix}$$

where $k_2 = 8\sqrt{2} \times 10^3/4\sqrt{2} = 2 \times 10^3$ kN/m. The structure stiffness matrix equation can now be assembled. We first write out the displacement vector and then enter the external loads in the corresponding positions in the load vector – note the negative sign for a downwards load.

We next take each member stiffness matrix in turn and transfer its elements into the appropriate locations in the structure stiffness matrix – if a stiffness coefficient relates the ith term in the load vector to the jth term in the displacement vector, then it is entered in row i and column j of the matrix. For example, the top left element of $\mathbf{K}_{AB}$ relates the X-direction force at B to the X-direction displacement at B; it is therefore entered in row 1, column 1 of the stiffness matrix. The top right element of $\mathbf{K}_{BD}$ relates the X-direction force at B to the Y-direction displacement at D; it therefore goes in row 1, column 5 of the stiffness matrix. If the location is already filled by a term from another member, then the two values are added. We therefore get:

$$\begin{bmatrix} 10 \\ 0 \\ 0 \\ -20 \\ 0 \end{bmatrix} = \begin{bmatrix} k_1 + 0.5k_2 & 0.5k_2 & 0 & 0 & -0.5k_2 \\ 0.5k_2 & k_1 + 0.5k_2 & 0 & -k_1 & -0.5k_2 \\ 0 & 0 & 0.5k_2 + k_1 & -0.5k_2 & 0 \\ 0 & -k_1 & -0.5k_2 & 0.5k_2 + k_1 & 0 \\ -0.5k_2 & -0.5k_2 & 0 & 0 & 0.5k_2 \end{bmatrix} \begin{bmatrix} u_{XB} \\ u_{YB} \\ u_{XC} \\ u_{YC} \\ u_{YD} \end{bmatrix}$$

Substituting numerical values for k_1 and k_2, this simplifies to:

$$\begin{bmatrix} 10 \\ 0 \\ 0 \\ -20 \\ 0 \end{bmatrix} = 10^3 \times \begin{bmatrix} 11 & 1 & 0 & 0 & -1 \\ 1 & 11 & 0 & -10 & -1 \\ 0 & 0 & 11 & -1 & 0 \\ 0 & -10 & -1 & 11 & 0 \\ -1 & -1 & 0 & 0 & 1 \end{bmatrix} \begin{bmatrix} u_{XB} \\ u_{YB} \\ u_{XC} \\ u_{YC} \\ u_{YD} \end{bmatrix}$$

This matrix equation can now be solved computationally – this aspect of the analysis will be discussed further in subsequent sections.

9.3 Stiffness matrices for beams and rigid-jointed frames

By far the most common application of the stiffness matrix approach is the analysis of redundant, rigid-jointed frames, in which the members carry axial, bending and shear loads. For 3D structures torsion may also be present.

The stiffness matrix equation for a rigid-jointed frame made up of beam elements can be assembled in much the same way as for a truss made of bar elements. We first generate the element matrices in local coordinates and apply a transformation to bring them into a single, global axis system. We then remove rows and columns corresponding to known zero displacements, and finally use the equilibrium and compatibility conditions to combine the member matrices into a single stiffness matrix equation for the structure.

The main difference is that there are now many more degrees of freedom. Whereas the simple bar elements of the previous section had only two degrees of freedom, a 2D beam element will have six (two perpendicular deflections and a rotation at each end) and a 3D member will have twelve (three deflections and rotations about three axes at each end). Clearly, large matrix equations will be generated even for quite simple structures.

9.3.1 Local stiffness matrices for 2D and 3D beam elements

We shall first derive the stiffness matrix for a 2D beam element, with the end deflections and rotations shown in Figure 9.14(a). The member has length l, area A, Young's modulus E and second moment of area about the local z-axis I. We make the assumption that the structural behaviour is independent of the level of axial load in the beam. We shall see in Chapter 11 that large axial loads cause buckling effects which significantly alter the behaviour, so that this assumption is not strictly true, but for most load levels it is a reasonable approximation.

The stiffness matrix is derived in the same way as for a bar, that is, we apply one displacement at a time, with all the others set to zero, then combine the results using linear superposition. If we first apply an axial deflection at end 1 (Figure 9.14(b)), the resulting forces are the same as those in the simple bar element of section 9.2.2:

$$F_{x1} = \frac{AE}{l} u_{x1}, \quad F_{x2} = -\frac{AE}{l} u_{x1}, \quad F_{y1} = F_{y2} = M_{z1} = M_{z2} = 0$$

Next apply a transverse deflection u_{y1} at end 1 (Figure 9.14(c)). To produce the deflected shape shown, both transverse forces and moments are required, but no

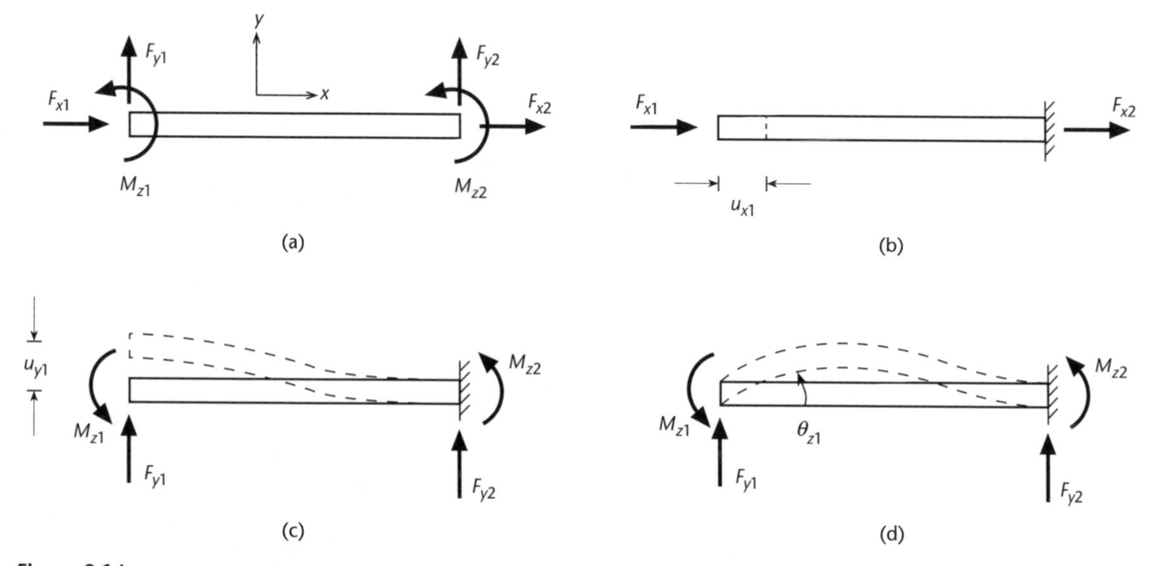

Figure 9.14
Forces and displacements in a 2D beam element.

axial forces, therefore $F_{x1} = F_{x2} = 0$. The moments can be found using the slope–deflection equations, (9.2) and (9.3):

$$M_{z1} = \frac{2EI}{l}\left(\frac{-3(-u_{y1})}{l}\right) = \frac{6EI}{l^2}u_{y1}, \quad M_{z2} = \frac{2EI}{l}\left(\frac{-3(-u_{y1})}{l}\right) = \frac{6EI}{l^2}u_{y1}$$

Note that the deflection u_{y1} is taken as negative in the slope–deflection equations because it involves a clockwise rotation between the member ends. The transverse forces can now be found by taking moments about end 2 and resolving vertically:

$$F_{y1} = \frac{M_{z1} + M_{z2}}{l} = \frac{12EI}{l^3}u_{y1}, \qquad F_{y2} = -F_{y1} = -\frac{12EI}{l^3}u_{y1}$$

Lastly we apply a rotation θ_{z1} (Figure 9.14(d)). Again, this requires transverse forces and moments at each end, but no axial forces. The slope–deflection equations give:

$$M_{z1} = \frac{2EI}{l}(2\theta_{z1}) = \frac{4EI}{l}\theta_{z1}, \quad M_{z2} = \frac{2EI}{l}\theta_{z1}$$

and then taking moments and resolving give:

$$F_{y1} = \frac{M_{z1} + M_{z2}}{l} = \frac{6EI}{l^2}\theta_{z1}, \quad F_{y2} = -F_{y1} = -\frac{6EI}{l^2}\theta_{z1}$$

A similar set of equations can be derived by applying displacements at end 2, and all the terms can then be combined to give the stiffness matrix equation:

$$\begin{bmatrix} F_{x1} \\ F_{y1} \\ M_{z1} \\ F_{x2} \\ F_{y2} \\ M_{z2} \end{bmatrix} = \left[\begin{array}{ccc|ccc} \frac{AE}{l} & 0 & 0 & -\frac{AE}{l} & 0 & 0 \\ 0 & \frac{12EI}{l^3} & \frac{6EI}{l^2} & 0 & -\frac{12EI}{l^3} & \frac{6EI}{l^2} \\ 0 & \frac{6EI}{l^2} & \frac{4EI}{l} & 0 & -\frac{6EI}{l^2} & \frac{2EI}{l} \\ \hline -\frac{AE}{l} & 0 & 0 & \frac{AE}{l} & 0 & 0 \\ 0 & -\frac{12EI}{l^3} & -\frac{6EI}{l^2} & 0 & \frac{12EI}{l^3} & -\frac{6EI}{l^2} \\ 0 & \frac{6EI}{l^2} & \frac{2EI}{l} & 0 & -\frac{6EI}{l^2} & \frac{4EI}{l} \end{array}\right] \begin{bmatrix} u_{x1} \\ u_{y1} \\ \theta_{z1} \\ u_{x2} \\ u_{y2} \\ \theta_{z2} \end{bmatrix} \qquad (9.10)$$

This is the general form of the stiffness matrix a for a 2D beam element in local coordinates, and may be used as the basis for the assembly of stiffness matrices for plane frames.

For a 3D beam element, forces and displacements can be generated in the x, y and z directions at each end, and moments applied about the x, y and z axes, as shown in Figure 9.15. M_z causes bending in the vertical plane, as in the 2D case above, M_y causes bending in the horizontal plane and M_x causes torsion. When working out the stiffness coefficients relating to these moments, we must take care to use the correct second moment of area in each case. Terms related to M_z or θ_z use the second moment of area about the z-axis, I_{zz} (this is what was simply called I above), those related to M_y or θ_y use I_{yy} and those related to M_x or θ_x use the polar second moment of area J.

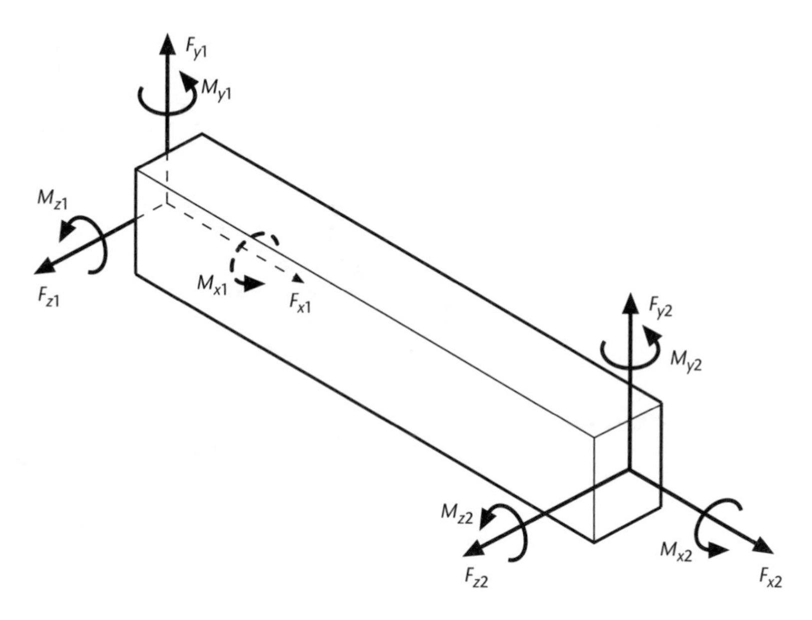

Figure 9.15
Forces in a 3D beam element.

By a similar process to that used above, we can show that the local stiffness matrix for a 3D beam element is:

$$\begin{bmatrix} F_{x1} \\ F_{y1} \\ F_{z1} \\ M_{x1} \\ M_{y1} \\ M_{z1} \\ F_{x2} \\ F_{y2} \\ F_{z2} \\ M_{x2} \\ M_{y2} \\ M_{z2} \end{bmatrix} = \left[\begin{array}{cccccc|cccccc}
\frac{AE}{l} & 0 & 0 & 0 & 0 & 0 & -\frac{AE}{l} & 0 & 0 & 0 & 0 & 0 \\
0 & \frac{12EI_{zz}}{l^3} & 0 & 0 & 0 & \frac{6EI_{zz}}{l^2} & 0 & -\frac{12EI_{zz}}{l^3} & 0 & 0 & 0 & \frac{6EI_{zz}}{l^2} \\
0 & 0 & \frac{12EI_{yy}}{l^3} & 0 & -\frac{6EI_{yy}}{l^2} & 0 & 0 & 0 & -\frac{12EI_{yy}}{l^3} & 0 & -\frac{6EI_{yy}}{l^2} & 0 \\
0 & 0 & 0 & \frac{GJ}{l} & 0 & 0 & 0 & 0 & 0 & -\frac{GJ}{l} & 0 & 0 \\
0 & 0 & -\frac{6EI_{yy}}{l^2} & 0 & \frac{4EI_{yy}}{l} & 0 & 0 & 0 & \frac{6EI_{yy}}{l^2} & 0 & \frac{2EI_{yy}}{l} & 0 \\
0 & \frac{6EI_{zz}}{l^2} & 0 & 0 & 0 & \frac{4EI_{zz}}{l} & 0 & -\frac{6EI_{zz}}{l^2} & 0 & 0 & 0 & \frac{2EI_{zz}}{l} \\
\hline
-\frac{AE}{l} & 0 & 0 & 0 & 0 & 0 & \frac{AE}{l} & 0 & 0 & 0 & 0 & 0 \\
0 & -\frac{12EI_{zz}}{l^3} & 0 & 0 & 0 & -\frac{6EI_{zz}}{l^2} & 0 & \frac{12EI_{zz}}{l^3} & 0 & 0 & 0 & -\frac{6EI_{zz}}{l^2} \\
0 & 0 & -\frac{12EI_{yy}}{l^3} & 0 & \frac{6EI_{yy}}{l^2} & 0 & 0 & 0 & \frac{12EI_{yy}}{l^3} & 0 & \frac{6EI_{yy}}{l^2} & 0 \\
0 & 0 & 0 & -\frac{GJ}{l} & 0 & 0 & 0 & 0 & 0 & \frac{GJ}{l} & 0 & 0 \\
0 & 0 & -\frac{6EI_{yy}}{l^2} & 0 & \frac{2EI_{yy}}{l} & 0 & 0 & 0 & \frac{6EI_{yy}}{l^2} & 0 & \frac{4EI_{yy}}{l} & 0 \\
0 & \frac{6EI_{zz}}{l^2} & 0 & 0 & 0 & \frac{2EI_{zz}}{l} & 0 & -\frac{6EI_{zz}}{l^2} & 0 & 0 & 0 & \frac{4EI_{zz}}{l}
\end{array}\right] \begin{bmatrix} u_{x1} \\ u_{y1} \\ u_{z1} \\ \theta_{x1} \\ \theta_{y1} \\ \theta_{z1} \\ u_{x2} \\ u_{y2} \\ u_{z2} \\ \theta_{x2} \\ \theta_{y2} \\ \theta_{z2} \end{bmatrix} \quad (9.11)$$

This is, of course, the most general case, from which simpler cases can be derived. For example, the stiffness matrix for the 2D beam given in equation (9.10) can be obtained from the 3D matrix by deleting rows and columns 3, 4, 5, 9, 10 and 11, which is equivalent to setting all the out-of-plane degrees of freedom to zero.

9.3.2 Global stiffness matrices for plane rigid-jointed frames

In this section we will look at the assembly of the stiffness matrix for a 2D rigid-jointed structure. The 3D case is not in principle any more difficult, but the matrix equations get rather too large to put down on paper.

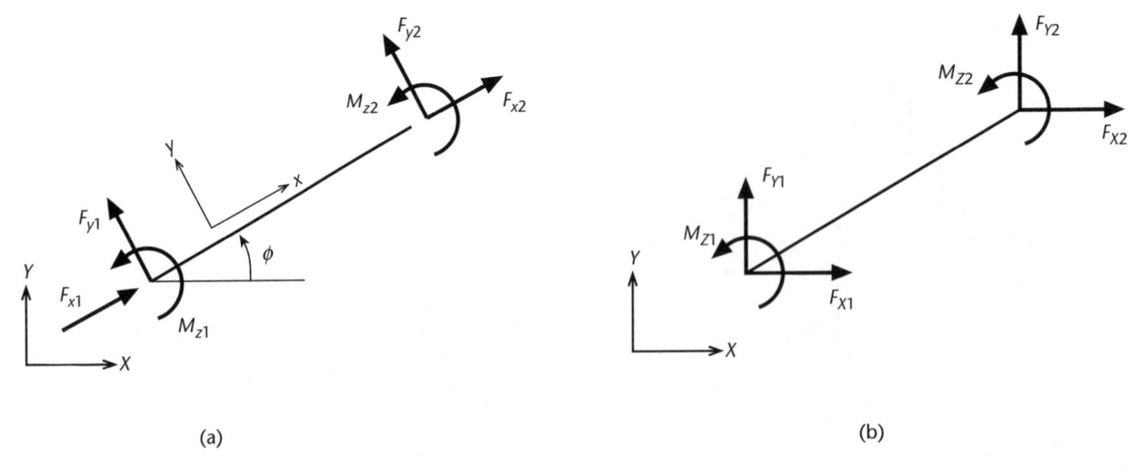

Figure 9.16
Axis transformation for a 2D beam element.

As with truss structures, the local element matrices can be written down directly using the standard result given in equation (9.10), and they must then be transformed into a common axis system. The transformation from local to global coordinates is given by equation (9.8):

$$\mathbf{K} = \mathbf{T}^{\mathrm{T}}\mathbf{k}\mathbf{T}$$

where **T** is the transformation matrix between the two axis systems. The transformation matrix used for a truss element will no longer be adequate, as more degrees of freedom are now being transformed. We must therefore derive a new one.

Consider the 2D beam element shown in Figure 9.16. The forces in the local axis system can be written in terms of their components in the global axis directions, and if the resulting equations are written in matrix form we get:

$$\begin{bmatrix} F_{x1} \\ F_{y1} \\ M_{z1} \\ F_{x2} \\ F_{y2} \\ M_{z2} \end{bmatrix} = \left[\begin{array}{ccc|ccc} \cos\phi & \sin\phi & 0 & 0 & 0 & 0 \\ -\sin\phi & \cos\phi & 0 & 0 & 0 & 0 \\ 0 & 0 & 1 & 0 & 0 & 0 \\ \hline 0 & 0 & 0 & \cos\phi & \sin\phi & 0 \\ 0 & 0 & 0 & -\sin\phi & \cos\phi & 0 \\ 0 & 0 & 0 & 0 & 0 & 1 \end{array}\right] \begin{bmatrix} F_{X1} \\ F_{Y1} \\ M_{Z1} \\ F_{X2} \\ F_{Y2} \\ M_{Z2} \end{bmatrix} \quad \text{or} \quad \mathbf{f} = \mathbf{TF} \tag{9.12}$$

The transformation matrix **T** is often written in shorthand as:

$$\mathbf{T} = \left[\begin{array}{c|c} \mathbf{R} & \mathbf{0} \\ \hline \mathbf{0} & \mathbf{R} \end{array}\right] \quad \text{where } \mathbf{R} = \begin{bmatrix} \cos\phi & \sin\phi & 0 \\ -\sin\phi & \cos\phi & 0 \\ 0 & 0 & 1 \end{bmatrix}$$

The stiffness matrix in global coordinates for a plane frame element is obtained by pre- and post-multiplying its local stiffness matrix (equation (9.10)), by **T**:

$$
\mathbf{K} = \frac{E}{l}\left[\begin{array}{ccc|ccc}
Ac^2 + \frac{12Is^2}{l^2} & Asc - \frac{12Isc}{l^2} & -\frac{6Is}{l} & -Ac^2 - \frac{12Is^2}{l^2} & -Asc + \frac{12Isc}{l^2} & -\frac{6Is}{l} \\
Asc - \frac{12Isc}{l^2} & As^2 + \frac{12Ic^2}{l^2} & \frac{6Ic}{l} & -Asc + \frac{12Isc}{l^2} & -As^2 - \frac{12Ic^2}{l^2} & \frac{6Ic}{l} \\
-\frac{6Is}{l} & \frac{6Ic}{l} & 4I & \frac{6Is}{l} & -\frac{6Ic}{l} & 2I \\
\hline
-Ac^2 - \frac{12Is^2}{l^2} & -Asc + \frac{12Isc}{l^2} & \frac{6Is}{l} & Ac^2 + \frac{12Is^2}{l^2} & Asc - \frac{12Isc}{l^2} & \frac{6Is}{l} \\
-Asc + \frac{12Isc}{l^2} & -As^2 - \frac{12Ic^2}{l^2} & -\frac{6Ic}{l} & Asc - \frac{12Isc}{l^2} & As^2 + \frac{12Ic^2}{l^2} & -\frac{6Ic}{l} \\
-\frac{6Is}{l} & \frac{6Ic}{l} & 2I & \frac{6Is}{l} & -\frac{6Ic}{l} & 4I
\end{array}\right] \tag{9.13}
$$

where $s = \sin\phi$ and $c = \cos\phi$. The stiffness matrix for a plane frame structure can now be assembled. We determine the stiffness matrix of each element in the global coordinate system, delete rows and columns corresponding to known zero displacements, and then insert the remaining elements in the appropriate locations in the structure stiffness matrix. A stiffness coefficient relating the ith force term to the jth displacement term goes in row i, column j of the matrix.

EXAMPLE 9.5

Find the stiffness matrix for the simple plane frame in Figure 9.17. Both members have cross-sectional area $A = 0.004$ m^2 and second moment of area about the z-axis $I = 100 \times 10^{-6}$ m^4, and are made of steel with Young's modulus $E = 205$ GPa.

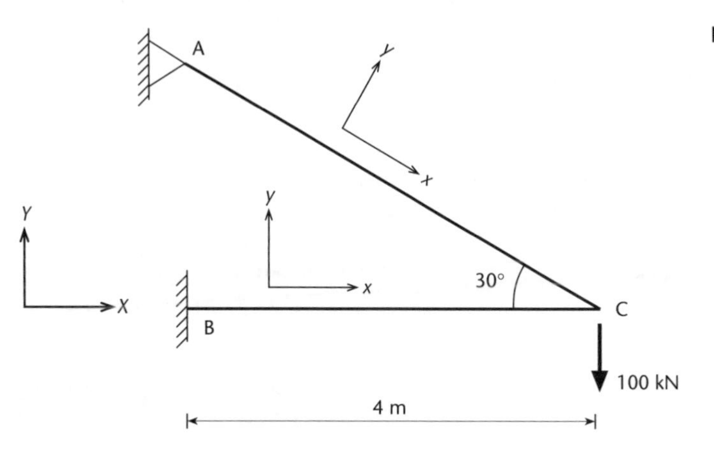

Figure 9.17

The active degrees of freedom are θ_{ZA}, u_{XC}, u_{YC} and θ_{ZC}. We must therefore determine a 4×4 stiffness matrix. The element stiffness matrices can be written in global coordinates by substituting the appropriate values in equation (9.13) and deleting rows and columns corresponding to inactive degrees of freedom.

First consider AC. The length $l = 8/\sqrt{3} = 4.619$ m and the anti-clockwise angle from the global to the local x-axis is $\phi = 330°$, giving $s = -1/2$, $c = \sqrt{3}/2$. With $u_{XA} = u_{YA} = 0$ we can delete the first two rows and columns from the 6×6 matrix, leaving (in units of kN and m):

$$\mathbf{K}_{AC} = \frac{205 \times 10^6}{4.619} \begin{bmatrix} 400 & -64.95 & -112.5 & 200 \\ -64.95 & 3,000 + 14.06 & -1,732 + 24.35 & -64.95 \\ -112.5 & -1,732 + 24.35 & 1000 + 42.19 & -112.5 \\ 200 & -64.95 & -112.5 & 400 \end{bmatrix} \times 10^{-6}$$

$$\rightarrow \mathbf{K}_{AC} = 10^3 \begin{bmatrix} 17.75 & -2.88 & -4.99 & 8.88 \\ -2.88 & 133.77 & -75.79 & -2.88 \\ -4.99 & -75.79 & 46.25 & -4.99 \\ 8.88 & -2.88 & -4.99 & 17.75 \end{bmatrix} \begin{matrix} \theta_{ZA} \\ u_{XC} \\ u_{YC} \\ \theta_{ZC} \end{matrix}$$

For BC the length $l = 4$ m. The local and global axes are coincident, so no transformation is needed and we can take terms directly from the local stiffness matrix equation, (9.10). B is fixed, so we delete the first three rows and columns, leaving:

$$\mathbf{K}_{BC} = 10^3 \begin{bmatrix} 205.0 & 0 & 0 \\ 0 & 3.85 & -7.69 \\ 0 & -7.69 & 20.5 \end{bmatrix} \begin{matrix} u_{XC} \\ u_{YC} \\ \theta_{ZC} \end{matrix}$$

The global stiffness matrix equation can now be written by transferring the element stiffness terms into the appropriate locations in the global matrix. The load vector consists of a single term corresponding to the degree of freedom u_{ZC}:

$$\begin{bmatrix} 0 \\ 0 \\ -100 \\ 0 \end{bmatrix} = 10^3 \begin{bmatrix} 17.75 & -2.88 & -4.99 & 8.88 \\ -2.88 & 338.77 & -75.79 & -2.88 \\ -4.99 & -75.79 & 50.10 & -12.68 \\ 8.88 & -2.88 & -12.68 & 38.25 \end{bmatrix} \begin{bmatrix} \theta_{ZA} \\ u_{XC} \\ u_{YC} \\ \theta_{ZC} \end{bmatrix}$$

Having assembled the stiffness matrix equation for a structure, we now wish to use it to ascertain the effects of the external loads. Typically, we require the structural deformations, the internal forces in the members and the support reactions. This first requires the direct solution of the structure stiffness matrix equation, then back-substitution into the member stiffness matrices, and finally some simple statics.

EXAMPLE 9.6

For the frame in Example 9.5, determine the joint displacements, internal forces and support reactions, and plot the bending moment diagram.

The structure stiffness matrix equation can be solved computationally to give (in units of radians and metres):

$$\begin{bmatrix} \theta_{ZA} \\ u_{XC} \\ u_{YC} \\ \theta_{ZC} \end{bmatrix} = \begin{bmatrix} -0.579 \\ -0.811 \\ -3.561 \\ -1.107 \end{bmatrix} \times 10^{-3}$$

The forces imposed on the individual members at the joints can be determined by back-substituting these values into the member stiffness equations:

$$\begin{bmatrix} M_{ZA} \\ F_{XC} \\ F_{YC} \\ M_{ZC} \end{bmatrix}_{AC} = \begin{bmatrix} 17.75 & -2.88 & -4.99 & 8.88 \\ -2.88 & 133.77 & -75.79 & -2.88 \\ -4.99 & -75.79 & 46.25 & -4.99 \\ 8.88 & -2.88 & -4.99 & 17.75 \end{bmatrix} \begin{bmatrix} -0.579 \\ -0.811 \\ -3.561 \\ -1.107 \end{bmatrix} = \begin{bmatrix} 0 \\ 166.2 \\ -94.8 \\ -4.7 \end{bmatrix}$$

$$\begin{bmatrix} F_{XC} \\ F_{YC} \\ M_{ZC} \end{bmatrix}_{BC} = \begin{bmatrix} 205.0 & 0 & 0 \\ 0 & 3.85 & -7.69 \\ 0 & -7.69 & 20.5 \end{bmatrix} \begin{bmatrix} -0.811 \\ -3.561 \\ -1.107 \end{bmatrix} = \begin{bmatrix} -166.2 \\ -5.2 \\ 4.7 \end{bmatrix}$$

These forces are shown on the exploded view of the structure in Figure 9.18(a). It can be seen that the horizontal forces and moments exerted on the members at C sum to zero, while the vertical forces sum to the applied load of 100 kN. Unsurprisingly, most of the vertical load is carried by AC, which has much greater vertical stiffness. It is now a simple matter to determine the reactions at A and B using statics, and these are also shown on the diagram. Lastly, the internal forces can be found by resolving forces on a segment of each element. Referring to Figure 9.18(b):

For AC:

$$T_{AC} = 166.2\cos 30^\circ + 94.8\sin 30^\circ = 191.3 \text{ kN}$$
$$S_{AC} = 166.2\sin 30^\circ - 94.8\cos 30^\circ = 1.0 \text{ kN}$$
$$M_{AC} = -1.0x$$

For BC:

$$T_{BC} = -166.2 \text{ kN (that is, compression)}$$
$$S_{BC} = -5.2 \text{ kN}$$
$$M_{BC} = 5.2x - 16.1$$

Clearly the frame carries the load primarily by developing axial forces in the two members, with bending moments and shear forces contributing comparatively little. The bending moment diagram can be plotted by considering whether the

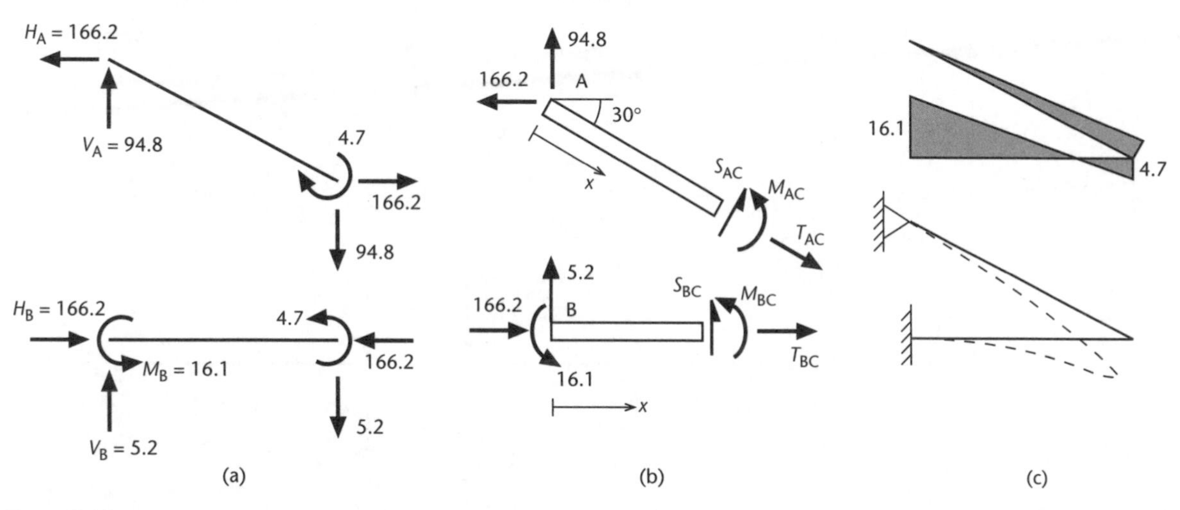

Figure 9.18

moments shown cause hogging or sagging of the members. For AC, the negative moment causes hogging (tension on the outside of the frame). For BC, the moment at B causes hogging and the moment at C sagging. The bending moment diagram and corresponding deflected shape are therefore as shown in Figure 9.18 (c).

Note that, if the forces at the member ends are transformed into the local axis system for the member, then the values obtained are automatically equal to the internal forces at the member ends, with no need for further resolution. This can be seen for member BC above, for which the local and global axes are the same.

9.4 Applications and special cases

9.4.1 Hand calculations using stiffness matrices

While it is essentially a computer method, the stiffness matrix approach can occasionally be used to give quick manual solutions for simple structures. It is particularly useful for cases for which axial deformations cannot be neglected, and which therefore cannot be analysed by traditional hand methods such as slope–deflection.

EXAMPLE 9.7

The beam ABC in Figure 9.19(a) has flexural stiffness EI and is supported on a central spring of stiffness k. Find the vertical deflection of B and the rotations at the pinned supports A and C.

Figure 9.19

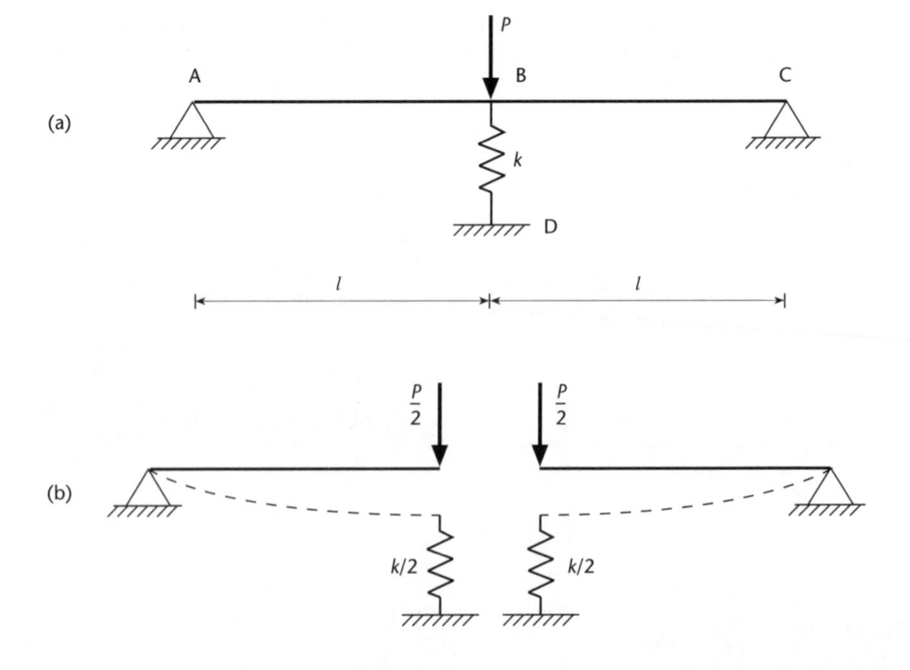

The structure and loading are both symmetrical (BC is the mirror image of AB), and so can be thought of as the superposition of the two identical substructures in Figure 9.19(b). The symmetry of the structure means that there is no rotation or horizontal deflection at B. The spring has only axial stiffness and so can be treated in the same way as an axially loaded bar. Considering just the left-hand substructure, the stiffness matrices in the global axis system can be written down straight away, using equations (9.9) and (9.10):

$$\mathbf{K}_{AB} = \begin{bmatrix} \frac{4EI}{l} & -\frac{6EI}{l^2} \\ -\frac{6EI}{l^2} & \frac{12EI}{l^3} \end{bmatrix} \begin{matrix} \theta_{ZA} \\ u_{YB} \end{matrix} \qquad \mathbf{K}_{BD} = \begin{bmatrix} \frac{k}{2} \end{bmatrix} u_{YB}$$

and the stiffness matrix equation for the substructure can then be written as:

$$\begin{bmatrix} 0 \\ -\frac{P}{2} \end{bmatrix} = \begin{bmatrix} \frac{4EI}{l} & -\frac{6EI}{l^2} \\ -\frac{6EI}{l^2} & \frac{12EI}{l^3} + \frac{k}{2} \end{bmatrix} \begin{bmatrix} \theta_{ZA} \\ u_{YB} \end{bmatrix}$$

This simple matrix equation can be easily solved by hand, giving:

$$\theta_{ZA} = -\frac{2}{3} \cdot \frac{Pl^2}{6EI + kl^3}, \qquad u_{YB} = -\frac{Pl^3}{6EI + kl^3}$$

and of course $\theta_{ZC} = -\theta_{ZA}$. An alternative, slightly longer, approach would be to assemble the 3×3 stiffness matrix for the full structure and then make the substitution $\theta_{ZC} = -\theta_{ZA}$ in the displacement vector. Two of the three equations then become identical and the same solution is achieved.

9.4.2 Members loaded between the joints

In the examples presented so far, the assembly of the load vector has been trivial – we have simply inserted each external load in the appropriate row, corresponding to the joint and direction at which it acts. A difficulty arises if loads are applied away from the joints, since the stiffness matrix equation is formulated entirely in terms of the forces and corresponding displacements at the joints. To get around this problem, we must apply a set of equivalent loads at the joints which would cause the same joint displacements as the real loads. The approach is similar to the use of fixed-end moments in a slope–deflection analysis.

Suppose we have a structure subjected to a variety of loads, some at the joints and some elsewhere in the structure. These loads give rise to a set of joint displacements **U** which we wish to calculate. We can easily formulate a load vector **F** consisting solely of the loads applied at the joints, but this will not give rise to the correct displacements, since the effect of the loads acting at other points in the structure is not included.

To deal with the loads away from the joints, we can split the loading and deformation of the structure into two loadcases:

- First clamp all the joints and apply the external loads. The loads acting at the joints will have no effect on the members and can be ignored, while those acting between the joints will cause a set of fixed-end reactions $\mathbf{F}^F$.

- Now remove the clamps and the external loads, and instead apply a set of loads $\mathbf{F}^D$ at the joints so as to cause the correct joint displacements $\mathbf{U}$. These forces must satisfy the stiffness matrix equation $\mathbf{F}^D = \mathbf{KU}$ and are therefore the forces we wish to use in our analysis.

The total forces acting at the joints are found by linear superposition of the fixed-end and displacement-related forces:

$$\mathbf{F} = \mathbf{F}^F + \mathbf{F}^D$$

Substituting for $\mathbf{F}^D$ from above:

$$\mathbf{F} - \mathbf{F}^F = \mathbf{KU} \tag{9.14}$$

Thus the correct load vector for use in the stiffness matrix analysis is obtained by first formulating the vector of nodal loads in the usual way and then *subtracting* the vector of fixed-end reactions due to the other loads. The fixed-end forces can usually be found very easily, and the fixed-end moments can be obtained using the formulae in section 8.1.2.

EXAMPLE 9.8

Find the load vector for the structure shown in Figure 9.20(a).

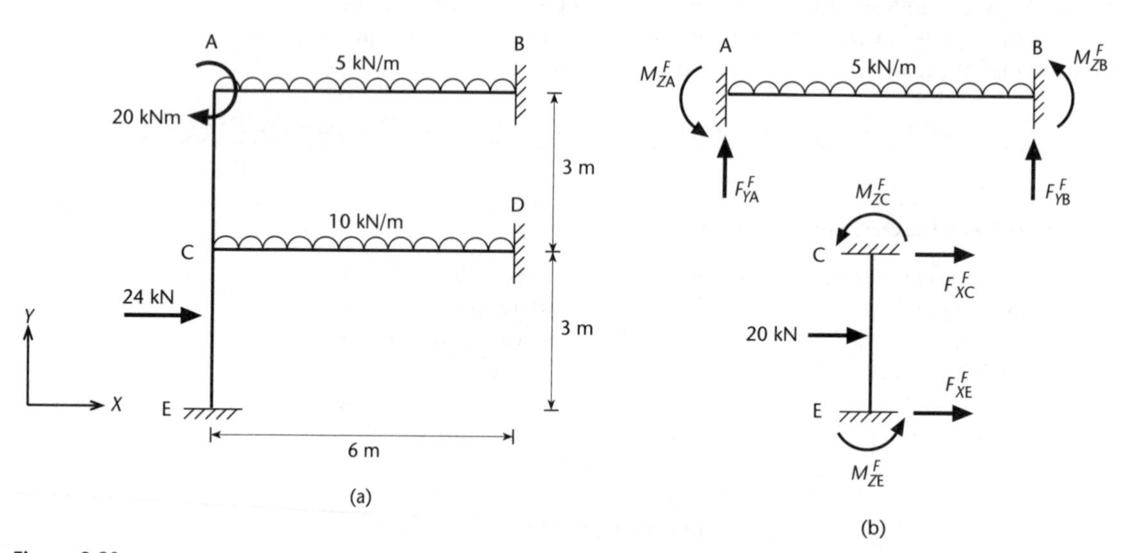

Figure 9.20

There are six active degrees of freedom: u_{XA}, u_{YA}, θ_{ZA}, u_{XC}, u_{YC} and θ_{ZC}. The vector of nodal loads corresponding to these degrees of freedom is:

$$\mathbf{F} = \begin{matrix} [0 & 0 & -20 & 0 & 0 & 0]^T \\ u_{XA} & u_{YA} & \theta_{ZA} & u_{XC} & u_{YC} & \theta_{ZC} \end{matrix}$$

We now need to find the fixed-end reactions at A and C for the members loaded between the joints. Figure 9.20(b) shows the forces on AB and CE; CD is similar to AB.

For AB: $F^F_{YA} = \frac{5 \times 6}{2} = 15$ kN, $M^F_{ZA} = \frac{5 \times 6^2}{12} = 15$ kNm

Similarly, for CD: $F^F_{YC} = 30$ kN, $M^F_{ZC} = 30$ kNm

For CE: $F^F_{XC} = -\frac{24}{2} = -12$ kN, $M^F_{ZC} = -\frac{24 \times 3}{8} = -9$ kNm

The fixed-end force vector is therefore:

$$\mathbf{F}^F = \underset{\begin{matrix} u_{XA} & u_{YA} & \theta_{ZA} & u_{XC} & u_{YC} & \theta_{ZC} \end{matrix}}{\begin{bmatrix} 0 & 15 & 15 & -12 & 30 & 21 \end{bmatrix}^T}$$

And the load vector to be used in the stiffness matrix equation is:

$$\mathbf{F} - \mathbf{F}^F = \underset{\begin{matrix} u_{XA} & u_{YA} & \theta_{ZA} & u_{XC} & u_{YC} & \theta_{ZC} \end{matrix}}{\begin{bmatrix} 0 & -15 & -35 & 12 & -30 & -21 \end{bmatrix}^T}$$

9.4.3 Frames with internal pins

If a frame comprises only pin-ended bars we can analyse it using the truss approach (section 9.2), and if it is made up of rigidly connected elements we use the frame approach (section 9.3). Quite often, however, a structure that is predominantly rigid-jointed will contain one or more joints which act as pins. Such cases are most easily dealt with by deriving a modified stiffness matrix for a beam element with one pinned end.

The stiffness matrix equation for a 2D beam element is given in equation (9.10). Suppose end 1 of the element is pinned to another member. We know that the pinned end can support no moment so, putting $M_{z1} = 0$, the third line of equation (9.10) becomes:

$$\frac{6EI}{l^2} u_{y1} + \frac{4EI}{l} \theta_{z1} - \frac{6EI}{l^2} u_{y2} + \frac{2EI}{l} \theta_{z2} = 0$$

Hence

$$\theta_{z1} = \frac{1}{2}\left(-\frac{3}{l} u_{y1} + \frac{3}{l} u_{y2} - \theta_{z2} \right) \tag{9.15}$$

All the terms in θ_{z1} in equation (9.10) can therefore be rewritten in terms of u_{y1}, u_{y2} and θ_{z2}. The third row and column of the matrix equation can then be eliminated, leaving:

$$\begin{bmatrix} F_{x1} \\ F_{y1} \\ F_{x2} \\ F_{y2} \\ M_{z2} \end{bmatrix} = \left[\begin{array}{cc|ccc} \frac{AE}{l} & 0 & -\frac{AE}{l} & 0 & 0 \\ 0 & \frac{3EI}{l^3} & 0 & -\frac{3EI}{l^3} & \frac{3EI}{l^2} \\ \hline -\frac{AE}{l} & 0 & \frac{AE}{l} & 0 & 0 \\ 0 & -\frac{3EI}{l^3} & 0 & \frac{3EI}{l^3} & -\frac{3EI}{l^2} \\ 0 & \frac{3EI}{l^2} & 0 & -\frac{3EI}{l^2} & \frac{3EI}{l} \end{array}\right] \begin{bmatrix} u_{x1} \\ u_{y1} \\ u_{x2} \\ u_{y2} \\ \theta_{z2} \end{bmatrix} \tag{9.16}$$

Similarly, for a member pinned at end 2:

$$\theta_{z2} = \frac{1}{2}\left(-\frac{3}{l}u_{y1} + \frac{3}{l}u_{y2} - \theta_{z1}\right) \tag{9.17}$$

and the stiffness matrix equation can be written as:

$$\begin{bmatrix} F_{x1} \\ F_{y1} \\ M_{z1} \\ F_{x2} \\ F_{y2} \end{bmatrix} = \left[\begin{array}{ccc|cc} \frac{AE}{l} & 0 & 0 & -\frac{AE}{l} & 0 \\ 0 & \frac{3EI}{l^3} & \frac{3EI}{l^2} & 0 & -\frac{3EI}{l^3} \\ 0 & \frac{3EI}{l^2} & \frac{3EI}{l} & 0 & -\frac{3EI}{l^2} \\ \hline -\frac{AE}{l} & 0 & 0 & \frac{AE}{l} & 0 \\ 0 & -\frac{3EI}{l^3} & -\frac{3EI}{l^2} & 0 & \frac{3EI}{l^3} \end{array}\right] \begin{bmatrix} u_{x1} \\ u_{y1} \\ \theta_{z1} \\ u_{x2} \\ u_{y2} \end{bmatrix} \tag{9.18}$$

The stiffness matrix for the structure can now be assembled, using equation (9.10) for ordinary members and either (9.16) or (9.18) for any member connected to an internal pin. If the rotation of the pinned end is required, this can be found by back-substitution into equation (9.15) or (9.17).

This approach to internal pins will work in all cases *except* when there is a concentrated moment applied at the pin; this rather more complex case will not be considered further here.

If a member is pinned at both ends then it can sustain no bending moments and therefore no shear forces. Thus it behaves as a simple bar element and its stiffness matrix is given by equation (9.5).

EXAMPLE 9.9

Find the stiffness matrix equation for the frame in Figure 9.21(a), in which there is a pin just to the right of B, that is, the joint between AB and DB is rigid, but BC is connected to this rigid joint via a pin. The member properties are $A = 0.004$ m^2, $I = 0.0001$ m^4 and $E = 205$ GPa.

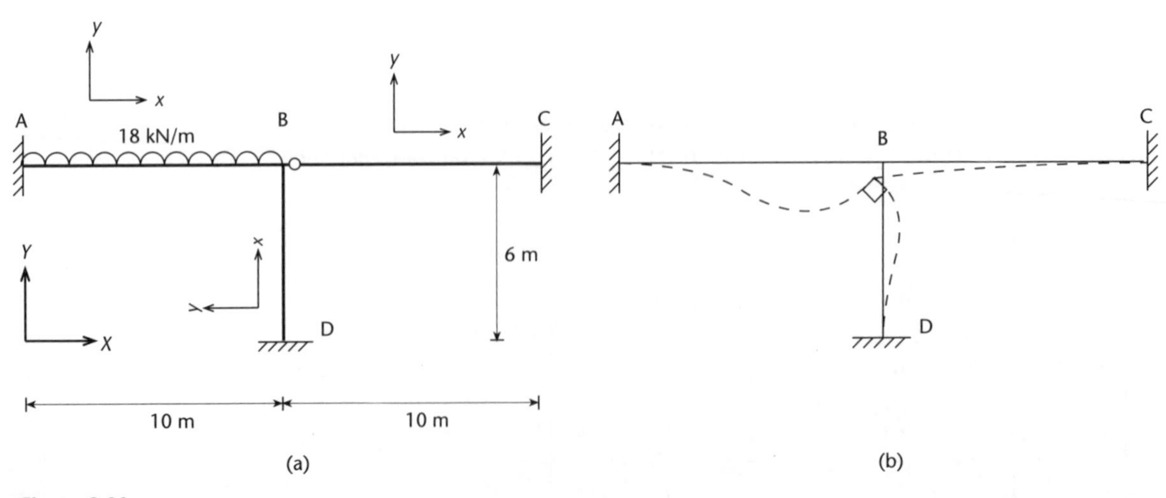

Figure 9.21

The required degrees of freedom are u_{XB}, u_{YB} and θ_{ZB}. There is a fourth unknown displacement – the rotation in BC just to the right of B – but this need not be included in the stiffness matrix equation. The element stiffness matrices are first found. For AB the local and global axes coincide, so equation (9.10) gives (using units of kilo-Newtons, metres and radians):

$$\mathbf{K}_{AB} = \begin{bmatrix} \frac{AE}{l} & 0 & 0 \\ 0 & \frac{12EI}{l^3} & -\frac{6EI}{l^2} \\ 0 & -\frac{6EI}{l^2} & \frac{4EI}{l} \end{bmatrix} = 10^3 \begin{bmatrix} 82.0 & 0 & 0 \\ 0 & 0.246 & -1.23 \\ 0 & -1.23 & 8.2 \end{bmatrix} \begin{matrix} u_{XB} \\ u_{YB} \\ \theta_{ZB} \end{matrix}$$

For BC, the local and global axes again coincide. There is a pin at end 1, so we use equation (9.16):

$$\mathbf{K}_{BC} = \begin{bmatrix} \frac{AE}{l} & 0 \\ 0 & \frac{3EI}{l^3} \end{bmatrix} = 10^3 \begin{bmatrix} 82.0 & 0 \\ 0 & 0.062 \end{bmatrix} \begin{matrix} u_{XB} \\ u_{YB} \end{matrix}$$

For DB, the angle from the global to the local axes is $\phi = 90°$, so use equation (9.13) with $s = 1$, $c = 0$:

$$\mathbf{K}_{DB} = \begin{bmatrix} \frac{12EI}{l^3} & 0 & \frac{6EI}{l^2} \\ 0 & \frac{AE}{l} & 0 \\ \frac{6EI}{l^2} & 0 & \frac{4EI}{l} \end{bmatrix} = 10^3 \begin{bmatrix} 1.14 & 0 & 3.42 \\ 0 & 136.67 & 0 \\ 3.42 & 0 & 13.67 \end{bmatrix} \begin{matrix} u_{XB} \\ u_{YB} \\ \theta_{ZB} \end{matrix}$$

The distributed load on member AB creates fixed-end reactions at B of $F^F_{YB} = 18 \times 10/2 = 90$ kN and $M^F_{ZB} = -18 \times 10^2/12 = -150$ kNm. The load vector is found by subtracting these terms from the vector of nodal loads, which is zero in this case. The matrix equation can therefore be written as

$$\begin{bmatrix} 0 \\ -90 \\ 150 \end{bmatrix} = 10^3 \begin{bmatrix} 165.14 & 0 & 3.42 \\ 0 & 136.98 & -1.23 \\ 3.42 & -1.23 & 21.87 \end{bmatrix} \begin{bmatrix} u_{XB} \\ u_{YB} \\ \theta_{ZB} \end{bmatrix}$$

This matrix equation can be inverted to give (in units of metres and radians):

$$\begin{bmatrix} u_{XB} \\ u_{YB} \\ \theta_{ZB} \end{bmatrix} = \begin{bmatrix} -0.14 \\ 0.59 \\ 6.85 \end{bmatrix} \times 10^{-3}$$

and, from equation (9.15), the rotation at the left-hand end of BC is $\theta_{ZB}(BC) = 0.09 \times 10^{-3}$ radians. The deflected shape is therefore as shown (greatly exaggerated) in Figure 9.21(b). The angle ABD remains a right angle, but DBC does not; the member BC is almost straight.

9.4.4 Grillages

A grillage is a frame in which all the members lie in a single (usually horizontal) plane and all the external loads are applied perpendicular to this plane. An example of a grillage would be a system of beams that forms the floor of a building or the deck of a bridge. In fact, bridge decks are often modelled as grillages even when they are actually solid slabs.

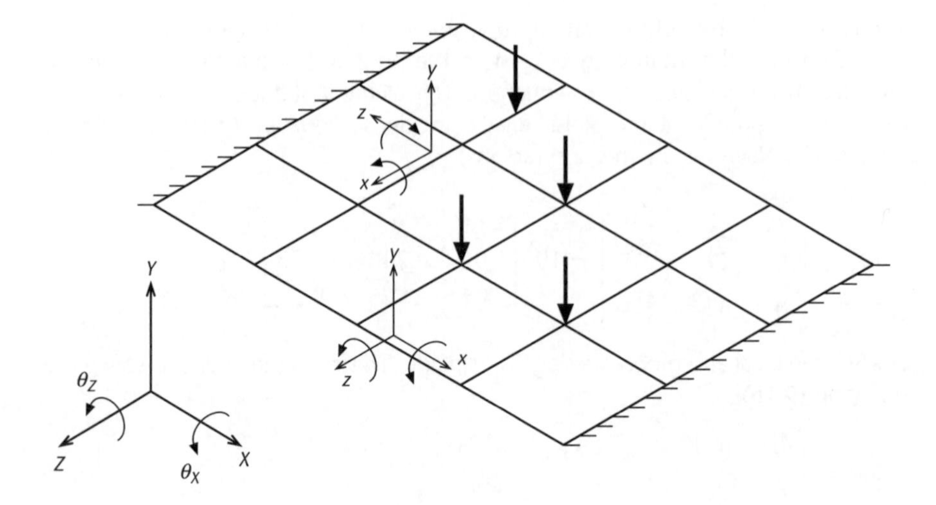

Figure 9.22
Grillage.

We will look briefly at the most common type of grillage, in which two rows of beams intersect at right angles (Figure 9.22). The structure has a very high in-plane stiffness and carries only out-of-plane loads, so it is reasonable to assume that the in-plane deformations are negligible. Therefore, at all joints we assume:

$$u_X = u_Z = \theta_Y = 0$$

The stiffness matrix of a grillage element in local coordinates can therefore be found from the general 3D element matrix (equation (9.11)), deleting rows and columns 1, 3, 5, 7, 9 and 11. For an element oriented in the X direction, the local and global axes are in the same directions and no further transformation is needed. The global stiffness matrix equation is therefore:

$$\begin{bmatrix} F_{Y1} \\ M_{X1} \\ M_{Z1} \\ F_{Y2} \\ M_{X2} \\ M_{Z2} \end{bmatrix} = \left[\begin{array}{ccc|ccc} \frac{12EI}{l^3} & 0 & \frac{6EI}{l^2} & -\frac{12EI}{l^3} & 0 & \frac{6EI}{l^2} \\ 0 & \frac{GJ}{l} & 0 & 0 & -\frac{GJ}{l} & 0 \\ \frac{6EI}{l^2} & 0 & \frac{4EI}{l} & -\frac{6EI}{l^2} & 0 & \frac{2EI}{l} \\ \hline -\frac{12EI}{l^3} & 0 & -\frac{6EI}{l^2} & \frac{12EI}{l^3} & 0 & -\frac{6EI}{l^2} \\ 0 & -\frac{GJ}{l} & 0 & 0 & \frac{GJ}{l} & 0 \\ \frac{6EI}{l^2} & 0 & \frac{2EI}{l} & -\frac{6EI}{l^2} & 0 & \frac{4EI}{l} \end{array}\right] \begin{bmatrix} u_{Y1} \\ \theta_{X1} \\ \theta_{Z1} \\ u_{Y2} \\ \theta_{X2} \\ \theta_{Z2} \end{bmatrix} \quad (9.19)$$

where $I = I_{zz}$. It is evident from this equation that grillage elements carry loads in torsion as well as in vertical bending and shear.

For an element oriented in the global Z direction the local matrix has the same form, but a transformation into global coordinates is now needed. The relation between the local and global degrees of freedom is:

$$\begin{bmatrix} u_{y1} \\ \theta_{x1} \\ \theta_{z1} \end{bmatrix} = \begin{bmatrix} 1 & 0 & 0 \\ 0 & 0 & 1 \\ 0 & -1 & 0 \end{bmatrix} \begin{bmatrix} u_{Y1} \\ \theta_{X1} \\ \theta_{Z1} \end{bmatrix} \quad \text{or} \quad \mathbf{u} = \mathbf{RU}$$

and the global stiffness matrix can be obtained by the transformation

$$\mathbf{K} = \mathbf{T}^{\mathrm{T}}\mathbf{k}\mathbf{T} \quad \text{where} \quad \mathbf{T} = \left[\begin{array}{c|c} \mathbf{R} & \mathbf{0} \\ \hline \mathbf{0} & \mathbf{R} \end{array}\right]$$

The global stiffness matrix equation for an element oriented in the Y direction is therefore:

$$\begin{bmatrix} F_{Y1} \\ M_{X1} \\ M_{Z1} \\ F_{Y2} \\ M_{X2} \\ M_{Z2} \end{bmatrix} = \left[\begin{array}{ccc|ccc} \frac{12EI}{l^3} & -\frac{6EI}{l^2} & 0 & -\frac{12EI}{l^3} & -\frac{6EI}{l^2} & 0 \\ -\frac{6EI}{l^2} & \frac{4EI}{l} & 0 & \frac{6EI}{l^2} & \frac{2EI}{l} & 0 \\ 0 & 0 & \frac{GJ}{l} & 0 & 0 & -\frac{GJ}{l} \\ \hline -\frac{12EI}{l^3} & \frac{6EI}{l^2} & 0 & \frac{12EI}{l^3} & \frac{6EI}{l^2} & 0 \\ -\frac{6EI}{l^2} & \frac{2EI}{l} & 0 & \frac{6EI}{l^2} & \frac{4EI}{l} & 0 \\ 0 & 0 & -\frac{GJ}{l} & 0 & 0 & \frac{GJ}{l} \end{array}\right] \begin{bmatrix} u_{Y1} \\ \theta_{X1} \\ \theta_{Z1} \\ u_{Y2} \\ \theta_{X2} \\ \theta_{Z2} \end{bmatrix} \tag{9.20}$$

The stiffness matrix for a right-angled grillage can now be assembled from equations (9.19) and (9.20) in the usual way.

9.4.5 Stiffness matrix computer programs

While structural engineers will almost always make use of the stiffness matrix method in the form of a pre-written computer program, rather than writing their own, it is important to have an idea of the steps that the program goes through. These can be summarised as follows:

1. Read in the data – we require:
   - Coordinates of joints.
   - Member connectivities – a list of members and the joints they span between.
   - Member properties – A, E, I_{zz} and (for 3D analyses) I_{yy} and J.
   - Support details – a list of supported joints and the type of restraint provided at each.
   - Any special conditions such as internal pins.
   - Loads – a list of loaded joints and/or members, with magnitudes and directions of loads. Of course, real structures are subjected to varying loads over their life span, and it is normal in design to come up with different load patterns in an attempt to represent possible worst cases. There may therefore be several separate loadcases.
2. Assemble the vector of unknown displacements **U**, using the list of joints and eliminating any inactive degrees of freedom at supports.
3. For each loadcase, assemble the load vector **F** corresponding to the degrees of freedom in the displacement vector. If any members are loaded between the joints, this will require the calculation of fixed-end force and moment terms.
4. For a given member, use the joint coordinates to calculate its length and its orientation with respect to the global axes. Hence calculate the member stiffness matrix in global coordinates. Transfer the terms to the appropriate locations in the structure stiffness matrix, remembering that a stiffness coefficient relating the ith load term to the jth displacement term goes in row i, column j, and that the new coefficient is added to any value already present in that location. Repeat for each member in the structure.

5. For a given loadcase, solve the stiffness matrix equation $\mathbf{F} = \mathbf{KU}$ to give the joint displacements. Note that the solution will not normally involve the inversion of $\mathbf{K}$, since this requires rather a large number of calculations. Usually a solution technique specially developed for banded matrices is used, and the efficiency of this is optimised by positioning all the non-zero terms in as narrow a band as possible around the leading diagonal. This in turn requires a re-ordering of the degrees of freedom – the re-ordering is, of course, internal to the program and would not be apparent to the user.
6. From the displacement vector, extract the joint displacements for a particular member and substitute into the member stiffness matrix equation to give the member forces. Transform these into the local axis system, that is, into axial force, transverse (shear) force etc. If the member is attached to a support, use statics to calculate the reactions. Repeat for each member.
7. Repeat steps 5 and 6 for each loadcase.

As an alternative to step 5, we could actually invert $\mathbf{K}$. The displacements for each loadcase can then be found very quickly simply by multiplying $\mathbf{K}^{-1}$ by the appropriate load vector. However, this is not normally economic unless $\mathbf{K}$ is quite small or the number of loadcases is very large.

9.4.6 Checking stiffness matrix analyses

The main potential pitfalls involved in analysing a structure by the stiffness matrix approach are those associated with any computer analysis technique:

- the risk of errors in the input data;
- the temptation to assume that the program is infallible and therefore to take an uncritical approach to the output;
- the possibility of violating the assumptions on which the method is based.

It is important to perform checks at each stage of an analysis in order to minimise the chances of errors creeping in.

Checking input data

At the input stage it is necessary to check that:

- all members have been included and are correctly connected;
- support conditions have been correctly specified;
- loads have been applied at the correct locations and in the right directions;
- the correct section properties have been assigned to the various members;
- dimensions, member properties and loads have been specified in consistent units.

Most analysis software now includes a graphical pre-processor, which simplifies the input phase and makes checking of the first three items on the list extremely easy. For three-dimensional structures it is always a good idea to view the frame from at least two different directions in order to establish that the geometry is correct in all three dimensions.

It is generally much harder to check that the numerical values assigned to loads and section properties are correct. In particular, great care must be taken to ensure that all quantities are entered in consistent units. For instance, if the loads are specified in kilo-Newtons and the dimensions in metres, then Young's modulus must be given in kN/m^2. It is very easy to introduce an error as large as a factor of 10^6 in the input data by forgetting what units you are working in. Such an error will often produce absurd results and thus be quite easy to detect and put right, but this is certainly not always the case.

Checking program output

Interpretation of the output requires some thought, for example in converting moments from the anti-clockwise positive to the hogging/sagging convention. However, many analysis packages have sophisticated post-processors that will do this conversion for you, and plot bending moment diagrams.

Checks on the output provide a further opportunity to confirm that the problem was correctly set up at the input stage, but are also necessary to ensure that the program is performing properly. Even the most well written and thoroughly debugged software may contain errors, and very often these can be detected by some very simple output checks. In general, you should aim to check that:

- there is overall equilibrium between the applied loads and support reactions;
- the reaction components correspond to the specified support conditions;
- there is equilibrium between the forces in members meeting at a joint (usually sufficient to check a random selection of joints);
- member forces are in agreement with the displacements of the end nodes.

Of course, a good analysis program will perform these checks internally, but it is nevertheless a good idea to conduct a full set of independent checks yourself – regardless of the sophistication of the software, it is advisable to treat the output with a healthy scepticism!

EXAMPLE 9.10

The results tabulated below were obtained from a stiffness matrix analysis of the plane frame shown in Figure 9.23(a). Perform random checks on the consistency of the results.

Nodal displacements:

Node	u_X mm	u_Y mm	θ 10^{-3}rads
A	30.2	0.06	−0.41
B	30.2	0	−0.28
C	30.2	−0.06	−0.42
D	28.0	0.06	−0.98
E	27.9	0	−0.91
F	27.9	−0.06	−0.99
G	21.9	0.04	−2.80
H	21.8	0	−2.03
I	21.8	−0.04	−2.79
J	0	0	−6.80
K	0	0	−7.16
L	0	0	−6.76

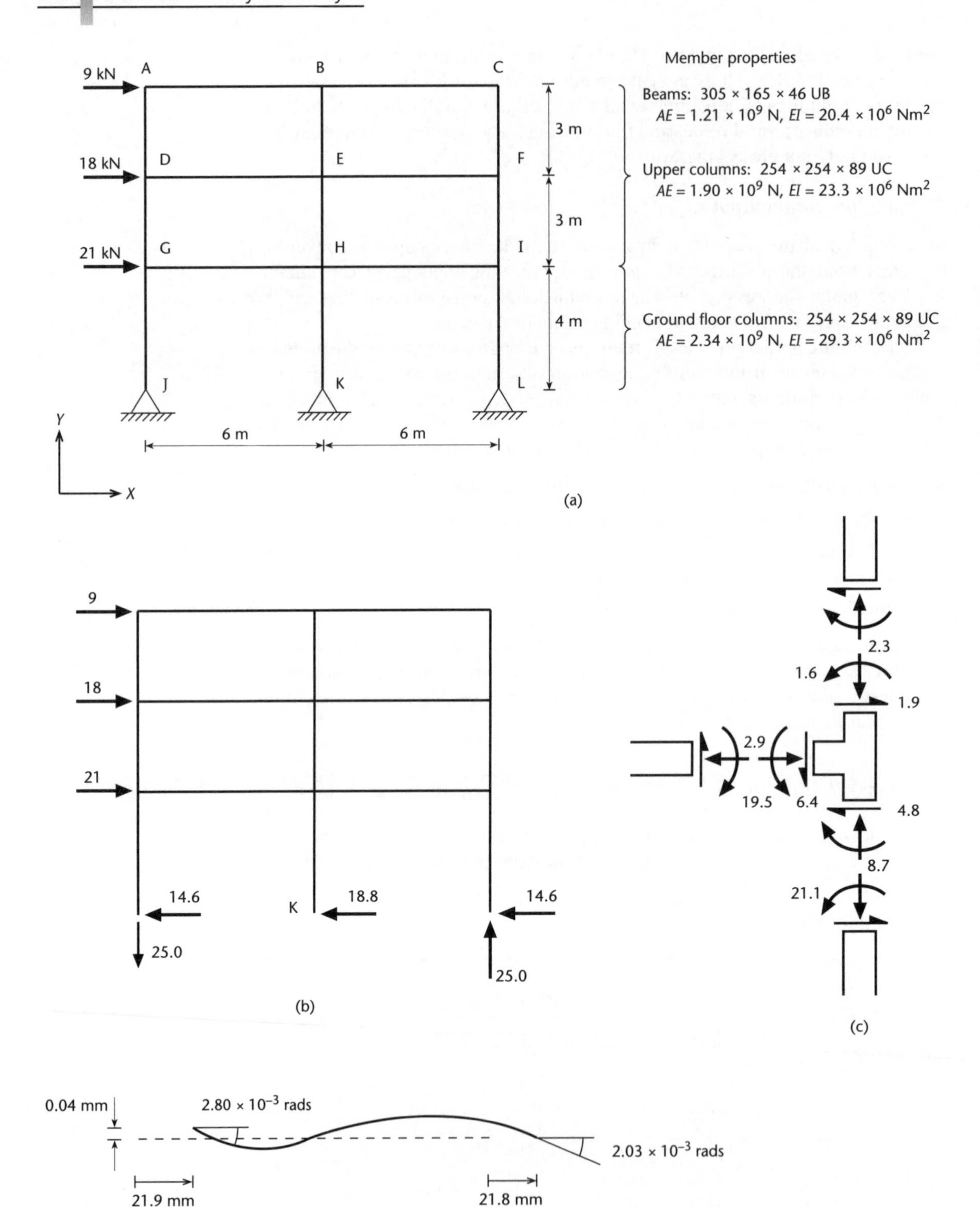
9 kN
A
B
C
18 kN
D
E
F
21 kN
G
H
I
J
K
L
3 m
3 m
4 m
6 m
6 m
Y
X
Member properties
Beams: 305 × 165 × 46 UB
AE = 1.21 × 10⁹ N, EI = 20.4 × 10⁶ Nm²
Upper columns: 254 × 254 × 89 UC
AE = 1.90 × 10⁹ N, EI = 23.3 × 10⁶ Nm²
Ground floor columns: 254 × 254 × 89 UC
AE = 2.34 × 10⁹ N, EI = 29.3 × 10⁶ Nm²
(a)
9
18
21
14.6
18.8
14.6
K
25.0
25.0
(b)
2.3
1.6
1.9
2.9
19.5
6.4
4.8
8.7
21.1
(c)
0.04 mm
2.80 × 10⁻³ rads
2.03 × 10⁻³ rads
21.9 mm
21.8 mm
(d)

Figure 9.23

Member forces (note: a positive axial force is tensile, a positive shear force acts in a positive axis direction and a positive moment is anti-clockwise):

Element	Axial kN	Shear (1) kN	Shear (2) kN	Moment (1) kNm	Moment (2) kNm
AB	−7.2	−2.3	2.3	−7.2	−6.4
BC	−1.9	−2.3	2.3	−6.4	−7.4
DE	−15.1	−6.4	6.4	−19.4	−18.9
EF	−2.9	−6.4	6.4	−19.0	−19.5
GH	−11.1	−16.3	16.3	−51.6	−46.4
HI	−9.8	−16.3	16.3	−46.3	−51.5
AD	2.3	1.8	−1.8	7.2	−1.8
BE	0	5.3	−5.3	12.8	3.0
CF	−2.3	1.9	−1.9	7.4	−1.6
DG	8.6	4.7	−4.7	21.1	−7.0
EH	0	17.5	−17.5	34.9	17.6
FI	−8.7	4.8	−4.8	21.1	−6.7
GJ	25.0	14.6	−14.6	58.6	0
HK	0	18.8	−18.8	75.2	0
IL	−25.0	14.6	−14.6	58.2	0

Reactions:

Node	Horiz. kN	Vert. kN	Moment kNm
J	−14.6	−25.0	0
K	−18.8	0	0
L	−14.6	25.0	0

First check overall equilibrium. Figure 9.23(b) shows the external forces and reactions on the structure.

Resolve horizontally: $\sum P_X = 9 + 18 + 21 - 14.6 - 18.8 - 14.6 = 0$

Resolve vertically: $\sum P_Y = 25 - 25 = 0$

Moments about K: $\sum M_Z = (25 \times 6) + (25 \times 6) - (21 \times 4) - (18 \times 7) - (9 \times 10) = 0$

Therefore overall equilibrium is satisfied. We can also see at a glance that the boundary conditions are satisfied, since the moments at the pinned supports are all zero.

Next we can check joint equilibrium. Consider for example joint F (Figure 9.23 (c)). Care needs to be taken with the directions of the internal forces. If we cut each member just away from the joint, then the forces in the above table are those acting at the ends of the members. It can be convenient to think of these as the forces imposed on the members by the joint, and of course there must be equal and opposite forces imposed on the joint by the members. So, for example, the shear force of +6.4 kN at end (2) of member EF acts in the positive Y direction (that is, upwards) at the right-hand end of EF and there is an equal and opposite (downwards) force imposed by the member on the joint F. Then, considering equilibrium of the forces on the joint:

Horizontally: $\sum P_X = 2.9 + 1.9 - 4.8 = 0$

Vertically: $\sum P_Y = 8.7 - 6.4 - 2.3 = 0$

Moments: $\sum M_Z = 19.5 + 1.6 - 21.1 = 0$

Therefore this joint is in equilibrium. Now consider a typical member, say GH. Figure 9.23(d) shows the end displacements of this member. Its properties are $AE = 1.21 \times 10^9$ N and $EI = 20.4 \times 10^6$ Nm^2. Using the axial force–extension relationship (in N and mm):

$$u_{X2} - u_{X1} = \frac{Tl}{AE} = \frac{-11,100 \times 6000}{1.21 \times 10^9} = -0.055 \text{ mm}$$

whereas the computed displacements give:

$$u_{X2} - u_{X1} = 21.8 - 21.9 = -0.1 \text{ mm}$$

which is in agreement, given the degree of accuracy to which the computational results are quoted, though obviously a greater number of decimal places would be desirable. Using the slope–deflection equations (in kN and m):

$$\begin{aligned} M_1 &= \frac{2EI}{l}\left[2\theta_1 + \theta_2 - \frac{3(u_{Y2} - u_{Y1})}{l}\right] \\ &= \frac{2 \times 20.4 \times 10^3}{6}\left[2 \times -2.80 - 2.03 + \frac{0.04}{6}\right] \times 10^{-3} = -51.8 \text{ kNm} \\ M_2 &= \frac{2EI}{l}\left[2\theta_2 + \theta_1 - \frac{3(u_{Y2} - u_{Y1})}{l}\right] \\ &= \frac{2 \times 20.4 \times 10^3}{6}\left[2 \times -2.03 - 2.80 + \frac{0.04}{6}\right] \times 10^{-3} = -46.6 \text{ kNm} \end{aligned}$$

These compare with computed values of $M_1 = -51.6$ kNm and $M_2 = -46.4$ kNm, again a tolerable level of agreement given the degree of accuracy to which the computed values have been given. It would, of course, be possible also to check the shear forces in this member, and to check other joints and members.

In addition to checking the consistency of the output, you should attempt to check that the analysis results make physical sense. For example, make sure that the directions and relative magnitudes of reactions, deflected shapes and bending moment diagrams accord with common sense – if possible, compare with a simple qualitative analysis. Also try to check that the displacements and internal forces are of the right order of magnitude – it is nearly always possible to do a simple hand calculation which gives an approximate check of the major results.

EXAMPLE 9.11

Perform qualitative and order-of-magnitude checks on the computer output given in Example 9.10.

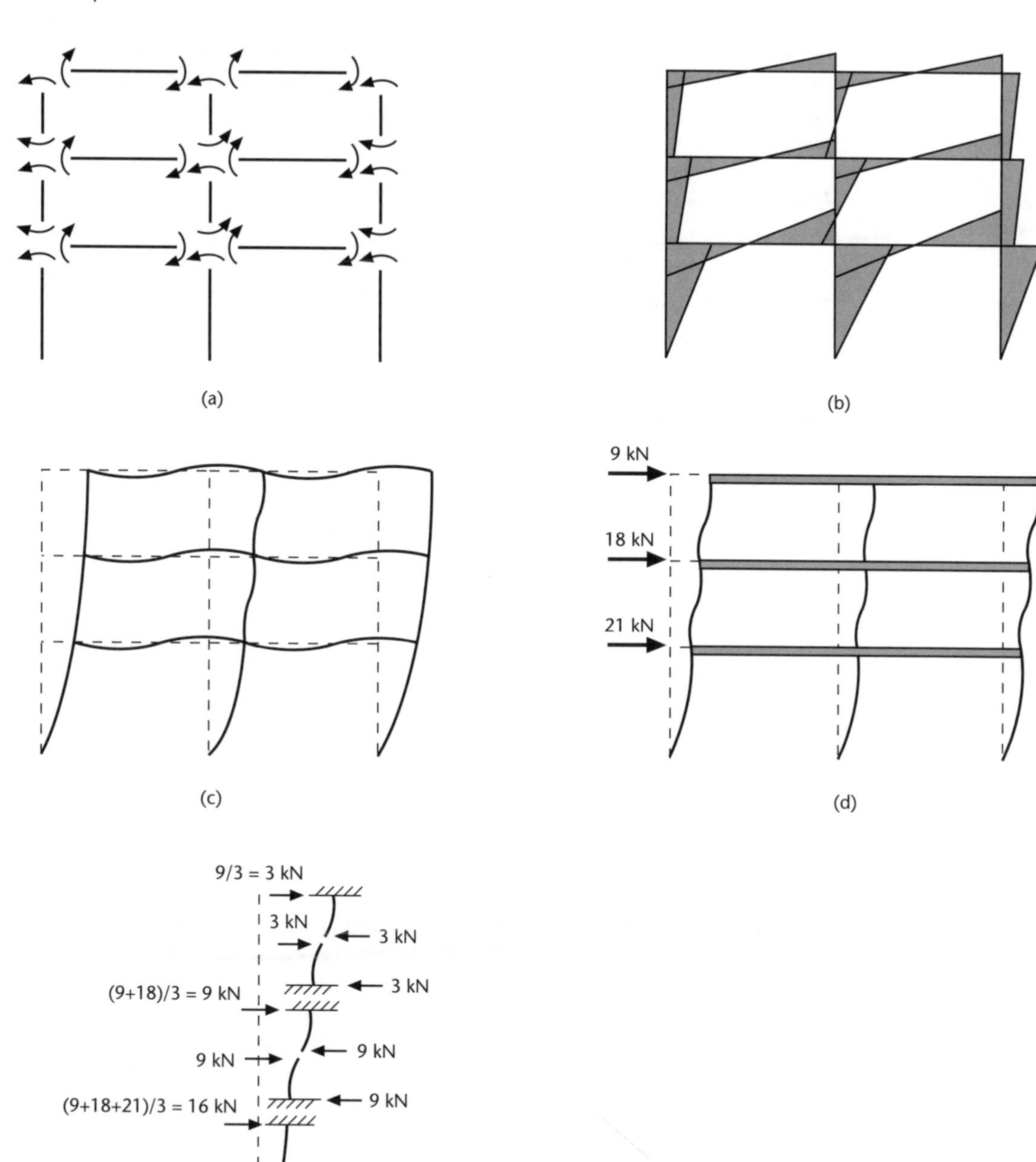

Figure 9.24

In terms of qualitative evaluation of the results, probably the most important check is that the bending moment distribution and deflected shape correspond to each other and to the applied loading. First consider the bending moments, which are tabulated in Example 9.10. Figure 9.24(a) shows the structure with the members separated at the joints and the directions of the moments indicated by arrows, remembering that a positive moment acts anti-clockwise. It is then possible to sketch the bending moment diagram (Figure 9.24(b)) – the lines are drawn on the tension sides of members and, since the only applied loads are at the joints, the moment distribution is linear within any member. Lastly, the deflected shape can be deduced from the bending moment diagram, and is shown in Figure 9.24(c). Note that:

- each beam is sagging at the left-hand end and hogging at the right;
- the curvature of the outer columns retains the same sign throughout the height, as do the corresponding bending moments;
- the central column undergoes several reversals of curvature, corresponding to the change in sign of the bending moments.

This sketched deflected shape is credible for a structure subjected to purely horizontal loading, and also agrees well with the computed deflections given in Example 9.10. In particular:

- every joint rotation is clockwise, that is, negative;
- the joint rotations reduce with height;
- the horizontal deflections increase with height.

We now have some confidence that the calculated moments and displacements make physical sense. Lastly, we need to check the magnitude of the results. It is not easy to do an accurate calculation by hand (otherwise we would not bother with the computer analysis) so we must accept a very rough approximation. Suppose that the beams were completely rigid, so that the frame deformed as shown in Figure 9.24(d) – this is of course, a much stiffer structure than the actual one, and so we would expect its lateral deflection to be lower.

We can find the deflection of each storey of our modified structure using simple beam formulae. As shown in Figure 9.24(e), each line of columns can be regarded as equivalent to a number of cantilevers. The bottom storey behaves just like a cantilever fixed at the top and loaded at the base. The other storeys, which are fixed against rotation at both ends, can each be treated as the sum of two cantilevers with length equal to half the storey height. Remember that each storey must transmit all the loads from its level and above down to the ground, so that the applied loads increase as we go down the structure. Note also that, with three identical lines of columns, each one carries one-third of the total applied load.

From Table 5.2 the deflection of a tip-loaded cantilever is $Pl^3/3EI$, so the deflections of each storey are (in units of N and mm):

Level GHI: $$\delta_{GHI} = \frac{16000 \times 4000^3}{3 \times 29.3 \times 10^{12}} = 11.7 \text{ mm}$$

Level DEF: $$\delta_{DEF} = 11.7 + 2 \times \frac{9000 \times 1500^3}{3 \times 23.3 \times 10^{12}} = 12.6 \text{ mm}$$

Level ABC: $$\delta_{ABC} = 12.6 + 2 \times \frac{3000 \times 1500^3}{3 \times 23.3 \times 10^{12}} = 12.9 \text{ mm}$$

These compare with computed values of $\delta_{GHI} = 21.8$ mm, $\delta_{DEF} = 27.9$ mm and $\delta_{GHI} = 30.2$ mm. As expected, the computed values are quite a bit higher because we have stiffened the structure considerably in order to simplify the hand calculation. Nevertheless, the results show that the displacements are of the right order of magnitude, and the rates of increase of displacement with height agree well. The hand analysis therefore gives us some additional confidence in the correctness of the stiffness matrix program output.

Lastly, when using the stiffness matrix approach, it is important always to be aware of the main assumptions:

- all members behave in a linear, elastic manner, with no yielding;
- members are assumed to be straight and to have constant properties between joints;
- deflections are sufficiently small so as not to affect the overall geometry of the structure;
- the axial forces are comparatively small, so that buckling effects are negligible;
- for all members the shear centre coincides with the centroid of the section, so that a load applied through the centroid produces no torsion.

An analysis may have been performed correctly, but if the results are such that the validity of any of these assumptions is in doubt, then they should be treated with great caution. For example, if the stresses resulting from axial and/or bending forces are close to the yield stress, then it is possible that yielding may occur. If large axial compressions are present, then buckling effects may greatly reduce member stiffnesses and load capacities (see Chapter 11).

Many of the issues raised here are also relevant to finite element computer programs; we shall therefore return to this discussion in Chapter 10.

9.5 Further reading

The following both include good accounts of stiffness matrix methods. The second one is written mainly as a programming guide. With so many stiffness matrix programs already available, it is very unlikely that you would want to write your own, but the book nevertheless has much to recommend it.

- R. K. Livesley, *Matrix Methods of Structural Analysis*, 2nd edn, Pergamon, 1975.
- P. Bhatt, *Programming the Matrix Analysis of Skeletal Structures*, Ellis Horwood, 1986.

9.6 Problems

Where appropriate, answers are given at the end of the book. The questions marked with an asterisk are a little more challenging.

9.1.* Sketch the deflected shapes and the bending moment diagrams for the frames in Figure 9.25 and state which unknown displacements would have to be considered in slope–deflection analyses of the frames. [*Hint: for (b), it may be helpful to consider this as the superposition of two loadcases, one of which involves introducing an artificial horizontal restraint at C.*]

Figure 9.25

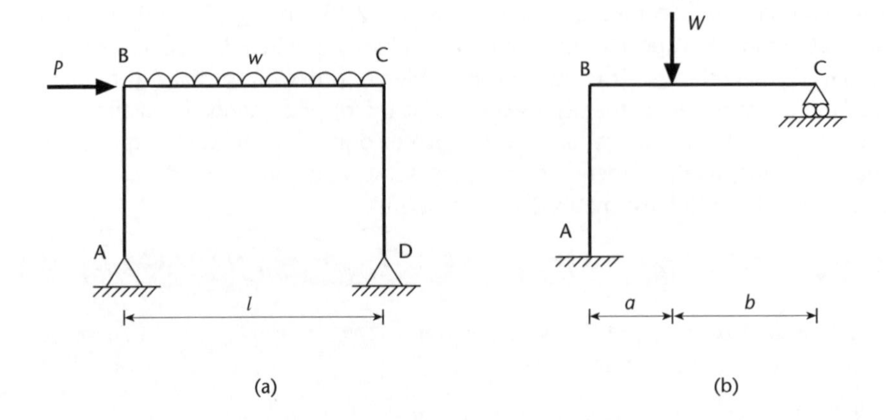

9.2. In the frame shown in Figure 9.26, AB has stiffness EI and BC has stiffness $2EI$. Find the rotations of the joints and draw the bending moment diagram.

Figure 9.26

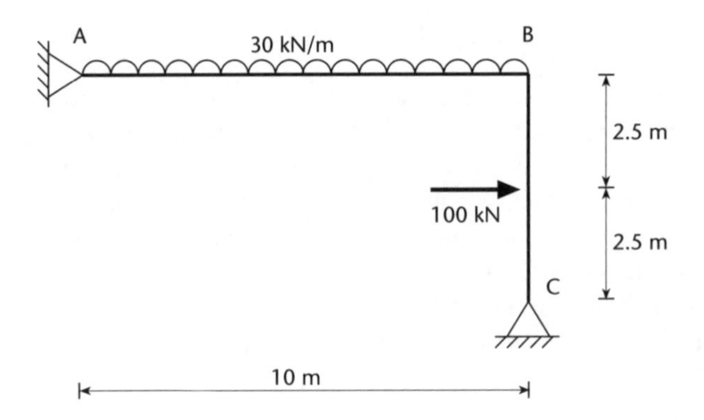

9.3. All the members of the portal frame shown in Figure 9.27 have the same properties. Use the slope–deflection method to calculate the moments at the joints and hence draw the bending moment diagram for the frame.

Figure 9.27

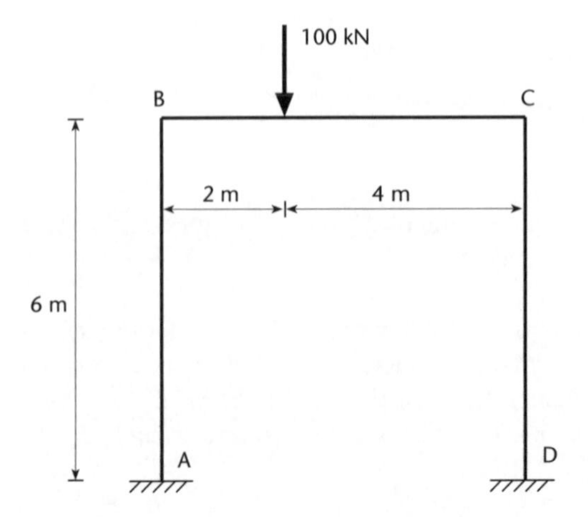

9.4.* Sketch the frame shown in Figure 9.28 when deformed by the action of the moment M applied to the central joint. Use the slope–deflection method to find the angle of rotation of this joint. All members have length l and flexural stiffness EI. [*Hints: (i) What does the way the structure deforms tell you about the moments in the members meeting at the pinned corners? (ii) It is only necessary to analyse a quarter of the structure.*]

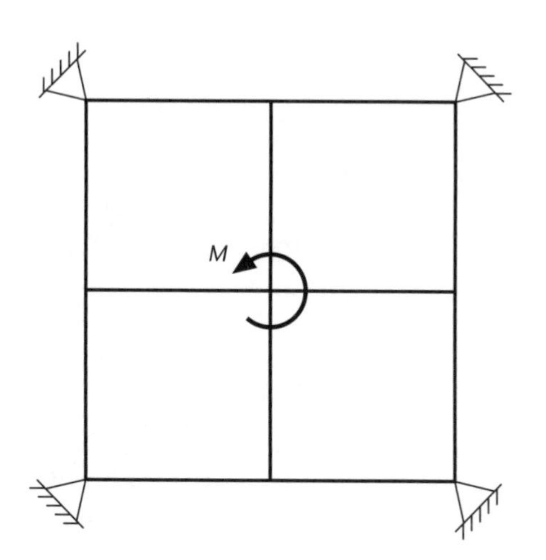

Figure 9.28

9.5. Explain why it would not be appropriate to analyse either of the structures shown in Figure 9.29 by the slope–deflection method, and in each case suggest a more appropriate method.

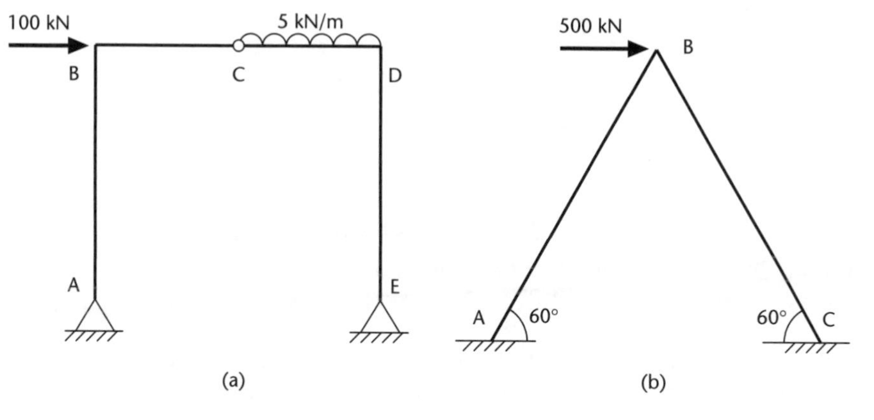

Figure 9.29

9.6. The bars AB and AC in Figure 9.30 have axial stiffnesses $2AE$ and AE respectively. Find the stiffness matrix equation in the global axis system shown. If $AE = 10,000$ kN and $h = 2.0$ m, find the deflections caused by:

(a) downwards load of 1.0 kN at A;
(b) load of 5.0 kN at A, acting at 45° to the X-axis.

9.7. Write down in algebraic form the stiffness matrix equation for the simple structure in Figure 9.31, in which both members have the same properties. If $l = 5$ m, the members are $203 \times 133 \times 30$ kg/m steel Universal Beams (that is,

Figure 9.30

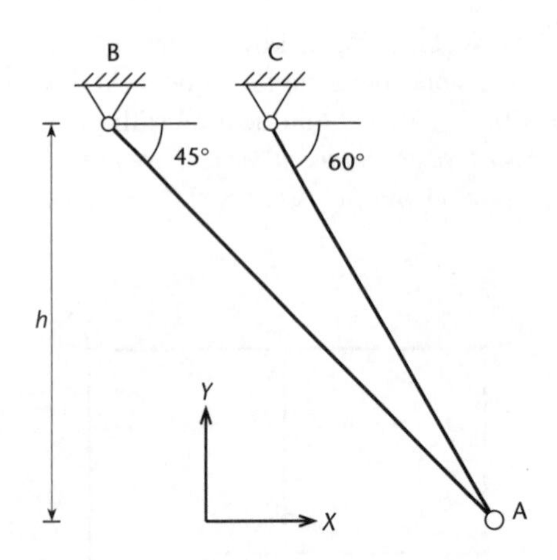

Figure 9.31

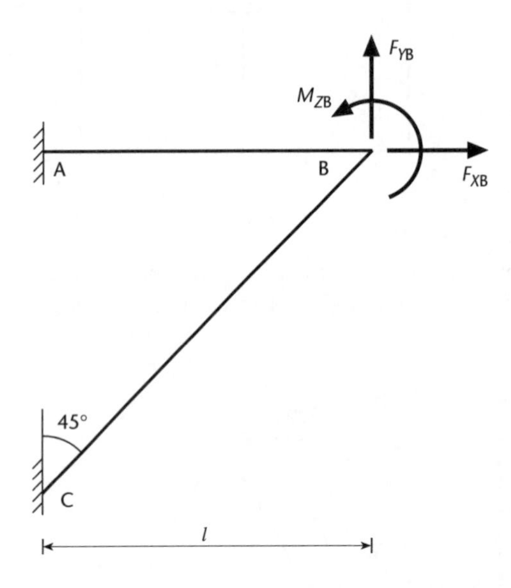

I-sections) with $A = 3.8 \times 10^{-3}$ m^2, $I = 28.87 \times 10^{-6}$ m^4 and $E = 205$ GPa, and the displacements of B are $u_{XB} = -0.732$ mm, $u_{YB} = 4.56$ mm and $\theta_{ZB} = 0.00734$ rads, what are the loads applied at B?

9.8. For each of the frames in Figure 9.25, write down the vector of equivalent joint loads which would be required for a stiffness matrix analysis.

9.9. Find the deflections of joint B of the structure in Figure 9.29(b) by the stiffness matrix approach. For both members take the length as 5 m, AE as 10^6 kN and EI as 10^4 kNm^2.

9.10. In Figure 9.32, AB and BC are both of length l and are joined together at right angles at A. The beam AB has a bending stiffness (EI) of $2K$ and torsional stiffness (GJ) of K, while for BC both the bending and torsional stiffnesses are K. Find the stiffness matrix equation in the global axis system shown.

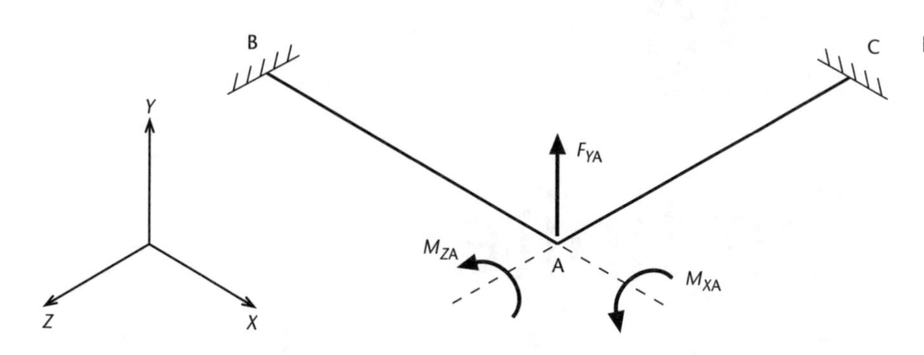

Figure 9.32

9.11.* The structure in Figure 9.33 is to be analysed by the stiffness matrix method. Identify the active degrees of freedom in the problem. Without detailed calculation, draw up the global stiffness matrix corresponding to these degrees of freedom, indicating by crosses which locations in the matrix will be filled by non-zero elements. [*Hint: the matrix should be* 21 × 21 *and must be symmetric.*]

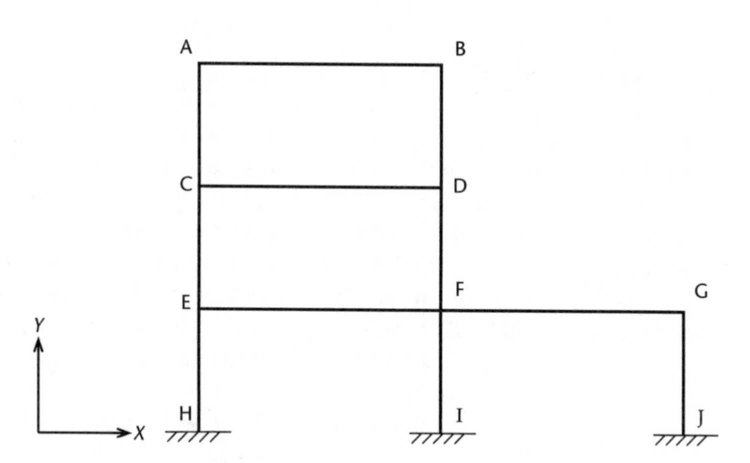

Figure 9.33

9.12. For the structure analysed in Examples 9.10 and 9.11, check for equilibrium of the internal forces at joint B and for agreement between the end forces and displacements of member AD.

CHAPTER 10

The finite element method

CHAPTER OUTLINE

The stiffness matrix method introduced in Chapter 9 provides a powerful, computerised approach to the analysis of *skeletal* structures, that is, ones whose elements are essentially one-dimensional and joined at their ends. While this is a reasonable approximation for many frame structures, there remain a large number of structures that cannot be treated in this way. For example, a beam whose depth is large compared to its length will not obey the simple beam theory of Chapter 5, and so must be analysed as a *continuum* structure, subjected to stresses and strains in two dimensions. Other examples of continua include plate and shell structures, gravity structures such as dams, mechanical components, or the ground underneath a building.

For very regular continua, it is sometime possible to determine internal stresses and deformations reasonably accurately by hand calculation. For anything other than the simplest cases, however, the analysis quickly becomes complex and it is preferable to use a computerised approach. By far the most versatile and popular approach is the *finite element method,* in which the continuum is represented as a *mesh* of discrete *elements* joined together at *nodes*. We can then proceed in a similar way to a skeletal structure, by establishing a stiffness matrix for each element, and then assembling these to give an overall stiffness matrix for the structure.

The finite element method is a huge topic, and it is not possible to provide a comprehensive account of it here. This chapter therefore aims to enable the reader to:

- understand the way in which the stiffness matrix for a finite element is formulated;
- understand how element matrices are combined to give a structure stiffness matrix;
- develop a feel for the way that factors such as element type and mesh density affect the accuracy and efficiency of the solution;
- get to grips with some of the terminology of finite elements;
- use finite element computer programs with an awareness of their strengths and limitations.

10.1 Outline of method

Suppose we wished to analyse the plate shown in Figure 10.1(a), in order to determine the displacements of the structure at any point, the internal stresses and the support reactions. The steps in a finite element analysis are:

(a) Split the structure up into a number of discrete elements joined together only at their nodes – this process is often referred to as *discretisation*. In Figure 10.1 (b) this has been done using four-noded quadrilateral elements, though many other element types could also be used. Note that the elements need not all be the same size. For economy, we want to use as few elements as possible, as otherwise the structure stiffness matrix becomes very large. However, the accuracy of the solution increases as the elements get smaller. It is therefore common to vary the mesh as shown, with a high density of elements in areas where the stresses and strains are expected to vary rapidly and a lower density elsewhere. The shape of the elements can also vary, so as to fit the shape of the structure being modelled. However, very large distortions of the basic element shape can cause numerical instabilities and so should be avoided.

(b) Define how the displacements within the element are related to the nodal displacements, and hence derive the stiffness matrix relating the nodal forces to the corresponding displacements. Matrices relating the element stresses and strains to the nodal displacements can also be derived. (This step will be explained in detail in the next section.) It is possible to develop finite elements based on many different displacement functions. In an analysis, it always advisable to use an element which deforms in much the same way as the structure being modelled, as this will enable an accurate solution to be obtained without resorting to a very high mesh density. For instance, if modelling an elastic beam, in which the variations of stress and strain over the cross-section would be expected to be close to linear, a linear strain element would be appropriate.

(c) Assemble the individual element matrices into the global stiffness matrix equation for the whole structure. This can be done in a similar way to the stiffness matrix approach, using equilibrium and displacement compatibility at the nodes.

(d) Solve the matrix equation to give the nodal displacements. Back-substitute into the element stress and strain matrices to give the stresses and strains at any point in each element.

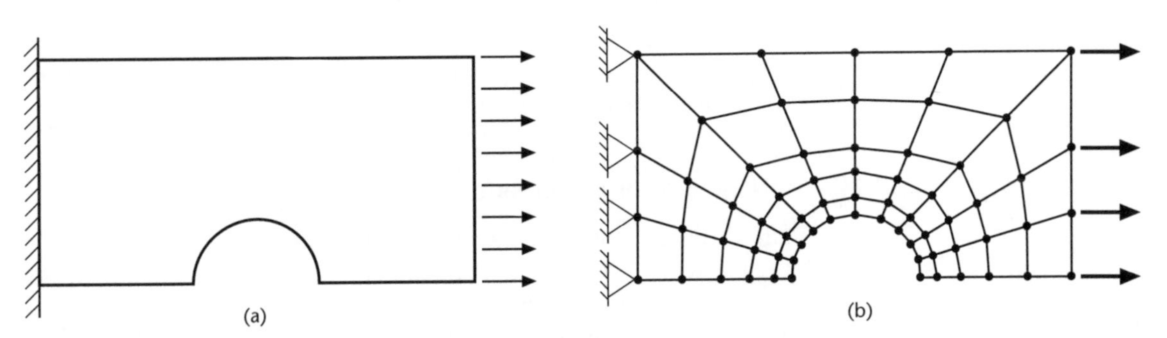

Figure 10.1
Finite element modelling of a plate.

Most users of finite elements will apply the technique using a commercial software package, of which there are many. The user will be required to input the geometry, material properties, boundary conditions and loading, and normally also to define a suitable finite element mesh and choose an appropriate element type (though many programs do have automatic meshing facilities). The remainder of the steps outlined above will be internal to the program, though often the user will be able to choose between alternative analysis options.

In this chapter we shall first look at steps (b) and (c). While these steps are usually carried out automatically by the program, an understanding of the underlying theory is essential if the program is to be used correctly. We shall then consider how to approach the discretisation process, step (a); essentially this always involves a trade-off between accuracy and efficiency. The solution phase, step (d), is important, but is really a topic in numerical analysis rather than structures. This will therefore be discussed only very briefly.

10.2 A simple plane stress finite element: the constant-strain triangle

It is possible to define 1D, 2D or 3D finite elements: 1D elements are similar to the bar and beam elements used in Chapter 9; 3D *brick* elements are used for modelling solid bodies which cannot be readily reduced to planar form. We shall concentrate on 2D *plate* elements, which are widely used for analysing continua which are, or can be reasonably approximated as, planar. In this section we will illustrate the procedure by reference to a particularly simple element known as the *constant strain triangle*. This is actually a very crude element and is rarely used in practice, but it serves to illustrate the derivation procedure without undue complexity. The same approach is then applicable to more sophisticated elements. The main steps of the derivation are:

- define the geometry of the element using *shape functions*;
- relate the displacements at points within the element to the nodal displacements using a *displacement function*;
- using the strain–displacement relationships of Chapter 4, find a matrix relating the strains within the element to the nodal displacements;
- using the stress–strain relationships for an elastic material, relate the stresses to the nodal displacements;
- relate the stresses within the element to the nodal forces (this is most conveniently done using virtual work);
- we are now able to write an equation relating the nodal force vector to the corresponding displacement vector, via a stiffness matrix.

10.2.1 Shape and displacement functions

Figure 10.2(a) shows a single triangular element, with nodes 1, 2 and 3 at the corners. The element could have any orientation in space, but for simplicity we shall assume it lies in the global (X, Y) plane.

A convenient way of defining the geometry of the element is by using *shape functions*. We first define a reference element in the (α, β) plane which has a very

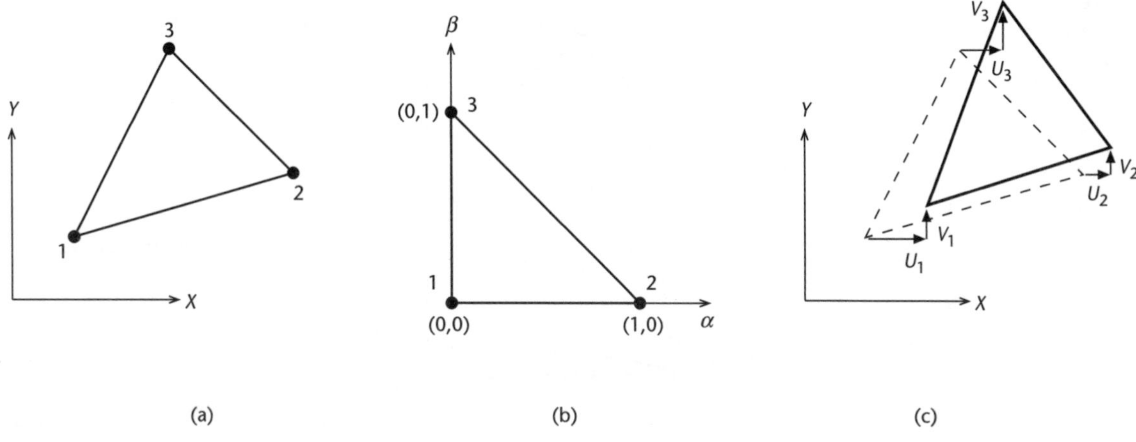

Figure 10.2
Three-noded triangular finite element.

simple geometry (Figure 10.2(b)). We then define one shape function for each node in the reference element, with each shape function S_i chosen so as to give a value of 1.0 at node i and zero at all other nodes. For our three-noded element, the shape functions have the general form:

$$S_i = A_1 + A_2\alpha + A_3\beta$$

The coefficients A_1–A_3 are determined by substituting the required function values at the nodes into this expression – it therefore follows that a shape function can only contain as many coefficients as there are nodes. The evaluation of the coefficients is very easy. For example, the function S_1 must equal 1.0 at $(\alpha, \beta) = (0, 0)$ and zero at $(1, 0)$ and $(0, 1)$, hence $A_1 = 1$, $A_2 = -1$, $A_3 = -1$. Repeating this simple exercise for S_2 and S_3 we get:

$$\begin{aligned} S_1 &= 1 - \alpha - \beta \\ S_2 &= \alpha \\ S_3 &= \beta \end{aligned} \qquad (10.1)$$

Note that each shape function varies linearly over the reference element, and over the real element. We can now define the coordinates (X, Y) of any point within the real element in terms of the coordinates (α, β) of the corresponding point in the reference element, and the coordinates of the nodes:

$$\begin{aligned} X &= S_1X_1 + S_2X_2 + S_3X_3 \\ Y &= S_1Y_1 + S_2Y_2 + S_3Y_3 \end{aligned}$$

This can be written in matrix form as:

$$\begin{bmatrix} X \\ Y \end{bmatrix} = \begin{bmatrix} S_1 & 0 & S_2 & 0 & S_3 & 0 \\ 0 & S_1 & 0 & S_2 & 0 & S_3 \end{bmatrix} \begin{bmatrix} X_1 \\ Y_1 \\ X_2 \\ Y_2 \\ X_3 \\ Y_3 \end{bmatrix} \quad \text{or} \quad \mathbf{x} = \mathbf{SX} \qquad (10.2)$$

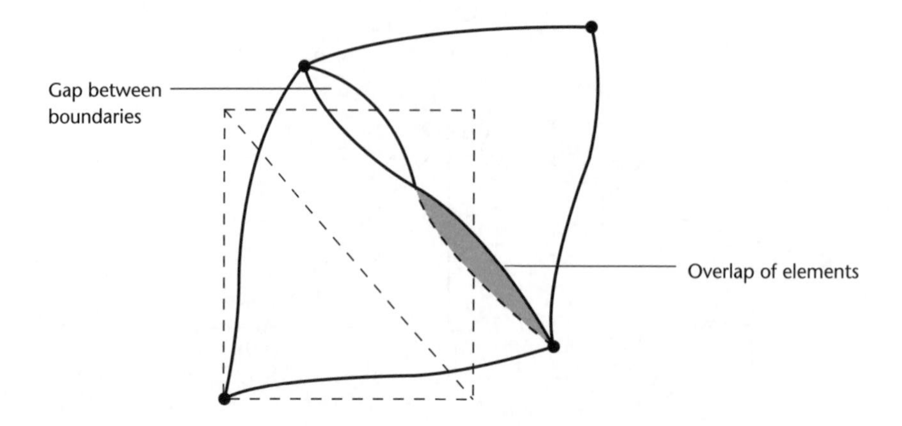

Figure 10.3
Deformations of non-conforming elements.

where **x** is the vector of coordinates of an arbitrary point within an element, **S** is the shape function matrix and **X** is the vector of nodal coordinates.

Having defined the geometry of the element, we now consider the displacements. We must define a *displacement function* relating the displacements at any point within the element to the nodal displacements shown in Figure 10.2(c). The displacement function is the key to how the element performs, since it determines whether the stress and strain variations over the element are linear, parabolic, or whatever. It must satisfy the following conditions:

(a) Both the function and its first derivatives must be continuous within the element.

(b) Nodal displacements corresponding to rigid-body movements of the element should not generate strain at any point within the element.

(c) The displacement function must satisfy the conditions of strain compatibility within the element.

(d) Compatibility of displacements between adjacent elements is also required. This is automatically assured at any node shared between two elements, but it is also desirable that adjacent edges of elements should undergo the same deformations, that is, that deformations of the kind shown in Figure 10.3 should not occur. A displacement function satisfying this condition is known as a *conforming function* and the resulting element as a *conforming element.*

There are many possible forms of displacement function, but the most common approach is simply to use the shape function matrix, so that the displacement at any point is given by:

$$\begin{bmatrix} U \\ V \end{bmatrix} = \begin{bmatrix} S_1 & 0 & S_2 & 0 & S_3 & 0 \\ 0 & S_1 & 0 & S_2 & 0 & S_3 \end{bmatrix} \begin{bmatrix} U_1 \\ V_1 \\ U_2 \\ V_2 \\ U_3 \\ V_3 \end{bmatrix} \quad \text{or} \quad \mathbf{u} = \mathbf{SU} \tag{10.3}$$

where **u** is the vector of displacements at the arbitrary point (X, Y) and **U** is the vector of nodal displacements. A finite element formulated using this approach is referred to as *isoparametric.*

EXAMPLE 10.1

Show that Equation (10.3) satisfies the conditions for a valid displacement function given above.

(a) Since the shape functions S_i are all linear functions of the reference coordinates α and β, both the displacements (U, V) and the coordinates (X, Y) are also linear functions of (α, β). It follows that the displacements must vary linearly with the real coordinates (X, Y). Therefore both the function and its first derivatives with respect to X and Y are continuous throughout the element.

(b) A rigid-body translation in the X direction requires $U_1 = U_2 = U_3 = \delta_X$, say, and $V_1 = V_2 = V_3 = 0$. From equation (10.3), the displacements of any point in the element are $U = \delta_X \cdot (S_1 + S_2 + S_3) = \delta_X$, since the shape functions sum to unity, and $V = 0$. All points in the element thus undergo the same displacement and so the element is not subjected to strain at any point.

Similarly, for a Y-translation $V_1 = V_2 = V_3 = \delta_Y$, say, and $U_1 = U_2 = U_3 = 0$, from which it follows that the displacements at any point are $U = 0$ and $V = \delta_Y$. Again, all points undergo the same displacement and so the element is unstrained.

Lastly consider a rigid-body rotation of θ about the origin (Figure 10.4). From diagram (b) it can be seen that the displacements are:

$$\begin{aligned}
U_1 &= -Y_1 \tan\theta, \quad & V_1 &= X_1 \tan\theta, \\
U_2 &= -Y_2 \tan\theta, \quad & V_2 &= X_2 \tan\theta, \\
U_3 &= -Y_3 \tan\theta, \quad & V_3 &= X_3 \tan\theta
\end{aligned}$$

Applying equation (10.3) and then substituting from (10.2) gives:

$$U = -\tan\theta \cdot (S_1 Y_1 + S_2 Y_2 + S_3 Y_3) = -Y \tan\theta$$

and $$V = \tan\theta \cdot (S_1 X_1 + S_2 X_2 + S_3 X_3) = X \tan\theta$$

All points thus undergo the same rotation about the origin and the element is unstrained. Any other rigid body movement can be generated by superposition of these three cases.

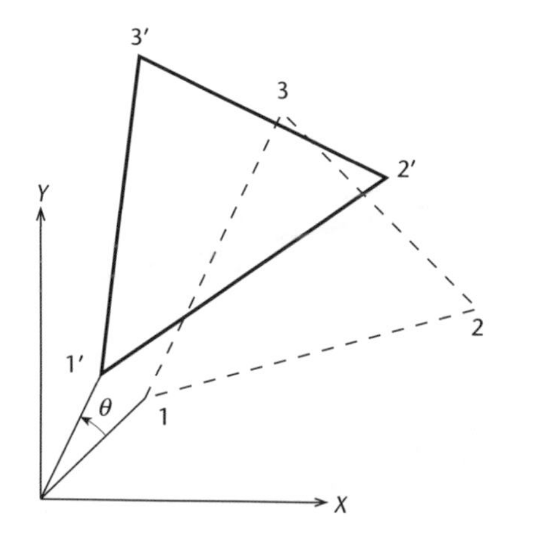

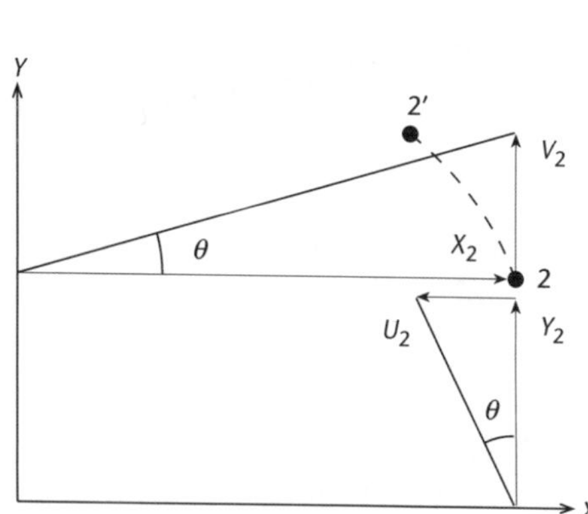

Figure 10.4

(c) The conditions required for local strain compatibility in two dimensions are not covered in detail in this book. It can be shown that compatibility is satisfied if a relationship between the partial second differentials of the direct and shear strains holds. In this case, since the displacements vary linearly within the element the strains are all constant. The partial second differentials are therefore all equal to zero and the relationship is satisfied.

(d) Consider a boundary between two adjacent elements, connected to the same two end nodes. Since the displacements within each element vary linearly, it follows that any initially straight line in an element must remain straight as the element deforms. The boundaries of both elements therefore remain as straight lines between the two end nodes, and are thus in contact over their full length. Hence equation (10.3) represents a conforming function.

10.2.2 Element strain and stress matrices

The next steps are to relate the nodal displacements to strains, and then to stresses. Strains at a point within the element can be calculated using the standard definitions (equations (4.9) and (4.10)):

$$\varepsilon_{XX} = \frac{\partial U}{\partial X}, \quad \varepsilon_{YY} = \frac{\partial V}{\partial Y}, \quad \gamma_{XY} = \frac{\partial U}{\partial Y} + \frac{\partial V}{\partial X}$$

The evaluation of these differentials is quite complex, because the displacements U and V are expressed as functions of α and β rather than X and Y. Consider for example ε_{XX}. Using the *chain rule* for partial differentiation we can write:

$$\varepsilon_{XX} = \frac{\partial U}{\partial X} = \frac{\partial U}{\partial \alpha} \cdot \frac{\partial \alpha}{\partial X} + \frac{\partial U}{\partial \beta} \cdot \frac{\partial \beta}{\partial X} \tag{10.4}$$

By combining equations (10.3) and (10.1) we get:

$$U = U_1 + \alpha(U_2 - U_1) + \beta(U_3 - U_1), \quad \text{so} \quad \frac{\partial U}{\partial \alpha} = U_2 - U_1 \quad \text{and} \quad \frac{\partial U}{\partial \beta} = U_3 - U_1$$

Combining equations (10.2) and (10.1) and rearranging gives:

$$\alpha = \frac{(X - X_1)(Y_3 - Y_1) - (Y - Y_1)(X_3 - X_1)}{(X_2 - X_1)(Y_3 - Y_1) - (Y_2 - Y_1)(X_3 - X_1)},$$

$$\text{so} \quad \frac{\partial \alpha}{\partial X} = \frac{Y_3 - Y_1}{(X_2 - X_1)(Y_3 - Y_1) - (Y_2 - Y_1)(X_3 - X_1)}$$

$$\beta = \frac{(X - X_1)(Y_2 - Y_1) - (Y - Y_1)(X_2 - X_1)}{(X_3 - X_1)(Y_2 - Y_1) - (Y_3 - Y_1)(X_2 - X_1)},$$

$$\text{so} \quad \frac{\partial \beta}{\partial X} = \frac{Y_1 - Y_2}{(X_2 - X_1)(Y_3 - Y_1) - (Y_2 - Y_1)(X_3 - X_1)}$$

In these last two partial differentials the denominator is in fact equal to twice the area A of the element. So, substituting all these terms into equation (10.4):

$$\varepsilon_{XX} = \frac{(U_2 - U_1)(Y_3 - Y_1) + (U_3 - U_1)(Y_1 - Y_2)}{2A}$$

Repeating this exercise for ε_{YY} and γ_{XY} and expressing the result in matrix form:

$$\begin{bmatrix} \varepsilon_{XX} \\ \varepsilon_{YY} \\ \gamma_{XY} \end{bmatrix} = \frac{1}{2A} \begin{bmatrix} Y_2 - Y_3 & 0 & Y_3 - Y_1 & 0 & Y_1 - Y_2 & 0 \\ 0 & X_3 - X_2 & 0 & X_1 - X_3 & 0 & X_2 - X_1 \\ X_3 - X_2 & Y_2 - Y_3 & X_1 - X_3 & Y_3 - Y_1 & X_2 - X_1 & Y_1 - Y_2 \end{bmatrix} \begin{bmatrix} U_1 \\ V_1 \\ U_2 \\ V_2 \\ U_3 \\ V_3 \end{bmatrix}$$

$$\text{or} \quad \boldsymbol{\varepsilon} = \mathbf{BU} \tag{10.5}$$

where ε is the strain vector and $\mathbf{B}$ is sometimes known as the element strain matrix. Note that, for this element, all the terms in $\mathbf{B}$ are simple expressions involving the nodal coordinates and so are constant throughout the element. Therefore, when a set of nodal displacements is applied, the resulting strains will be the same at every point within the element. This is why the element is usually referred to as the *constant strain triangle*.

There are various ways of arriving at equation (10.5); the method used here has been chosen for its simplicity rather than its mathematical elegance. For other element types this approach may not be suitable and a more sophisticated approach is usually adopted. The reader should consult specialist finite element texts for details of such methods (see section 10.5).

We now wish to convert the strains to stresses. For direct stresses and strains in two dimensions we use Hooke's law, (equation (4.13)), with σ_{ZZ} set to zero:

$$\varepsilon_{XX} = \frac{1}{E}(\sigma_{XX} - \nu\sigma_{YY})$$

$$\varepsilon_{YY} = \frac{1}{E}(\sigma_{YY} - \nu\sigma_{XX})$$

and the shear stress and strain are related by equation (4.14):

$$\gamma_{XY} = \frac{\tau_{XY}}{G}$$

Inverting these expressions and writing in matrix form:

$$\begin{bmatrix} \sigma_{XX} \\ \sigma_{YY} \\ \tau_{XY} \end{bmatrix} = \begin{bmatrix} E/(1-\nu^2) & E\nu/(1-\nu^2) & 0 \\ E\nu/(1-\nu^2) & E/(1-\nu^2) & 0 \\ 0 & 0 & G \end{bmatrix} \begin{bmatrix} \varepsilon_{XX} \\ \varepsilon_{YY} \\ \gamma_{XY} \end{bmatrix} \quad \text{or} \quad \boldsymbol{\sigma} = \mathbf{D}\boldsymbol{\varepsilon} \tag{10.6}$$

where $\boldsymbol{\sigma}$ is the stress vector and $\mathbf{D}$ is the element stress matrix. Of course, equations (10.5) and (10.6) can be combined to give:

$$\boldsymbol{\sigma} = \mathbf{DBU}$$

EXAMPLE 10.2

A three-noded triangular finite element is used to model a steel plate having $E = 205$ GPa and $\nu = 0.3$. The nodal coordinates are $(0,0)$, $(150,20)$ and $(100,100)$. Calculate the stresses and strains corresponding to the nodal displacement vector $[-0.014 \quad 0.025 \quad -0.050 \quad 0.028 \quad 0.075 \quad 0.092]^{\mathrm{T}}$. (Coordinates and displacements are both in units of millimetres.)

Figure 10.5

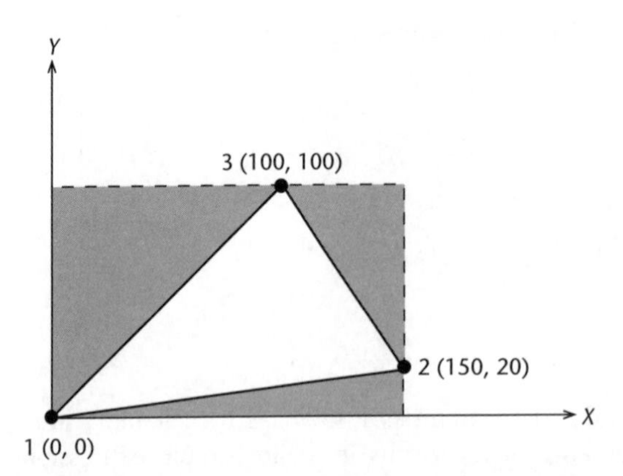

The area of the element is most easily found by drawing a rectangle enclosing the element (Figure 10.5), and then subtracting the shaded triangles around the outside:

$$A = 150 \times 100 - \tfrac{1}{2}(150 \times 20 + 50 \times 80 + 100 \times 100) = 6{,}500 \text{ mm}^2$$

The strains can now be found from equation (10.5):

$$\begin{bmatrix} \varepsilon_{XX} \\ \varepsilon_{YY} \\ \gamma_{XY} \end{bmatrix} = \frac{1}{2 \times 6{,}500} \begin{bmatrix} -80 & 0 & 100 & 0 & -20 & 0 \\ 0 & -50 & 0 & -100 & 0 & 150 \\ -50 & -80 & -100 & 100 & 150 & -20 \end{bmatrix} \begin{bmatrix} -0.014 \\ 0.025 \\ -0.050 \\ 0.028 \\ 0.075 \\ 0.092 \end{bmatrix}$$

$$= \begin{bmatrix} -414 \\ 750 \\ 1224 \end{bmatrix} \times 10^{-6}$$

The shear modulus is $G = E/2(1 + \nu) = 79$ GPa. Equation (10.6) then gives the stresses:

$$\begin{bmatrix} \sigma_{XX} \\ \sigma_{YY} \\ \tau_{XY} \end{bmatrix} = 10^3 \begin{bmatrix} 225.6 & 67.6 & 0 \\ 67.6 & 225.6 & 0 \\ 0 & 0 & 79 \end{bmatrix} \begin{bmatrix} -414 \\ 750 \\ 1224 \end{bmatrix} \times 10^{-6} = \begin{bmatrix} -42.7 \\ 141.2 \\ 96.7 \end{bmatrix} \text{MPa}$$

10.2.3 Element stiffness matrix

We are now very close to being able to formulate the stiffness matrix for our element. The last step is to convert the element stresses into equivalent nodal forces and relate these to the nodal displacements. This can be done using the virtual work method introduced in Chapter 7.

Let the vector of equivalent nodal forces be $\mathbf{F}$. Suppose we impose on our triangular element an arbitrary set of nodal displacements $\hat{\mathbf{U}}$, which generates corresponding strains $\hat{\boldsymbol{\varepsilon}}$. The principle of virtual work states that the external virtual work done by the real nodal loads $\mathbf{F}$ moving through the virtual displacements $\hat{\mathbf{U}}$ is equal to the internal virtual work due to the real stresses $\boldsymbol{\sigma}$ undergoing virtual strains $\hat{\boldsymbol{\varepsilon}}$, that is:

$$\hat{\mathbf{U}}^{\mathrm{T}}\mathbf{F} = \int \hat{\boldsymbol{\varepsilon}}^{\mathrm{T}}\boldsymbol{\sigma}\mathrm{d}V$$

where the integral is performed over the volume of the element. We know from the previous section that the vector of real stresses is related to the real nodal displacements by $\boldsymbol{\sigma} = \mathbf{DBU}$. Also, the virtual strains and displacements must satisfy $\hat{\boldsymbol{\varepsilon}} = \mathbf{B}\hat{\mathbf{U}}$, which can be transposed to give $\hat{\boldsymbol{\varepsilon}}^{\mathrm{T}} = \hat{\mathbf{U}}^{\mathrm{T}}\mathbf{B}^{\mathrm{T}}$. Substituting into the virtual work equation:

$$\hat{\mathbf{U}}^{\mathrm{T}}\mathbf{F} = \int (\hat{\mathbf{U}}^{\mathrm{T}}\mathbf{B}^{\mathrm{T}}) \cdot (\mathbf{DBU})\mathrm{d}V \tag{10.8}$$

We now note two important facts about $\hat{\mathbf{U}}^{\mathrm{T}}$. First, it is a vector of nodal displacements and does not vary within the element, so it may be taken outside the integral. Second, it is an *arbitrary* vector, that is, its elements may take any value (so long as they are compatible with each other) and need bear no relation to the real forces $\mathbf{F}$ and real displacements $\mathbf{U}$ – this is a fundamental aspect of the principle of virtual work. Thus equation (10.8) must be true for any possible vector $\hat{\mathbf{U}}^{\mathrm{T}}$; this condition is only satisfied if:

$$\mathbf{F} = \left(\int \mathbf{B}^{\mathrm{T}}\mathbf{DB}\mathrm{d}V\right)\mathbf{U} \quad \text{or} \quad \mathbf{F} = \mathbf{KU} \tag{10.9}$$

Equation (10.9) gives us a relationship between the nodal forces and displacements, in which the stiffness matrix for the element is obtained by pre- and post-multiplying the element stress matrix $\mathbf{D}$ by the strain matrix $\mathbf{B}$, and then integrating the result over the volume of the element.

For the constant strain triangle all the terms in $\mathbf{B}$ and $\mathbf{D}$ are constant so, if the element has area A and thickness t, the stiffness matrix is simply:

$$\mathbf{K} = \int \mathbf{B}^{\mathrm{T}}\mathbf{DB}\mathrm{d}V = \mathbf{B}^{\mathrm{T}}\mathbf{DB}\int \mathrm{d}V = \mathbf{B}^{\mathrm{T}}\mathbf{DB}At$$

Carrying out the matrix multiplication will result in an element stiffness matrix equation of the form:

$$\begin{bmatrix} F_{X1} \\ F_{Y1} \\ F_{X2} \\ F_{Y2} \\ F_{X3} \\ F_{Y3} \end{bmatrix} = \begin{bmatrix} K_{11} & K_{12} & K_{13} & K_{14} & K_{15} & K_{16} \\ K_{21} & K_{22} & K_{23} & K_{24} & K_{25} & K_{26} \\ K_{31} & K_{32} & K_{33} & K_{34} & K_{35} & K_{36} \\ K_{41} & K_{42} & K_{43} & K_{44} & K_{45} & K_{46} \\ K_{51} & K_{52} & K_{53} & K_{54} & K_{55} & K_{56} \\ K_{61} & K_{62} & K_{63} & K_{64} & K_{65} & K_{66} \end{bmatrix} \begin{bmatrix} U_{X1} \\ U_{Y1} \\ U_{X2} \\ U_{Y2} \\ U_{X3} \\ U_{Y3} \end{bmatrix} \tag{10.10}$$

Even for this extremely crude finite element, the stiffness terms are quite complicated, for example

$$K_{11} = \frac{E}{(1 - \nu^2)}(Y_2 - Y_3)^2 + G(X_3 - X_2)^2$$

It would be extremely cumbersome to write the matrix out in full. This is, in any case, not a particularly useful thing to do, since the method is never used for hand calculations, and numerous finite element computer programs are available which can calculate the coefficients very quickly. Instead, let us just note a few salient points about the form of the matrix:

- it is a 6×6 matrix, corresponding to the two degrees of freedom at each of the three nodes;
- in its general form it does not contain any zero elements (though of course for a particular set of nodal coordinates some terms may turn out to be zero) – unlike the 1D beam elements in Chapter 9, there is a direct coupling between all the degrees of freedom in a plane stress element;
- the terms are functions of both the element geometry (through **B**) and the material properties (through **D**);
- like the matrices derived in Chapter 9, it is symmetric, that is, $K_{ij} = K_{ji}$;
- because the derivation for this element is based entirely on the global coordinates of the nodes, the stiffness matrix achieved is automatically in global form, with no need for any transformation.

For most elements, the integration required to generate the stiffness matrix is much more complex than in this simple case. For a planar element with uniform thickness t, the volume integral can be expressed as:

$$\mathbf{K} = t \int \int \mathbf{B}^{\mathrm{T}} \mathbf{D} \mathbf{B} \, \mathrm{d}X \mathrm{d}Y$$

However, because the element may have any orientation, the appropriate integration limits in (X, Y) coordinates can be difficult to determine. It is therefore normal to transform into an integral over the reference element, in (α, β) coordinates:

$$\mathbf{K} = t \int_0^1 \int_0^{1-\beta} \mathbf{B}^{\mathrm{T}} \mathbf{D} \mathbf{B} \bar{A} \, \mathrm{d}\alpha \mathrm{d}\beta$$

where $\bar{A}$ is the ratio of the area of the real element to that of the reference element ($= 2A$ for our constant strain triangle). Even after transformation, the integral is likely to be complicated and is usually evaluated using a numerical method such as Gaussian integration.

10.2.4 Assembly of structure stiffness matrix

As with the stiffness matrix method, once we have found the stiffness matrices of the individual elements in the global axis system, the next stage is to assemble them into an overall matrix equation for the whole structure. As before, this requires considerations of:

(a) displacement compatibility – two elements connected at a node must undergo the same displacement at that node; and

(b) force equilibrium – the total force acting at a node is the sum of the nodal forces in all the elements meeting at that node.

In practice, the procedure is quite simple and methodical. We first identify all the active degrees of freedom; for plane stress elements there are two at each node. As with the stiffness matrix method, we omit any degrees of freedom that are known to be zero because of the support conditions. We can now write load and displacement vectors, and then assemble the structure stiffness matrix by transferring the element stiffness terms into the appropriate locations. A stiffness term relating the *i*th term in the load vector to the *j*th term in the displacement vector goes in row *i*, column *j* of the stiffness matrix. If a location in the matrix is already occupied by a stiffness term, the new term is added to the existing one.

EXAMPLE 10.3

Assemble the structure stiffness matrix for the finite element model of a pin-supported beam in Figure 10.6. (Note: This is an extremely poor model, which would give very inaccurate results. A better model would use more sophisticated elements, or a higher mesh density, or preferably both. It is used here purely to illustrate the assembly procedure.)

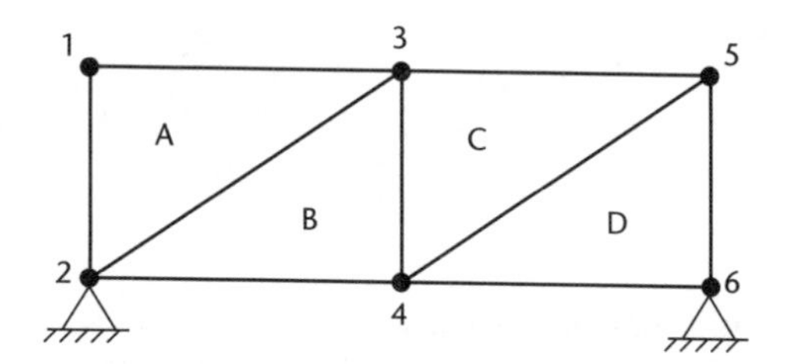

Figure 10.6

There are eight degrees of freedom in the problem (two at each unrestrained node), so an 8×8 matrix is required. For each element, we can write down a matrix equation of the form of equation (10.10). We must be careful when transferring terms from these element equations, as they have been derived on the basis of a particular node numbering system – the nodes on each element were numbered in *anti-clockwise* order. We therefore define a global node numbering system and then establish which global nodes are used to make up each element. The node numbering is shown in Figure 10.6, in which the elements are identified by the letters A–D. We shall use the same letter to denote the terms in each element stiffness matrix, that is, A_{11} is the first element of the stiffness matrix for element A, etc. It makes sense to number nodes by working across the short dimension of the structure, so as to keep the difference between the highest and lowest node number on any adjoining elements as small as possible. This helps to minimise the bandwidth of the final matrix, which in turn enables the matrix equation to be solved as efficiently as possible.

The information on which global nodes make up each element can be conveniently tabulated:

Element	Node 1	Node 2	Node 3
A	1	2 (pinned)	3
B	2 (pinned)	4	3
C	3	4	5
D	4	6 (pinned)	5

We can now transfer terms to the appropriate locations in the stiffness matrix. For element A, the second node is pinned, so we can delete rows and columns 3 and 4 from the element stiffness matrix. The remaining 4×4 matrix goes in the first four rows and columns of the global matrix, corresponding to global nodes 1 and 3. This element is particularly straightforward as the element and global node numbers are the same.

For element B, the first node is pinned so rows and columns 1 and 2 can be deleted. The remaining terms are inserted in the global matrix noting that element node 2 corresponds to global node 4 and element node 3 corresponds to global node 3. The term B_{35}, for example, relates the X-direction force at element node 2 to the X-displacement at element node 3. From the table above, this is equivalent to relating the X-force at global node 4 to the X-displacement at global node 3. This element therefore goes in row 5, column 3 of the stiffness matrix. Continuing in this way for all elements, we arrive at the stiffness matrix equation for the structure:

$$\begin{bmatrix} F_{X1} \\ F_{Y1} \\ F_{X3} \\ F_{Y3} \\ F_{X4} \\ F_{Y4} \\ F_{X5} \\ F_{Y5} \end{bmatrix} = \left[\begin{array}{cc|cc|cc|cc} A_{11} & A_{12} & A_{15} & A_{16} & 0 & 0 & 0 & 0 \\ A_{21} & A_{22} & A_{25} & A_{26} & 0 & 0 & 0 & 0 \\ \hline A_{51} & A_{52} & A_{55}+B_{55}+C_{11} & A_{56}+B_{56}+C_{12} & B_{53}+C_{13} & B_{54}+C_{14} & C_{15} & C_{16} \\ A_{61} & A_{62} & A_{65}+B_{65}+C_{21} & A_{66}+B_{66}+C_{22} & B_{63}+C_{23} & B_{64}+C_{24} & C_{25} & C_{26} \\ \hline 0 & 0 & B_{35}+C_{31} & B_{36}+C_{32} & B_{33}+C_{33}+D_{11} & B_{34}+C_{34}+D_{12} & C_{35}+D_{15} & C_{36}+D_{16} \\ 0 & 0 & B_{45}+C_{41} & B_{46}+C_{42} & B_{43}+C_{43}+D_{21} & B_{44}+C_{44}+D_{22} & C_{45}+D_{25} & C_{46}+D_{26} \\ \hline 0 & 0 & C_{51} & C_{52} & C_{53}+D_{51} & C_{54}+D_{52} & C_{55}+D_{55} & C_{56}+D_{56} \\ 0 & 0 & C_{61} & C_{62} & C_{63}+D_{61} & C_{64}+D_{62} & C_{65}+D_{65} & C_{66}+D_{66} \end{array}\right] \begin{bmatrix} U_{X1} \\ U_{Y1} \\ U_{X3} \\ U_{Y3} \\ U_{X4} \\ U_{Y4} \\ U_{X5} \\ U_{Y5} \end{bmatrix}$$

The non-zero terms are located in a band around the leading diagonal – the banded nature of the stiffness matrix would be much more apparent for a larger structure, so long as a sensible numbering scheme is adopted.

10.2.5 Elements loaded between the nodes

We have seen how to generate a stiffness matrix equation for an assemblage of finite elements subjected to point loads at the nodes. If the structure is loaded only at a few discrete points then it is usually quite simple to define the finite element mesh so that there are nodes at all the loaded points. Often, however, we will want to model problems involving distributed loads acting on the edges of elements. As with the stiffness matrix method previously, we must replace such loads by equivalent point loads acting at the nodes.

The equivalent nodal loads can be found using virtual work. Suppose we have a vector of distributed loads $\mathbf{w} = [w_X \quad w_Y]^T$ acting along the edge of an element between nodes 1 and 2. We wish to replace this by a vector of point loads $\mathbf{F} = [F_{X1} \quad F_{Y1} \quad F_{X2} \quad F_{Y2}]^T$ acting at the end nodes.

We apply a set of virtual displacements $\hat{\mathbf{U}}$ at nodes 1 and 2 with other nodes fixed, and these in turn generate virtual displacements $\hat{\mathbf{u}}$ at some point within the element. The virtual work done by the distributed load must equal that done by the equivalent point loads, so

$$\int \hat{\mathbf{u}}^T \mathbf{w} \, dl = \hat{\mathbf{U}}^T \mathbf{F}$$

where the integration is performed over the length of the element edge. Now from equation (10.3) the internal displacements are related to the nodal displacements by the shape functions. If we define a reduced shape function matrix $\mathbf{s}$, comprising only those terms related to nodes 1 and 2, then:

$$\int (\hat{\mathbf{U}}^{\mathrm{T}}\mathbf{s}^{\mathrm{T}})\mathbf{w}\,\mathrm{d}l = \hat{\mathbf{U}}^{\mathrm{T}}\mathbf{F}$$

As before, since $\hat{\mathbf{U}}$ is arbitrary, this can only be true if:

$$\mathbf{F} = \int \mathbf{s}^{\mathrm{T}}\mathbf{w}\,\mathrm{d}l \qquad (10.11)$$

Note that the equivalent loads are related to the shape function, and so will vary for different element types.

For a constant strain triangle subjected to a uniform load w per unit length normal to an edge of length L, the resulting unit loads are $wL/2$ normal to the edge and zero parallel to the edge at each end node.

EXAMPLE 10.4

Calculate the equivalent nodal loads for the finite element mesh in Figure 10.7(a).

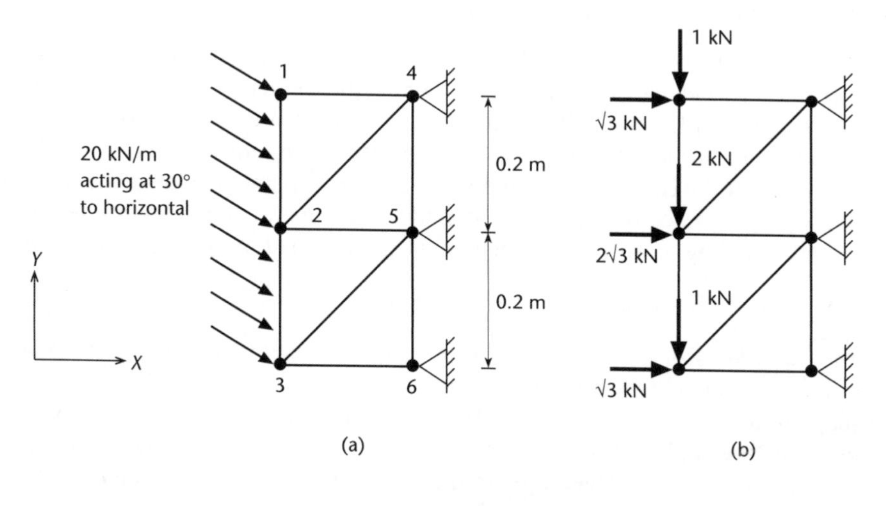

Figure 10.7

Consider edge 12 on the top element. The distributed load vector is

$$\mathbf{w} = \begin{bmatrix} 10\sqrt{3} \\ -10 \end{bmatrix} \text{kN/m}$$

The shape function matrix relating to nodes 1 and 2 is

$$\mathbf{s} = \begin{bmatrix} S_1 & 0 & S_2 & 0 \\ 0 & S_1 & 0 & S_2 \end{bmatrix} = \begin{bmatrix} 1-\alpha & 0 & \alpha & 0 \\ 0 & 1-\alpha & 0 & \alpha \end{bmatrix}$$

since $\beta = 0$ along edge 12 (from the definition of the reference element, Figure 10.2). We need to integrate the product of $\mathbf{s}^{\mathrm{T}}$ and $\mathbf{w}$ along edge 12 from 0 to 0.2 m. However, because the shape functions are in terms of α, we must transform this into an integral along edge 12 of the reference element. Now the element of length along the edge is $dl = 0.2d\alpha$, so equation (10.11) gives:

$$\mathbf{F} = \int \mathbf{s}^T \mathbf{w} dl = \int_0^1 \begin{bmatrix} 10\sqrt{3}(1-\alpha) \\ -10(1-\alpha) \\ 10\sqrt{3}\alpha \\ -10\alpha \end{bmatrix} 0.2d\alpha = \begin{bmatrix} \sqrt{3} \\ -1 \\ \sqrt{3} \\ -1 \end{bmatrix} \text{kN}$$

The loading and geometry of the lower element are identical, so the same nodal forces will result. The total nodal loads to be used in the analysis are therefore as shown in Figure 10.7(b).

10.3 Other element types

We illustrated the procedure by which a finite element is derived with reference to a very crude element, which is in fact rarely used in practice. We shall now look briefly at some of the more common element types and their shape functions, without going into detail on the matrix derivation.

10.3.1 Other plane stress triangles

An obvious improvement to the constant strain triangle would be to make it capable of modelling variations in strain across the element. To do this we need to use a higher-order displacement function and, if we are using an isoparametric formulation, this means also using higher-order shape functions. Since the shape functions for the constant strain triangle are linear in α and β, a logical next step is to use second-order, or quadratic, shape functions, that is, ones involving terms up to and including α^2 and β^2. The general form of a quadratic shape function is:

$$S_i = A_1 + A_2\alpha + A_3\beta + A_4\alpha^2 + A_5\alpha\beta + A_6\beta^2$$

The coefficients A_1 to A_6 are found by putting $S_i = 1.0$ at node i and zero at all the others. It follows that in order to determine all the coefficients for a quadratic shape function we need six nodes. We can create a six-noded triangle by putting nodes at the midpoint of each side. A typical six-noded triangular element is shown in Figure 10.8, together with the corresponding reference element.

Following the procedure outlined above, the shape functions are:

$$\begin{aligned} S_1 &= 1 - 3\alpha - 3\beta + 2\alpha^2 + 4\alpha\beta + 2\beta^2 \\ S_2 &= -\alpha + 2\alpha^2 \\ S_3 &= -\beta + 2\beta^2 \\ S_4 &= 4\alpha - 4\alpha\beta - 4\alpha^2 \\ S_5 &= 4\alpha\beta \\ S_6 &= 4\beta - 4\alpha\beta - \beta^2 \end{aligned} \qquad (10.12)$$

These can be used to formulate the required matrices for the finite element in much the same way as before. The element strain matrix **B** will now contain terms

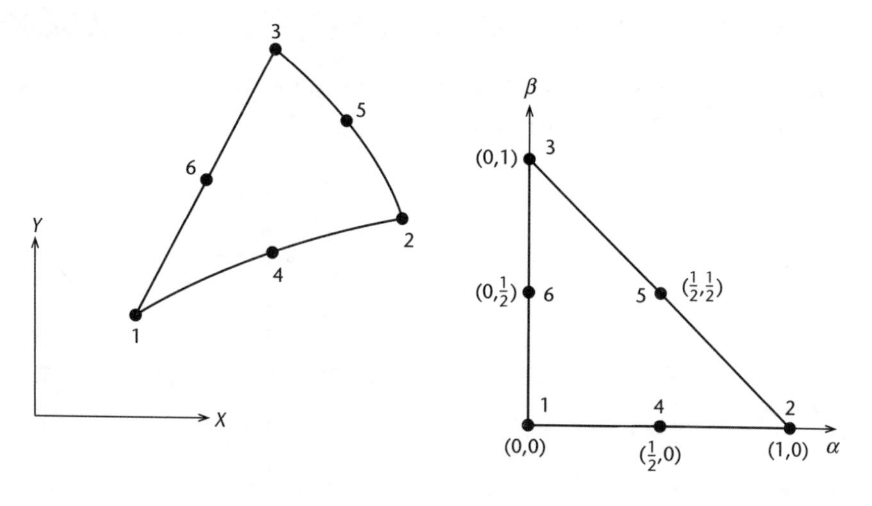

Figure 10.8
Six-noded triangular finite element.

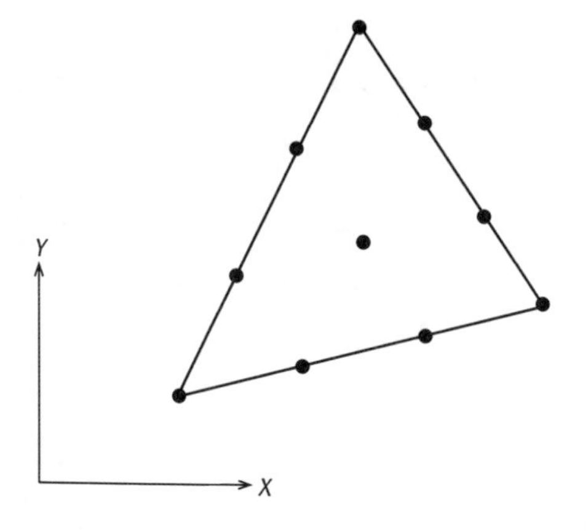

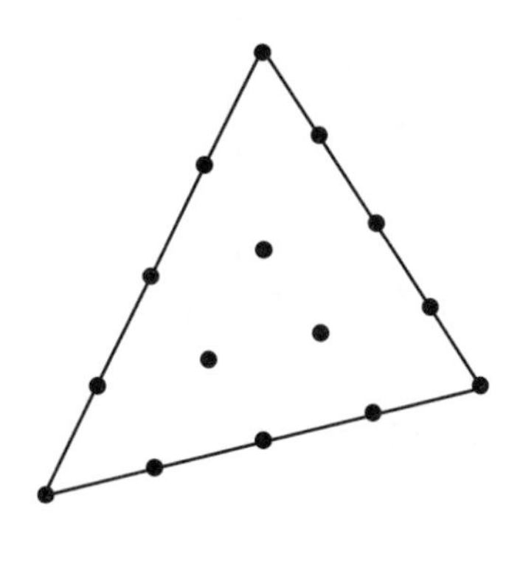

Figure 10.9
Ten- and fifteen-noded triangular elements.

that are linear functions of the coordinates of a point within the element. This element is therefore often known as the *linear strain triangle*.

The linear strain triangle is the most popular of the many triangular elements, since it is capable of modelling quite accurately the large number of problems in which the strain variation within a structure is linear. For instance, it would provide a very good representation of the direct stresses and strains in an elastic beam under bending. It can also fit higher-order strain distributions reasonably well without the need for very high mesh densities. A further advantage of higher-order elements such as the linear strain triangle is that the element edges can be curved, enabling easy meshing of structures with curved boundaries.

Even more sophisticated triangular elements can be formulated. Figure 10.9 shows ten- and fifteen-noded triangles, in which nodes have been introduced within the element as well as along the edges. These allow, respectively, quadratic and cubic strain variations. However, their use is very rare.

One further difference between higher-order elements and the constant strain triangle is the way that distributed loads are lumped at the nodes. We saw in section 10.2.5 that the equivalent loads are related to the shape functions, and so will vary for different element types.

EXAMPLE 10.5

Repeat the equivalent nodal load calculation of Example 10.4, but using six-noded triangles instead of three-noded triangles.

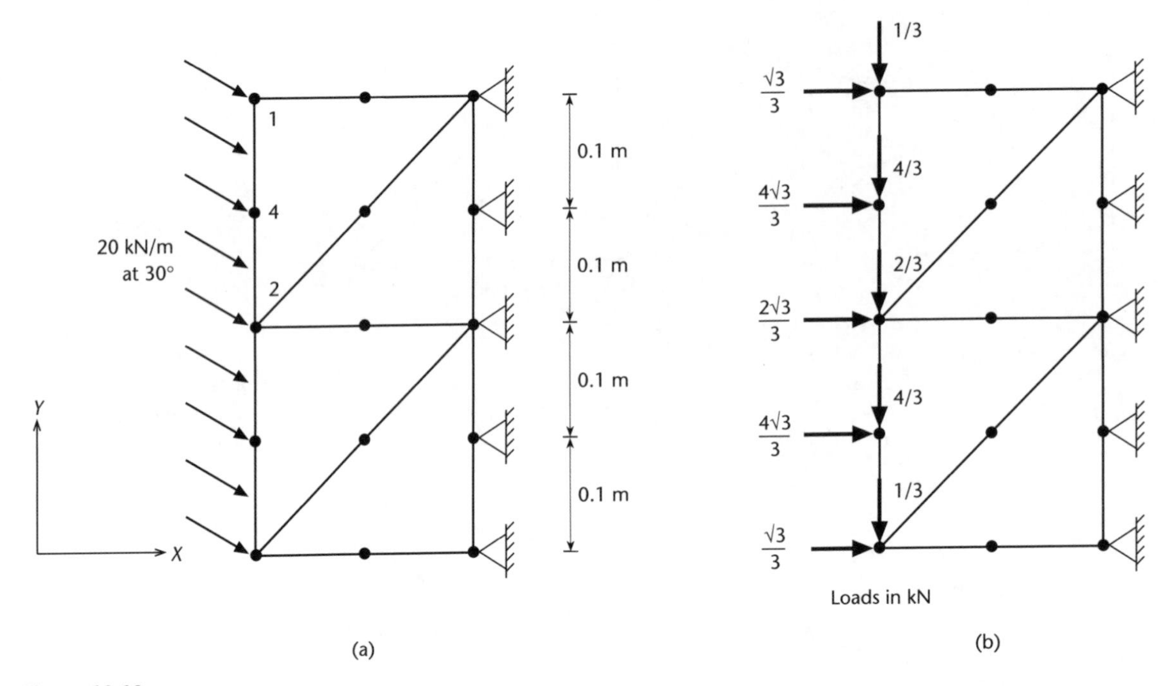

Figure 10.10

The new mesh is shown in Figure 10.10(a). Consider edge 142 on the upper element. As before, the distributed load vector is:

$$\mathbf{w} = \begin{bmatrix} 10\sqrt{3} \\ -10 \end{bmatrix} \text{ kNm}$$

The shape function matrix relating to nodes 1, 4 and 2 is:

$$\mathbf{s} = \begin{bmatrix} S_1 & 0 & S_4 & 0 & S_2 & 0 \\ 0 & S_1 & 0 & S_4 & 0 & S_2 \end{bmatrix}$$
$$= \begin{bmatrix} 1-3\alpha+2\alpha^2 & 0 & 4\alpha-4\alpha^2 & 0 & -\alpha+2\alpha^2 & 0 \\ 0 & 1-3\alpha+2\alpha^2 & 0 & 4\alpha-4\alpha^2 & 0 & -\alpha+2\alpha^2 \end{bmatrix}$$

since $\beta = 0$ along edge 142. Equation (10.11) gives:

$$\mathbf{F} = \int \mathbf{s}^{\mathrm{T}}\mathbf{w}\,\mathrm{d}l = \int_0^1 \begin{bmatrix} 10\sqrt{3}(1-3\alpha+2\alpha^2) \\ -10(1-3\alpha+2\alpha^2) \\ 10\sqrt{3}(4\alpha-4\alpha^2) \\ -10(4\alpha-4\alpha^2) \\ 10\sqrt{3}(-\alpha+2\alpha^2) \\ -10(-\alpha+2\alpha^2) \end{bmatrix} 0.2\mathrm{d}\alpha = \begin{bmatrix} \sqrt{3}/3 \\ -1/3 \\ 4\sqrt{3}/3 \\ -4/3 \\ \sqrt{3}/3 \\ -1/3 \end{bmatrix} \text{ kN}$$

Whereas for the constant strain triangle the uniform load was divided equally between the two end nodes, it is now distributed in the ratio $\frac{1}{6} : \frac{2}{3} : \frac{1}{6}$ between nodes 1, 4 and 2. The results for the lower element will be the same, and the total nodal loads are therefore as shown in Figure 10.10(b).

10.3.2 Plane stress quadrilaterals

Besides the triangle, the other element shape commonly used for plane stress problems is the quadrilateral, or quad for short. The simplest such element has nodes only at the four corners. Figure 10.11 shows a four-noded quad and the corresponding reference element.

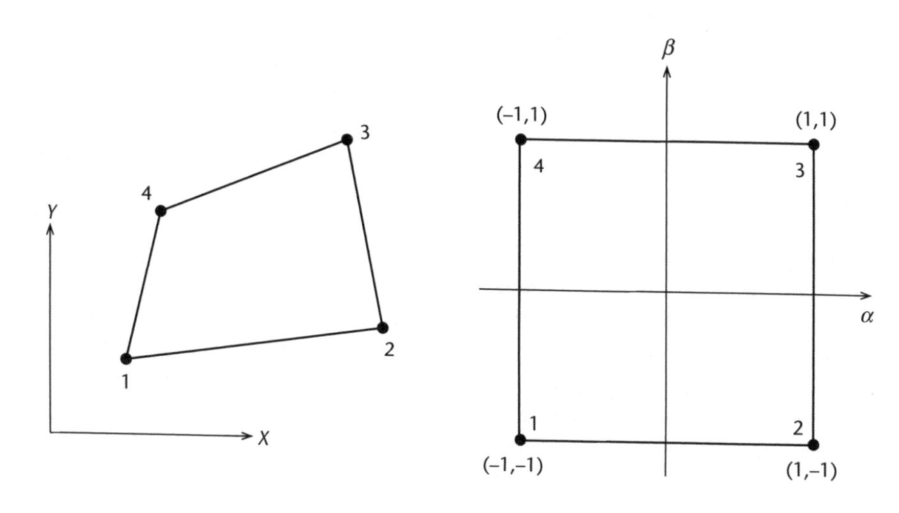

Figure 10.11
Four-noded quadrilateral.

For this element we take a shape function of the general form:

$$S_i = A_1 + A_2\alpha + A_3\beta + A_4\alpha\beta$$

Again, the coefficients are found by setting each shape function to 1.0 at one node and zero at the others. The shape functions are therefore:

$$\begin{aligned} S_1 &= \frac{1}{4}(1 - \alpha - \beta + \alpha\beta) \\ S_2 &= \frac{1}{4}(1 + \alpha - \beta - \alpha\beta) \\ S_3 &= \frac{1}{4}(1 + \alpha + \beta + \alpha\beta) \\ S_4 &= \frac{1}{4}(1 - \alpha + \beta - \alpha\beta) \end{aligned} \qquad (10.13)$$

The behaviour of this element is rather more complex than the triangles considered above, since the shape functions are neither purely linear nor fully quadratic. For example, Figure 10.12 shows a 3D plot of S_1. Since it contains only single powers of α and β, the variation of displacement within the element is linear along any line of constant α or constant β (for example, lines AB, CD and GH).

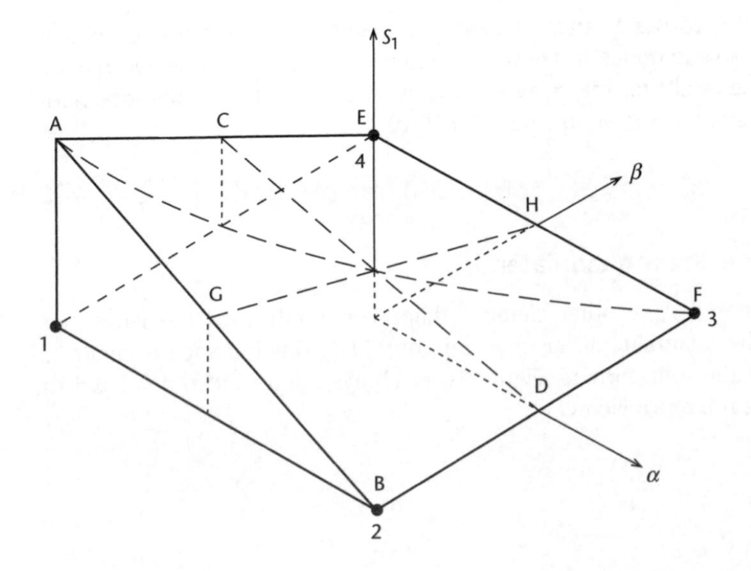

Figure 10.12
Plot of shape function for a four-noded quadrilateral.

Because of the presence of the $\alpha\beta$ terms, however, the displacement varies quadratically along diagonal lines such as $\alpha = \beta$ (AF in the diagram).

Using an isoparametric formulation, the displacements exhibit the same characteristics as the shape functions. When we come to find the strains within the element, differentiation of the $\alpha\beta$ terms results in strains that are linear functions of α and β. Thus the strain may vary linearly over the element, even though the displacement function is not fully quadratic (no α^2 and β^2 terms). This four-noded element thus provides a degree of sophistication somewhere between the three- and six-noded triangles.

As with the triangles, it is possible to generate a range of elements, ranging from the simple four-noded quad to higher-order elements. The most popular is the eight-noded quad (Figure 10.13), which has shape functions of the form:

$$S_i = A_1 + A_2\alpha + A_3\beta + A_4\alpha^2 + A_5\alpha\beta + A_6\beta^2 + A_7\alpha^2\beta + A_8\alpha\beta^2$$

This gives a displacement function that is not fully cubic, but does contain some third-order terms ($\alpha^2\beta, \alpha\beta^2$), and the strain matrix **B** will therefore contain some quadratic terms.

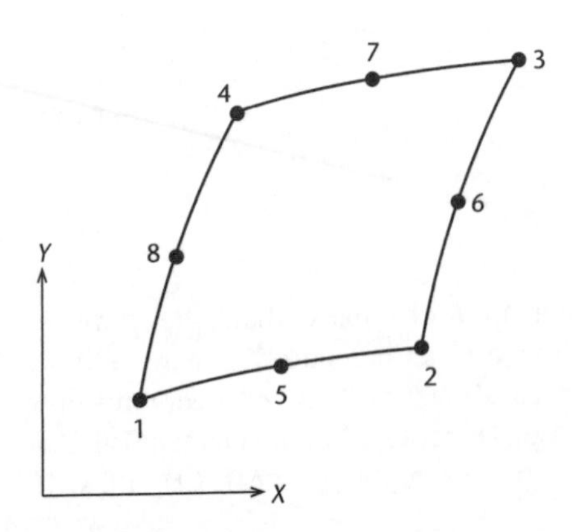

Figure 10.13
Eight-noded quadrilateral element.

Even higher-order quads are of course possible, but these are rarely used and will not be considered here.

10.3.3 Plate bending elements

The plane stress elements discussed so far are suitable for plate structures in which all the loads act in the plane of the plate. The deformations can be adequately described by X and Y translations of the nodes, and there are therefore just two degrees of freedom at each node. These can be thought of as the 2D equivalent of the 1D bar elements in Chapter 9.

When a plate is subjected to out-of-plane loads, bending moments and rotations may be developed about two perpendicular axes in the plane of the plate (Figure 10.14(a)). Plane stress elements will not be suitable for modelling such behaviour, and it will be necessary to formulate a special plate finite element, which can be thought of as the 2D equivalent of the beam elements in Chapter 9. Figure 10.14 (b) shows a four-noded plate element – again, higher-order elements are possible. For plate elements it is normal to neglect the in-plane deformations, so that the active degrees of freedom at each node are the vertical deflection and the rotations about the two in-plane axes. The four-noded element therefore has a total of twelve degrees of freedom.

Formulation of the plate finite element requires knowledge of plate theory, which is not covered in this book. We shall therefore not proceed beyond this simple introduction.

10.3.4 Three-dimensional solid elements

For problems that are fully three-dimensional, it may necessary to model the structure using 3D solid elements. This is avoided where possible, because the resulting models are likely to have an extremely high number of degrees of freedom and so be very demanding of computational resources. Some of the common 3D elements are shown in Figure 10.15. The most popular types are tetrahedra and cuboid or *brick* elements, and again different levels of sophistication are possible depending on the number of nodes used. Note that there are three

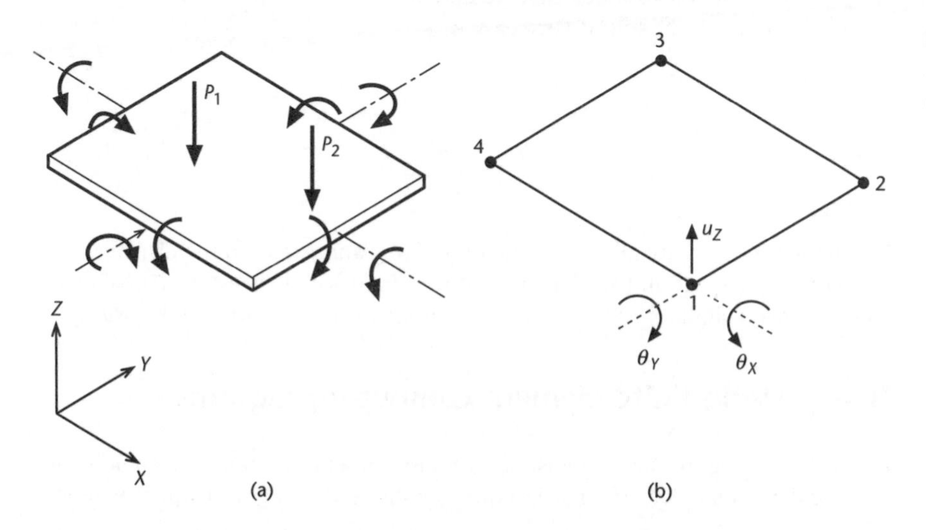

Figure 10.14
Four-noded element for plate bending.

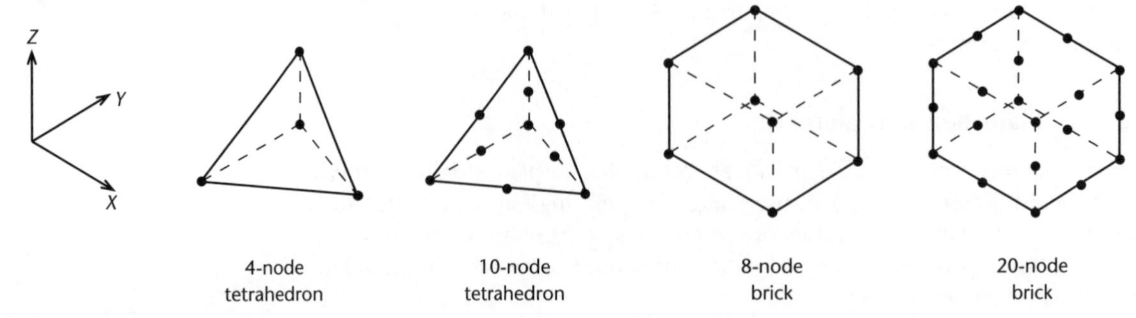

Figure 10.15
Three-dimensional solid finite elements.

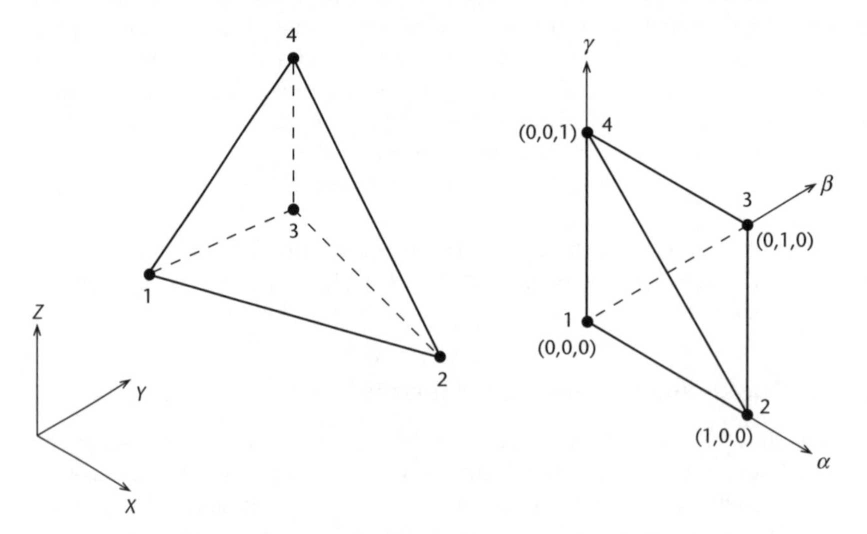

Figure 10.16
Constant strain tetrahedron.

degrees of freedom (X, Y and Z deflections) at each node, so that the 20-node brick element, for example, has a total of 60 degrees of freedom!

As a very simple example, consider the four-noded tetrahedron in Figure 10.16. This must be related to a reference element in (α, β, γ) space. The appropriate shape functions can be found very easily:

$$S_1 = 1 - \alpha - \beta - \gamma$$
$$S_2 = \alpha$$
$$S_3 = \beta$$
$$S_4 = \gamma$$

The shape functions are all linear functions of α, β and γ, and so the displacements will vary linearly within the element and the strain will be constant. The element thus has many similarities with the constant strain triangle discussed earlier.

10.4 Using finite element computer programs

There are many extremely sophisticated finite element programs available which can calculate and solve the matrix equations for assemblages of finite elements.

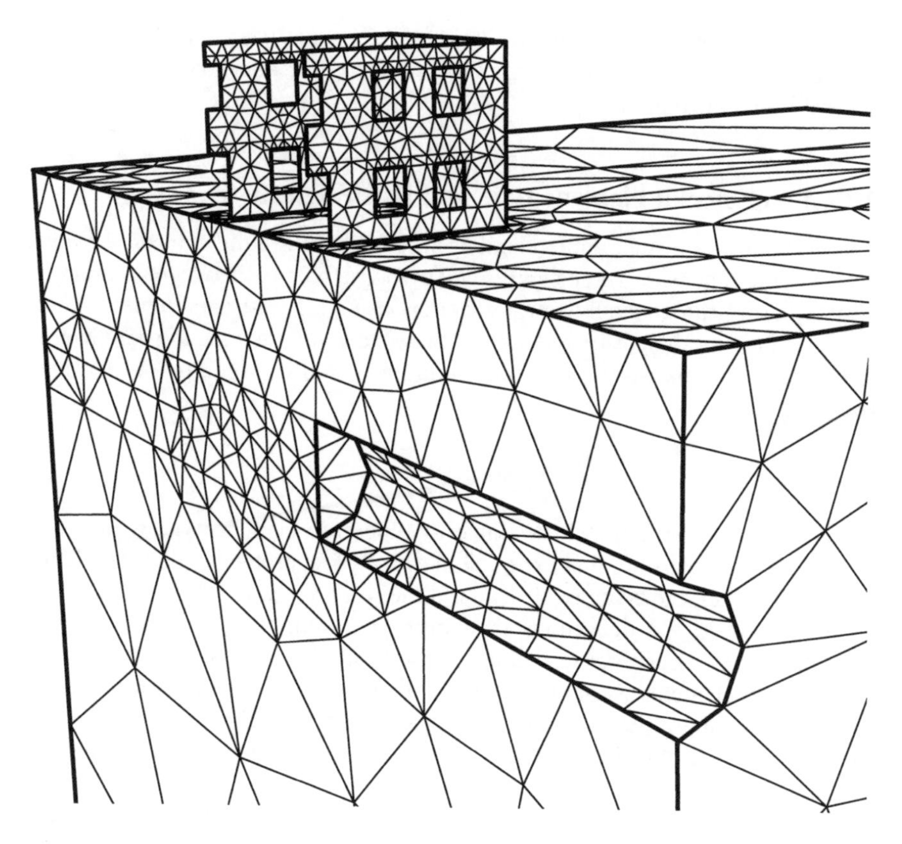

Figure 10.17
A typical finite element model.

Most nowadays have advanced graphics pre-processors to facilitate the generation of the finite element mesh, and post-processors to give a visual display of the results. Once the mesh has been set up and support conditions and material properties entered, the steps that the program goes through are similar to those for a stiffness matrix program discussed in section 9.4.5.

Figure 10.17 shows an example of a finite element mesh for the analysis of the effects of tunnelling operations beneath masonry buildings. A model of a building made up of triangular plane stress elements rests on a large block of soil, discretised using solid, tetrahedral elements. Note that:

- While the model may be made up of differing element types, it is normal to use elements having shape functions of the same order. Problems would arise if we attempted to join a linear strain element to a constant strain element, for example, as the adjacent boundaries would deform in different ways, violating the displacement compatibility conditions.
- Because the problem being modelled is symmetrical, only half of the structure and soil block need be modelled, so long as appropriate boundary conditions are applied at the nodes lying on the plane of symmetry.

The boring of the tunnel is modelled by removing finite elements along the line of the tunnel in a predetermined sequence – at the instant shown, the tunnelling has advanced to a point directly below the near wall of the building. Analyses performed at different stages during the tunnel construction will result in very different patterns of movement at the ground surface, and hence in quite different stresses being generated in the masonry building. A mesh such as the one shown would have many thousand degrees of freedom and one analysis would take several hours of computer time.

10.4.1 Choice of element type and mesh density

The major issues facing a finite element program user are the choices of element type and mesh density. These choices will involve a balance between accuracy and cost. Accuracy can generally be improved by using higher-order elements and/or increasing the mesh density, but both of these steps will increase the bandwidth of the stiffness matrix and hence the computer resources required to solve the problem.

Accuracy

For an accurate solution, it is desirable to choose an element type whose strain pattern gives a close fit to that expected in the physical problem – this may depend on both the structure being modelled, and the results required. For instance, if we are interested in the deflections of a uniform beam, these are dominated by bending effects, which cause a linear strain variation across the cross-section. Linear strain elements would therefore be appropriate. If, on the other hand, we require detailed information on the shear stresses in the beam, these vary parabolically over the depth of the section, and so a quadratic strain element would provide a more accurate representation.

If the appropriate order of element is used, then it is possible to obtain good results without using a very fine mesh. Alternatively, it is possible to obtain acceptable results using a lower-order element, but a significant increase in mesh density will be required.

Cost

The analysis of a large finite element model requires the solution of a matrix equation which may have several thousand degrees of freedom. The structure stiffness matrix is generally banded, that is, all the non-zero terms are concentrated in a band around the leading diagonal. Considerable research effort has been put into developing efficient methods of solving banded matrix equations, and numerous different solution techniques are available. For most such techniques, the number of calculations required is proportional to the square of the bandwidth, and we therefore wish to make this as small as possible.

Increasing the mesh density and using higher-order elements will both have the effect of increasing the bandwidth. For all but very simple models, it will therefore be necessary to compromise so as to achieve an acceptable level of accuracy at a reasonable cost. As we saw earlier, the bandwidth can be kept to a minimum by careful numbering of the nodes, so as to minimise the largest difference in node number between nodes on adjoining elements. For a regular mesh, this is achieved by starting at one corner, numbering nodes across the shorter dimension of the mesh, and then repeating for each line of nodes along the length of the mesh. For irregular meshes no such simple rule exists. However, most software now includes automated routines for reordering the nodes in an optimal way.

EXAMPLE 10.6

Figure 10.18 shows two finite element models of a simply supported beam, using three- and six-noded triangles. Both meshes have the same number of degrees of freedom. Make a qualitative assessment of the accuracy and cost of analyses using these meshes.

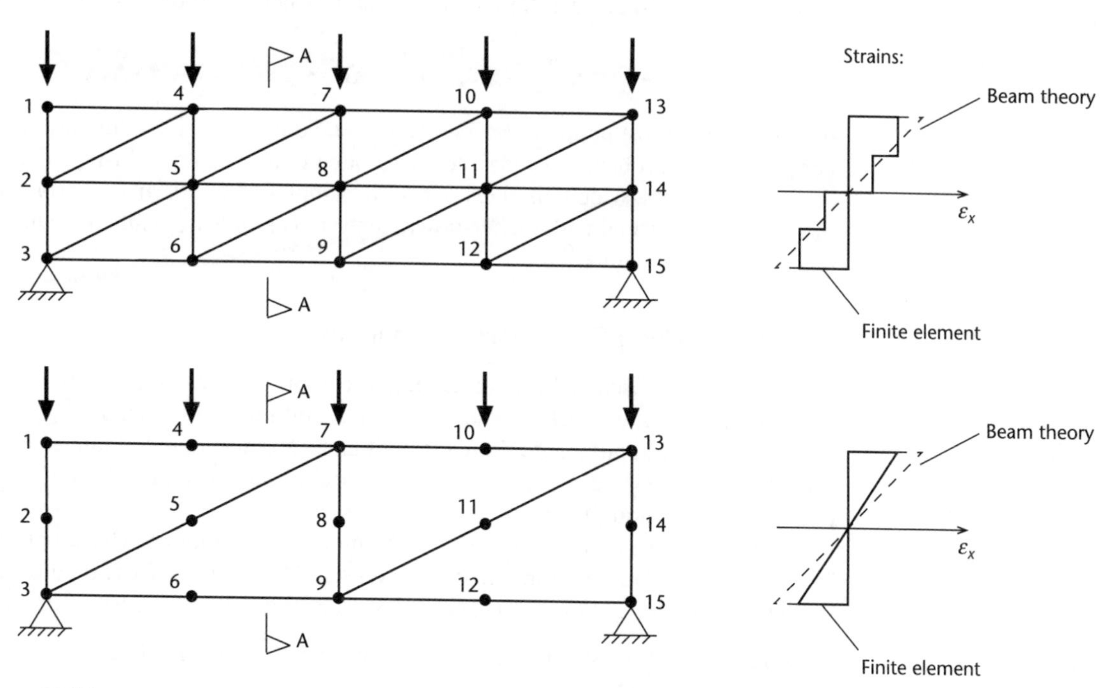

Figure 10.18

In the figure, the distributions of axial strain along the section A–A are shown. It can be seen that the three-noded triangles (Figure 10.18(a)) give a very crude, stepwise distribution which compares poorly with the linear distribution predicted by simple beam theory. The linear strain triangle mesh (Figure 10.18(b)) gives a much more accurate strain distribution and so would be expected to give far better estimates of bending stresses and deformations, even though fewer elements are used.

(Note: Beam theory would itself not be completely accurate in this case, since it takes account only of axial (X-direction) stresses. For a beam such as the one shown, which is quite deep compared to its length, both Y-direction and shear stresses would be expected to contribute to the deformation. Nevertheless, simple beam theory provides a reasonable approximation with which to compare the finite element results.)

The cost of an analysis can be assessed by considering the *semi-bandwidth* of the stiffness matrix, defined as the distance (measured as a number of columns) from the leading diagonal to the furthest non-zero element on any row. The semi-bandwidth is closely related to the difference between the highest and lowest node numbers on any elements sharing a common node. If we were to write out the stiffness matrices for the two meshes (not shown here because of the high number of degrees of freedom) we would find that the semi-bandwidths are 8 for the three-noded mesh and 14 for the six-noded mesh. For the most common solution techniques, the number of calculations required is proportional to the square of the bandwidth, hence the solution phase for the six-noded mesh would take $(14/8)^2$, or about three times as long as for the three-noded mesh.

In spite of the longer solution time, it would be preferable to use the mesh of six-noded triangles for this analysis, as a very large increase in mesh density would be required in order to achieve comparable accuracy using the three-noded elements.

In practice, the most popular elements tend to be one order above the most basic. For plane stress problems, for instance, six-noded triangles and eight-noded quads are by far the most widely used elements; the simpler elements are too crude and the expense associated with higher-order elements can only be justified when very high accuracy is required.

10.4.2 Checking finite element analyses

In Chapter 9 we discussed the limitations of computer analysis methods and the rigorous approach to checking that is required, and most of the comments made there are equally applicable to finite element analysis. Indeed, with finite elements it is particularly easy to develop huge, complicated models and lose sight of the underlying structural principles.

As with a stiffness matrix analysis of a frame, it is important to check both the input to a finite element program and the results. Most of the checks required are similar to those for a stiffness matrix analysis. For the input we must ensure that:

- the geometry of the structure has been correctly defined and meshed;
- support conditions have been correctly specified;
- loads have been applied at the correct locations and in the right directions;
- the correct properties have been assigned to the various members;
- all quantities have been specified in consistent units.

The geometrical checks can be greatly simplified by the availability of an advanced graphical pre-processor and a high-resolution graphics terminal. Two-dimensional finite element meshes are relatively easy to check, but for three-dimensional structures it may be necessary to view the model from several different directions in order to ensure correct meshing. Techniques such as surface shading and hidden line removal can facilitate this process. For example, Figure 10.19(a) shows the finite element mesh for a simple two-storey plated structure. With so many intersecting lines it is extremely difficult to see whether the model is geometrically correct, but with hidden line removal (Figure 10.19(b)), the structural form becomes much clearer. Another useful graphical technique is element shrinking, in which the outline of each element is drawn separately (Figure 10.19(c)). This enables one to see at a glance whether any elements have been omitted.

Care needs to be taken with the support conditions for a finite element mesh – in order to restrain the edge or end of a member it may be necessary to apply restraints to several nodes. Figure 10.20 shows a beam modelled by plane stress elements, which have only X and Y degrees of freedom at each node. To achieve complete fixity at an edge we must restrain both degrees of freedom at each node

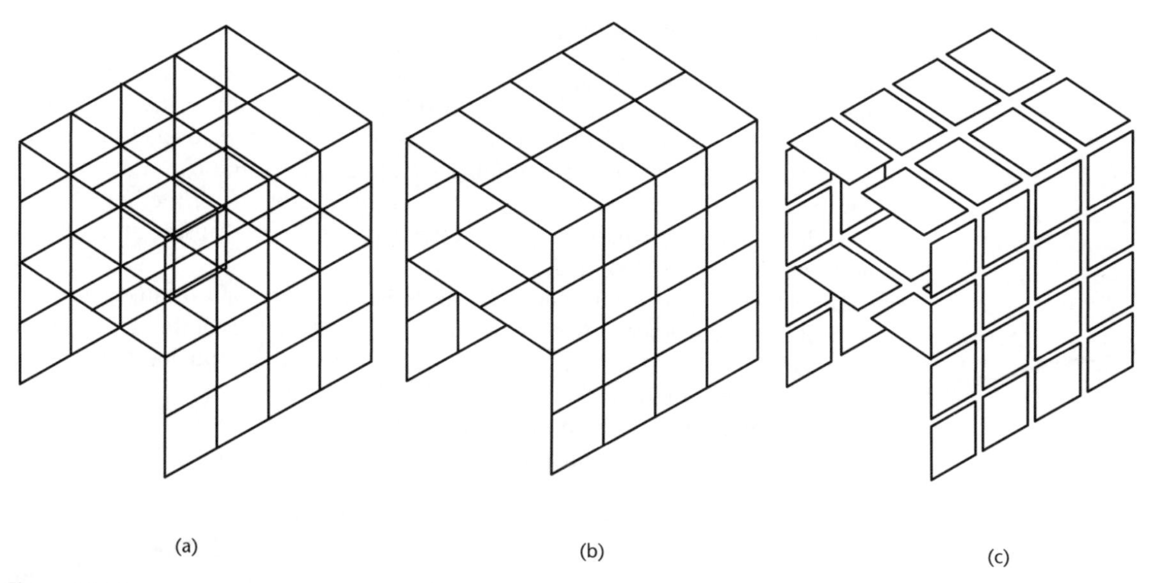

Figure 10.19
Three views of a finite element mesh: (a) basic view; (b) with hidden lines removed; (c) with hidden line removal and element shrinking.

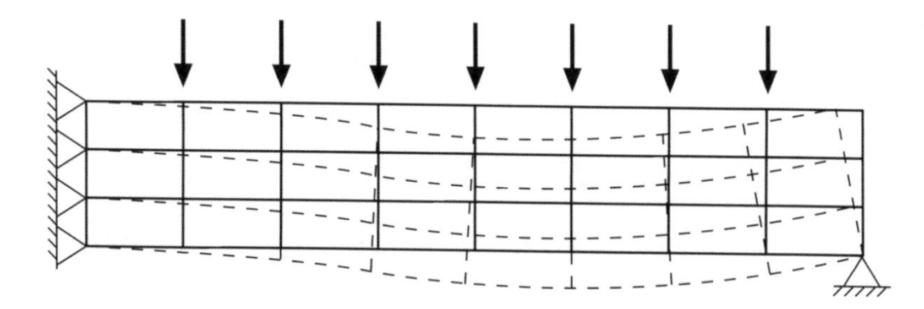

Figure 10.20
Support conditions for a finite element model of a fixed-pinned beam.

along that edge. To provide a pinned support, on the other hand, only one node should be restrained.

The other checks listed above simply require care, though a good graphical pre-processor can still be helpful. For instance, colour coding can be used to display groups of elements sharing the same material or section properties.

Turning to the output of results, one should check that:

- there is overall equilibrium between the applied loads and support reactions;
- the reaction components and deformed shape correspond to the specified support conditions – for example, it can be seen that the deformation of the mesh in Figure 10.20 satisfies the requirement for fixity at the left-hand end and a pinned support at the right-hand end;
- the directions and relative magnitudes of reactions, deflected shapes and bending moment diagrams accord with common sense;
- the displacements and internal forces are of the right order of magnitude.

These aspects were covered in some detail for stiffness matrix analysis in section 9.4.6, and will not be discussed further here.

10.4.3 A cautionary tale

A spectacular recent example of the limitations of computer analysis and design methods was the loss of the Sleipner A offshore oil platform in the Norwegian sector of the North Sea. This enormous concrete structure, over 100 m high (comparable to a 30-storey building), took three years to design and construct. During installation in 1991, a shear failure in the wall of one of the huge hollow cells that made up the base initiated a catastrophic collapse, causing an estimated economic loss of US$700 million.

The structure had been analysed and designed using sophisticated computer software which first performed a global finite element analysis of the structure and then carried out detailed strength checks at several thousand locations under several hundred different loadcases. Unfortunately, because of quite small errors in the assumptions made in both the global analysis and the strength calculations, the wall that failed was not identified as a critical location and so was not checked!

Faced with enormous financial pressure to build a replacement, and the obvious need for a major revision to their computer software, the engineers took the decision to design the new structure using simple hand calculations. By the time the new computer results were available, the design was complete and most of the structure had been built. It has now been operating safely for several years.

The lesson is obvious: no matter how large and complex the structure, or how sophisticated the computer software available, it is *always* possible to obtain the most important design parameters from relatively simple hand calculations. Such calculations should always be performed, both as a check on computer output and to improve the engineer's understanding of the structural behaviour.

10.5 Further reading

Good introductory texts on finite elements, which introduce the method without too much mathematical detail, include:

- R. J. Astley, *Finite Elements in Solids and Structures*, Chapman & Hall, 1992.
- C. T. F. Ross, *Finite Element Methods in Engineering Science*, Ellis Horwood, 1990.

While most engineers use finite elements in the form of a commercially available software package, those involved in very specialised fields of work may wish to write their own programs. A useful programming guide is:

- I. M. Smith and D. V. Griffiths, *Programming the Finite Element Method*, 3rd edn, Wiley, 1998.

Those wishing to study the technique in greater depth are referred to the following extremely comprehensive books, written by the pioneers in the field:

- K. J. Bathe, *Finite Element Procedures in Engineering Analysis*, Prentice-Hall, 1982.
- O. C. Zienkiewicz and R. L. Taylor, *The Finite Element Method*, 4th edn, McGraw-Hill, Vol. 1, 1989 and Vol. 2, 1991.

10.6 Problems

It is difficult to set realistic problems on finite element analysis that can be tackled without computer software. There are therefore only a few problems for hand solution in this chapter. The reader is strongly encouraged to try to get a feel for finite elements by gaining hands-on experience with a computer program, experimenting with different mesh densities and element types, and looking at the variations in accuracy and solution speed. Where appropriate, answers to problems are given at the end of the book.

10.1. Find the shape functions S_3 and S_8 for the 8-noded quad in Figure 10.13. The corresponding nodes in the reference element have (α, β) coordinates $(1, 1)$ and $(-1, 0)$ respectively.

10.2. Write down the **B** matrices for the two triangular finite elements in Figure 10.21. If the only non-zero nodal displacements are (in mm) $U_1 = 1.0$, $U_2 = 1.5$, $V_2 = V_3 = 1.2$, calculate the strains in the two elements.

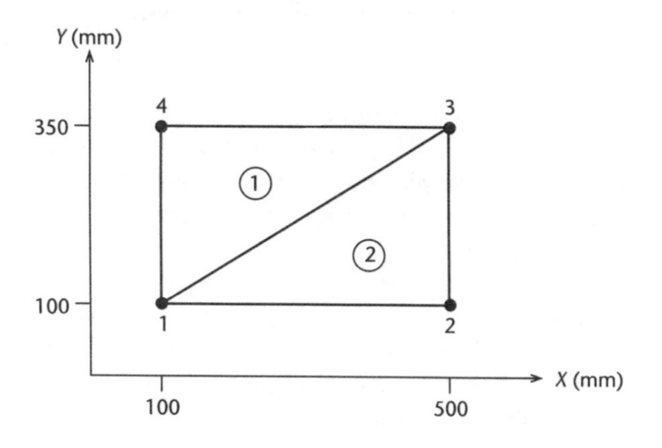

Figure 10.21

10.3. Calculate the equivalent nodal loads corresponding to the distributed loading shown in Figure 10.22, which varies linearly from zero at node 1 to intensity w per unit length at node 2.

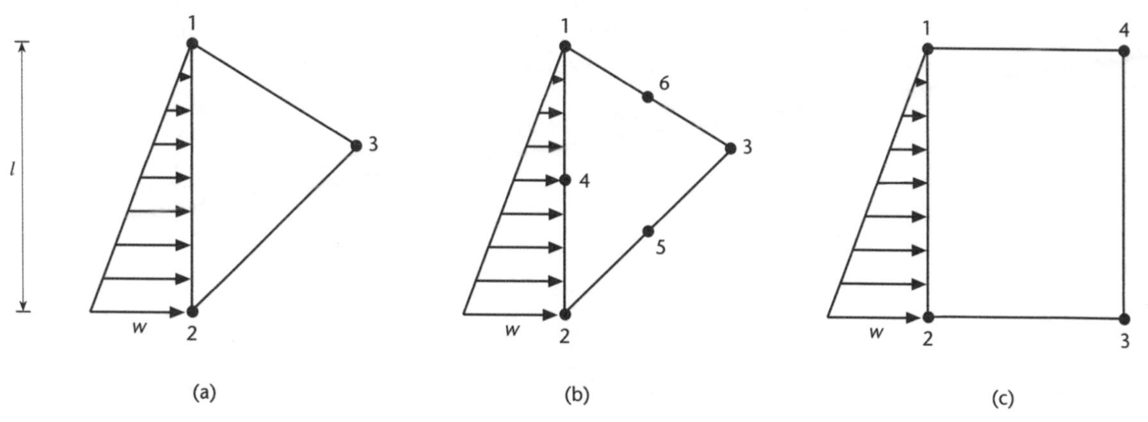

Figure 10.22

10.4. The wall of a water-tank is modelled by six-noded triangular elements as shown in Figure 10.23. When the tank is full, the hydraulic pressure on the wall

Figure 10.23

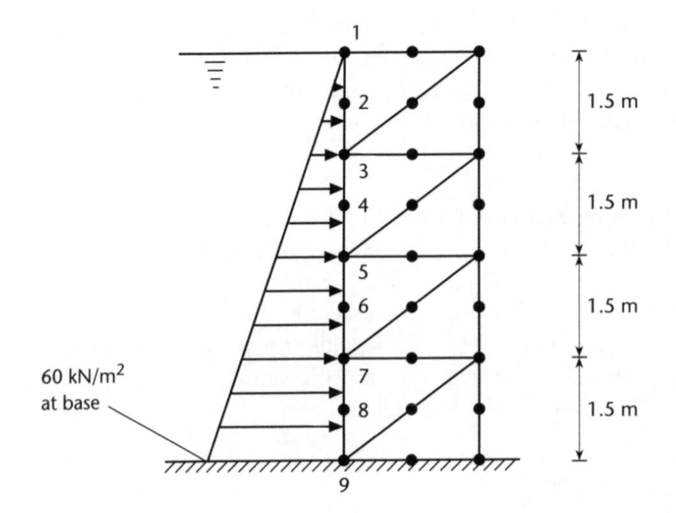

varies linearly from zero at the top to 60 kN/m^2 at the base. Assuming a unit thickness into the page, what nodal loads should be used in the finite element analysis? [*Hint: the pressure distribution on each element can be thought of as the sum of a uniform component and a component of the form shown in Figure 10.22.*]

10.5. *If you have access to a finite element computer program:* use the program to calculate the maximum deflection of a simply supported beam of length 5.0 m, depth 0.5 m and thickness 1.0 m, subjected to a uniform vertical load over the top face of 100 kN/m^2. Take the material properties of the beam to be $E = 10$ GPa and $\nu = 0.25$. Use the following meshes:

(a) three-noded triangles based on a square grid of nodes spaced at 0.25 m;
(b) six-noded triangles based on the same nodes as in (a);
(c) three-noded triangles based on a square grid of nodes spaced at 0.125 m;
(d) six-noded triangles based on the same nodes as in (c).

The maximum deflection given by simple beam theory is 7.81 mm, while a more rigorous 2D elastic calculation gives a deflection of 7.99 mm. Compare your finite element results with these figures and comment on the differences.

CHAPTER 11

Buckling and instability

CHAPTER OUTLINE

In some instances structures and solid bodies may become unstable; if they are subjected to a small movement or force then their subsequent movements are large and uncontrolled.

In structures by far the most common and important form of instability is buckling, which occurs primarily in members in compression. If the compressive load on a relatively slender member is increased to some critical value then the member remains in equilibrium but ceases to be stable – a very small lateral load or imperfection will cause it to bow out sideways, or *buckle*.

For a slender element, buckling is likely to occur at a load substantially lower than that which would cause yielding of the material, and thus it will be buckling rather than yielding which will govern the maximum force the element can carry. Buckling is therefore crucial to the design of any structure in which relatively slender elements must carry compressive loads. It can also arise in elements that are not in pure compression. For example, the compressive stresses developed in the cross-section of a beam in bending can give rise to a form of instability known as lateral-torsional buckling. This will be briefly discussed towards the end of this chapter.

This chapter should enable the reader to:

- appreciate the difference between stable, neutral and unstable equilibria;
- recognise situations in which instability may arise;
- determine the buckling loads of idealised, perfect struts;
- analyse the compressive behaviour of struts with initial imperfections;
- understand how practical design guidelines have been developed from the theoretical analysis of the buckling and yielding behaviour of struts;
- apply simple energy methods for the determination of buckling loads, and appreciate the possible sources of inaccuracy in such methods;
- understand the basic principles of lateral-torsional buckling of beams.

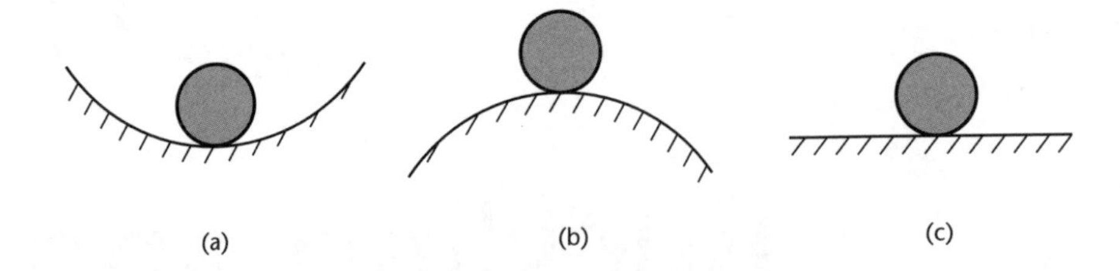

Figure 11.1
Stable, unstable and neutral equilibria.

11.1 Stable and unstable equilibria

The concepts of stable and unstable equilibria are most simply illustrated by the example of a ball at rest on a surface. Three different cases are shown in Figure 11.1. In each case the ball is in equilibrium; its weight is balanced by an equal and opposite reaction from the ground, both forces acting vertically through the ball's centre of gravity.

- Case (a) is an example of a stable equilibrium. If the ball is displaced sideways by a small amount and let go it returns to its original position, after a small amount of oscillation.
- The ball in (b) is in unstable equilibrium. If it is given a small disturbance, it continues to move, that is, the displacement increases without bound.
- The intermediate case (c) is known as a *neutral* equilibrium. If the ball is given a sideways displacement it neither returns to its original position nor keeps going, but remains in the displaced position. In this case the equilibrium is independent of the horizontal position, or, put another way, *there is an infinite number of positions in which the ball can be in equilibrium*. This is in contrast to cases (a) and (b), which have a unique equilibrium position.

We can use this simple example to illustrate how equilibrium and stability are related to the potential energy of the system. The stable equilibrium in (a) occurs at the position where the potential energy of the ball is a minimum, the unstable equilibrium in (b) occurs at the position of maximum potential energy and in (c) the potential energy is the same regardless of the position of the ball. Expressing these concepts mathematically, we can state that, if the potential energy V of the system varies with some parameter x (the horizontal position in this case) then the system is in equilibrium if:

$$\frac{dV}{dx} = 0 \tag{11.1}$$

The equilibrium is stable if:

$$\frac{d^2V}{dx^2} > 0 \tag{11.2a}$$

it is unstable if:

$$\frac{d^2V}{dx^2} < 0 \tag{11.2b}$$

and it is neutral if:

$$\frac{d^2V}{dx^2} = 0 \tag{11.2c}$$

EXAMPLE 11.1

As an introduction to the use of potential energy in stability analysis, consider the simple assembly in Figure 11.2(a). A rigid bar of length l is supported at its base by a torsion spring of stiffness k (that is, a rotation θ of the bar gives rise to a restoring moment $k\theta$). If the bar is initially vertical ($\theta = 0$), analyse the behaviour of the assembly as the load P is increased.

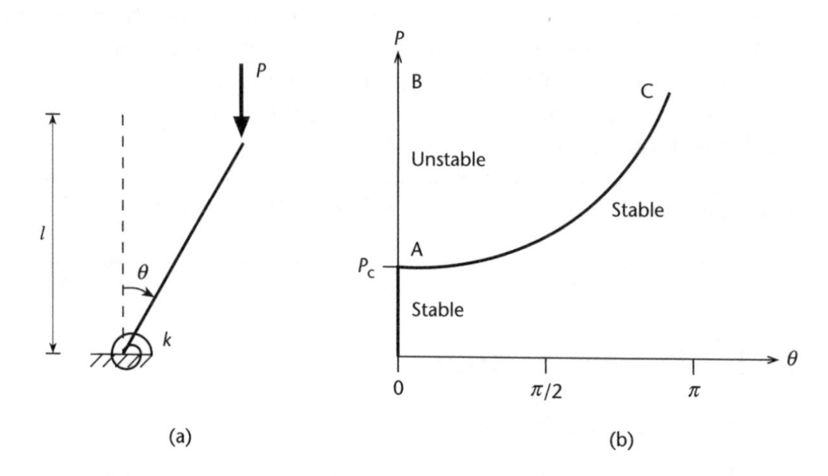

Figure 11.2

To analyse the system, first assume the bar has undergone a rotation θ from its initial position. The potential energy of the system has two parts – the strain energy in the spring is $\frac{1}{2}k\theta^2$ and the loss of gravitational potential energy of the load is $Pl(1 - \cos\theta)$. Therefore

$$V = \tfrac{1}{2}k\theta^2 - Pl(1 - \cos\theta)$$

$$\frac{dV}{d\theta} = k\theta - Pl\sin\theta$$

$$\frac{d^2V}{d\theta^2} = k - Pl\cos\theta$$

For equilibrium $dV/d\theta = 0$, from which either $\theta = 0$ or $\dfrac{\theta}{\sin\theta} = \dfrac{Pl}{k}$.

First consider the case $\theta = 0$. This gives $\cos\theta = 1$ and hence $d^2V/d\theta^2 = k - Pl$. Therefore, from equation (11.2), the equilibrium is:

- stable if $P < k/l$;
- neutral if $P = k/l$; and
- unstable if $P > k/l$.

The transitional value of P is known as the *critical load* P_c. If we plot this result on a graph of P against θ (Figure 11.2(b)), we get the line OAB, with the point A representing the transition from stable to unstable behaviour.

Now consider the case where $\dfrac{\theta}{\sin\theta} = \dfrac{Pl}{k}$. This can be plotted as the curve AC in Figure 11.2(b). Using this expression to eliminate P from the expression for the second differential, we get:

$$\frac{d^2V}{d\theta^2} = k(1 - \theta/\tan\theta)$$

This is positive for all values of θ between 0 and π, so that the line AC represents a stable equilibrium.

Now consider the physical behaviour of the assembly as the load P is increased from zero. Initially, the bar remains vertical and the system is in stable equilibrium (the line OA on the graph). At A the critical load is reached and the equilibrium becomes neutral. A is known as a *bifurcation point* because there are now two equilibrium paths that the structure can follow. If the bar remains vertical then it becomes unstable (the line AB) – a small lateral disturbance would then result in large, uncontrolled deformations. Alternatively, the bar may rotate so as to take up a stable equilibrium position (the line AC).

Normally, the effect of small imperfections will be to make the bar deflect as it approaches the critical load, so that it follows path AC rather than AB.

11.2 Buckling of struts

11.2.1 Buckling behaviour

We now go on to consider the most important form of instability in civil engineering – the buckling of *struts*, that is, elements under compressive loads.

Imagine first a perfectly straight, slender element subjected to an axial tension. If a small lateral load is briefly applied at some point along the length of the element, it bends slightly but returns to its original straight state as soon as the load is removed (so long as no yielding occurs at any point). This is an example of *stable* behaviour.

Now suppose the same slender element is loaded in compression. We shall see in this chapter that there is a critical compressive load at which the equilibrium ceases to be stable. If the compressive force is increased to this critical load and a small lateral force applied then the element *buckles*, that is, it bends outwards and does not straighten when the lateral load is removed. In practice, it is impossible to achieve a perfectly straight member. Any small imperfection will have the same effect as the small lateral load, so that a compression member will never be able to exceed its critical load without buckling.

This behaviour can easily be demonstrated using an ordinary 300 mm perspex ruler. Stand the ruler vertically on a hard surface such as a desk and press down on the top with your hand (Figure 11.3). If the ruler is initially quite straight then it will bow out suddenly. If, as is often the case, it has a significant initial curvature then the buckling effect will be rather less dramatic.

11.2.2 Euler buckling load for a perfect pin-ended strut

We shall return later to the application of potential energy to stability problems. First we will see that simple buckling formulae can be derived by considering equilibrium of a member in its buckled position. This approach yields results known as the Euler formulae, after the Swiss mathematician who first derived them.

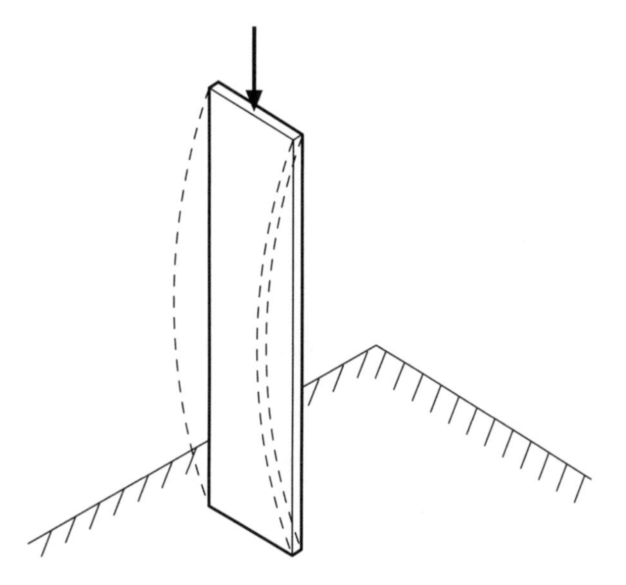

Figure 11.3
Buckling of a perspex ruler.

Consider first a uniform strut of flexural rigidity EI and length l, with pinned ends. The strut is assumed to be perfect in that it is initially completely straight and free from flaws. If an axial compressive load P is applied and steadily increased then eventually a critical load will be reached. If a small lateral disturbance is now applied at some point along the member, then it will bow out sideways, or *buckle*. We shall assume that the buckling causes lateral deflections that are small compared to the overall dimensions of the element, and that no yielding occurs at any point.

Consider the buckled strut shown in Figure 11.4. If we consider equilibrium of a length x from one end and then apply the moment–curvature relationship, (equation (5.12)) we get:

$$M = -Pv = EI\frac{d^2v}{dx^2}$$

Therefore

$$\frac{d^2v}{dx^2} + \alpha^2 v = 0 \quad \text{where} \quad \alpha^2 = \frac{P}{EI} \tag{11.3}$$

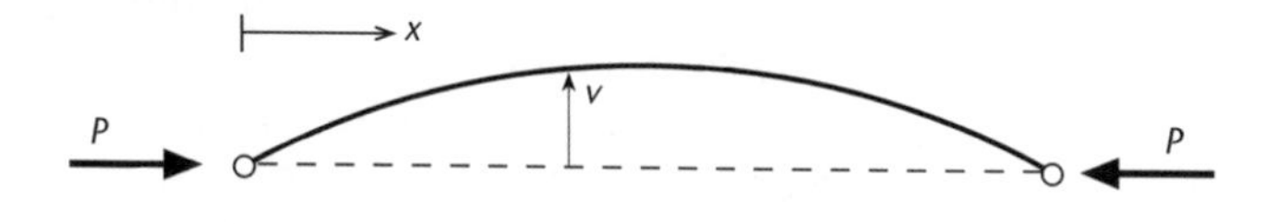

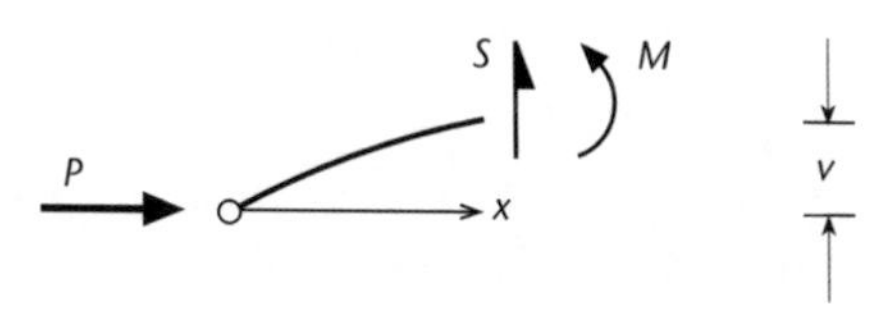

Figure 11.4
Buckling of a pin-ended strut.

The solution to this differential equation is

$$v = a\sin\alpha x + b\cos\alpha x$$

where the constants a and b must be determined from the boundary conditions. Putting $v = 0$ at $x = 0$ gives $b = 0$ and then putting $v = 0$ at $x = l$ gives

$$a\sin\alpha l = 0 \tag{11.4}$$

So either $a = 0$, in which case the strut remains straight, or $\sin\alpha l = 0$, which requires that $\alpha l = n\pi$, where n is any positive integer. Substituting for α from equation (11.3) gives the critical load:

$$P_c = n^2\pi^2\frac{EI}{l^2} \tag{11.5}$$

and the corresponding buckled shape is:

$$v = a\sin\frac{n\pi x}{l} \tag{11.6}$$

There is thus an infinite number of solutions to the governing equation (11.3), corresponding to the possible values of n. The higher the value of n, the more complex the buckled shape and the higher the corresponding critical load. The higher modes of buckling are really more of mathematical than practical interest, since the most important instability from a designer's point of view is the one that occurs at the lowest load ($n = 1$). Putting $n = 1$ in equation (11.5) gives the lowest critical load, or *Euler load* for the strut:

$$P_E = \pi^2\frac{EI}{l^2} \tag{11.7}$$

and this corresponds to the buckled shape:

$$v = a\sin\frac{\pi x}{l} \tag{11.8}$$

Equation (11.7) shows that the buckling load is proportional to the second moment of area I of the section, but the axis about which I is calculated has not so far been defined. Normally buckling will occur at the lowest possible load, and this will be achieved by using the smallest possible I-value for the section. (Remember the example of the ruler in Figure 11.3 – it buckled outwards in the direction of its smallest dimension.) Buckling calculations should therefore be based on the second moment of area about the *minor* axis, and the resulting Euler load will cause the member to buckle by bending about the minor axis. This is illustrated for the example of an I-shaped cross-section in Figure 11.5; Y–Y is the minor axis and the corresponding second moment of area is termed I_{yy}.

The behaviour of a pin-ended strut as the load P is increased can be illustrated by plotting the load against the lateral deflection v_{max} at the midpoint (Figure 11.6). Initially the strut is in stable equilibrium and the lateral deflection is zero (the line OA). When the magnitude of the load reaches P_E a bifurcation point occurs and two forms of behaviour are possible:

- either the strut remains straight, in which case the load increases above P_E but the equilibrium becomes unstable (the line AB); or

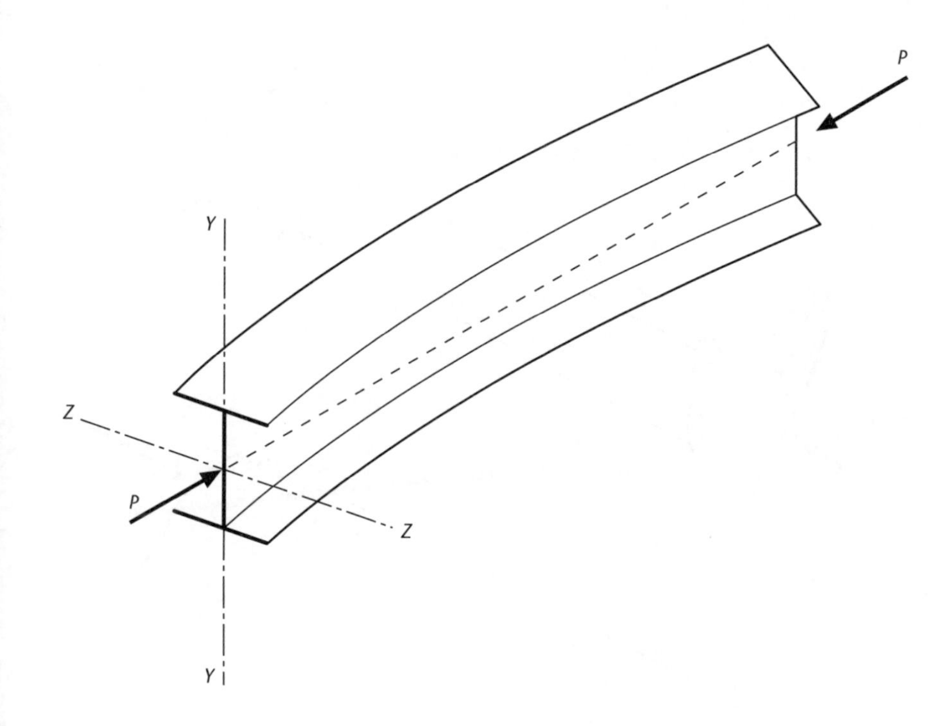

Figure 11.5
Minor-axis buckling for an I-section strut.

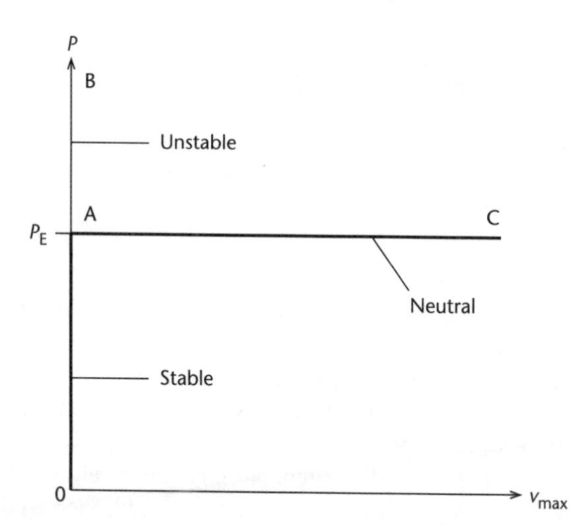

Figure 11.6
Axial load vs lateral deflection for a perfect pin-ended strut.

- the strut buckles into the shape given by equation (11.8) – line AC. In this case the load remains constant at P_E and the amplitude of the buckled shape (a in equation (11.8)) may take any value. Thus when the strut buckles it goes into *neutral* equilibrium, in which there are infinite possible equilibrium positions.

11.2.3 Struts with other end conditions

Struts will frequently be encountered which have end conditions other than simple pins. In general, if some rotational restraint is provided at the member ends this will tend to increase the buckling load. This can again be demonstrated using a perspex ruler. If you place the ruler between the palms of your hands and press

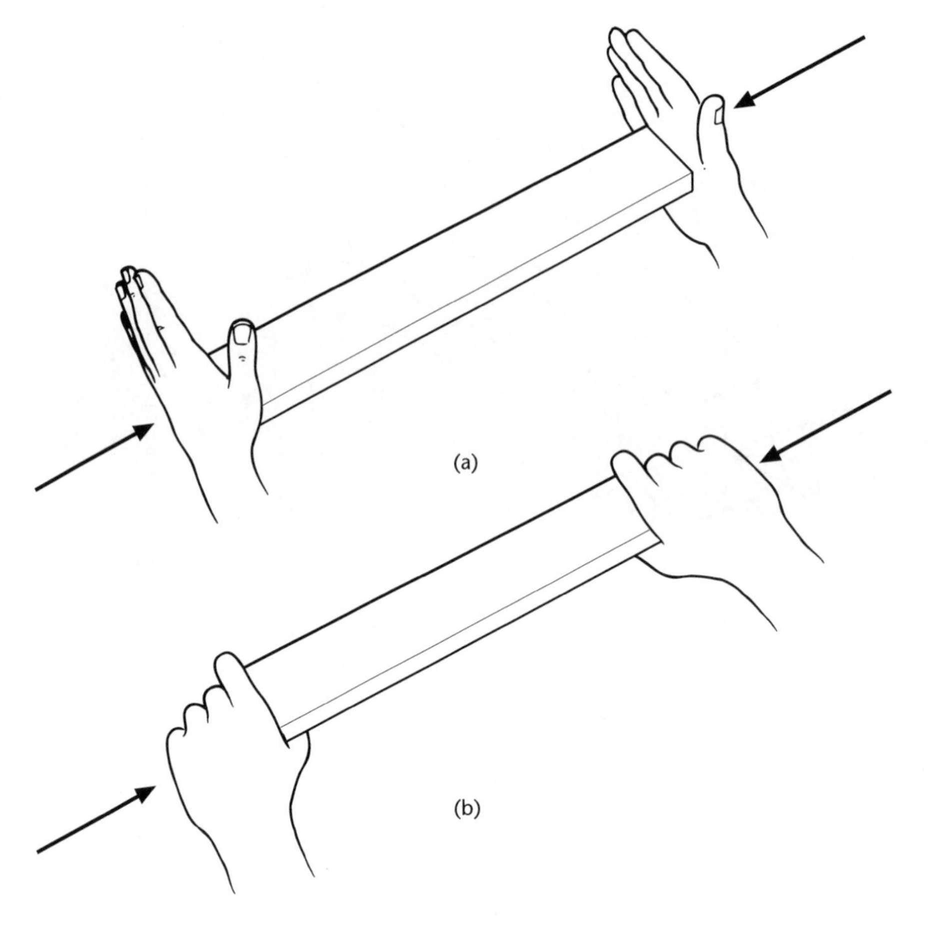

Figure 11.7
Practical demonstration of the effect of end restraint on buckling load.

inwards (Figure 11.7(a)) it buckles quite easily. If you now grip the ends of the ruler and attempt to prevent them from rotating (Figure 11.7(b)) it becomes much more difficult to force the ruler to buckle.

In practice all ends are usually restrained by supports having some finite stiffness, but since this stiffness is difficult to assess it is common to use the simple approximations of pinned or fixed ends. Figure 11.8 shows some of the most widely used idealisations of support conditions together with the resulting buckled shapes.

The critical loads for these cases are:

Cantilever: $$P_c = \frac{\pi^2}{4}\frac{EI}{l^2} = 0.25P_E \tag{11.9}$$

Fixed-fixed: $$P_c = 4\pi^2\frac{EI}{l^2} = 4P_E \tag{11.10}$$

Fixed-pinned: $$P_c = 2.04\pi^2\frac{EI}{l^2} = 2.04P_E \tag{11.11}$$

These can be derived in a similar way to the pin-ended result above, but care must be taken to incorporate the correct boundary conditions and restraining forces at the member ends, as illustrated in the following example.

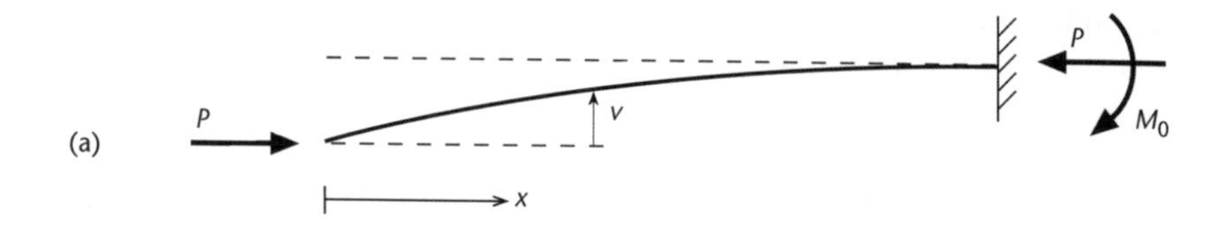

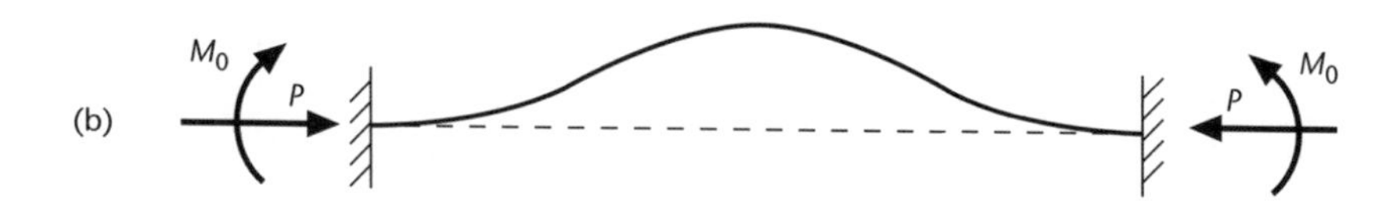

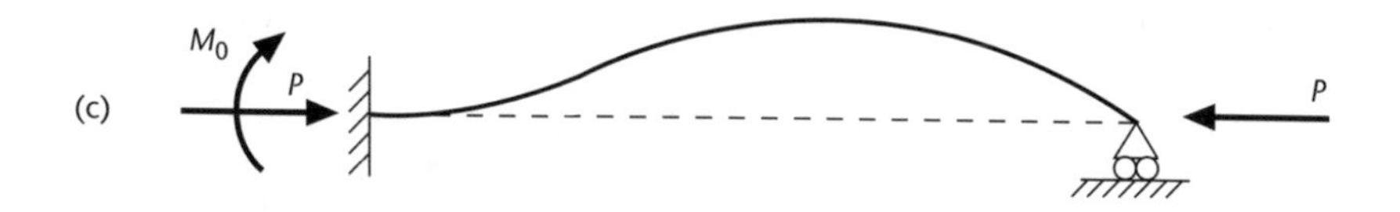

Figure 11.8
Buckling of struts with various end conditions.

EXAMPLE 11.2

Find the critical load for the cantilever strut in Figure 11.8(a) above.

The problem is most easily set up by defining the origin to be at the tip of the cantilever as shown in the diagram. The moment at a distance x from one end is then:

$$M = -Pv = EI\frac{d^2v}{dx^2}$$

This is identical to the equation for the pin-ended strut derived earlier, and so the solution also has the same form:

$$v = a\sin\alpha x + b\cos\alpha x$$

which can be differentiated to give:

$$v' = a\alpha\cos\alpha x - b\alpha\sin\alpha x$$

Applying the boundary conditions for this case:

$$x = 0, v = 0 \quad \rightarrow b = 0$$
$$x = l, v' = 0 \quad \rightarrow a\alpha\ \cos\alpha l = 0$$

Therefore either $a = 0$ and the strut remains straight, or $\cos\alpha l = 0$. The latter case has an infinite number of solutions, the lowest of which is $\alpha l = \pi/2$. Remembering that $\alpha^2 = P/EI$, this gives a critical load of:

$$P_c = \frac{\pi^2}{4}\frac{EI}{l^2}$$

and the corresponding deflected shape is

$$v = a\sin\frac{\pi x}{2l}$$

So the cantilever buckles into the shape of a quarter-sine wave at a quarter of the load required to cause buckling of a pin-ended strut having the same length and properties.

A useful concept in the analysis of buckling problems is the *effective length*. Consider, for example, the fixed-ended strut of Figure 11.8(b). The buckled shape of the strut is a full sine wave, so that there are points of contraflexure at $l/4$ from each end. The central portion of length $l/2$ between the points of contraflexure behaves exactly as a pin-ended strut of length $l/2$ – it has the same buckled shape and buckling occurs at the same load. We therefore say that the fixed-ended strut of length l has an effective length l_e of $l/2$. If we use effective length rather than actual length, then buckling loads for all end conditions can be found using the one formula (equation (11.7)).

Effective lengths for struts having various end conditions are given in Figure 11.9. It is easy to see that using the effective length together with equation (11.7) gives the same buckling load as using the actual length with equations (11.9)–(11.11). Note that for the cantilever the only point of contraflexure is at the tip; the distance between points of contraflexure is therefore found by extending the strut symmetrically beyond its fixed end as shown.

In design, it can be convenient to express the critical load in a member in terms of the stress in the material rather than the overall force. To do this, we first write the second moment of area I_{yy} as

$$I_{yy} = Ar_y^2 \tag{11.12}$$

where A is the cross-sectional area and r_y is known as the *radius of gyration* about the minor axis. (Similarly, for the major axis we could write $I_{zz} = Ar_z^2$, but since we are interested in minor-axis buckling this need not concern us.) Then, if we denote the stress in the material at buckling by σ_c we can write equation (11.7) as

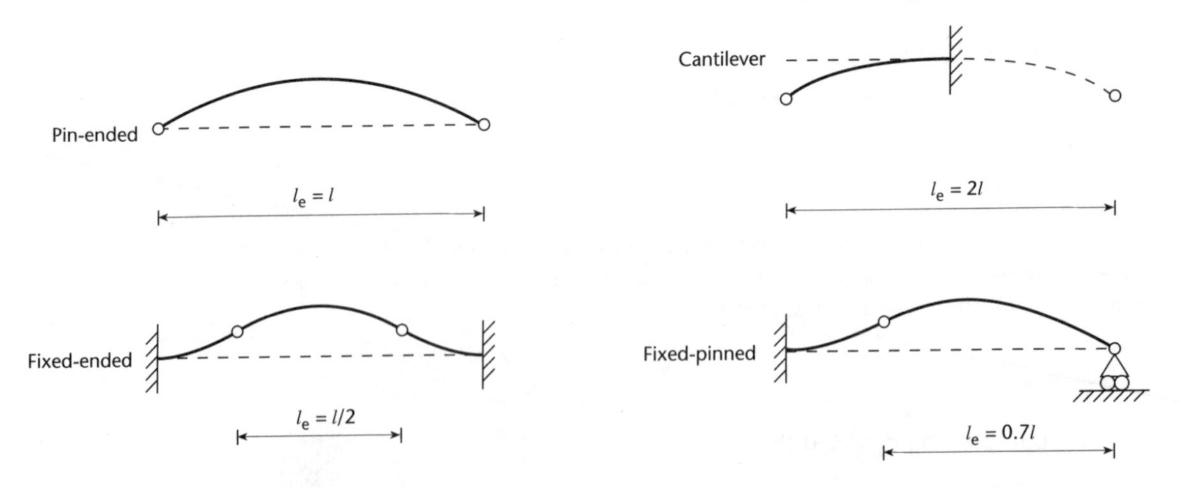

Figure 11.9
Effective lengths of struts.

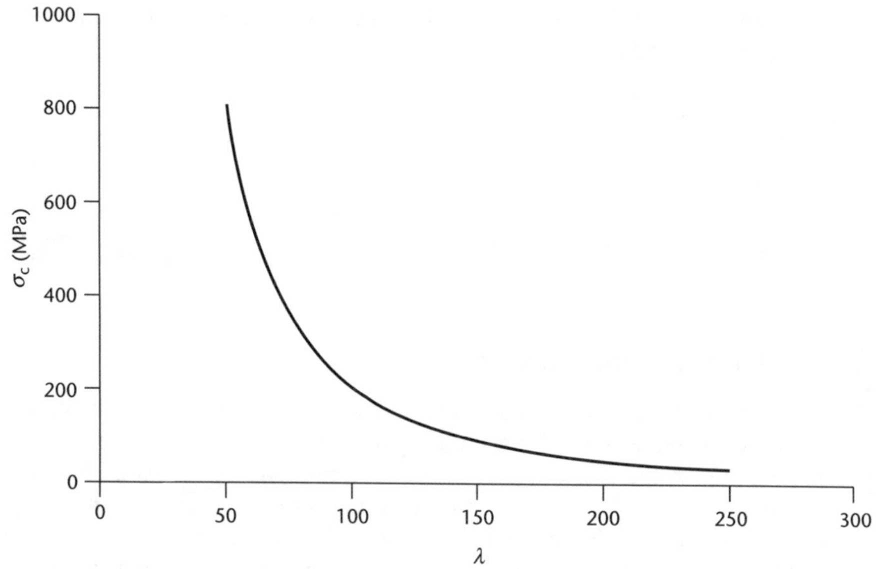

Figure 11.10
Critical stress vs slenderness ratio for steel.

$$A\sigma_c = \pi^2 \frac{EAr_y^2}{l_e^2}$$

or

$$\sigma_c = \pi^2 \frac{E}{\lambda^2} \quad \text{where} \quad \lambda = \frac{l_e}{r_y} \tag{11.13}$$

λ is known as the *slenderness ratio*. This is an extremely important property of a compression member, since it is the only geometric parameter which influences the stress at which buckling occurs. Equation (11.13) is plotted for steel ($E = 205$ GPa) in Figure 11.10. Clearly the critical stress reduces very rapidly as the slenderness ratio increases.

EXAMPLE 11.3

A steel 254 × 254 × 89 kg/m Universal Column section has cross-sectional area 11,400 mm^2 and second moment of area about the minor axis 48.49×10^6 mm^4. If the column carries an axial load of 1600 kN and is fixed at the base and pinned at the top, what is the greatest length of column that can be used without buckling?

The axial stress in the column is $1600 \times 10^3/11,400 = 140.4$ MPa. From equation (11.13) this stress will be sufficient to cause the column to become unstable if the slenderness ratio is:

$$\lambda = \pi\sqrt{\frac{205 \times 10^3}{140.4}} = 120$$

(Alternatively, this value could have been read approximately from Figure 11.10.) The radius of gyration r_y is

$$r_y = \sqrt{\frac{I_{yy}}{A}} = \sqrt{\frac{48.49 \times 10^6}{11,400}} = 65.2 \text{ mm}$$

and the effective length of a fixed-pinned member is $l_e = 0.7l$. So the slenderness ratio can be expressed as

$$\lambda = \frac{0.7l}{r_y} \rightarrow 120 = \frac{0.7l}{65.2}$$

from which $l = 11.18$ m.

11.2.4 Effect of initial imperfections

The preceding sections have been based on the assumption of a perfect strut, that is, initially completely straight, axially loaded, and remaining linear and elastic throughout the loading. In practice these perfect conditions are virtually impossible to achieve. For instance, most steel is produced by a process known as hot-rolling, in which the desired section is formed by passing lengths of heated steel through profiled rollers. The hot-rolling process and the subsequent rather uneven cooling of the steel tend to cause a slight initial curvature of the member, and also result in patterns of *residual stresses* being locked into the section. In addition, so-called axial loads are rarely applied exactly along the axis of a member; a small eccentricity is practically impossible to avoid. Each of these small variations from the assumed ideal characteristics can have a significant effect on the overall behaviour.

Probably the most important of these imperfections is the initial curvature of a member. This will therefore be considered in some depth here. Figure 11.11 shows a pin-ended strut which has a small, sinusoidal initial curvature of the form

$$v_0 = V \sin \frac{\pi x}{l}$$

where V is the initial midspan deflection. When an axial load is applied bending moments are immediately set up in the strut, causing the curvature to increase by an amount u so that the total deviation of the strut from a straight line is

$$v = v_0 + u$$

The moment–curvature equation for the strut is:

$$M = -Pv = -P(v_0 + u) = EI \frac{d^2u}{dx^2}$$

Figure 11.11
Strut with initial curvature.

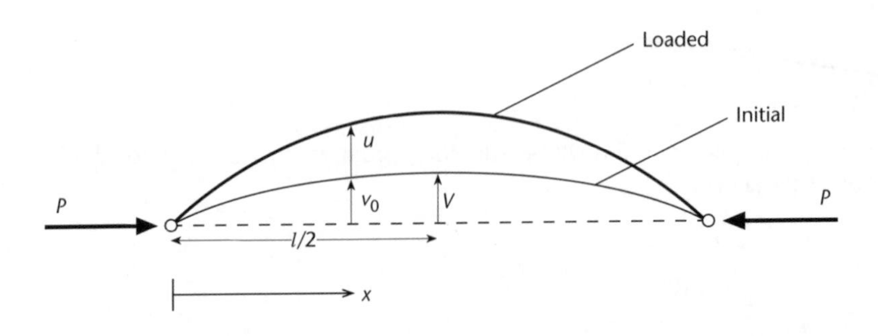

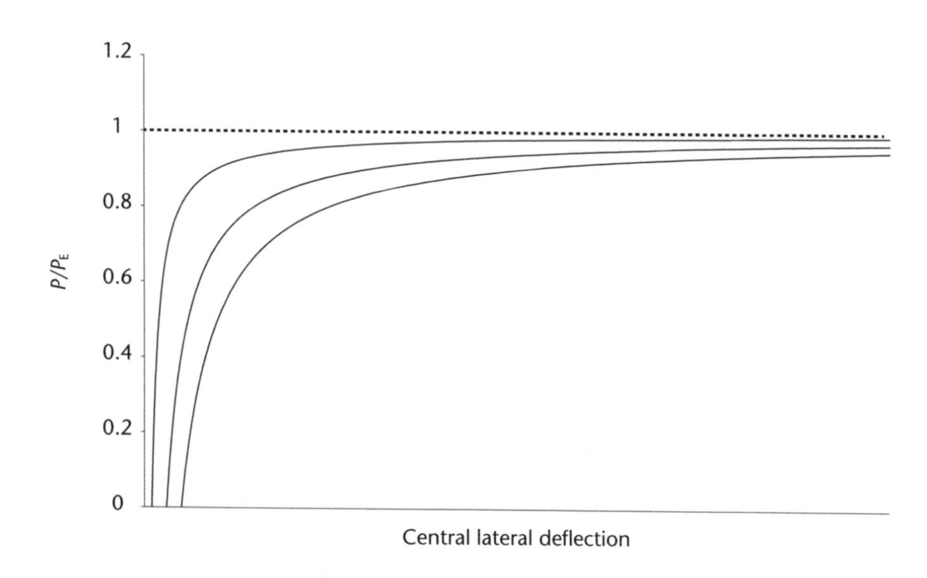

Figure 11.12
Load vs lateral deflection for struts with an initial sinusoidal curvature.

Note that the moment causes only the *additional* curvature, not the total curvature (that is, u'' not v''). Rearranging gives

$$\frac{d^2u}{dx^2} + \alpha^2 u = -\alpha^2 v_0 = -\alpha^2 V \sin\frac{\pi x}{l}$$

This has a similar solution to equation (11.3) for the pin-ended strut, but with the addition of a particular integral corresponding to the non-zero right-hand side:

$$u = a \sin \alpha x + b \cos \alpha x + \frac{\alpha^2}{(\pi/l)^2 - \alpha^2} V \sin\frac{\pi x}{l}$$

As before, the constants a and b can be found from the boundary conditions. Putting $u = 0$ at $x = 0$ gives $b = 0$ and putting $u = 0$ at $x = l$ gives $a = 0$, so that we are left with only the particular integral term. Remembering that $\alpha^2 = P/EI$, this becomes:

$$u = \frac{P/EI}{(\pi/l)^2 - P/EI} V \sin\frac{\pi x}{l} = \frac{P}{\pi^2 EI/l^2 - P} V \sin\frac{\pi x}{l} = \frac{P}{P_E - P} V \sin\frac{\pi x}{l}$$

The total lateral deflection at any time is then found by adding u to the initial shape:

$$v = v_0 + u = V \sin\frac{\pi x}{l}\left(1 + \frac{P}{P_E - P}\right) = \frac{P_E}{P_E - P} V \sin\frac{\pi x}{l} \qquad (11.14)$$

So the initial curvature is multiplied by the factor $P_E/(P_E - P)$. As P increases, the deflection initially increases very slowly, but then it tends towards infinity as P approaches the Euler load P_E.

Figure 11.12 shows the relationship between the axial load and the central lateral deflection for various values of the initial central deflection V. For an initially straight bar the deflection is zero until the Euler load is reached, and the bar then buckles sideways into a neutral equilibrium position (the dashed line). For an initially curved bar the load–deflection path follows one of the solid lines, with the exact path depending on the initial deflection V. Note in particular that:

- The Euler buckling load for a perfectly straight strut represents an upper bound on the load capacity; an initially curved strut will tend towards but never reach this load. Knowledge of the Euler load is therefore extremely valuable even though it represents idealised rather than real behaviour.
- There is no discontinuity in the behaviour of a curved member.
- For a curved member there is always a unique equilibrium position, that is, every value of P corresponds to a particular value of lateral deflection. There is therefore no load at which neutral equilibrium is reached and the bar remains stable throughout.

It should be remembered that Figure 11.12 is plotted using the assumption of fully elastic material behaviour throughout. In reality, the large increase in lateral deflection would be accompanied by an increase in the bending moment, which would eventually cause yielding. The resulting behaviour will be considered in more detail in the next section.

EXAMPLE 11.4

A pin-ended strut of length 4.0 m is made from a circular hollow steel section with outer and inner radii 60 mm and 50 mm respectively. The manufacturing process causes an initial sinusoidal curvature, with the midspan point lying a distance of 2.0 mm from a straight line between the two ends. Calculate the load required to increase the midspan deflection to 10 mm and compare this to the Euler buckling load of an initially straight strut.

The second moment of area about any axis in the plane of the section and passing through its centroid is:

$$I = \frac{\pi}{4}(0.06^4 - 0.05^4) = 5.27 \times 10^{-6}\ \text{m}^4$$

The Euler load can then be found from equation (11.7):

$$P_E = \frac{\pi^2 \times 205 \times 10^6 \times 5.27 \times 10^{-6}}{4^2} = 666\ \text{kN}$$

Putting $x/l = 0.5$ in equation (11.14):

$$v_{max} = V\frac{P_E}{P_E - P}$$

Rearranging:

$$P = P_E\frac{(v_{max} - V)}{v_{max}}$$

With $v_{max} = 10$ mm and $V = 2$ mm this gives:

$$P = 0.8P_E = 533\ \text{kN}$$

Note that even this very small initial curvature results in a significant reduction in load capacity – the limiting deflection is achieved at 80% of the Euler load.

11.2.5 Interaction of buckling and yielding

In the preceding sections we considered only perfectly elastic behaviour. Of course, most materials have a stress limit at which yielding will occur, and in design it will often be necessary to consider buckling and yielding behaviour together.

Take first the idealised case of an initially straight, pin-ended strut made from a perfectly elasto-plastic material with yield stress σ_y. If we apply a steadily increasing axial compressive load P, then one of two things will eventually happen – either the strut will become unstable and buckle, or the material from which it is made will yield. Both of these events are forms of failure, limiting the load that can be carried by the strut, and we therefore wish to avoid them. We can determine the range of acceptable loads and the most likely mode of failure by plotting the axial stress in the strut against the slenderness ratio λ.

We have already seen in equation (11.13) that the stress at which instability occurs is $\sigma = \pi^2 E/\lambda^2$. Obviously yielding will occur at a stress $\sigma = \sigma_y$, independent of the slenderness. These two lines are plotted in Figure 11.13. Obviously for a given value of λ the critical failure mechanism will be the one which occurs at the lower stress. So for relatively short, *stocky* members (low values of λ) the material is likely to fail by reaching its yield stress, whereas more slender elements are likely to fail by buckling at stresses well below σ_y. The value of λ at which the transition between the two failure modes occurs is found by equating the two expressions for σ, giving:

$$\lambda = \pi\sqrt{\frac{E}{\sigma_y}} \tag{11.15}$$

Now consider the more realistic case of a pin-ended strut with a small initial curvature $v_0 = V \sin \pi x/l$ (Figure 11.14(a)). The behaviour is now rather more complex. We have already seen that, for elastic behaviour, an axial load amplifies the initial deflection by a factor $P_E/(P_E - P)$ where P_E is the Euler buckling load for an initially straight member. The deflection gives rise to a bending moment in

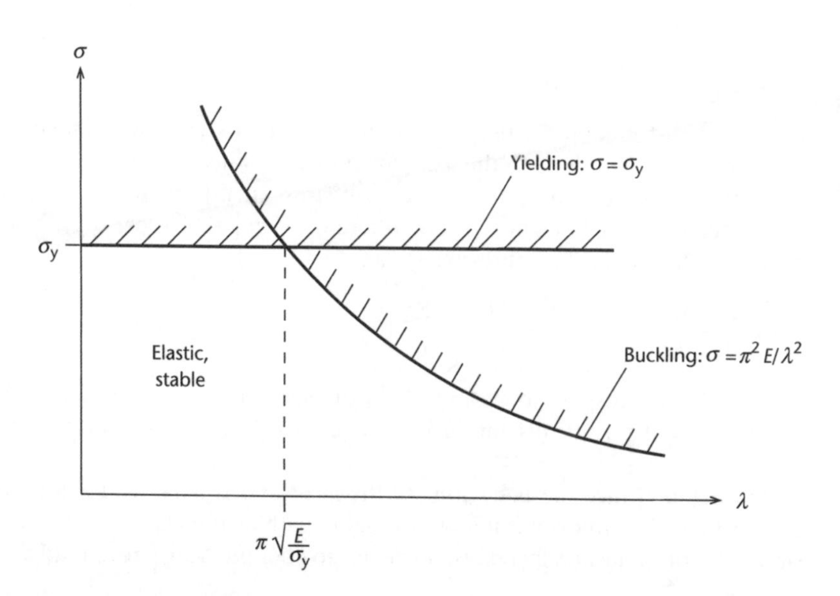

Figure 11.13
Interaction of buckling and yielding for an ideal strut.

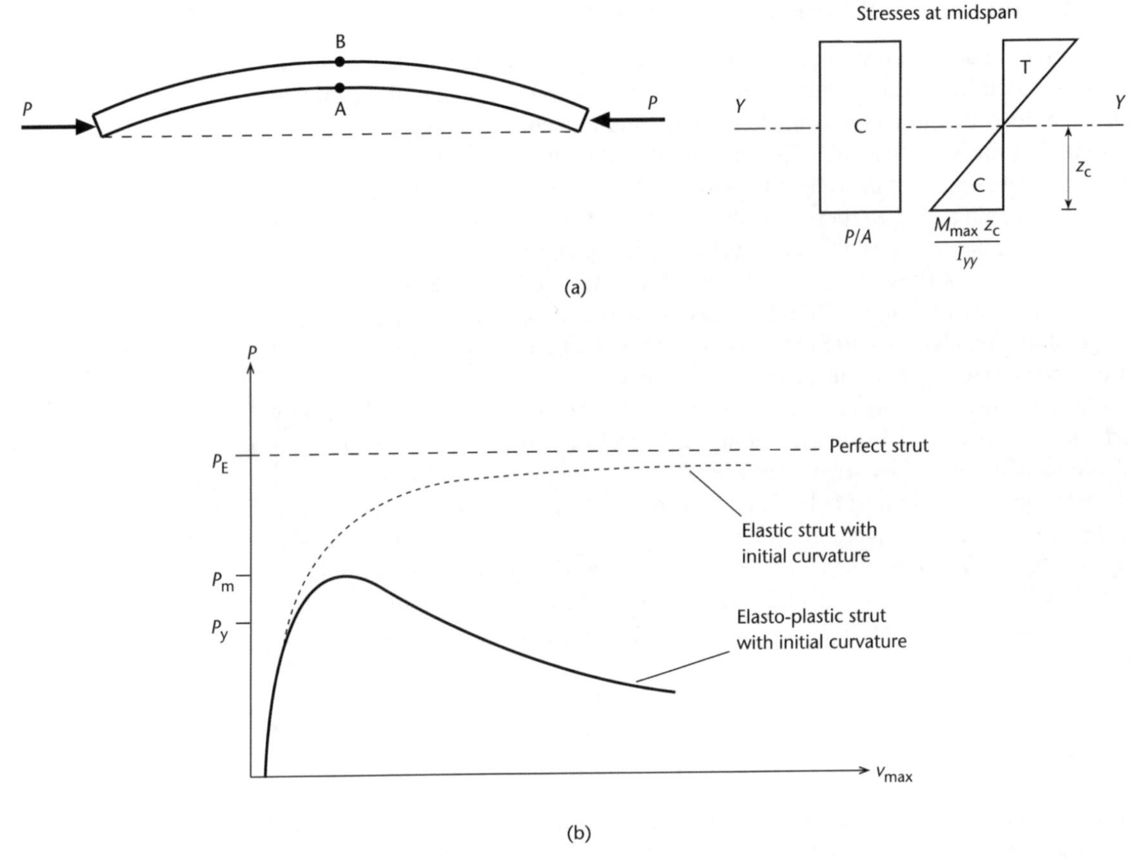

Figure 11.14
Behaviour of a perfectly elasto-plastic strut with initial curvature.

addition to the axial load. This takes its maximum magnitude at the midpoint of the strut:

$$M_{max} = Pv_{max} = \frac{PP_E}{P_E - P} V$$

The stress distribution in the cross-section therefore has two components: a uniform compression due to the axial force P and a linearly varying stress due to the bending moment M_{max}. The maximum stress in the member occurs on the inside face (point A in Figure 11.14), where the compressive part of the bending stress adds to the axial compressive stress:

$$\sigma = \frac{P}{A} + \frac{M_{max} z_c}{I_{yy}} = P\left(\frac{1}{A} + \frac{P_E}{P_E - P} \cdot \frac{V z_c}{I_{yy}}\right) \tag{11.16}$$

where A is the cross-sectional area, I_{yy} is the second moment of area about the minor axis, and z_c is the distance from the minor axis to the extreme compression fibre.

We can now trace the behaviour of the strut as the axial load P is increased (Figure 11.14(b)). Initially it follows the same path as the elastic strut considered earlier. As the deflection increases, however, so does the peak stress, until a load P_y

is reached where the stress given by equation (11.16) equals the yield stress of the material, and yielding occurs at the extreme compression fibre, point A. The yielding causes a reduction in the lateral stiffness of the strut, so that from this point onwards the elasto-plastic curve has a more shallow gradient than the fully elastic one. As the lateral deflection increases, so the yielding spreads both through the cross-section and along the member and the gradient of the load–deflection curve reduces further. Eventually, at a load P_m, the moment caused by the lateral deflection increases to the point where the peak tensile bending stress is bigger than the axial compressive stress, and the member goes into tension at point B. Beyond this point, the load carried by the member reduces as the lateral deflection increases.

The behaviour of most real compression members is similar to the solid line in Figure 11.14(b).

EXAMPLE 11.5

In Example 11.4 we looked at the elastic behaviour of a strut with an initial lack of straightness of amplitude 2 mm. If the strut is made from steel with yield strength $\sigma_y = 275$ MPa, determine the load required to cause first yield and the corresponding peak lateral deflection.

The cross-sectional area of the member is $A = \pi(60^2 - 50^2) = 3456\ \text{mm}^2$. In Example 11.4 we calculated $I = 5.27 \times 10^6\ \text{mm}^4$, and the Euler buckling load of an initially straight element having the same properties was found to be 666 kN. From equation (11.16) first yield occurs when

$$\sigma_y = P\left(\frac{1}{A} + \frac{P_E}{P_E - P} \cdot \frac{Vz_c}{I_{yy}}\right)$$

Rearranging this expression gives a quadratic for the axial load P:

$$P^2 - P(A\sigma_y + P_E + P_E V \cdot Az_c/I_{yy}) + A\sigma_y P_E = 0$$

which, on substituting numerical values for the various parameters (in units of N and mm), becomes:

$$P^2 - 1.643 \times 10^6 P + 633.0 \times 10^9 = 0$$

Solving and taking the lower of the two roots gives $P = 617$ kN (that is, 93% of the Euler load) and the corresponding lateral deflection can then be found from equation (11.14):

$$v_y = V\frac{P_E}{P_E - P} = 27.1\ \text{mm}$$

The load calculated here is a realistic upper limit for this strut. The calculation also serves to reassure us that the analysis performed in Example 11.4 was valid, that is, that our assumption of linear behaviour up to a lateral displacement of 10 mm was correct.

11.2.6 Application to design

When designing elements carrying compressive loads, it is desirable to have design charts of the form of Figure 11.13, which enable us to relate the load capacity of a member to a single, easily calculated geometric parameter. Unfortunately, Figure 11.13 is not itself suitable because it is based on the assumption of an ideal strut, whereas we know that real struts have initial imperfections. However, we can derive similar curves based on the more realistic analyses in sections 11.2.4 and 11.2.5. We know that first yield will occur when the stress given by equation (11.16) becomes equal to σ_y. Working in terms of stresses rather than forces, we can write (11.16) as:

$$\sigma_y = \sigma_c\left(1 + \frac{\sigma_E}{\sigma_E - \sigma_c}\cdot\frac{AVz_c}{I_{yy}}\right) \qquad (11.17)$$

where $\sigma_c = P/A$ and $\sigma_E = P_E/A$. The ratio AVz_c/I_{yy} gives a non-dimensional measure of how the initial curvature compares to the cross-sectional dimensions, and is usually termed the *initial curvature parameter*, denoted by η. Equation (11.17) can then be further rearranged to give a quadratic for σ_c , the average stress at first yield:

$$\sigma_c^2 - \sigma_c\left[\sigma_y + \sigma_E(1+\eta)\right] + \sigma_y\sigma_E = 0$$

and the lower of the two roots of this equation is:

$$\sigma_c = \frac{\sigma_y + \sigma_E(1+\eta)}{2} - \left[\left(\frac{\sigma_y + \sigma_E(1+\eta)}{2}\right)^2 - \sigma_y\sigma_E\right]^{1/2} \qquad (11.18)$$

Equation (11.18) can be used to determine the limiting stress σ_c that an initially curved member can carry without yielding, so long as we know what value to use for the curvature parameter η. Since this is a function of the manufacturing process, it will tend to vary from member to member. As a general rule, however, very slender members will tend to curl up rather more than stocky ones, so that it is reasonable to assume that an approximate relationship exists between η and the slenderness ratio λ. Various formulae have been used over the years, but the one currently favoured in the UK is:

$$\begin{aligned} \eta &= 0.001a(\lambda - \lambda_0) \quad & \lambda \geq \lambda_0 \\ \eta &= 0 & \lambda < \lambda_0 \end{aligned} \qquad (11.19)$$

where $\lambda_0 = 0.2(\pi E/\sigma_y)$. The parameter a takes one of four values (2.0, 3.5, 5.5 or 8.0) depending on the section shape and dimensions. This formulation for η in fact accounts for rather more than the initial curvature effect, since the values of a include an allowance for the effects of residual stresses caused by the manufacturing process.

The above formulae are named after the engineers who did much of the pioneering work in this field. Equation (11.18) is known as the Perry–Robertson formula, the form of η given in (11.19) is often referred to as the Perry factor and the coefficient a is the Robertson constant.

Figure 11.15 shows plots of the Perry–Robertson formula for struts made of mild steel ($E = 205$ GPa, $\sigma_y = 275$ MPa), together with the limiting stress curves for an ideal strut. It can be seen that, at very low slenderness ratios, both real and ideal struts fail by simple compressive yielding. For real struts, however, there is no

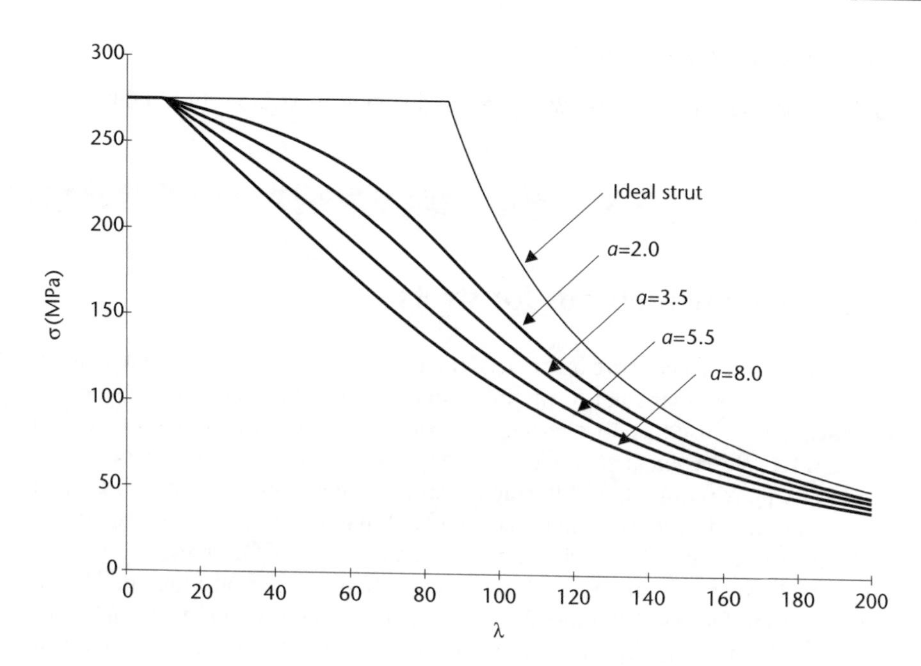

Figure 11.15
Strut design curves for steel with $\sigma_y = 275$ MPa.

sharp cut-off between yielding and buckling behaviour. Instead, the load capacity reduces gradually as the slenderness increases, and the stress is always limited by the onset of yielding. For intermediate values of λ the reduction in load capacity caused by the initial curvature is very significant. At very high slenderness ratios the real and ideal curves again become very similar.

EXAMPLE 11.6

A $254 \times 254 \times 89$ kg/m Universal Column section is made from steel with $\sigma_y = 275$ MPa. This section has $A = 11,400$ mm^2, $I_{yy} = 48.49 \times 10^6$ mm^4, and its Robertson constant for buckling about the minor axis is $a = 5.5$. If the column is 3.25 m high, fixed at its base and free at the top, calculate the maximum axial load it can carry and compare this with the load capacity of the corresponding ideal strut.

The radius of gyration is:

$$r_y = \sqrt{\frac{48.49 \times 10^6}{11,400}} = 65.2 \text{ mm}$$

and the slenderness ratio is:

$$\lambda = \frac{l_e}{r_y} = \frac{2 \times 3250}{65.2} = 100$$

Therefore, using the $a = 5.5$ curve in Figure 11.15, the limiting stress is 122 MPa and the axial load capacity is

$$P = 11,400 \times 122 = 1,391 \text{ kN}$$

Using the top curve in Figure 11.13, an ideal strut having this slenderness ratio would buckle at a stress of 202 MPa, giving a load capacity of

$$P = 11,400 \times 202 = 2,306 \text{ kN}$$

So the imperfections in the real strut have reduced its load capacity to 60% of the ideal value.

11.3 Energy methods for struts

At the start of this chapter we introduced the concepts of buckling and instability in terms of the potential energy of the system under consideration. An energy approach can be used as an alternative to the calculation methods presented in section 11.2. However, a major disadvantage is that a prior knowledge of the buckled shape is required, and if a significantly wrong buckled shape is assumed then large errors in the calculated buckling load may result. Nevertheless, energy methods can be useful in providing approximate solutions for cases where more precise methods prove unwieldy, for instance when the second moment of area varies over the length of the member. The method presented here is suitable for struts free of initial imperfections – as we have seen, this idealised case provides an upper bound to the behaviour of real struts.

11.3.1 Potential energy of a buckled strut

The first step in formulating an energy analysis is to determine a general expression for the potential energy of a buckled strut. When a strut is loaded it gains potential energy in the form of strain energy, but there is also a loss of energy due to the reduction in the distance between the two end loads.

Consider first the strain energy. The axial force, shear force and bending moment developed in the strut all generate strain energy, but the first two are negligible compared to the bending term. The strain energy due to bending is given by equation (5.10), and since $M = EI\mathrm{d}^2v/\mathrm{d}x^2$ this can be written:

$$U = \int_0^l \frac{M^2}{2EI}\mathrm{d}x = \int_0^l \frac{EI}{2}\left(\frac{\mathrm{d}^2v}{\mathrm{d}x^2}\right)^2 \mathrm{d}x$$

Now consider the loss of potential energy due to the deformation. When the strut buckles the large curvature developed will cause the distance between the two ends to reduce by an amount δ as shown in Figure 11.16. Using calculus it is possible to obtain an approximate expression for δ:

$$\delta = \int_0^l \frac{1}{2}\left(\frac{\mathrm{d}v}{\mathrm{d}x}\right)^2 \mathrm{d}x$$

There will also be a small change in distance between the two ends due to the elastic shortening of the strut under the compressive load, but this again may be treated as negligible.

The total potential energy is then equal to the gain in bending strain energy minus the loss in potential energy due to the load P moving through the distance δ:

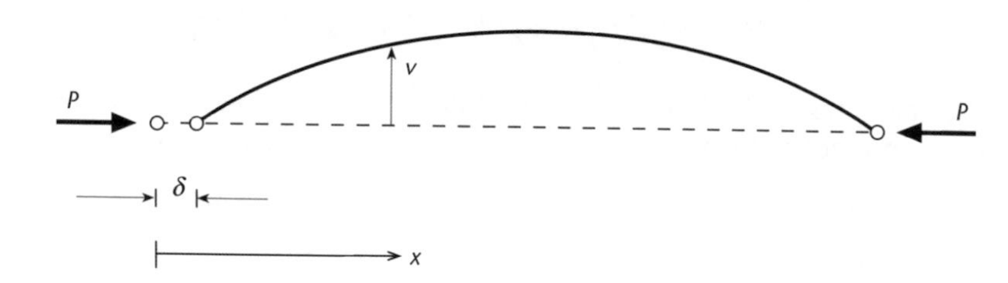

Figure 11.16
Reduction in length of a strut due to buckling.

$$V = \int_0^l \frac{EI}{2}\left(\frac{d^2v}{dx^2}\right)^2 dx - \frac{P}{2}\int_0^l \left(\frac{dv}{dx}\right)^2 dx \tag{11.20}$$

11.3.2 Rayleigh's method

Obviously the evaluation of the potential energy requires knowledge of the deflected shape v of the strut. Let us assume this to be of the form

$$v = af(x)$$

where a represents the amplitude. Since a is just a constant, the first two differentials of v are af' and af'' and equation (11.20) can now be written:

$$V = \frac{a^2}{2}\int_0^l EI(f'')^2 dx - \frac{Pa^2}{2}\int_0^l (f')^2 dx$$

Putting $dV/da = 0$ gives Rayleigh's formula for the critical load, usually denoted by P_R:

$$P_R = \frac{\int_0^l EI(f'')^2 dx}{\int_0^l (f')^2 dx} \tag{11.21}$$

This is often written as:

$$P_R = \frac{\int_0^l EI\left(\frac{d^2v}{dx^2}\right)^2 dx}{\int_0^l \left(\frac{dv}{dx}\right)^2 dx} \tag{11.22}$$

If the deflected shape of a strut is known then Rayleigh's method will give the true value of critical load. Except for very simple cases, however, the correct deflected shape will not be known and some approximation will have to be used. Rayleigh's formula will then give a critical load that is *higher* than the correct value. This is because the assumption of an incorrect buckled shape implies the imposition of additional, artificial forces and restraints, which will have the effect of increasing the stiffness of the structure, and hence increasing its critical load.

EXAMPLE 11.7

Use Rayleigh's method to calculate the critical load of a pin-ended strut of length l and constant flexural rigidity EI, assuming a buckled shape of the form:

(a) $v = a\sin(\pi x/l)$;

(b) $v = 4a(lx - x^2)/l^2$.

The assumed buckled shape has the form $v = af(x)$ where $f(x) = \sin(\pi x/l)$. Differentiating:

$$f' = \frac{\pi}{l}\cos\frac{\pi x}{l}, \quad f'' = -\frac{\pi^2}{l^2}\sin\frac{\pi x}{l}$$

Therefore equation (11.21) gives:

$$P_R = \frac{EI\int_0^l \frac{\pi^4}{l^4}\sin^2\frac{\pi x}{l}\,dx}{\int_0^l \frac{\pi^2}{l^2}\cos^2\frac{\pi x}{l}\,dx}$$

Both integrals can be evaluated by rewriting them using the double angle formulae, therefore:

$$P_R = \frac{\pi^2 EI}{l^2}\cdot\frac{\int_0^l \frac{1}{2}\left(1 - \cos\frac{2\pi x}{l}\right)dx}{\int_0^l \frac{1}{2}\left(1 + \cos\frac{2\pi x}{l}\right)dx} = \frac{\pi^2 EI}{l^2}\cdot\frac{l/2}{l/2} = \frac{\pi^2 EI}{l^2}$$

This agrees exactly with the Euler load for a pin-ended strut (equation (11.7)) since the correct deflected shape has been used.

(b) Following the same procedure we have $f(x) = lx - x^2$ and hence:

$$f' = l - 2x, \quad f'' = -2$$

and equation (11.21) gives:

$$P_R = \frac{EI\int_0^l 4\,dx}{\int_0^l (l^2 - 4lx + 4x^2)dx} = 4EI\cdot\frac{1}{[l^2x - 2lx^2 + 4l^3/3]_0^l} = \frac{12EI}{l^2}$$

The use of a quadratic deflected shape has therefore produced a result that is approximately 21% higher than the true value determined in (a).

The source of the inaccuracy in Example 11.7(b) can be seen clearly if we plot the two deflected shapes and their differentials. This has been done in Figure 11.17, with the values non-dimensionalised by the deflection amplitude a and the length l.

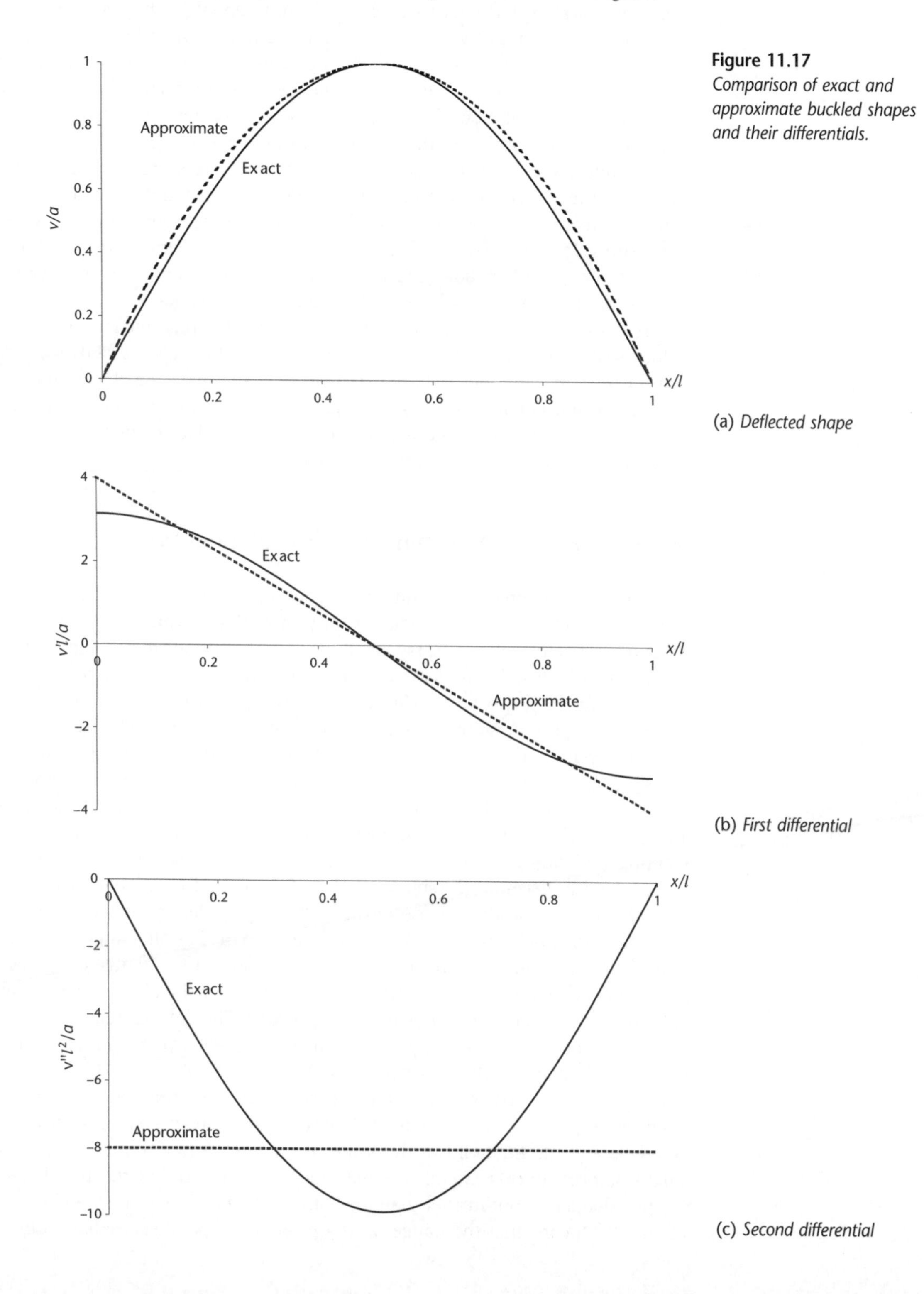

Figure 11.17 *Comparison of exact and approximate buckled shapes and their differentials.*

(a) *Deflected shape*

(b) *First differential*

(c) *Second differential*

Looking first at the deflected shapes (Figure 11.17(a)), the approximate, quadratic curve follows the exact, sinusoidal curve quite closely. We might intuitively feel that two such similar curves should give approximately the same value of critical load. However, if we plot the differentials of the two curves (Figure 11.17(b)), we see that there are some quite significant variations between them, particularly at the ends. If we differentiate again to give the member curvatures (Figure 11.17(c)), then the differences between the two curves are even more striking; the quadratic buckled shape results in a constant curvature, which is a very poor approximation to the correct, sinusoidally varying curvature.

The trends shown in Figure 11.17 are typical of most such approximations; errors in the deflected shape are amplified each time the deflection curve is differentiated, so that even quite small errors in the deflection function can give rise to gross inaccuracies, particularly in the second differential. Since the Rayleigh formula makes use of both the first and second differentials of the estimated deflection function, it is highly sensitive to any inaccuracies.

There are other energy formulations, not covered here, which are rather more robust than Rayleigh's method. For example, Timoshenko's method, a modification of the Rayleigh approach, is based on only the deflected shape and its first differential and so tends to give smaller errors when an incorrect deflected shape is used. However, it cannot be applied to all cases and can often be difficult to evaluate, and as a result is not widely used.

11.4 Lateral-torsional buckling of beams

Buckling does not only occur under axial compressive loads. Indeed, elements that carry pure compression are comparatively rare. Beams carry loads primarily in bending and many columns carry bending moments and shear forces in addition to axial loads – such elements are often referred to as *beam-columns*. Bending elements may fail by a form of instability known as *lateral-torsional buckling*. This extremely important failure mechanism will be briefly discussed here.

In all bending members, some part of the cross-section will be in compression. Consider, for instance, the relatively straightforward case of a simply supported I-section beam subjected to a constant bending moment M over its length (Figure 11.18). The elevation shows the beam bending, with the top flange going into compression and the bottom into tension. Like a strut, the compression flange can be expected to become unstable if the load is increased to some critical value. It cannot buckle vertically because of the restraint provided by the web, so it must bow out horizontally, as shown in the plan view. In the bottom flange, meanwhile, quite different conditions prevail. The tensile stresses in this flange cause it resist any tendency to move laterally, rather like a taut string. The differing behaviour of the two flanges therefore causes the beam to twist. The deformation of any cross-section can be characterised by a rotation ϕ about some point remote from the member, but lying on the y-axis of the section.

There is one further important aspect to the deformation that must be emphasised. Consider the length AB indicated on the plan view in Figure 11.18, where A is at the midpoint of the beam. If we look at how this initially straight length deforms during lateral-torsional buckling, we see that the top flange, in compression, bows out further than the bottom, tension flange, as shown in Figure 11.19. This means that the flanges at B are no longer parallel, so that an initially

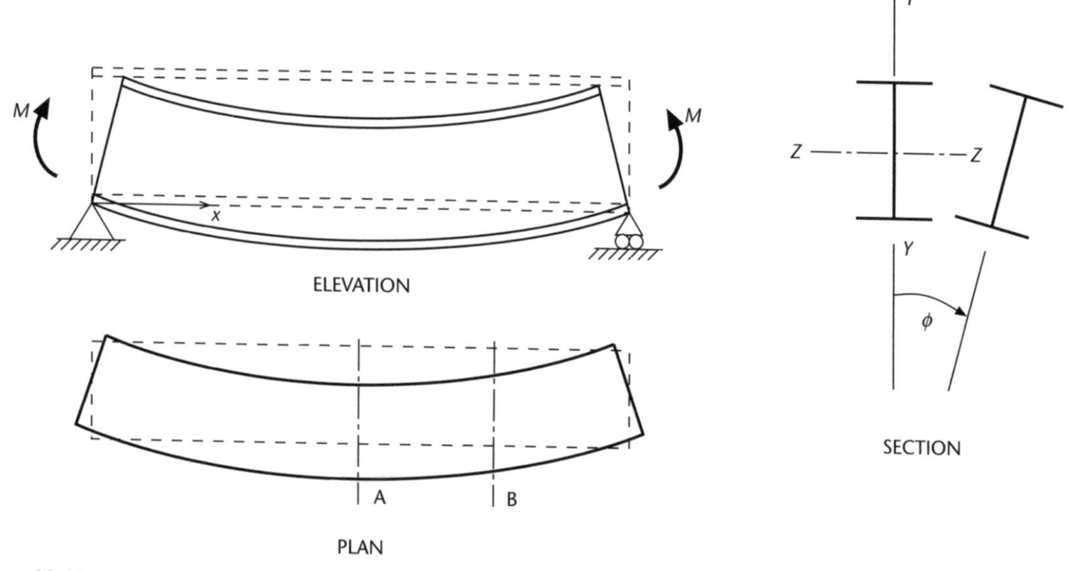

Figure 11.18
Lateral-torsional buckling of a simply supported beam.

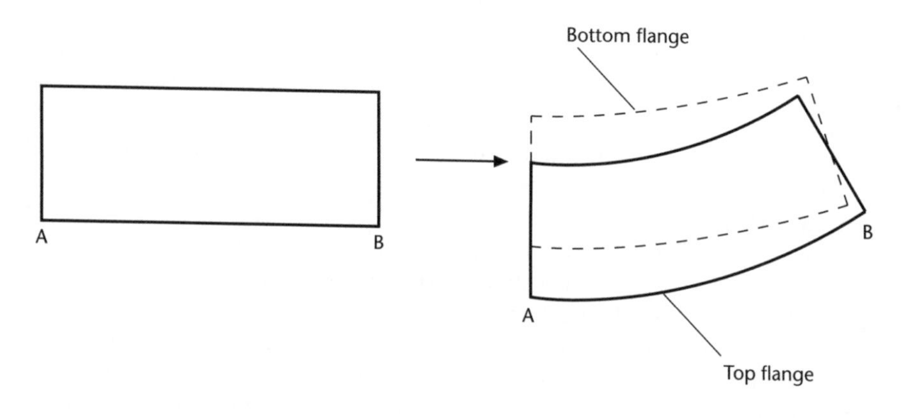

Figure 11.19
Warping of a beam during lateral-torsional buckling.

planar cross-section has distorted out-of-plane, or *warped*. In most members, there is some restraint to warping at the supports, so that the beam is not free to warp throughout its length and additional, longitudinal stresses are created.

The analysis of a buckled beam including warping effects is highly complex and only the main results will be presented here. For the beam in Figure 11.18, subjected to a constant bending moment M, a fourth-order differential equation in terms of the angle of twist ϕ can be obtained:

$$EH\frac{d^4\phi}{dx^4} - GJ\frac{d^2\phi}{dx^2} + \frac{M^2\phi}{EI_{yy}} = 0 \tag{11.23}$$

in which H is the *warping constant*, given by $H = I_{yy}h^2/4$ where h is the depth of the section. Equation (11.24) thus comprises a warping term, a twisting term and a bending term. This can be solved to give the critical moment M_c at which lateral-torsional buckling occurs. For a beam restrained from twisting at its ends the solution is:

$$M_c = \frac{\pi}{l}\sqrt{EI_{yy}GJ\left(1 + \frac{\pi^2 EH}{l^2 GJ}\right)} = \frac{\pi}{\lambda}\sqrt{EAGJ\left(1 + \frac{\pi^2 EH}{l^2 GJ}\right)} \quad (11.24)$$

where $\lambda = l/r_y$. So the critical moment for a beam is inversely proportional to its slenderness ratio. This is similar to buckling of a strut, where the critical load is proportional to $1/\lambda^2$.

Equation (11.25) is the basis for the design of steel beams, and can be thought of as the equivalent of the Euler buckling load for a strut (equation (11.7)). Of course, like (11.7) it needs modifications to account for the effects of initial imperfections, and the interaction between buckling and yielding modes of failure needs to be considered. These refinements are, however, beyond the scope of this text.

11.5 Other forms of instability

In this chapter we have looked briefly at some of the most common and straightforward forms of buckling and instability, concentrating particularly on struts carrying axial compressive loads. There are, however, many other instances in which buckling is important, most of which are quite mathematically complex. They will therefore not be covered in any detail here, but the reader should be aware of their existence.

First, there is the important case of beam-columns, that is, members carrying both axial and transverse loads. In these cases, both axial and lateral-torsional buckling are possible. Also, if the axial loads are large they have the effect of significantly reducing the bending stiffness of the member. In such cases it would be quite unrealistic to use, for example, the stiffness matrix for a bending element derived in Chapter 9. Instead, it is necessary to develop a new matrix in which the flexural terms are functions of the axial load. This makes the analysis considerably more complicated, because the stiffness terms are no longer constant, but must be recalculated whenever the loading is changed.

Whereas we have looked at single elements in isolation, most real elements are part of a frame. In pin-jointed frames the members carry only axial loads and so the methods presented here can be applied to them quite straightforwardly. Rigid-jointed structures present a much more difficult problem since the members will be subject to end moments in addition to axial loads. In addition, it may be necessary to consider the overall stability of an entire frame as well as the stability of the individual elements.

11.6 Further reading

Three of the most well-known and authoritative books on buckling and instability are:

- S. P. Timoshenko and J. M. Gere, *Theory of Elastic Stability*, 2nd edn, McGraw-Hill, 1961.
- M. R. Horne and W. Merchant, *The Stability of Frames*, Pergamon, 1965.
- J. M. T. Thompson and G. W. Hunt, *A General Theory of Elastic Stability*, Wiley 1973.

All three books are quite theoretical, especially the first one; Timoshenko was certainly the greatest structural theoretician of the twentieth century and all of his books are regarded as classics, but they are not generally considered easy reading! For those looking for a rather more applied account of buckling and particularly its role in the design of steel structures, an excellent text is:

- P. J. Dowling, P. Knowles and G. W. Owens, *Structural Steel Design*, Butterworth, 1988.

11.7 Problems

Where appropriate, answers are given at the end of the book. The questions marked with an asterisk are a little more challenging.

11.1. Figure 11.20 shows a light, rigid bar pinned to a rigid support. The free end is supported by a spring which is attached to a slider, so that the spring always remains vertical as the bar rotates. By considering the potential energy of the system, determine the three possible equilibrium positions of the bar and examine the stability of each. If $\theta = 0$ when the load is first applied, describe the behaviour as the load is steadily increased.

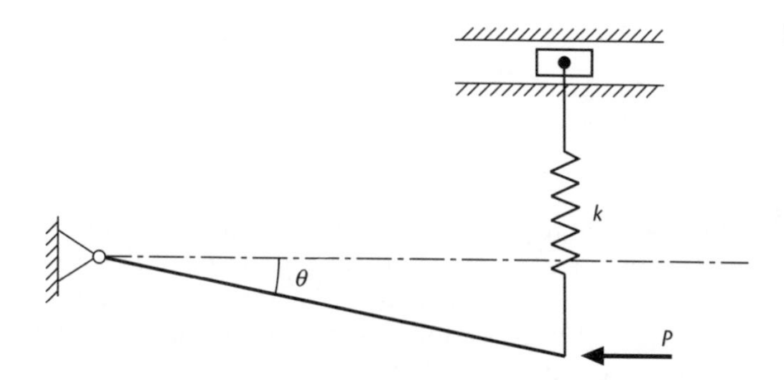

Figure 11.20

11.2. A steel ($E = 205$ GPa, $\sigma_y = 275$ MPa) rectangular hollow section has outer dimensions 200 × 120 mm and wall thickness 10 mm. It is used as a pin-ended column of length 5.0 m and is initially perfectly straight. Calculate the load capacity and identify the mode of failure if:

(a) the column is free to buckle in any direction;
(b) the column is restrained in such a way that it can only buckle about the major axis.

11.3. The strut in Problem 11.2 is now taken to be a real strut, with initial imperfections, and is free to buckle about any axis. If the Robertson constant for this cross-section is $a = 2.0$, show that the design critical load is approximately 79% of the Euler load calculated in 11.2(a). Show also that the value of a is equivalent to an initial sinusoidal curvature of the strut of amplitude 7.0 mm.

11.4. Figure 11.21 shows the buckled shape of a non-uniform, pin-ended strut. The left-hand half has flexural rigidity EI, while the right-hand half may be taken as rigid. Show that the critical load may be found from the equation $\tan(\alpha l/2) + \alpha l/2 = 0$, where $\alpha^2 = P/EI$.

Figure 11.21

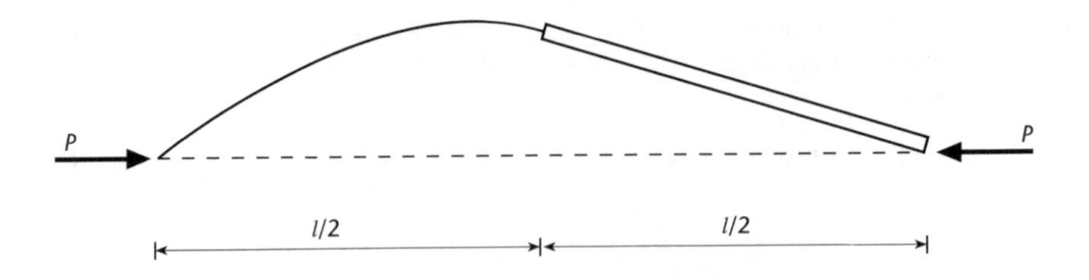

11.5.* Figure 11.22 shows a strut of length l and flexural rigidity EI, fixed at one end and at the other restrained laterally by a spring of stiffness k. Show that the buckling load P is given by the solution of the equation

$$\frac{\tan \alpha l}{\alpha l} = 1 - \frac{P}{kl}$$

where $\alpha^2 = P/EI$. Verify that this equation reduces to a standard solution when $k = 0$ and $k \to \infty$.

Figure 11.22

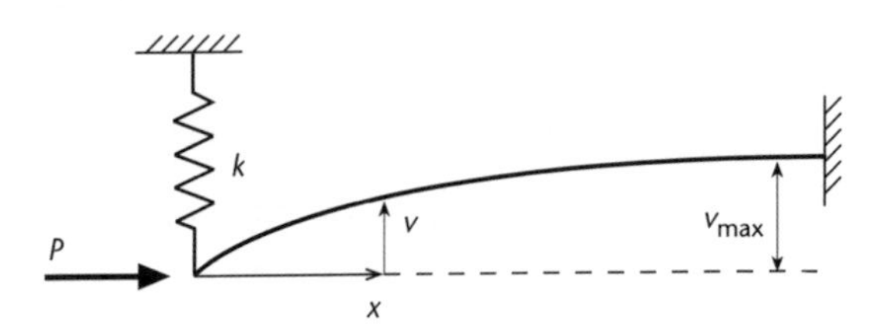

11.6.* A uniform pin-ended strut has length l and flexural rigidity EI. It carries an axial compressive load P and a transverse load W at midspan. Show that buckling occurs when $P = \pi^2 EI/l^2$ (that is, the buckling load is unaffected by the presence of a lateral load). If P takes a value equal to one-quarter of the buckling load, find the maximum bending moment in the strut.

11.7. A uniform cantilever of length 2.0 m has a solid rectangular cross-section with dimensions 150 × 75 mm, and is made from timber with $E = 12$ GPa. The member has an initial curvature about the minor axis which can be approximated as a quarter-sinewave of amplitude 8 mm. Find the maximum lateral deflection when an axial load of 25 kN is applied to the cantilever. Find also the maximum compressive stress and compare this with the stress that would be experienced by an initially straight member having the same properties.

11.8. An initially straight pin-ended strut has a second moment of area which varies linearly from I_0 at one end to $2I_0$ at the other. Use Rayleigh's method to estimate the buckling load, assuming a buckled shape of the form $v = a \sin \pi x/l$. Calculate the Euler load for a uniform strut with second moment of area $2I_0$ and comment on the comparison between the two results.

11.9.* Repeat Problem 11.5 using Rayleigh's method, assuming a deflected shape $v = a \sin(\pi x/2l)$. Compare the result with standard solutions for the cases $k = 0$ and $k \to \infty$. Compare the Rayleigh solution with the exact solution determined in 11.5 by sketching both solutions on a graph of critical load P against spring stiffness k.

[Hint: remember that there is strain energy in the spring as well as in the strut – this extra term must be added into the Rayleigh formula.]

11.10. A 254 × 146 × 37 kg/m Universal Beam has the cross-sectional properties: $A = 4740\ \text{mm}^2$, $r_y = 33.4$ mm, $r_z = 108$ mm, $J = 0.149 \times 10^6\ \text{mm}^4$. It is made from steel with properties $E = 205$ GPa, $G = 79$ GPa, $\sigma_y = 275$ MPa. The beam is simply supported and carries a uniformly distributed load w per unit length, causing it to bend about its major (ZZ) axis. Determine the maximum value of w the beam can carry and the mode of failure if its length is (a) 4.0 m and (b) 6.0 m.

CHAPTER 12

Plastic analysis of structures

CHAPTER OUTLINE

The traditional approach to the design of structures has been emphasised thus far in this book. In this approach, we first determine the stresses caused by a set of loads assuming fully elastic behaviour, and then size the members to ensure that the largest stress does not exceed the yield stress of the material. The ratio of the yield stress to the largest stress in the structure is called the *safety factor*. If the behaviour is always linear, then the safety factor is also the ratio of the yield load to the largest applied load.

We will now introduce an alternative approach known as plastic analysis. This is based on a recognition that most structures, particularly redundant ones, have considerable reserves of strength, so that the loads required to cause them to collapse are quite a lot larger than the loads which cause the onset of yielding. Therefore, instead of designing on the basis of no yield anywhere in the structure, we calculate the loading required to cause complete collapse, and then make sure that this exceeds the largest applied load by an appropriate margin. The ratio of the collapse load to the maximum applied load is called the *load factor*. The plastic design approach generally results in a more economical structure, though obviously this depends on the values chosen for the load factor and safety factor in the two approaches.

In this chapter we will look at how a plastic collapse mechanism develops and how the collapse load can be calculated. On completion the reader should be able to:

- understand the process of development of a plastic hinge and appreciate the approximations inherent in the analytical approach;
- predict likely collapse mechanisms for beams and plane frames;
- determine collapse loads using either the statical or virtual work approach;
- check the correctness of a collapse mechanism;
- understand the terms *load factor*, *shape factor* and *safety factor*, and the relationships between them.

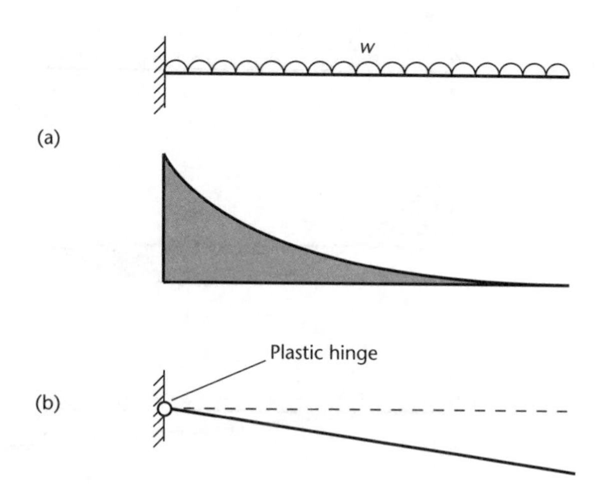

Figure 12.1 *Plastic collapse of a cantilever.*

12.1 Plastic collapse of structures

In order to perform a plastic analysis, we need to understand the sequence of events that leads to collapse of a structure.

Consider first the statically determinate, uniformly loaded cantilever in Figure 12.1(a). The elastic bending moment diagram is drawn below the structure, with the largest moment occurring at the fixed end. If the magnitude of the load w is steadily increased then a load will be reached at which yielding commences at the extreme fibres at the fixed end. As the load is increased further the plasticity spreads inwards until an entire cross-section has yielded. The yielded section has now lost all resistance to rotation and is known as a *plastic hinge*. As we saw in Chapter 5, the moment required to cause the plastic hinge to form is typically 1.1 to 1.2 times the moment which causes first yield at the extreme fibres; this ratio is known as the *shape factor*, since it depends on the shape of the cross-section.

The effect of creating a plastic hinge is equivalent to introducing an additional pin joint into the structure. In this case, since the structure is statically determinate, the creation of one plastic hinge is sufficient to reduce it to a mechanism, and it collapses as shown in Figure 12.1(b).

Now consider a structure with a single redundancy. Figure 12.2(a) shows a propped cantilever carrying a uniform load. Again the largest elastic bending moment occurs at the fixed end and so yielding will again commence at this location as the load is increased. In Figure 12.2(b) a plastic hinge has formed at this point. Introducing the plastic hinge effectively reduces the number of restraints on the structure by one, since there is no longer any resistance to rotation at the root. However, since it was initially redundant, the structure has not yet been reduced to a mechanism and can still carry more load.

As additional load is applied, no further moment can be carried at the plastic hinge, but other points in the beam do still have reserve capacity and so the bending moment at those locations can increase. The development of the hinge has thus caused a *redistribution* of the bending moments across the structure. It is this redistribution that enables a redundant structure to carry more load after the first hinge has formed.

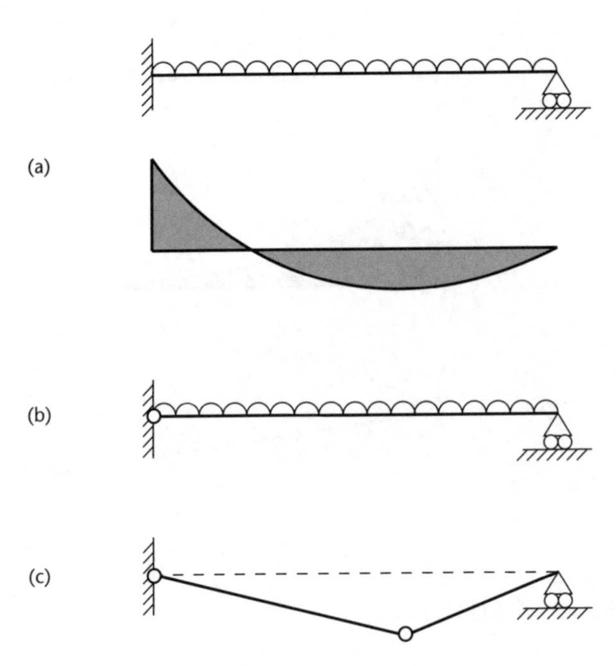

Figure 12.2
Plastic collapse of a propped cantilever.

Eventually the moment at some point along the span will become large enough to cause another plastic hinge to develop. The structure has now become a mechanism and collapses as shown in Figure 12.2(c).

12.2 The development of a plastic hinge

Before going on to look at the loads required to form plastic collapse mechanisms, we must consider in a little more detail how a plastic hinge is formed, and how hinge behaviour is idealised in plastic analysis. This, of course, involves yielding in flexure, a topic which was introduced in section 5.3.

A plastic hinge forms when yielding has spread through the entire cross-section of a member, reducing the rotational stiffness to zero. Consider, for example, the I-section beam subjected to a bending moment M in Figure 12.3. Yielding occurs at the extreme fibres at a moment M_y, defined by equation (5.16):

$$M_y = \frac{I_{zz}}{y_{max}}\sigma_y = Z_e\sigma_y$$

where Z_e is the elastic modulus of the section. As the moment is increased, yielding spreads inwards from the extreme fibres until the entire section has become plastic. The plastic moment capacity M_p can be found by taking moments of the stresses in the section about the neutral axis. For the section shown we can simply take moments of the total forces in the web and in the flange giving:

$$M_p = 2\left(\frac{\sigma_y h_w t_w}{2} \cdot \frac{h_w}{4} + \sigma_y b_f t_f \cdot \frac{(h_w + t_f)}{2}\right) \tag{12.1}$$

More generally we can write M_p as, equation (5.17):

$$M_p = Z_p\sigma_y$$

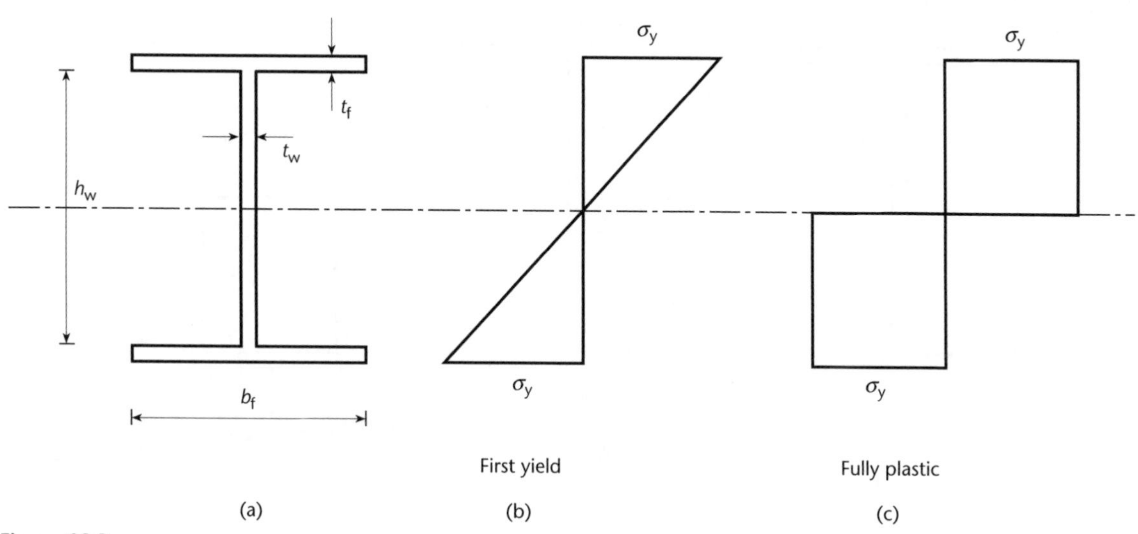

Figure 12.3
Stress distributions in an I-beam at first yield and full plasticity.

where Z_p is known as the plastic modulus and represents the area × lever arm terms in equation (12.1). The shape factor α of the section is then defined by equation (5.18) as:

$$\alpha = \frac{M_p}{M_y} = \frac{Z_p}{Z_e}$$

α gives a measure of the amount by which the moment on a section can be increased between first yield and the formation of a plastic hinge, and normally takes a value between 1.1 and 1.2 for a thin-walled section.

If we plot the moment–curvature relationship for a section undergoing yield (Figure 12.4(a)), the behaviour is initially linear until the moment reaches M_y. Thereafter, the flexural stiffness reduces as yielding spreads through the section and the stiffness tends to zero as the moment approaches M_p. It can be seen from this graph that the fully plastic moment is a theoretical concept which is never quite achieved in practice, since it implies an infinite curvature.

For a plastic analysis, we can idealise the behaviour in two stages. First, since we are concerned only with the formation of plastic hinges we are not particularly interested in the transition phase between the elastic and fully plastic states. We can therefore simplify the moment–curvature graph as shown in Figure 12.4(b). This idealised curve implies a sudden transition from elastic to plastic behaviour, and is equivalent to taking the shape factor to be unity. Next, we note that the curvature at an elastic section will be far smaller than that at a plastic hinge, and can therefore be neglected. This results in the *rigid-plastic* moment–curvature characteristic shown in Figure 12.4(c).

For beam-columns, which carry axial forces in addition to bending loads, we showed in section 5.3 that the plastic moment capacity is reduced in proportion to the square of the axial load. It can also be demonstrated that the presence of

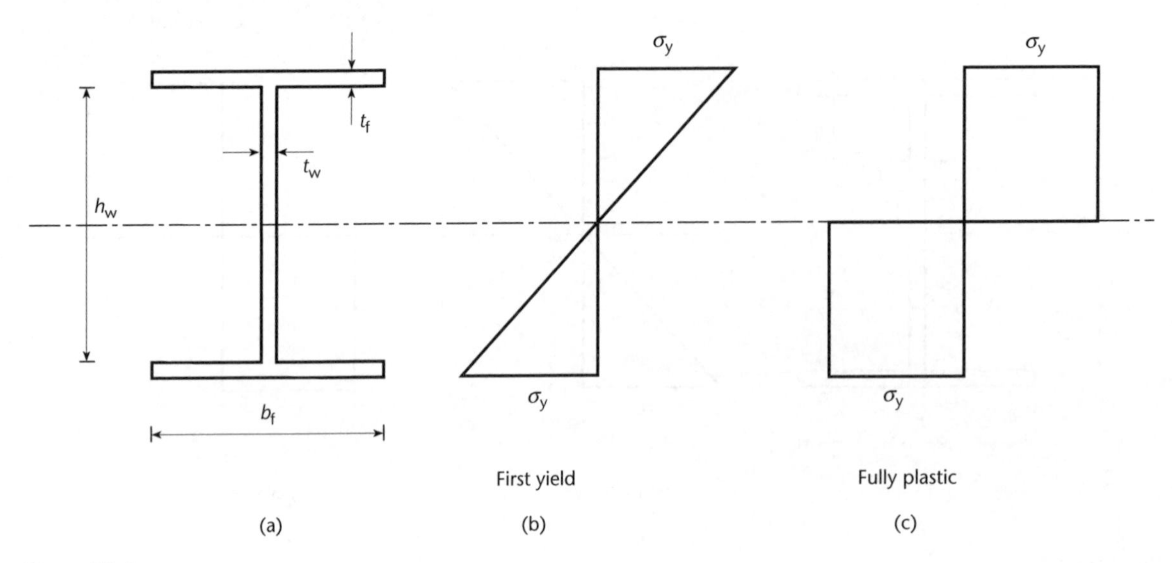

Figure 12.4
Real and idealised moment–curvature relations.

significant shear stresses reduces the moment capacity of a section. Such effects are, however, normally neglected in plastic analysis, with the plastic moment calculated from equation (5.17), regardless of any axial load or shear force that may be present.

Thus far we have considered the behaviour of a single cross-section in isolation, but in reality yielding spreads along the length of the beam at the same time as it spreads through the section. Consider for example the case of a simply supported beam of symmetrical cross-section carrying a uniformly distributed load w (Figure 12.5).

As w is increased, the bending moment increases until the yield moment M_y is reached at midspan. Yielding then occurs at the extreme top and bottom fibres (Figure 12.5(a)). If the load is increased further, then the plasticity at midspan spreads inwards from the edges of the section. In addition, the moments at points either side of midspan reach M_y and so yielding commences at these locations. This results in the pattern shown in Figure 12.5(b), with the plastic zones deepest at midspan and receding to the extreme fibres at the point where $M = M_y$. By the time a plastic hinge forms at midspan (Figure 12.5(c)), yielding in other parts of the beam is quite widespread.

With the assumption of rigid-plastic behaviour introduced above, this spread of yielding along a member is not modelled; a cross-section is treated as either fully elastic or fully plastic, with no partial yielding permitted.

In the remainder of this chapter, the following assumptions will be made:

- The plastic moment capacity M_p of a member may be calculated from equation (5.17), neglecting the influence of any axial load or shear force that may be present.
- A cross-section is assumed elastic and rigid until the plastic moment is reached, at which point a hinge forms and the rotational stiffness becomes zero.
- A plastic hinge is assumed to have infinitesimal length; the effects of partial yielding at points away from the hinges are neglected.

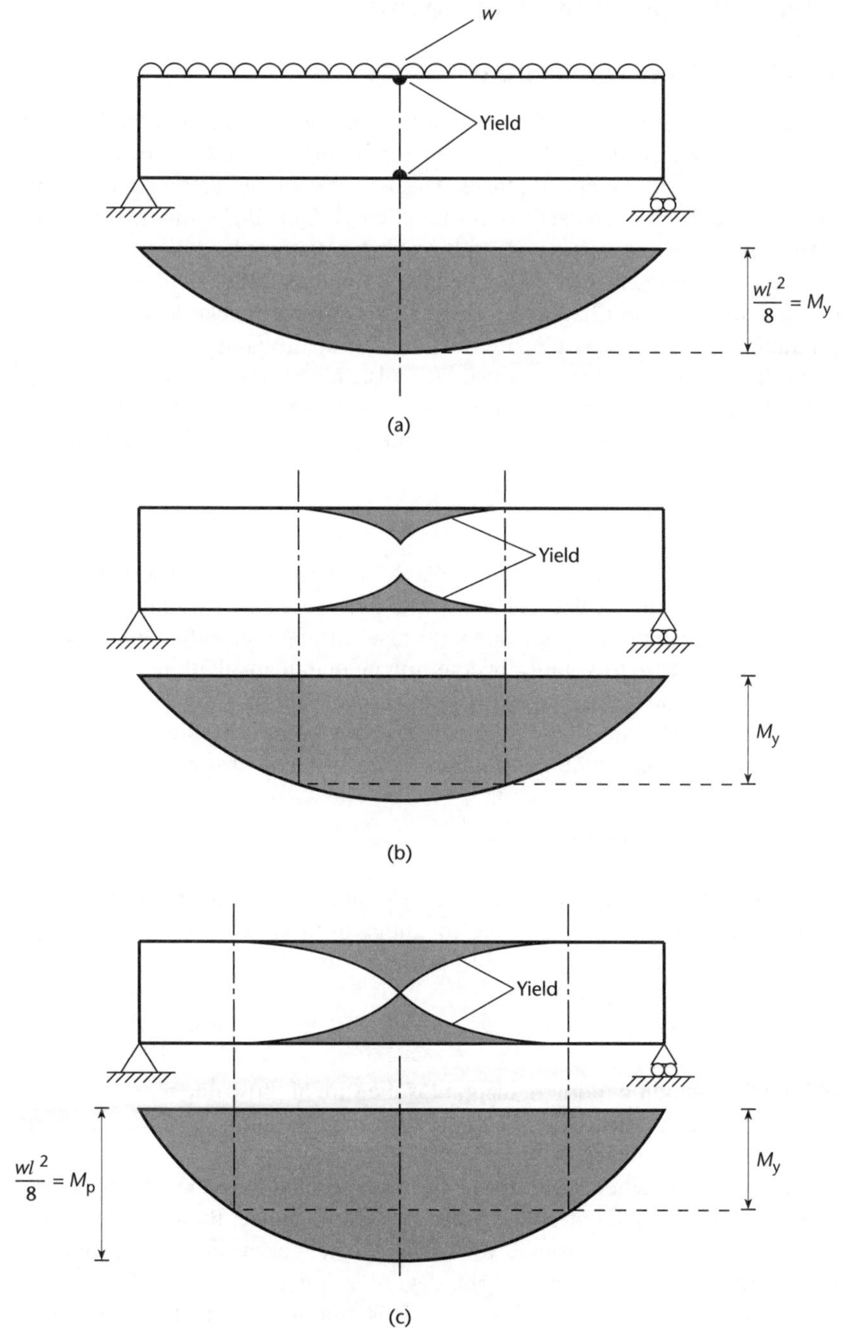

Figure 12.5
Spreading of plasticity in a simply supported beam.

- For simplicity, cross-sections are assumed symmetrical about the major axis, so that plastic moment capacity takes the same value whether the bending moment is hogging or sagging.
- It is assumed that other modes of failure such as buckling and shear are prevented, so that the structure is everywhere able to achieve its full plastic moment capacity.

12.3 Plastic analysis of beams

12.3.1 Collapse mechanisms

The development of a collapse mechanism was described in qualitative terms in section 12.1. We shall now look in a little more detail at collapse mechanisms and how they relate to the applied loads. This is best done by reference to a particular example. Consider a fixed-ended beam of length l, having a uniform cross-section with plastic moment capacity M_p, subjected to a uniform load of intensity w. If the structure remains elastic, then the bending moments can be determined using the methods described in Chapter 5. Figure 12.6(a) shows the elastic bending moment diagram, in which the largest moments occur at the ends of the beam, and are twice the magnitude of the moment at midspan.

The load is now increased to some value w_1 which causes plastic hinges to form at the two ends (Figure 12.6(b)). The value of w_1 can be easily determined since:

$$\frac{w_1 l^2}{12} = M_p \quad \text{hence} \quad w_1 = \frac{12M_p}{l^2}$$

At this point the midspan moment is $M_p/2$. The structure has now lost all stiffness at the two ends, but is still capable of sustaining more load without collapsing. The creation of the hinges is equivalent to inserting pins at either end, reducing a redundant structure to a statically determinate one. If an additional load w_2 is now applied, this is carried as though the beam was simply supported, as shown in Figure 12.6(c). The total bending moment at any point is the sum of the moments in (b) and (c). A plastic hinge therefore forms at the centre when:

$$\frac{M_p}{2} + \frac{w_2 l^2}{8} = M_p \quad \text{hence} \quad w_2 = \frac{4M_p}{l^2}$$

This third hinge reduces the structure to a mechanism, which cannot carry any further load. The structure is free to undergo gross deformations as shown in Figure 12.6(d). The load required to cause collapse is:

$$w = w_1 + w_2 = \frac{16M_p}{l^2}$$

and the final bending moment diagram is the sum of those due to w_1 and w_2. Note that the moments must have magnitude M_p at all the hinge locations, and cannot exceed M_p at any point.

Collapse occurs when sufficient hinges have formed to reduce the structure to a mechanism. For the fixed-ended beam considered above, there are six unknown reactions (vertical, horizontal and moment at each end) and only three equilibrium equations available. The degree of indeterminacy is therefore three. For vertically loaded beams, however, it is common to neglect the horizontal reactions and the horizontal equilibrium equation since these do not affect the bending behaviour; if this is done the degree of indeterminacy becomes two. Three hinges are therefore required to reduce the structure to a mechanism.

Based on the above, it is tempting to assume that if the degree of statical indeterminacy is D then the number of hinges required to cause a collapse mechanism is $(D+1)$. In practice this rule only holds when the collapse mechanism involves the entire structure. Very often collapse can occur by the formation of a *partial mechanism*, in which part of the structure forms a mechanism

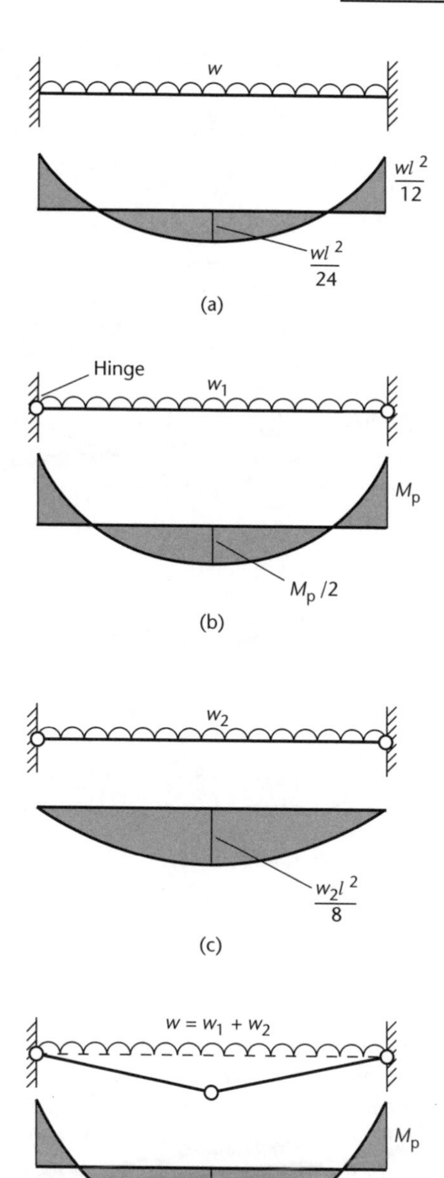

Figure 12.6
Development of collapse mechanism for a fixed-ended beam.

while another part remains indeterminate. In these cases, collapse may occur with fewer than $(D + 1)$ hinges.

EXAMPLE 12.1

Sketch possible collapse mechanisms for the three-span continuous beam in Figure 12.7(a).

Figure 12.7

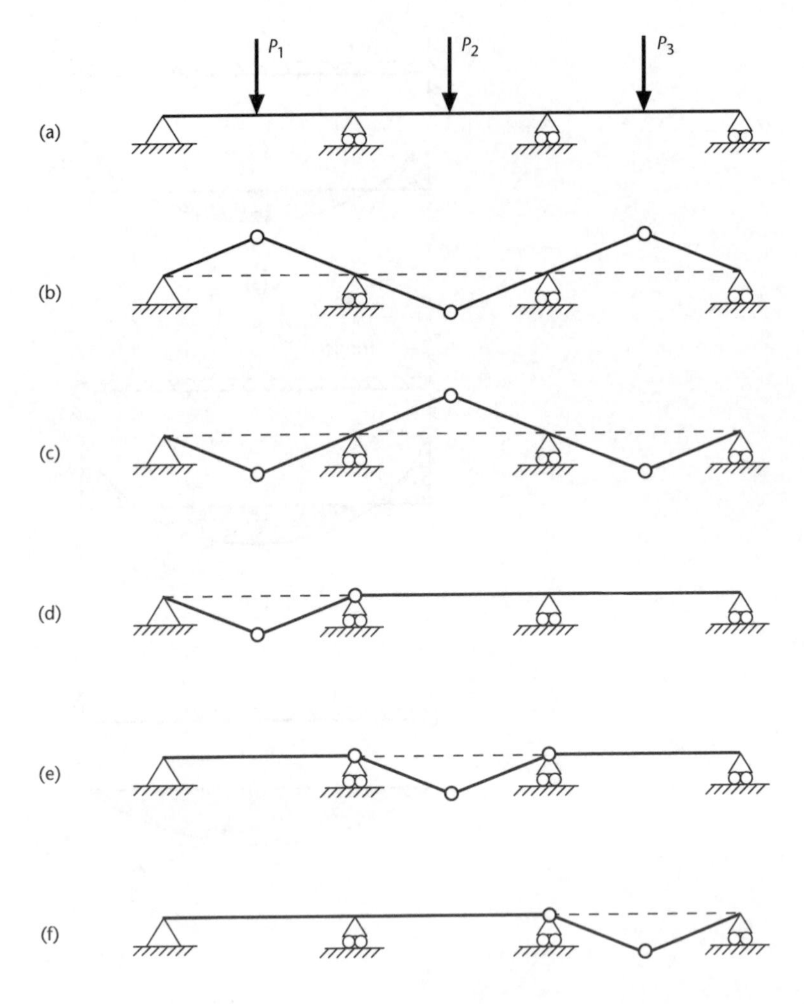

There are many possible collapse mechanisms for this structure, the correct one depending on the relative magnitudes of the three loads. The unknowns are the vertical reactions at each pinned support and the horizontal reaction at the left-hand pin, and the equations available are vertical, horizontal and moment equilibrium. The degree of statical indeterminacy is therefore $D = 5 - 3 = 2$ and three hinges would be required to form a global mechanism.

Five possible collapse mechanisms are shown in Figure 12.7(b)–(f). Mechanism (b) is likely only if P_2 is much larger than P_1 and P_3, since hogging moments are required in the outer spans in order to form the hinges shown. Similarly, (c) requires P_2 to be much smaller than the other loads.

If the three loads have similar magnitudes then the partial mechanisms (d)–(f) are much more likely, with the collapse confined to the most heavily loaded span. While (e) requires three hinges, like the global mechanisms, (d) and (f) each require only two and therefore do not conform to the $(D + 1)$ rule.

Lastly, note that other mechanisms are possible besides those shown in Figure 12.7. In particular, collapse could occur over two spans, with the third remaining intact.

12.3.2 Statical method of analysis

As we saw in Example 12.1, there are often many possible collapse mechanisms for a structure. As the load is increased from zero the structure will collapse into the first possible mechanism, that is, the one which occurs at the lowest load, and it is this we wish to identify.

The correct collapse mechanism can be identified by tracing the loading history and the sequence of hinge formation, as was done for the fixed-ended beam in Figure 12.6. However, this can be a difficult and long-winded process and is not usually necessary, since the collapse load can be determined directly from the final collapse mechanism, without reference to the load history. The disadvantage of this approach is that the correct mechanism may not be obvious, so that it may be necessary to analyse several in order to identify the one that occurs at the lowest load. This is, however, usually quite a simple matter and preferable to the full step-by-step analysis given above.

The first method we shall look at is known as the statical method. This requires a sketch of the general form of the bending moment diagram, from which the likely hinge locations can be deduced. The approach will be illustrated by example.

EXAMPLE 12.2

A propped cantilever of length 4 m and plastic moment capacity 150 kNm carries a concentrated load P at midspan (Figure 12.8(a)). Find the value of P required to cause collapse.

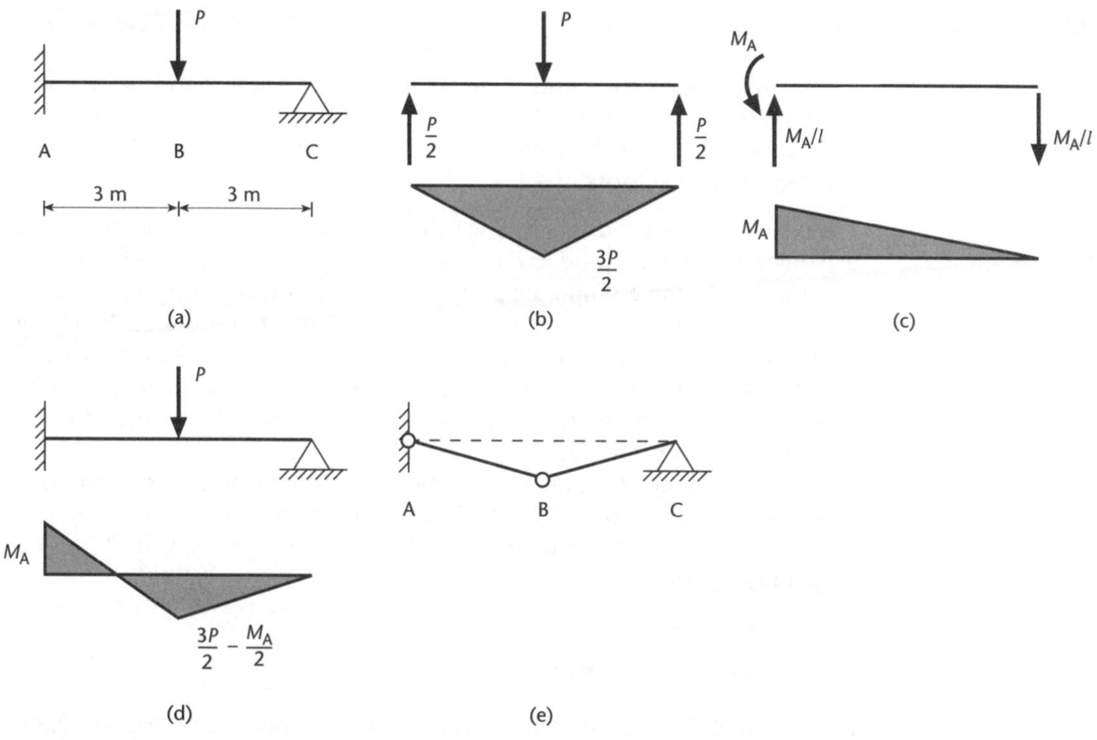

Figure 12.8

First we need to sketch the general form of the bending moment diagram. For a redundant structure, the following simple procedure can be used. First, reduce the structure to a statically determinate one by ignoring the appropriate number of rotational restraints. For the propped cantilever, this can be done by inserting an imaginary pin at the fixed end A, reducing the structure to a simply supported beam. The bending moment diagram for this case can be easily calculated, and is shown in (b).

Now we need to add back in the effect of the rotational restraint at A. This has the effect of imposing a moment of unknown magnitude at A, which reduces linearly to zero at the pinned end C, as shown in (c). This is often referred to as the *reactant* moment diagram. The total moment is then found by summing (b) and (c), and is shown in (d).

For collapse we need two hinges, since the structure has one redundancy. From the bending moment diagram these must occur at A and B, giving the collapse mechanism shown in (e). The moment must equal the plastic moment capacity at each hinge, therefore:

At A: $M_A = 150$ kNm

At B: $3P/2 - M_A/2 = 150$ hence $P = 150$ kN

Example 12.2 is relatively simple since there is only one possible collapse mechanism. Next, consider a case where several mechanisms are likely.

EXAMPLE 12.3

A continuous beam comprises three equal spans of length l and has a uniform section with plastic moment M_p. It carries point loads of P at the centre of the two outer spans and $2P$ at the centre of the middle span. Determine the value of P required to cause collapse.

The bending moment diagram is sketched in Figure 12.9(a). First, the statically determinate diagram is produced by inserting imaginary pins at B and C, reducing the structure to three simply supported spans. The reactant moment diagram is then found. Moments must be introduced at B and C to restore continuity, and from symmetry these must be equal. Let them be, say, M_0. The moments at the pinned ends must be zero and the reactant moment diagram simply comprises straight lines between these four points. The total moments are then found by summing the two diagrams.

From examination of the bending moment diagram, collapse must occur by one of two partial mechanisms. Consider first collapse of the centre span (Figure 12.9 (b)). This requires that the moments at B, C and the midpoint of BC all have magnitude M_p, so:

$$M_0 = M_p$$

$$Pl/2 - M_0 = M_p \qquad \text{hence} \quad P = 4M_p/l$$

Alternatively, collapse occurs simultaneously in the two outer spans (Figure 12.9(c)), for which moments of M_p must be achieved at B, C and the midpoints of AB and

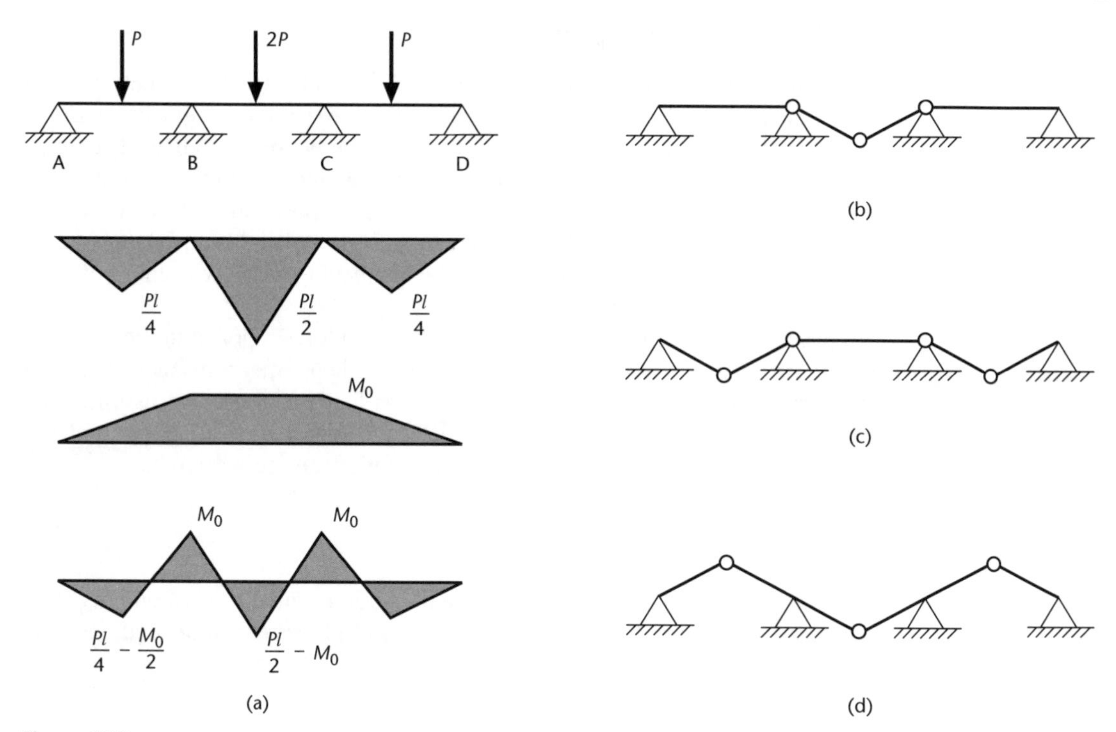

Figure 12.9

CD, therefore:

$$M_0 = M_p$$
$$Pl/4 - M_0/2 = M_p \quad \text{hence} \quad P = 6M_p/l$$

The mechanism in (b) occurs at the lower load and is therefore more critical, so collapse will occur in span BC at $P = 4M_p/l$.

Note that the global collapse mechanism shown in (d) cannot occur, since it would require the midspan moments in the outer and central spans to be of opposite sign. Since one is twice the other, this is not possible.

One further point can be made regarding alternative collapse mechanisms. Suppose in Example 12.3 we had failed to spot the correct collapse mechanism and instead had investigated only the mechanism in Figure 12.9(c). We would then have deduced that $P = 6M_p/l$ and $M_0 = M_p$. The moment at the midspan of BC is then:

$$M_{BC} = \frac{Pl}{2} - M_0 = \frac{6M_p}{l} \cdot \frac{l}{2} - M_p = 2M_p$$

This is an impossibility since the section cannot sustain a moment greater than M_p. Calculating the moments at key points away from the hinges thus provides a useful check that the mechanism arrived at is physically possible. This point will be discussed further in section 12.3.5.

12.3.3 Virtual work approach

An alternative to the statical method is the use of virtual work. If the likely collapse mechanisms can be recognised by eye, then this approach is even quicker than the statical method, since there is no need to sketch the bending moment diagram.

The principle of virtual work was introduced in Chapter 7. If we take a structure subjected to a set of forces in equilibrium and impose on it a set of virtual displacements, then the virtual work done by the internal and external forces must be equal. The displacements chosen must be compatible with each other but need not be related to the loading.

In plastic analysis, our force set comprises the real forces applied to the structure and the virtual displacement sets are the possible collapse mechanisms, only one of which can actually be created by the applied loading. Since we assume that internal deformation occurs only at the plastic hinges, with other parts of the structure remaining rigid, the virtual work equation can be written as:

$$\sum_i P_i\delta_i = \sum_j M_j\theta_j \qquad (12.2)$$

where P_i are the applied forces, which are in equilibrium with the moments M_j at the hinges, and δ_i and θ_j are the corresponding displacements and rotations. The subscript i refers to the load points and j to the hinge locations.

EXAMPLE 12.4

Repeat Example 12.3 using virtual work.

Figure 12.10

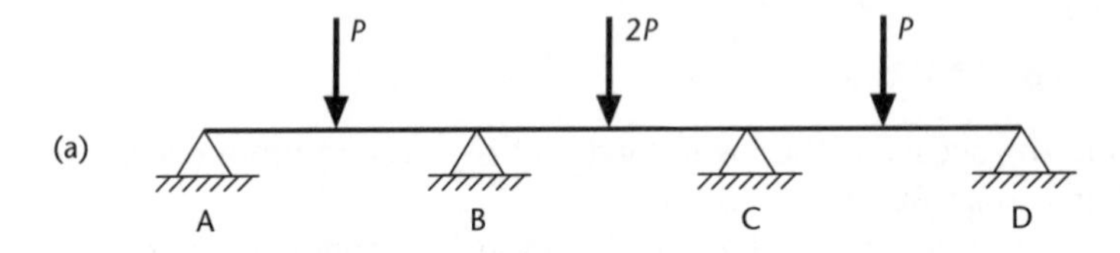

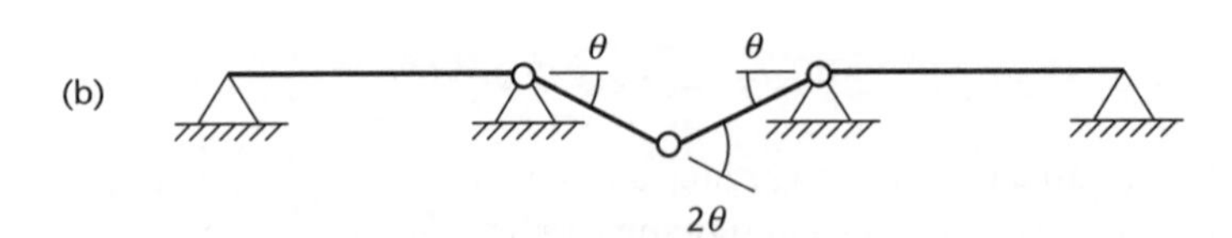

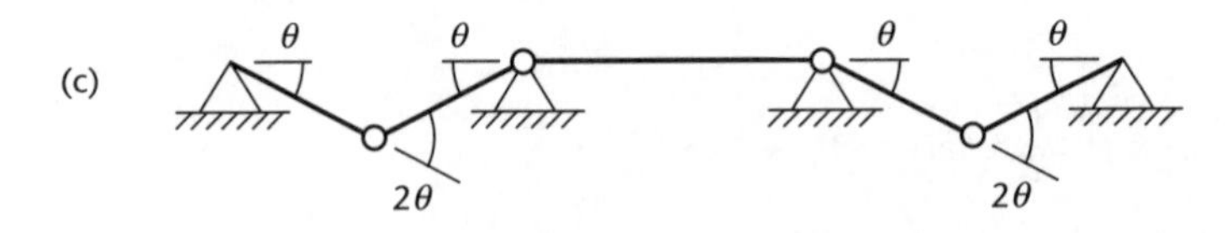

The loading is shown in Figure 12.10(a) and the two possible mechanisms in (b) and (c). Consider first (b). Let the rotation at B be θ. From symmetry the rotation at C must also be θ and the total rotation at the central hinge is the sum of the rotations of the two beam segments meeting there, that is, 2θ. Using small angle assumptions, the vertical displacement of the central pin is $\theta l/2$. Applying virtual work (equation (12.2)):

$$2P \cdot \frac{\theta l}{2} = M_p \cdot \theta + M_p \cdot 2\theta + M_p \cdot \theta \qquad \text{hence} \quad P = \frac{4M_p}{l}$$

Similarly for the mechanism in (c):

$$P \cdot \frac{\theta l}{2} + P \cdot \frac{\theta l}{2} = M_p \cdot 2\theta + M_p \cdot \theta + M_p \cdot \theta + M_p \cdot 2\theta \qquad \text{hence} \quad P = \frac{6M_p}{l}$$

These answers are, of course, the same as in Example 12.3 and again the mechanism in (b) is the critical one.

12.3.4 Load factor

In general we wish to ensure that the load to which a structure is subjected is lower than the value which would cause collapse by a fairly large margin; we can define this margin using a *load factor*. Suppose we have a structure subjected to a known working load P. We can calculate that the onset of yielding would be caused by a load P_y and complete collapse by a load P_c.

The load factor against collapse is:

$$\lambda = \frac{P_c}{P} \tag{12.3}$$

This contrasts with the conventional elastic approach, in which we use a safety factor against yielding, defined as:

$$\gamma = \frac{P_y}{P} \tag{12.4}$$

For a statically determinate structure, only one plastic hinge is required to cause collapse. A hinge will form when the load reaches αP_y, where α is the shape factor (see section 12.1). Therefore the load factor is:

$$\lambda = \frac{P_c}{P} = \frac{\alpha P_y}{P} = \alpha\gamma$$

For a redundant structure, additional hinges will be required to cause collapse, therefore $\lambda > \alpha\gamma$.

Alternatively, load and safety factors can be approached in the opposite way. If we can calculate P_y and P_c then by specifying a required value of γ or λ the maximum permissible load can be established. Suppose we performed two alternative analyses of the same structure, one using elastic theory and the other plastic, with γ and λ both assigned the same value. Then, from equations (12.3) and (12.4), we can see that the plastic analysis would result in a greater permissible load, since $P_c > P_y$. In practice, it would be normal to set the value of λ rather higher than that of γ, since it is guarding against a more catastrophic event (complete collapse as opposed to localised yielding). Even so, a plastic analysis would normally result in a higher permissible load.

12.3.5 Theorems used in plastic analysis

We now introduce three fundamental structural theorems. The theorems are valid for elasto-plastic structures in which displacements are small compared to the overall dimensions and instability does not occur.

The uniqueness theorem

The following conditions must be satisfied by a structure in its collapse state:

- the *equilibrium* condition – the system of bending moments must be in equilibrium with the applied loads;
- the *yield* condition – the bending moment may not exceed the plastic moment capacity at any point;
- the *mechanism* condition – sufficient plastic hinges must have formed to reduce all or part of the structure to a mechanism.

If all three of these conditions are satisfied, then a critical mechanism has been formed and the applied loading is the true collapse load. The corresponding load factor is usually denoted by $\lambda = \lambda_c$.

The lower bound theorem, or safe theorem

If a distribution of moments can be found which satisfies the equilibrium and yield conditions, then the structure is either safe or just on the point of collapse, that is $\lambda \leq \lambda_c$.

The upper bound theorem, or unsafe theorem

If a loading is found which causes a collapse mechanism to form, then it must be greater than or equal to the actual collapse load, that is $\lambda \geq \lambda_c$.

It can be seen that the uniqueness theorem is satisfied only when both the safe and unsafe theorems apply simultaneously.

In plastic analysis, we principally make use of the unsafe theorem. We formulate possible mechanisms and calculate the corresponding collapse load, knowing that all mechanisms must give a result greater than or equal to the true collapse load. We therefore deduce that the critical mechanism is the one giving the lowest load factor. There is, however, a risk that we have simply overlooked a mechanism that would give a lower load factor. Therefore, having determined what we think to be the correct mechanism, we must then apply the safe theorem by checking that it satisfies the equilibrium and yield conditions.

EXAMPLE 12.5

Figure 12.11 shows a collapse mechanism for a propped cantilever of length l and plastic moment M_p carrying a uniformly distributed load w. The load required to form this mechanism is $w = 12M_p/l^2$. Use the safe theorem to check whether a critical mechanism has been found.

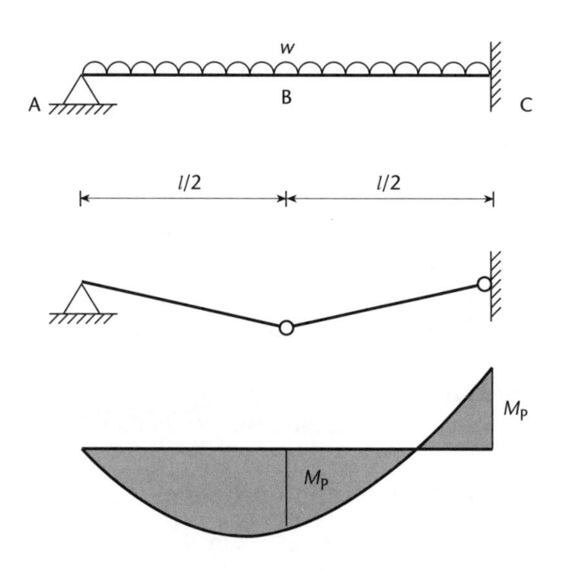

Figure 12.11

We must check that the structure is in equilibrium, and that the moment does not exceed M_p at any point. First, use equilibrium to calculate the reactions. The formation of the hinge at C requires a hogging moment of magnitude M_p so, taking moments about C:

$$M_p = \frac{wl^2}{2} - V_A l = \frac{12M_p}{l^2} \cdot \frac{l^2}{2} - V_A l \quad \text{hence} \quad V_A = \frac{5M_p}{l}$$

Similarly, taking moments about A gives $V_B = 7M_p/l$ and resolving vertically we see that these reactions sum to the applied load $wl = 12M_p/l$. Splitting the structure at B and taking moments for the left hand side gives:

$$M_B = \frac{V_A l}{2} - \frac{w(l/2)^2}{2} = \frac{5M_p}{l} \cdot \frac{l}{2} - \frac{12M_p}{l^2} \cdot \frac{l^2}{8} = M_p$$

and a similar result can be obtained taking moments for the right-hand half. The moment distribution therefore satisfies the equilibrium condition. The moment at a distance x from A is:

$$M = V_A x - \frac{wx^2}{2} = \frac{5M_p x}{l} - \frac{12M_p}{l^2} \cdot \frac{x^2}{2} = M_p\left(\frac{5x}{l} - \frac{6x^2}{l^2}\right)$$

Differentiating to find the location of the maximum moment:

$$\frac{dM}{dx} = M_p\left(\frac{5}{l} - \frac{12x}{l^2}\right) = 0 \quad \text{hence} \quad x = \frac{5l}{12}$$

and substituting back into the moment expression gives $M_{max} = 25M_p/24$. This is greater than M_p, so the yield condition is violated and the safe theorem is not satisfied. The mechanism given is therefore not the critical one and a lower collapse load can be found (see Example 12.6 for the correct solution).

In contrast to the plastic approach, elastic design methods are based exclusively on the safe theorem. We find a moment distribution that is in equilibrium with the applied loads and satisfies the yield condition, but is insufficient to produce a collapse mechanism. Using the safe theorem, we can then be sure that the structure will not collapse under the applied loading; this is true *even if the bending moment distribution calculated is not the one that actually occurs in practice.*

This last point is crucial to the elastic designer, since the real moments in a structure may differ hugely from the elastically calculated values used in design, particularly if the structure is redundant. Discrepancies may arise owing, for example, to the approximations inherent in using idealised support conditions. According to the safe theorem, however, this will not affect the safety of the structure. So long as the structure has been given adequate strength to resist an equilibrium bending moment distribution it will not collapse.

12.3.6 Minimisation problems

In some instances, even when the general form of the collapse mechanism is obvious the exact locations of the hinges may be uncertain. This is particularly likely when members carry distributed loads. Such cases can be analysed by setting up the problem in terms of a variable dimension and then using differentiation to minimise the collapse load.

EXAMPLE 12.6

Find the critical collapse load for the uniformly loaded propped cantilever of Example 12.5.

Figure 12.12

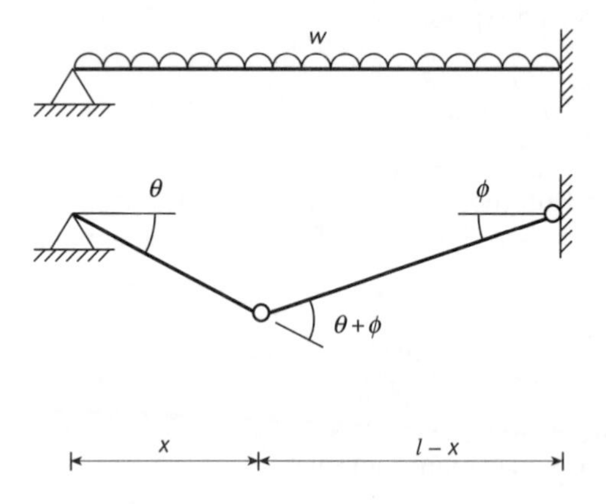

The mechanism investigated in Example 12.5 was found not to be the critical one. It is clearly of the right general form, but the location of the sagging hinge at midspan is incorrect. Instead, let it be at some distance x from A. The collapse mechanism is now unsymmetrical, with the two beam segments undergoing rotations of θ and ϕ (Figure 12.12). Using small angle assumptions, the vertical deflection δ of the hinge can be written as:

$$\delta = \theta x = \phi(l - x) \quad \text{hence} \quad \phi = \theta \frac{x}{l - x}$$

The virtual work equation can now be written. Note that the vertical distance moved by the centroid of the distributed load on each beam segment is $\delta/2 = \theta x/2$.

$$wl \cdot \frac{\theta x}{2} = M_p(\theta + \phi) + M_p\phi = M_p\theta\left(1 + \frac{2x}{l - x}\right)$$

Rearranging:

$$w = \frac{2M_p}{lx}\left(\frac{l + x}{l - x}\right)$$

The value of w can now be minimised by differentiating with respect to the variable distance x:

$$\frac{dw}{dx} = \frac{2M_p}{l}\left(\frac{x(l - x) - (l + x)(l - 2x)}{x^2(l - x)^2}\right) = 0$$

This reduces to a quadratic equation for x:

$$x^2 + 2lx - l^2 = 0$$

and taking the positive root gives $x = l(\sqrt{2} - 1) = 0.414l$. Substituting back into the expression for w gives the minimum collapse load:

$$w = \frac{11.66M_p}{l^2}$$

Strictly speaking, we ought to check that this mechanism satisfies the safe theorem, so as to reassure ourselves that we have found the critical collapse load.

12.4 Plastic analysis of frames

Plastic collapse analysis can be performed for frame structures in much the same way as for beams. However, since the structures are now two- or three-dimensional, the possible collapse mechanisms may be harder to formulate and to analyse. The virtual work approach will normally provide the simplest solution. As an introduction to the topic, we shall start with the relatively simple case of plane, rectangular portal frames and then look briefly at some more complex frame structures.

12.4.1 Rectangular portal frames

The rectangular portal frame with fixed feet in Figure 12.13(a) carries a vertical point load V at the centre of the beam and a horizontal load H at the top of one column. For simplicity, we shall initially assume that all members have the same plastic moment capacity M_p, though this is not usually the case. We can sketch the elastic bending moment diagram for this structure by superposition of cases (a) and (c) in Figure 9.2; the result is shown in Figure 12.13(b). Obviously the exact form will depend on the relative magnitudes of V and H. In particular, if H is large the moment at A may reverse in sign.

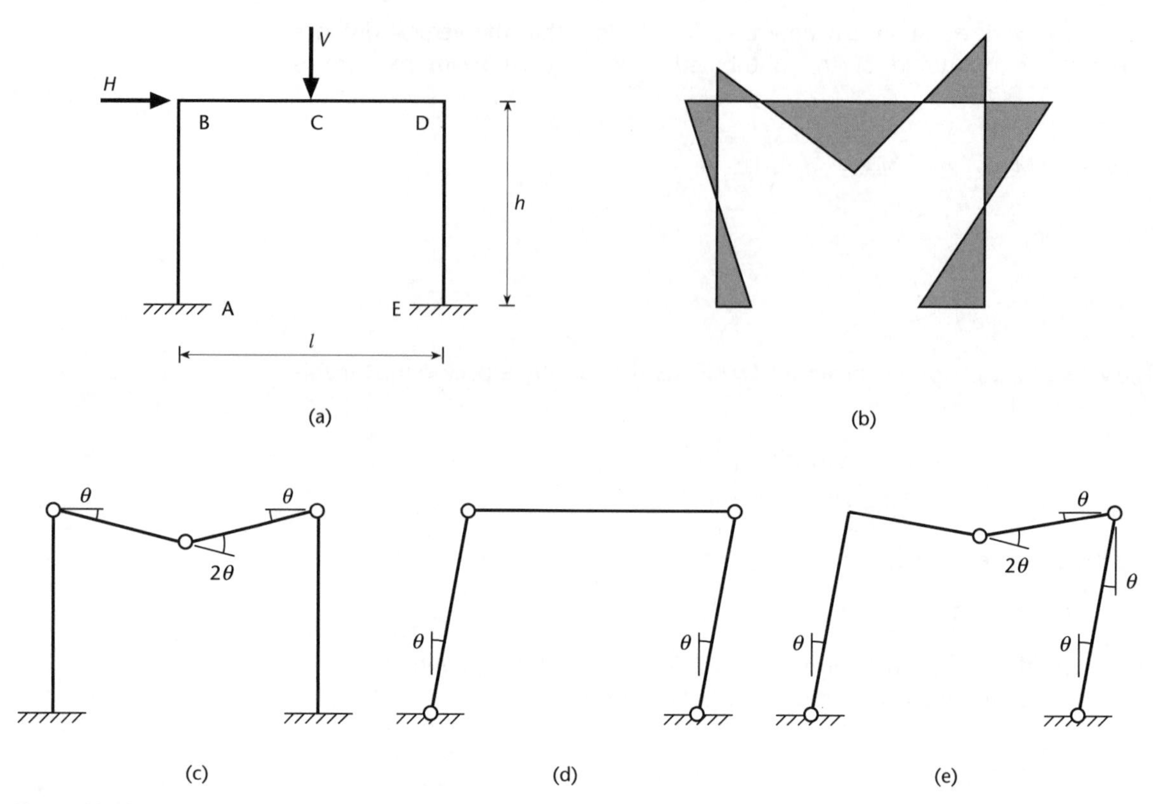

Figure 12.13
Collapse mechanisms for a rectangular portal frame.

We can now identify three possible collapse mechanisms; which is the most critical will depend on the values of V, H, l and h:

Beam mechanism – If V is much larger than H then the moments at B, C and D are greater than those at the fixed feet, and a partial mechanism forms in the beam BCD (Figure 12.13(c)) – the behaviour is similar to a fixed-ended beam.

Sway mechanism – If H is much larger than V then the four corner moments are larger than the sagging moment at C. Collapse therefore occurs by overall sway of the frame (Figure 12.13(d)).

Combined mechanism – If H and V have comparable magnitudes then the smallest moment is at B, where the moments due to H and V oppose each other. Hinges therefore form at the other four possible locations, giving the mechanism in Figure 12.13(e), in which the angle ABC remains a right angle.

The displaced shape in Figure 12.13(e) is in fact simply the sum of those in (c) and (d), with the two hinges at B 'cancelling out' because the moments are in opposite directions. There are thus only two independent mechanisms for this frame, the third being a linear combination of the first two. Nevertheless, for certain values of V and H, the combined mechanism may be more critical than the beam and sway mechanisms from which it is derived. We shall consider combination of mechanisms in more detail in section 12.4.3.

EXAMPLE 12.7

For the case where $h = 2l/3$, analyse the three mechanisms in Figure 12.13 and determine the ranges of values of V and H for which each mechanism is critical.

Beam mechanism – from symmetry the rotations of BC and CD are equal, so using virtual work:

$$V \cdot \frac{\theta l}{2} = M_p(\theta + 2\theta + \theta) \quad \text{hence} \quad V = \frac{8M_p}{l}$$

Sway mechanism – AB and DE have both rotated by θ giving:

$$H \cdot \frac{2\theta l}{3} = M_p(\theta + \theta + \theta + \theta) \quad \text{hence} \quad H = \frac{6M_p}{l}$$

Combined mechanism – both AB and BC rotate through θ, because joint B remains rigid. CD must also rotate by θ so as to give the same vertical displacement at C and DE rotates by θ since all points on BCD must undergo an equal lateral sway. Alternatively, these rotations could have been determined by summing the values for the beam and sway mechanisms. The virtual work equation is:

$$V \cdot \frac{\theta l}{2} + H \cdot \frac{2\theta l}{3} = M_p(\theta + 2\theta + 2\theta + \theta) \quad \text{hence} \quad 3V + 4H = \frac{36M_p}{l}$$

We can determine which is the most critical mechanism by plotting these three results on a graph of V against H (non-dimensionalised by the factor l/M_p) – see Figure 12.14.

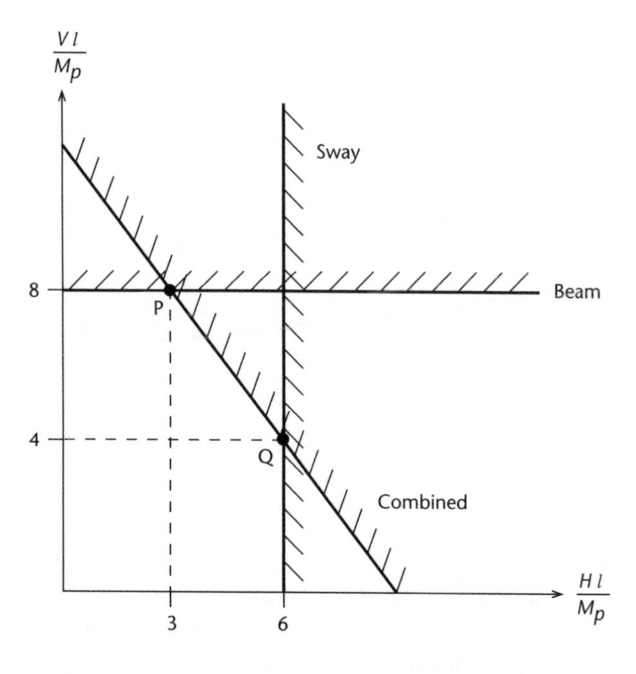

Figure 12.14

For each mechanism, points on the line represent collapse load combinations and the range of safe load values is the area enclosed by the three lines and the two axes. As V and H are increased, failure occurs by the mechanism corresponding to the first line reached. Transition between failure modes therefore occurs at the intersection points P and Q.

At P: $V/H = 8/3$
At Q: $V/H = 4/6 = 2/3$

Therefore if: $V < 2H/3$ – sway mechanism is critical,
$2H/3 < V < 8H/3$ – combined mechanism is critical,
$V > 8H/3$ – beam mechanism is critical.

12.4.2 Analysis of more complex structures

Thus far the mechanisms have been drawn with plastic hinges at the beam–column joints. In practice the hinge would normally form a very small distance away from the joint, in one of the two members meeting there. This distinction does not significantly affect the overall geometry of the problem, and only becomes important when the plastic moment capacities of the two members are different; in this case the hinge would, of course, form in the weaker element.

EXAMPLE 12.8

Repeat Example 12.7, but this time taking the plastic moment capacity of the cap beam to be $2M_p$ and that of the columns to be M_p.

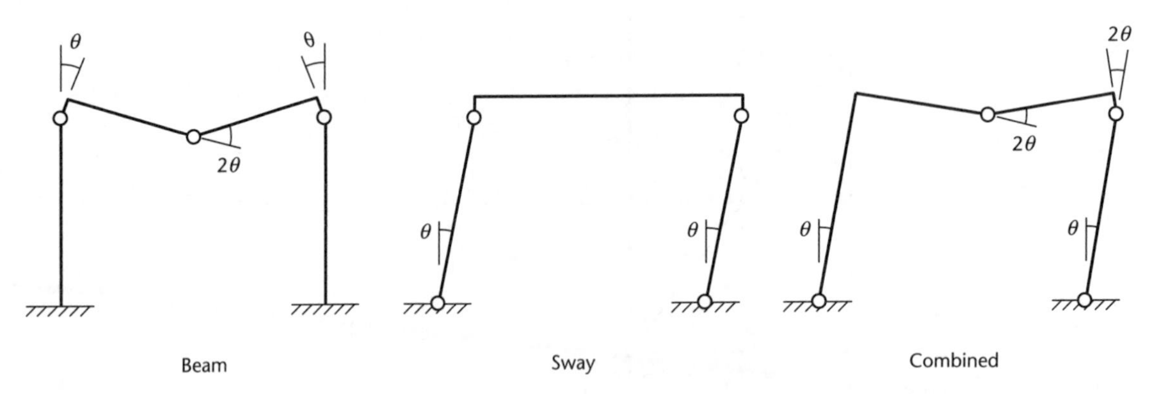

Figure 12.15

The basic collapse mechanisms are the same as previously, but hinges previously drawn at the beam-column joints must now occur at the tops of the columns (Figure 12.15). This change will not significantly affect the mechanism geometry. The collapse equations are therefore:

Beam: $V \cdot \frac{\theta l}{2} = M_p\theta + 2M_p \cdot 2\theta + M_p\theta$ hence $V = \frac{12M_p}{l}$

Sway: $H \cdot \frac{2\theta l}{3} = M_p(\theta + \theta + \theta + \theta)$ hence $H = \frac{6M_p}{l}$

Combined: $V \cdot \frac{\theta l}{2} + H \cdot \frac{2\theta l}{3} = M_p\theta + 2M_p \cdot 2\theta + M_p \cdot 2\theta + M_p\theta$

hence $3V + 4H = \frac{48M_p}{l}$

Note that the sway mechanism is unaffected because it involves only column hinges. As in Example 12.7, it is now a simple matter to determine the ranges of values of V and H for which each mechanism is critical:

$V < 4H/3$ – sway mechanism is critical,
$4H/3 < V < 4H$ – combined mechanism is critical,
$V > 4H$ – beam mechanism is critical.

While simple rectangular portal frames are useful for illustrating the principles of collapse analysis, more complex structures are likely to be encountered in practice. As an example, Figure 12.16 shows the possible collapse modes for a pitched-roof portal frame. Because the members do not meet at right angles, their rotations can be difficult to determine. The easiest way of doing this is to use the concept of the *instantaneous centre of rotation*.

Figure 12.17 shows mechanism (c) in more detail. Each of the members BC, CD and DE rotates as a rigid body about a point. If the angles are small, all points on a member can be assumed to move at right angles to a line from the centre of rotation. For example, BC rotates through an angle θ about B. C therefore moves in a direction normal to BC to a new position C′. Similarly, DE rotates through an angle β about E, so that D moves horizontally to D′.

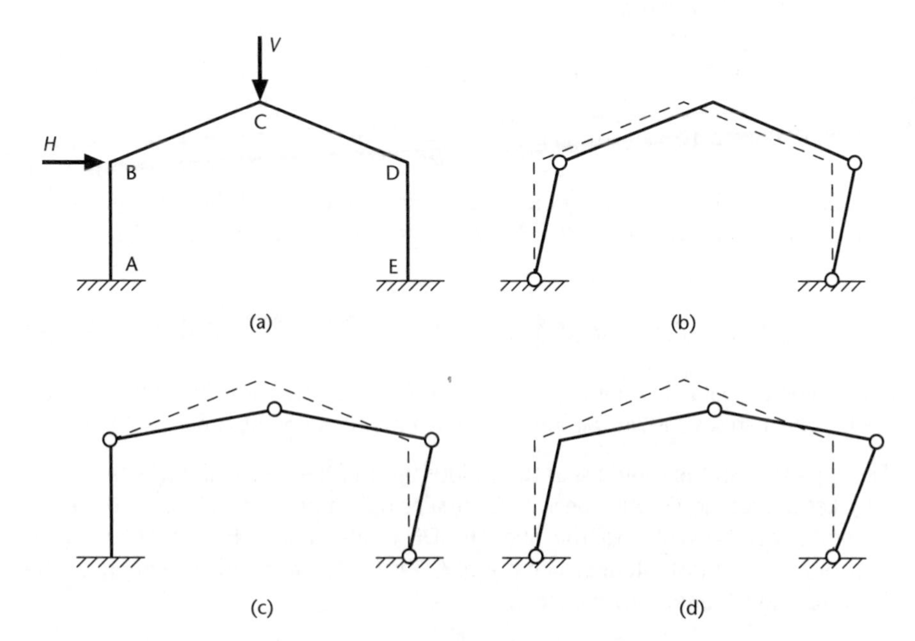

Figure 12.16
Collapse mechanisms for pitched-roof portal frames.

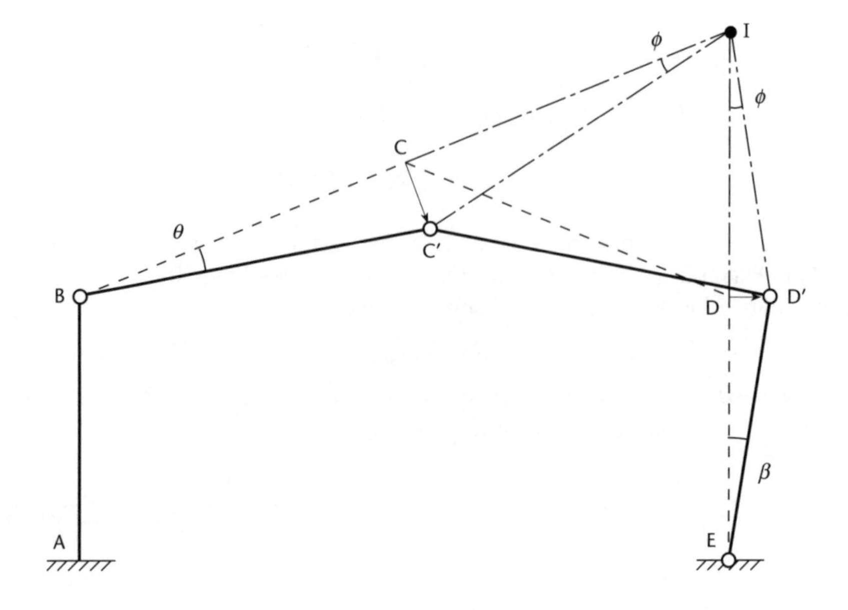

Figure 12.17
Instantaneous centre of a pitched-roof portal frame member.

Having defined the directions of motion of C and D, it is now possible to find the centre of rotation of CD. We note that C is moving at right angles to the line BC and D is moving at right angles to DE. It follows that CD must be rotating about the point I, where the extensions of BC and DE intersect. I is known as the instantaneous centre of CD. Since CD rotates as a rigid body, all points on the member must undergo the same rotation about I. So both the lines IC and ID turn through the same angle ϕ.

For a virtual work analysis, we need to express all deformations in terms of one displacement variable, say θ. Having located I, we can now do this by considering the deflections of the hinges C and D:

$$\mathrm{CC'} = \mathrm{BC}\cdot\theta = \mathrm{IC}\cdot\phi \quad \text{so} \quad \phi = \frac{\mathrm{BC}}{\mathrm{IC}}\theta$$

$$\mathrm{DD'} = \mathrm{DE}\cdot\beta = \mathrm{ID}\cdot\phi \quad \text{so} \quad \beta = \frac{\mathrm{ID}}{\mathrm{DE}}\phi = \frac{\mathrm{ID}}{\mathrm{DE}}\cdot\frac{\mathrm{BC}}{\mathrm{IC}}\theta$$

Finally, the total angle of rotation at each joint is the sum of the rotations of the two members meeting there: θ at B, $(\theta+\phi)$ at C, $(\phi+\beta)$ at D and β at E.

EXAMPLE 12.9

Determine the value of P required to cause collapse of the frame in Figure 12.18(a) by the mechanism shown. All members have plastic moment capacity 200 kNm.

First we establish the rotations of the various parts of the mechanism (Figure 12.18 (b)). Let the rotation of ABC be θ. ABC remains rigid and rotates about A, therefore C moves perpendicular to the line AC. DE rotates about E so that D moves horizontally. The instantaneous centre of CD is therefore found by extending the lines AC and DE until they intersect.

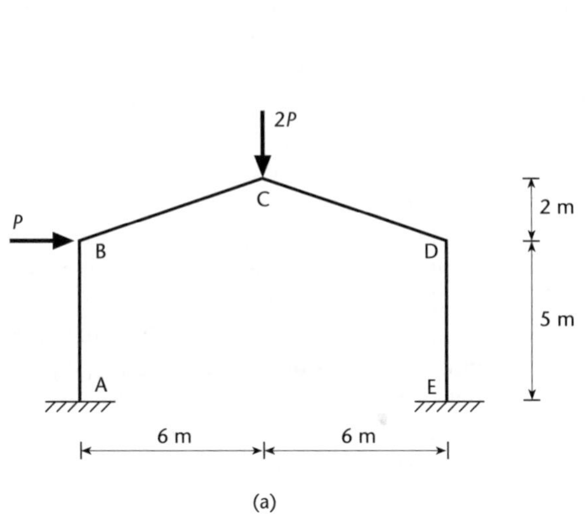

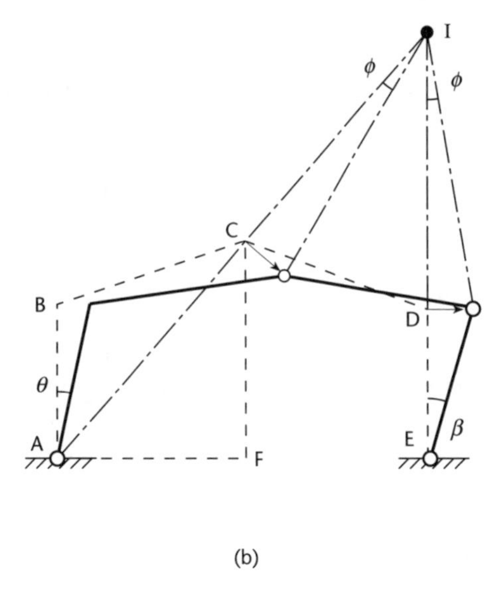

Figure 12.18

If we define a point F midway between A and E, we can see that ACF and AIE are similar triangles, the latter having dimensions of twice the former. It follows that:

$$\text{IC} = \text{AC}$$

$$\text{ID} = \text{IE} - \text{DE} = 14 - 5 = 9\text{ m}$$

Therefore:

$$\phi = \frac{\text{AC}}{\text{IC}}\theta = \theta, \quad \beta = \frac{\text{ID}}{\text{DE}}\phi = \frac{9}{5}\phi = \frac{9}{5}\theta$$

We must also find the deflections corresponding to the external loads. B moves horizontally by a distance $\text{AB} \cdot \theta = 5\theta$ and C moves vertically by a distance $\text{AF} \cdot \theta = 6\theta$. We can now write the virtual work equation for the mechanism:

$$P \cdot 5\theta + 2P \cdot 6\theta = 200[\theta + (\theta + \phi) + (\phi + \beta) + \beta] = 200\theta\left[1 + 2 + \frac{14}{5} + \frac{9}{5}\right]$$

hence $P = 89.4\text{ kN}$

12.4.3 Combination of mechanisms

When more complicated structures or load configurations are analysed, there may be many possible collapse mechanisms. Investigating them all can be tedious and time-consuming, and there is a greater risk of missing the critical mechanism. In such situations, it can be beneficial to use the method of combination of mechanisms. This concept was briefly introduced in section 12.4.1, where we saw that the beam and sway modes of a rectangular portal frame could be combined to give a third mechanism.

It is first necessary to establish how many independent mechanisms exist. For a structure under a given loading, it is normally possible to identify a finite number of possible hinge locations. If this number is H then it can be shown that there are $(H - D)$ independent mechanisms, where D is the degree of statical indeterminacy. For example, the rectangular portal frame of Figure 12.13 has $D = 3$ (six unknown reactions, three equilibrium equations available). Hinges can only form at A, B, C, D or E, since the moments at these locations will always be greater than at points in between. So $H = 5$ and there are two independent mechanisms.

Having identified and analysed the independent mechanisms, we can then start combining them in an attempt to find the critical mechanism. When two mechanisms are combined, it is possible that a hinge that is common to the two mechanisms will disappear. The virtual work equation for the combined mechanism can be obtained by summing the equations for the two independent mechanisms, omitting the internal virtual work terms for any 'disappeared' hinges.

As a simple illustration, consider again the rectangular portal frame of Example 12.7. For the beam mechanism the virtual work equation can be written:

$$V \cdot \frac{\theta l}{2} = 3M_p\theta + \{M_p\theta\}$$

and for the sway mechanism:

$$H \cdot \frac{2\theta l}{3} = 3M_p\theta + \{M_p\theta\}$$

where the terms in curly brackets represent the internal virtual work in the plastic hinge at B. When the mechanisms are combined the rotations at B cancel out and so this hinge disappears. Therefore, to find the virtual work equation for the combined mechanism we simply sum these two equations, omitting the bracketed terms, giving:

$$V \cdot \frac{\theta l}{2} + H \cdot \frac{2\theta l}{3} = 6M_p\theta \quad \text{or} \quad 3V + 4H = \frac{36M_p}{l}$$

which is the same result as was obtained in Example 12.7.

EXAMPLE 12.10

All the members of the frame in Figure 12.19(a) have a plastic moment capacity of 125 kNm. Find the minimum load factor and the corresponding collapse mechanism.

Under the loading shown, it is possible that hinges could form at A, B, C, D, E and F, therefore $H = 6$. The degree of indeterminacy $D = 3$ and so $6 - 3 = 3$ independent mechanisms can be found. Three suitable mechanisms are shown in Figure 12.19 (b), and these can easily be shown to be independent. Mechanism **1** has a hinge at D which cannot be created by combination of mechanisms **2** and **3**; mechanism **2** has a hinge at B and which cannot be created from **1** and **3**; and mechanism **3** has a hinge at F, which cannot be generated from **1** and **2**. Analysis of these three mechanisms by virtual work is straightforward – for collapse to occur the loads must be increased by a load factor λ, so:

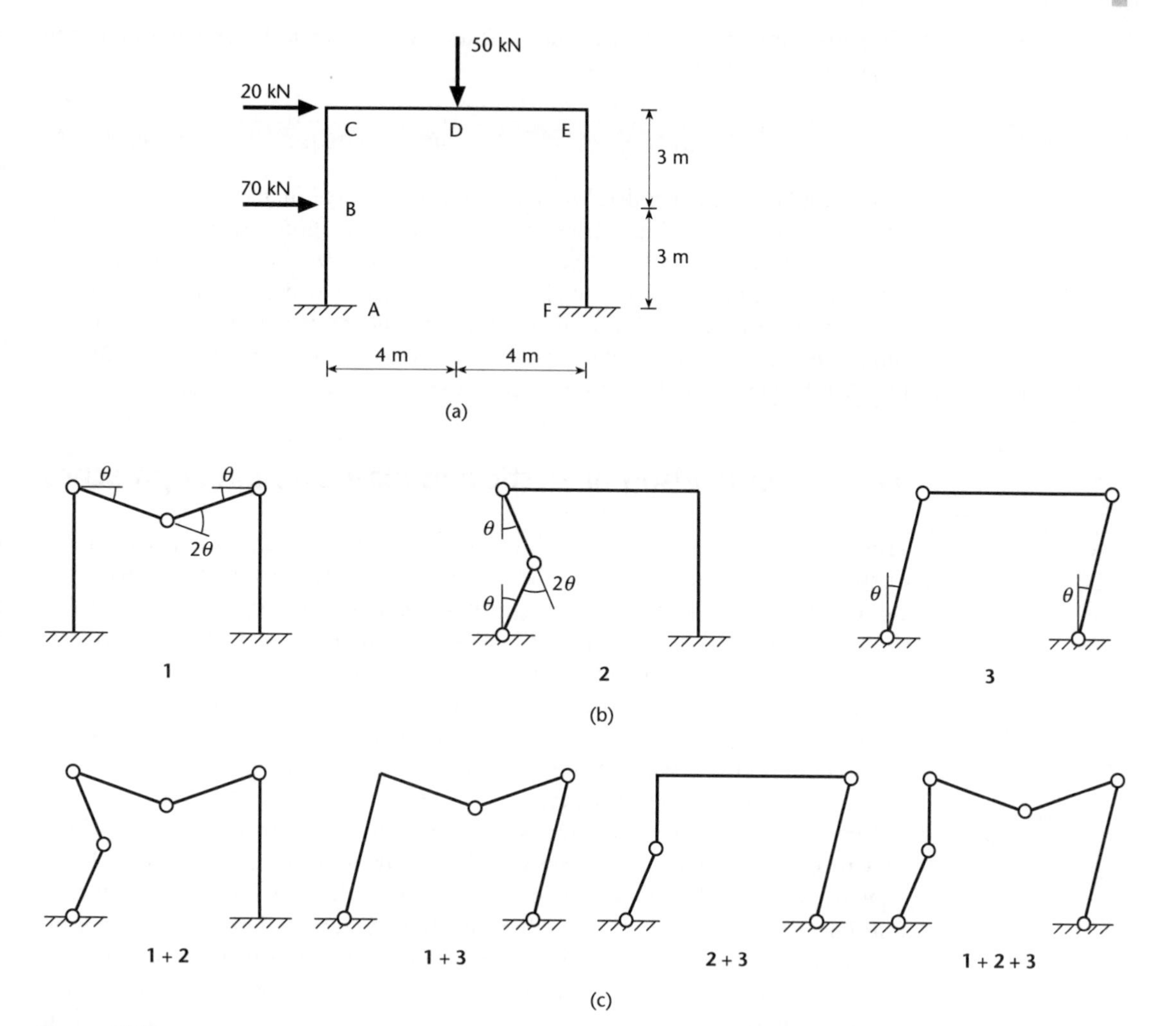

Figure 12.19

$$
\begin{aligned}
&\mathbf{1}: \quad 50\lambda \cdot 4\theta = 125 \cdot 4\theta && \text{or } 200\lambda\theta = 500\theta \quad \rightarrow \lambda = 2.50 \\
&\mathbf{2}: \quad 70\lambda \cdot 3\theta = 125 \cdot 4\theta && \text{or } 210\lambda\theta = 500\theta \quad \rightarrow \lambda = 2.38 \\
&\mathbf{3}: \quad 70\lambda \cdot 3\theta + 20\lambda \cdot 6\theta = 125 \cdot 4\theta && \text{or } 330\lambda\theta = 500\theta \quad \rightarrow \lambda = 1.52
\end{aligned}
$$

So **3** is clearly the most critical of the independent mechanisms. In addition, four combined mechanisms can be formed (three by combining pairs of independent mechanisms, one by combining all three). The combined mechanisms are shown in Figure 12.19(c) and are analysed below:

1 + 2 – no hinges disappear so: $200\lambda\theta + 210\lambda\theta = 500\theta + 500\theta$
$\rightarrow \lambda = 2.44$

1 + 3 – the hinge at C disappears so: $200\lambda\theta + 330\lambda\theta = 500\theta + 500\theta - 250 \cdot \theta$
$\rightarrow \lambda = 1.42$

2 + 3 – the hinge at C disappears so: $210\lambda\theta + 330\lambda\theta = 500\theta + 500\theta - 250 \cdot \theta$
$\rightarrow \lambda = 1.39$

1 + (**2** + **3**) – no further hinges disappear so: $200\lambda\theta + 540\lambda\theta = 500\theta + 750\theta$
$\rightarrow \lambda = 1.69$

Thus the combination of mechanisms **2** and **3** is the critical mechanism and the collapse load factor is $\lambda_c = 1.39$.

If the number of independent mechanisms is quite large then there may be many combinations to consider. However, it is usually quite easy to see which combinations are likely to be critical, and the others need not be investigated. A combined mechanism can only give a lower collapse load than the independent mechanisms from which it is formed if the combination results in the removal of a hinge. In the above example, therefore, the combined mechanisms $(\mathbf{1}+\mathbf{2})$ and $(\mathbf{1}+\mathbf{2}+\mathbf{3})$ were bound not to be critical and need not have been analysed.

12.5 Comparison of elastic and plastic design approaches

From the material presented in this chapter, it might seem that the use of plastic analysis in the design of structures would be extremely attractive. By allowing the full collapse mechanism to develop, a more economical design can be achieved than one based on the prevention of yielding at any point. For simple structures, the plastic approach provides an extremely quick means of hand analysis, requiring only the formulation of an appropriate set of collapse mechanisms and some elementary virtual work equations. The approach also forces the designer to think about how and where failure will occur, thus concentrating attention on the most critical parts of the structure.

There are, however, some drawbacks. For large and complex structures there may be many possible mechanisms and these may not be simple to analyse. Hand analysis therefore becomes cumbersome. Plastic behaviour involves substantial non-linearity, and the usual analysis approaches are rather non-methodical, involving an element of trial and error. Both of these aspects make plastic analysis difficult to computerise. Elastic methods, on the other hand, can be very easily applied to all sorts of structures in the forms of stiffness matrix and finite element computer programs; the use of such programs is now extremely widespread.

A further problem with plastic methods is that it can be difficult to provide sufficient restraint to members to enable them to achieve their full plastic moment capacity. Particularly in the case of steel structures, members will often fail by buckling at a load well below the plastic collapse load. For these reasons design methods based on elastic analysis remain widespread.

12.6 Further reading

A good recent account of plastic theory and its application to the design of civil engineering structures is:

- S. S. J. Moy, *Plastic Methods for Steel and Concrete Structures*, 2nd edn, Macmillan, 1996.

The following rather earlier books are the classic texts on plastic analysis, written by the pioneers in the field:

- J. F. Baker and J. Heyman, *Plastic Design of Frames, Volume 1: Fundamentals*, Cambridge University Press, 1969.

- J. Heyman, *Plastic Design of Frames, Volume 2: Applications*, Cambridge University Press, 1971.
- M. R. Horne, *Plastic Theory of Structures*, 2nd edn, Pergamon, 1979.

12.7 Problems

Where appropriate, answers are given at the end of the book. The questions marked with an asterisk are a little more challenging.

12.1. For each of the four beams shown in Figure 12.20, sketch the elastic bending moment diagram and all the possible collapse mechanisms.

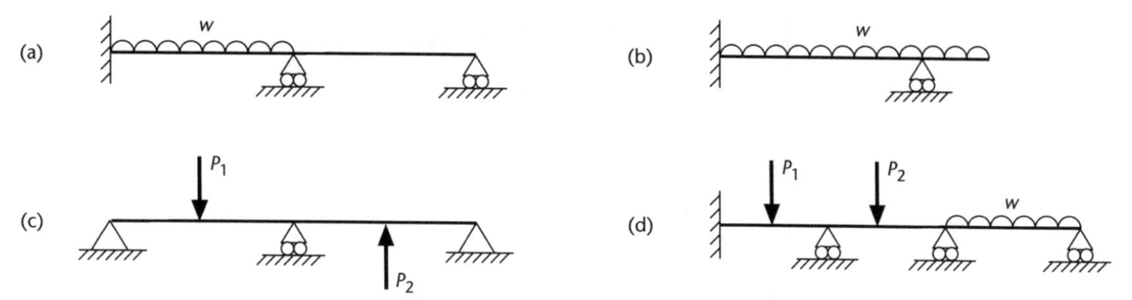

Figure 12.20

12.2. A beam of uniform cross-section with plastic moment capacity 100 kNm and length 10 m rests on simple supports at its ends and on a central prop. If equal concentrated loads P are applied at the centre of each span, use the statical method to find the value of P required to cause collapse. How does the result change if the ends are built-in rather than simply supported?

12.3. Repeat Problem 12.2 using the virtual work approach.

12.4.* The continuous beam in Figure 12.21 has plastic moment capacity $2M_p$ in span AB and M_p in spans BC and CD. Analyse all the possible collapse mechanisms and hence show that the minimum collapse load is $w = 1.5M_p/l^2$.

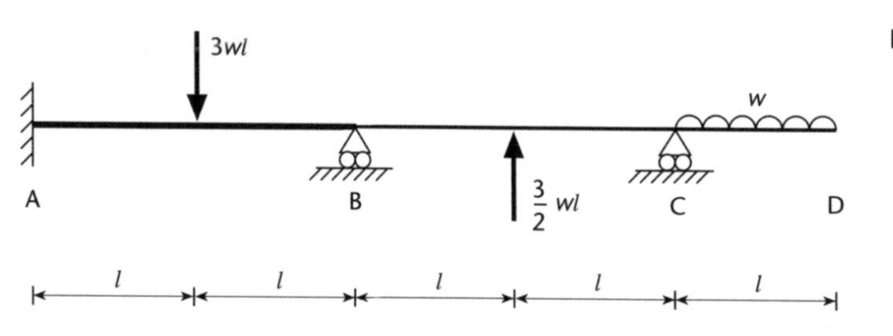

Figure 12.21

12.5. A propped cantilever of length l and plastic moment M_p carries a concentrated load W at a distance x from the propped end, where x may take any value between 0 and l. Show that collapse occurs at a load factor $\lambda = 5.83M_p/Wl$.

12.6.* A uniform, continuous beam has length l and plastic moment M_p. It rests on simple supports at each end and at two intermediate points such that the end spans are of equal length. A uniformly distributed load is applied over the entire length. Determine the positions of the intermediate supports which maximise the critical collapse load. [*Hint: the optimum support location is one which will cause two partial mechanisms to occur at the same load – why is this?*]

12.7. The structure shown in Figure 12.22 has a plastic moment capacity of 135 kNm throughout. Find the value of P required to cause collapse.

Figure 12.22

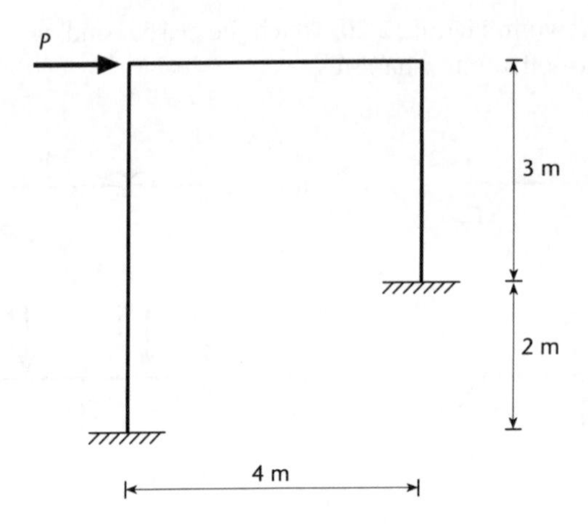

12.8. A portal frame made from members of equal plastic moment of resistance $Pl/2$ is loaded as shown in Figure 12.23. Find the critical collapse mechanism and the corresponding value of λ. Verify that this mechanism satisfies the yield condition by drawing the bending moment diagram.

Figure 12.23

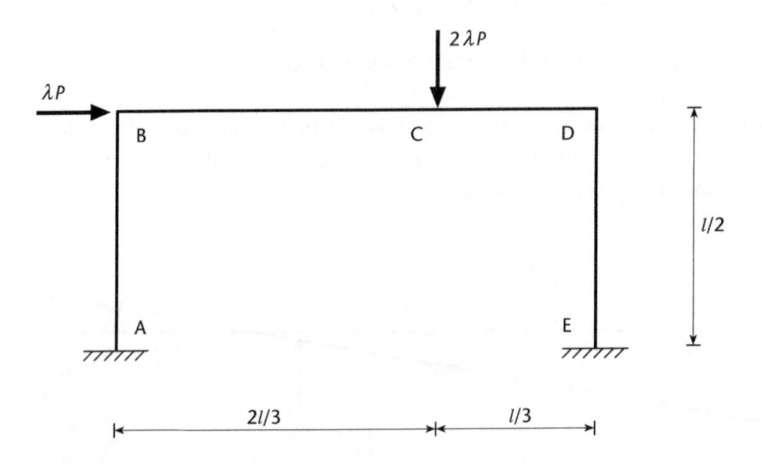

12.9.* The symmetrical frame in Figure 12.24 has constant plastic moment capacity M_p. It carries a vertical load P at C and a horizontal load kP at B, where the scalar k may take any positive value. Show that if $k > 1/\sqrt{3}$ collapse occurs by sway to the right, whereas if $k < 1/\sqrt{3}$ collapse is by sway to the left, and find the collapse load P in each case. What happens if $k = 1/\sqrt{3}$?

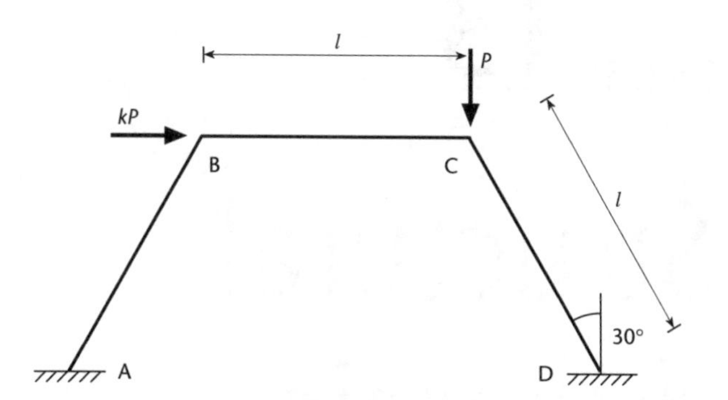

Figure 12.24

12.10.* Find the collapse load factor for the frame in Figure 12.25, in which members AB and DE have $M_p = 50$ kNm and BCD has $M_p = 80$ kNm. [*Hint: in order for the load P to do positive work, C must move downwards – for the sway and combined mechanisms this requires the frame to sway to the left.*]

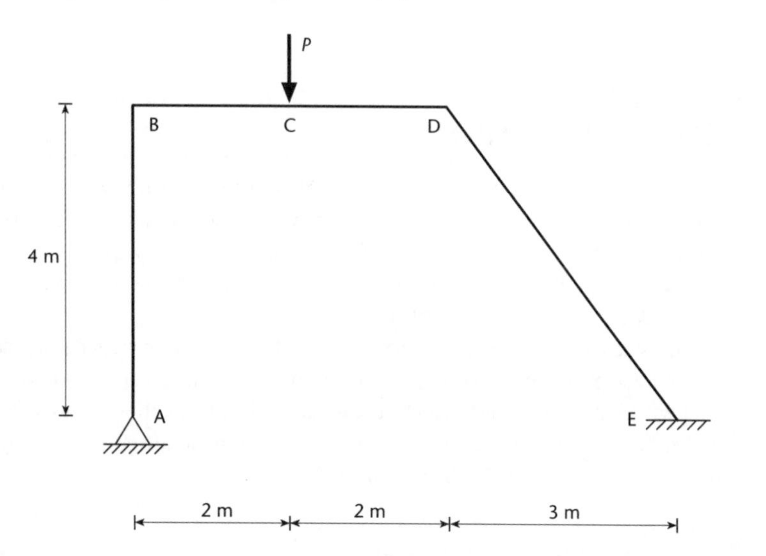

Figure 12.25

12.11. Sketch the elastic bending moment diagrams for the frames in Figure 12.26 and hence determine the possible hinge locations and the number of independent collapse mechanisms. In each case sketch suitable sets of independent mechanisms and all their possible combinations.

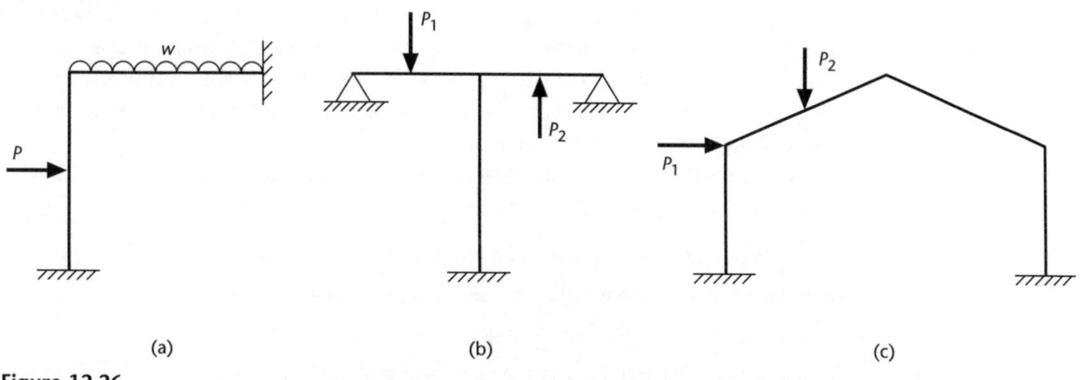

Figure 12.26

CHAPTER 13

Structural dynamics

CHAPTER OUTLINE

The science of dynamics deals with how and why bodies move. Obviously all structures move from time to time, as loads are applied and removed, but in dynamics we are concerned with motions that are continuous over a period of time. Dynamic loadings can arise from many different sources, for example wave loads on offshore oil platforms, ground shaking due to earthquakes and wind gusts on tall buildings.

In the static analysis methods covered in the rest of this book, we have frequently made use of the principle of equilibrium of the forces on a structure. In dynamics this principle no longer applies – it is the lack of equilibrium that causes a structure to move. We therefore make use of Newton's second law of motion, which states that the resultant force on a body causes a proportional acceleration. This is normally written as:

$$F = ma \tag{13.1}$$

where F is the resultant force, m the mass and a the acceleration.

As with statics, the complexity of a dynamic analysis will increase as the number of degrees of freedom in the structure increases. In reality, all structures have distributed mass and stiffness and therefore have infinite numbers of degrees of freedom. However, in most cases it is possible to obtain reasonably accurate estimates of the dynamic behaviour using *lumped parameter* models, in which the structure is represented by a relatively small number of discrete masses, thus reducing the number of degrees of freedom. We shall concentrate here on the dynamics of lumped parameter systems, and shall also restrict ourselves to linear elastic behaviour; those interested in the dynamics of continuous or non-linear systems are referred to the specialist texts recommended at the end of the chapter.

Only a limited introduction to structural dynamics is possible in the space available here. This chapter therefore aims to enable the reader to:

- understand the basic principles and terminology of structural dynamics;
- recognise which properties of a structure affect its dynamic behaviour;
- derive lumped-parameter models of simple structures;
- analyse free and forced vibrations of single degree of freedom systems;

- understand some of the methods employed in the analysis of multi degree of freedom structures;
- recognise some of the types of load which are likely to cause a structure to respond dynamically.

This chapter requires slightly more advanced mathematics than the remainder of the book. In particular, some familiarity with the solution of second-order ordinary differential equations is assumed.

13.1 Single-degree-of-freedom systems

We shall start by considering the simplest possible system. A single-degree-of-freedom (SDOF) system is one whose deformation can be completely defined by a single displacement. Obviously most real structures have many degrees of freedom, but a surprisingly large number of structures can be modelled approximately as SDOF systems.

Two examples of SDOF systems are shown in Figure 13.1. In (a) the mass m rests on frictionless rollers and is attached to a rigid support by a spring of stiffness k. Under the action of a horizontal force $P(t)$ which varies with time, the mass undergoes dynamic horizontal displacements $u(t)$. We shall see that it is extremely convenient in dynamic analysis to be able to represent structures in this simple mass–spring form.

In (b) a rectangular portal frame with pinned feet is loaded by a time-varying horizontal force $P(t)$ at the roof level. The structure is not strictly a single-degree-of-freedom system, since it has distributed mass and stiffness, but it can quite easily be modelled as a mass–spring system of the form shown in (a). First we must make some simplifying assumptions about the overall mode of deformation. If we assume that the masses of the columns are small compared to that of the beam, and that the beam is rigid, then the structure deforms by swaying horizontally as shown. Its deformation can now be defined entirely by the lateral deflection $u(t)$ of the roof beam.

We now see that the structure has been reduced to a rigid mass (the roof beam) supported on a light, flexible substructure (the columns). We can therefore represent this diagrammatically as a mass–spring system of the form shown in Figure 13.1(a), where the lumped mass represents the mass of the roof beam and the spring stiffness k equals the sway stiffness of the two columns. Now each column deforms as a cantilever (Figure 13.1(c)), so that the lateral deflection due to a force F is, from Table 5.2:

$$\delta = \frac{Fl^3}{3EI}$$

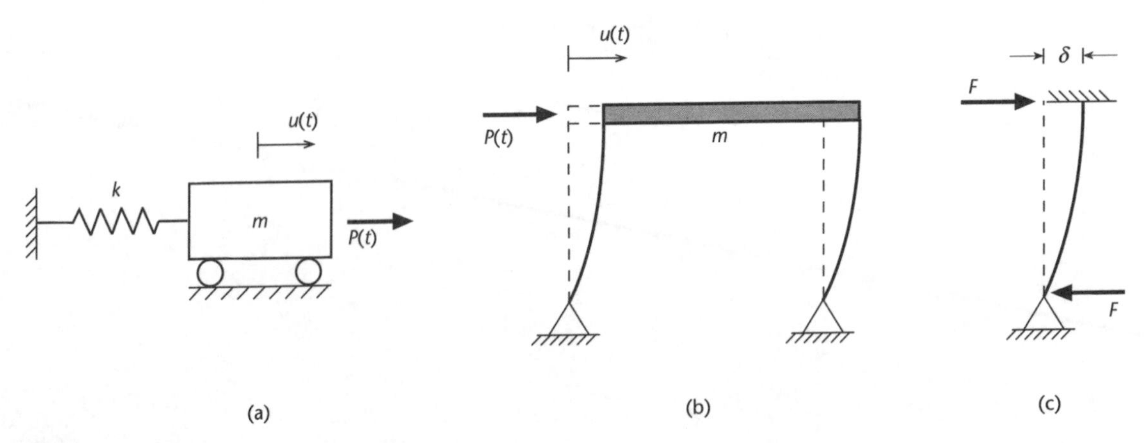

Figure 13.1
Modelling of a single-degree-of-freedom system.

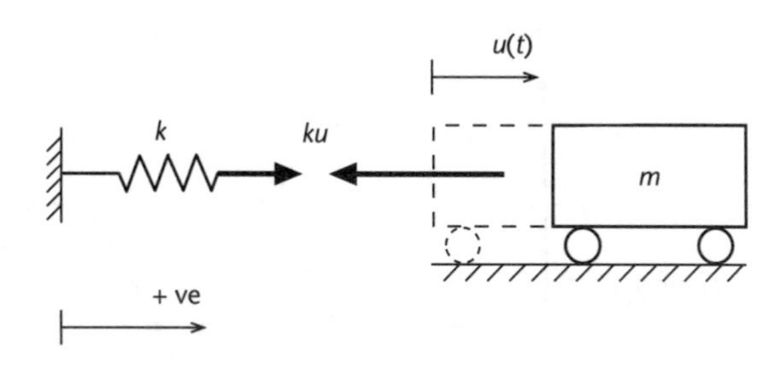

Figure 13.2
Free vibration of an undamped mass–spring system.

and the total stiffness provided by two columns is therefore:

$$k = 2\frac{F}{\delta} = \frac{6EI}{l^3}$$

The SDOF model thus created is, of course, an approximation. In reality the flexibility of the roof beam would tend to reduce the overall sway stiffness and the mass of the columns may not be negligible. This latter problem could be crudely accounted for by lumping half of the column mass at roof level and half at the base.

There is one other important parameter that has not been included in our model. Most structures gradually dissipate energy as they move, through a variety of internal mechanisms. For instance, in concrete energy can be expended by, among other things, the repeated opening and closing of micro-cracks. The various energy dissipation mechanisms are normally grouped together and known as *damping*. This will be discussed in more detail later.

13.1.1 Undamped free vibrations

Consider first the theoretical case of a simple mass–spring system with no damping and no exciting force. The mass is given a small displacement U from its equilibrium position, held stationary for a moment and then released. The subsequent behaviour can be analysed using Newton's second law of motion.

Consider the system at some time t after release, with the mass a distance $u(t)$ from its equilibrium position. Figure 13.2 shows the free body diagrams for the mass and the spring; it is the forces acting on the mass which are of interest. With the mass displaced to the right, as shown, the spring goes into tension and so imposes a force ku on the mass to the left. Conversely, if the mass moves to the left of its equilibrium position then the spring is compressed and exerts a force to the right. Clearly the spring force always opposes the displacement which causes it, and is the only force acting on the mass. Therefore, applying Newton's second law (equation (13.1)) to the mass gives:

$$m\ddot{u} = -ku$$

or

$$m\ddot{u} + ku = 0 \tag{13.2}$$

where each dot represents one differentiation with respect to time, so that $\dot{u}$ is the velocity and $\ddot{u}$ is the acceleration. This second-order differential equation is called the *equation of motion* of the system, and has the general solution:

$$u(t) = A\sin\omega_{\mathrm{n}}t + B\cos\omega_{\mathrm{n}}t \tag{13.3}$$

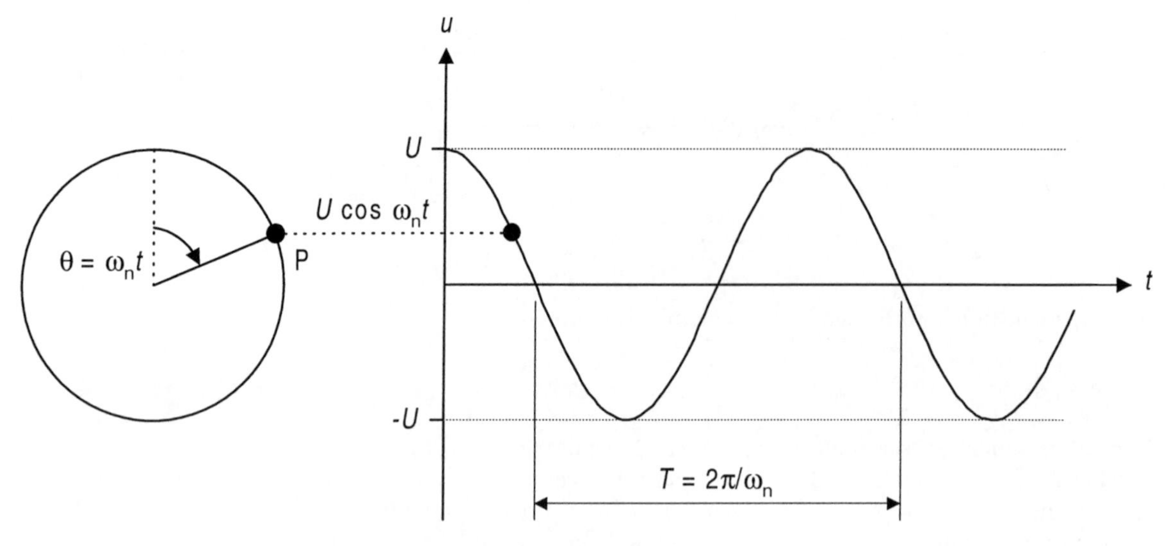

Figure 13.3
Relationship between circular and harmonic motion.

where

$$\omega_n = \sqrt{\frac{k}{m}} \tag{13.4}$$

The sinusoidal variation of displacement with time is known as *harmonic* motion. Differentiating (13.3) gives the velocity of the mass:

$$\dot{u}(t) = A\omega_n \cos \omega_n t - B\omega_n \sin \omega_n t \tag{13.5}$$

The constants A and B can now be found from the initial conditions. Substituting the initial velocity $\dot{u}(0) = 0$ into equation (13.5) gives $A = 0$ and then putting the initial displacement $u(0) = U$ into equation (13.3) gives $B = U$, so the motion of our system simplifies to:

$$u(t) = U \cos \omega_n t$$

In order to understand the meaning of ω_n it can be helpful to consider the relationship between harmonic motion and simple circular motion. Imagine a point P moving on a circle of radius U at constant angular velocity ω_n, as shown in Figure 13.3. The displacement u of our mass from its equilibrium position is equal to the vertical component of the displacement of P.

The time taken to complete one forwards-and-backwards cycle of motion of the mass m is called the *natural period* of the system, T. Using the circular motion analogy, one full cycle of motion corresponds to one complete revolution of P, so:

$$\omega_n T = 2\pi \rightarrow T = \frac{2\pi}{\omega_n} = 2\pi\sqrt{\frac{m}{k}} \tag{13.6}$$

An alternative way of expressing the rate of motion is the *natural frequency* of the system f_n, defined as the number of complete oscillations occurring in a unit time, measured in cycles per second, or Hertz (Hz). This is, of course, just the reciprocal of the natural period, therefore:

$$f_n = \frac{\omega_n}{2\pi} = \frac{1}{2\pi}\sqrt{\frac{k}{m}} \tag{13.7}$$

For reasons that should be obvious from the preceding discussion, ω_n is known as the *circular natural frequency*, measured in radians per second.

Whether it is defined as a natural frequency or as a period, the rate at which free vibration occurs is a property of a structure which depends only on its mass and stiffness and is not affected by the external loading or the initial conditions.

EXAMPLE 13.1

A structure is idealised as a single mass of 500 kg supported on a spring of stiffness 200 kN/m. When at its equilibrium position an impulse is applied to the mass, giving it an initial velocity of 0.2 m/s. Describe the subsequent motion.

From equation (13.4) the circular natural frequency is

$$\omega_n = \sqrt{\frac{200 \times 10^3}{500}} = 20 \text{ rad/s}$$

and dividing by 2π gives $f_n \approx 3.2$ Hz. Since the system has the same general form as the one considered above, the initial analysis will be identical and the same general solution will result:

$$u(t) = A \sin \omega_n t + B \cos \omega_n t$$
$$\dot{u}(t) = A\omega_n \cos \omega_n t - B\omega_n \sin \omega_n t$$

Substituting the initial conditions into these equations:

$$u(0) = 0 \quad \rightarrow B = 0$$
$$\dot{u}(0) = 0.2 \quad \rightarrow A = 0.2/20 = 0.01 \text{ m}$$

So the displacement of the mass is:

$$u = 0.01 \sin(20t)$$

This means that in every second the mass completes approximately 3.2 cycles of harmonic motion of amplitude 0.01 m about its equilibrium position.

13.1.2 Damped free vibrations

Whereas the harmonic motion of the idealised system in Figure 13.2 will carry on indefinitely at constant amplitude, we know that in a real structure the motion would gradually reduce due to energy dissipation, or damping, within the structural system. There are many different forms of damping in structures. For instance, energy may be dissipated by friction within the structure as it deforms; this is known as Coulomb damping. However, the vast majority of analytical

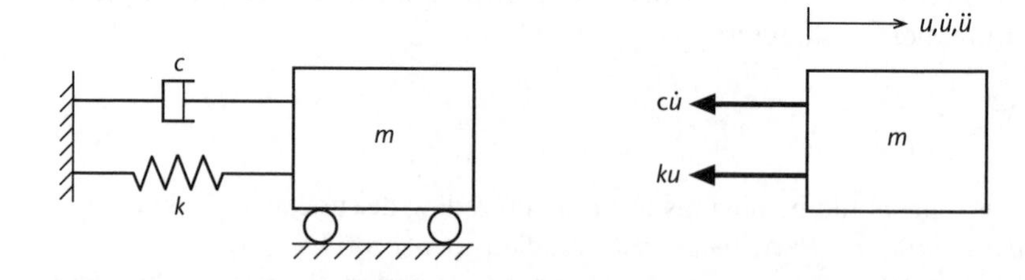

Figure 13.4
Damped single-degree-of-freedom system.

methods make the assumption that all energy dissipation is caused by linear viscous damping, in which the damping force is linearly related to the velocity. This is a reasonable approximation so long as the damping is not too great, and leads to a much simpler mathematical formulation than other damping models.

Returning to our SDOF system, damping can be represented by the inclusion of a viscous dashpot with damping coefficient c (Figure 13.4). Whereas the spring generates a force proportional to its extension, the dashpot gives rise to a force proportional to the change in velocity across it. For the SDOF system shown, the velocity of points to the left of the dashpot is zero and to the right is $\dot{u}$, so that the damping force is $c\dot{u}$. If the force is measured in N and the velocity in m/s, it follows that c has units of Ns/m. Like the spring force, the damping force always opposes the motion of the mass. So applying Newton's second law to the mass:

$$m\ddot{u} = -ku - c\dot{u}$$

or

$$m\ddot{u} + c\dot{u} + ku = 0 \tag{13.8}$$

The inclusion of damping therefore results in an additional, velocity-dependent term in the equation of motion. Solving this differential equation in the normal way leads us to a solution of the form:

$$u = e^{-ct/2m}\left(Ae^{\alpha_1 t} + Be^{\alpha_2 t}\right) \tag{13.9}$$

where

$$\alpha_{1,2} = \frac{\pm\sqrt{c^2 - 4km}}{2m} \tag{13.10}$$

The behaviour defined by equations (13.9) and (13.10) may take one of three possible forms, depending on the relative magnitudes of m, c and k. These are illustrated in Figure 13.5.

(i) Underdamped

If the damping coefficient is relatively small, such that $c^2 < 4km$, then the square root term in equation (13.10) is negative and the two exponents α_1 and α_2 are complex. It can be shown mathematically that a complex exponential is equivalent to a sine function. The term in brackets in equation (13.8) is therefore oscillatory, while the term outside the brackets causes the amplitude of the oscillations to decay exponentially with time. This chapter will be concerned principally with underdamped systems, which will be considered in more detail once we have introduced some more terminology.

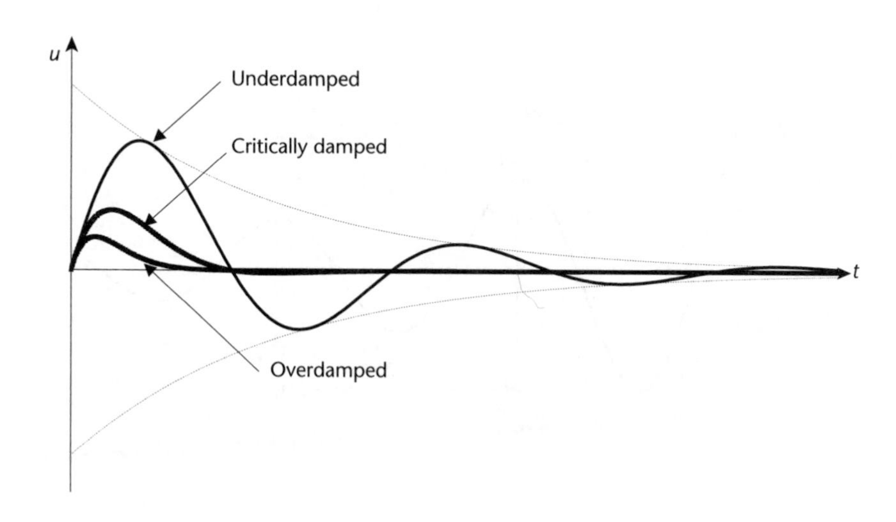

Figure 13.5
Effect of damping on free vibrations.

(ii) Overdamped

If $c^2 > 4km$, then α_1 and α_2 are both real. Under these circumstances the displacement of the mass simply decays exponentially, with no oscillations.

(iii) Critically damped

Critical damping occurs when $c^2 = 4km$ so that, from equation (13.10), both α_1 and α_2 are zero. This results in non-oscillatory behaviour similar to the overdamped case. The value of c required to cause this is known as the *critical damping coefficient* c_c:

$$c_c = 2\sqrt{km} \tag{13.11}$$

c_c is the lowest value of damping that will prevent oscillatory motion.

The critical damping condition occurs very rarely in practice, and is useful mainly as a reference case against which others can be scaled. Rather than using c, it is often convenient to work in terms of the *damping ratio* ζ, defined as the fraction of the critical damping value:

$$\zeta = \frac{c}{c_c} = \frac{c}{2\sqrt{km}} \tag{13.12}$$

This parameter is generally preferred to the damping coefficient because it is non-dimensional and because it is simple to use in practice. In general, it is much easier to estimate a damping ratio, which will not vary much between structures of a similar type, than a dashpot coefficient. For civil engineering structures damping is usually quite light, with ζ typically taking a value between 0.02 and 0.1, the exact figure depending on a range of parameters including the structural material, any non-structural elements present and the magnitude of the motion.

Remembering that $\omega_n = \sqrt{(k/m)}$, we can rewrite the damped equation of motion (13.8) as:

$$m(\ddot{u} + 2\zeta\omega_n\dot{u} + \omega_n^2 u) = 0 \tag{13.13}$$

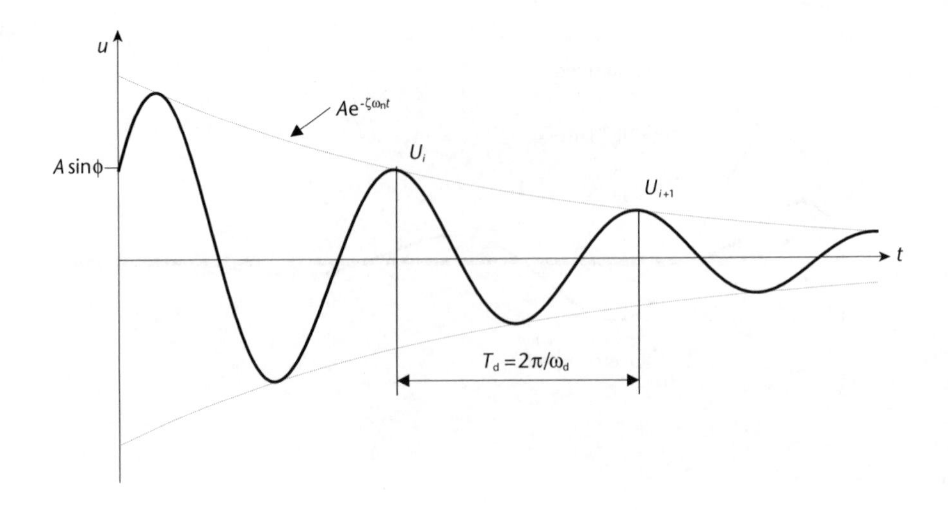

Figure 13.6
Logarithmic decrement.

and its solution (equation (13.9)) can be written in either of the following forms:

$$u = e^{-\zeta\omega_n t}\left(A\sin\sqrt{1-\zeta^2}\omega_n t + B\cos\sqrt{1-\zeta^2}\omega_n t\right) \tag{13.14}$$

$$u = Ce^{-\zeta\omega_n t}\sin\left(\sqrt{1-\zeta^2}\omega_n t + \phi\right) \tag{13.15}$$

where A and B, or C and ϕ are determined from the initial conditions. From these equations it is apparent that the damping results in a slight reduction of natural frequency compared to the undamped case. The damped circular natural frequency can be written as:

$$\omega_d = \omega_n\sqrt{1-\zeta^2} \tag{13.16}$$

For most realistic damping values, the difference between the damped and undamped frequencies is very small. For example, with a damping of 10% of critical ($\zeta = 0.1$) we get $\omega_d = 0.995\omega_n$. Nevertheless, we shall maintain the distinction between the two in this chapter.

Lastly in this section, we shall introduce one more way of quantifying damping. Figure 13.6 shows the general form of the damped vibration defined by equation (13.15). Since the curve is enveloped by an inverse exponential, the ratio of any two successive peaks must be the same. We define the *logarithmic decrement* δ as the natural log of the ratio between any two successive peaks:

$$\delta = \ln\frac{U_i}{U_{i+1}} = \ln\frac{e^{-\zeta\omega_n t}}{e^{-\zeta\omega_n(t+T_d)}} = \zeta\omega_n T_d = \zeta\omega_n\frac{2\pi}{\omega_d}$$

Substituting for ω_d from equation (13.16), we get:

$$\delta = \frac{2\pi\zeta}{\sqrt{1-\zeta^2}} \tag{13.17}$$

The logarithmic decrement is mostly used as a convenient way of assessing damping from experimental measurements of structural vibrations. For the majority of analytical work, the damping ratio ζ remains the more convenient parameter.

EXAMPLE 13.2

A simple mass–spring–damper system has $m = 10$ kg and $k = 25$ kN/m. When the mass is given an initial disturbance and then left to oscillate, it is found that the amplitude of the vibrations reduces by 50% each cycle. Calculate the damping ratio ζ, the viscous damping coefficient c and the damped natural period T_d. How long will it take for the amplitude of vibration to reduce to 1% of its initial value?

The logarithmic decrement is the natural log of the ratio of one peak to the next, therefore $\delta = \ln 2$. Then, taking the square of equation (13.17):

$$(\ln 2)^2 = \frac{4\pi^2\zeta^2}{1-\zeta^2} \quad \rightarrow \zeta^2 = \frac{(\ln 2)^2}{4\pi^2 + (\ln 2)^2} \quad \rightarrow \zeta = 0.110$$

and from equation (13.12) the damping coefficient is:

$$c = \zeta \cdot 2\sqrt{km} = 0.22\sqrt{10 \times 25{,}000} = 110 \text{ Ns/m}$$

The damped circular natural frequency is:

$$\omega_\mathrm{d} = \sqrt{\frac{k}{m}} \cdot \sqrt{1-\zeta^2} = 49.7 \text{ rad/s}$$

and the damped natural period is then simply given by:

$$T_\mathrm{d} = \frac{2\pi}{\omega_\mathrm{d}} = 0.126 \text{ s}$$

Now if the amplitude of vibration U reduces by a factor β in each cycle then we can say that the second cycle has an amplitude of β times the first, that is $U_2 = \beta U_1$. Similarly for the third cycle we have $U_3 = \beta U_2 = \beta^2 U_1$. Repeating for n cycles gives $U_\mathrm{n} = \beta^n U_1$. In our case $\beta = 0.5$, so the number of cycles required to reduce the amplitude to 1% of its initial value is given by:

$$\frac{U_\mathrm{n}}{U_1} = (0.5)^n = 0.01$$

Taking logs gives $n = \ln(0.01)/\ln(0.5) = 6.64$ and the time required for this number of cycles is simply $nT_\mathrm{d} = 6.64 \times 0.126 = 0.84$ s. It can be seen that this level of damping reduces the vibrations to a negligible level very quickly.

13.1.3 Forced vibration

While free vibrations are important in establishing the basic dynamic characteristics of a system, we are generally interested in the motion caused by some time-dependent exciting force $P(t)$ (Figure 13.7). Resolving the forces acting on the mass and applying Newton's second law in the usual way leads to the equation of motion:

$$m\ddot{u} + c\dot{u} + ku = P(t) \qquad (13.18)$$

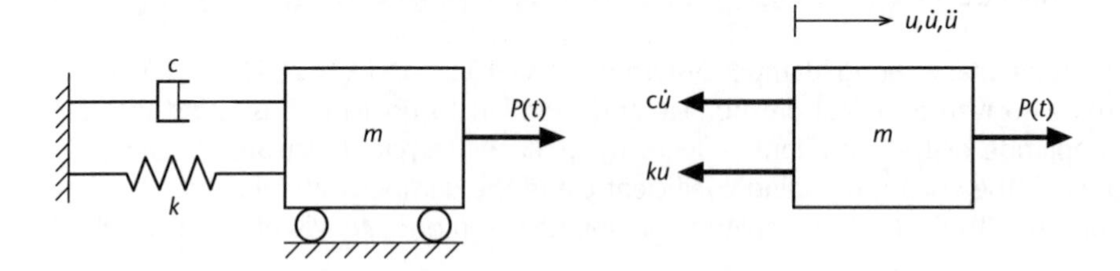

Figure 13.7
SDOF system with external exciting force.

or, using the form introduced in equation (13.13):

$$\ddot{u} + 2\zeta\omega_n\dot{u} + \omega_n^2 u = \frac{P(t)}{m} \tag{13.19}$$

The full solution to this differential equation now has two parts: a complementary function u_t and a particular integral u_s:

$$u = u_t + u_s$$

The complementary function u_t is obtained by putting the right-hand side of equation (13.19) equal to zero, giving the free vibration solution (equation (13.14)). As we have seen, damping causes this part of the motion to die out after quite a short time, and it is therefore often referred to as the *transient* part of the response. The particular integral u_s is related to the external loading and will continue so long as the loading continues. It is therefore known as the *steady-state* part of the response.

EXAMPLE 13.3

An undamped SDOF system having mass m and stiffness k is initially stationary at its equilibrium position. If a constant load P is suddenly applied to the mass, describe the subsequent motion.

We have omitted damping here in order to keep the mathematics relatively simple. The solution of the damped case is not in principle any more difficult, but the algebra becomes lengthy. The equation of motion of the system is:

$$m\ddot{u} + ku = P$$

The solution consists of a complementary function given by equation (13.3) and a particular integral. The latter is found by assuming a solution having a similar form to the loading function and then substituting this into the equation of motion. In this case, since the loading does not vary with time, we try a solution of the form $u_s = C$ where C is a constant. Substituting this into the above equation gives $C = P/k$ and so the general solution is:

$$u = A\sin\omega_n t + B\cos\omega_n t + \frac{P}{k}$$

Note that this particular integral is simply the displacement that would occur if the load P were applied statically. Applying the initial conditions $u(0) = 0$ and $\dot{u}(0) = 0$ gives $A = 0$ and $B = -P/k$, so the solution becomes:

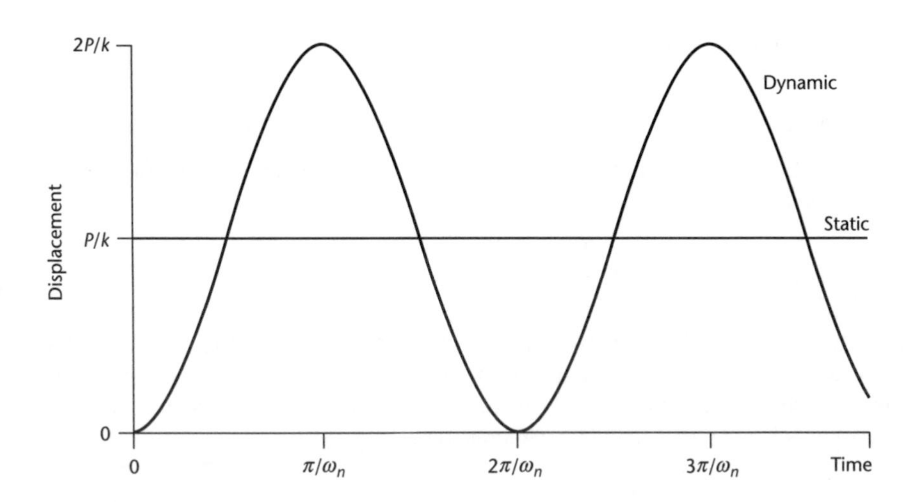

Figure 13.8

$$u = \frac{P}{k}(1 - \cos \omega_n t)$$

The dynamic and static responses to the load P are shown in Figure 13.8, from which it can be seen that the dynamic displacement oscillates between zero and twice the static displacement. If the system were damped then the oscillations would die away over time, leaving only the static displacement P/k.

For damped systems it is common to consider only the steady-state component and neglect the transient response, which quickly disappears.

We shall now consider the steady-state response to a *harmonic* excitation – one which has long duration and whose magnitude varies sinusoidally with time. This case arises quite often in practice, and provides a clear illustration of some of the most important principles of dynamic behaviour. The mathematics involved in the analysis is quite lengthy and only a condensed version is given here.

Let the loading function in (13.19) be

$$P(t) = P_0 \sin \omega t$$

where ω is known as the *forcing frequency*. Now whereas the transient response will take place at the natural frequency of the system, the steady-state response will occur at the forcing frequency. We therefore assume a steady-state solution of the form:

$$u_s = A \sin \omega t + B \cos \omega t$$

If we differentiate this expression and substitute into equation (13.19) then we get a mixture of sine and cosine terms on the left-hand side, while the right-hand side is simply $(P_0/m) \sin \omega t$. It follows that the coefficients of the cosine terms must sum to zero, while the sine coefficients sum to P_0/m. This gives us two simultaneous equations which we can solve for A and B and the following expression for the displacement is then achieved:

$$u_s = \frac{(1 - \Omega^2) \sin \omega t - 2\zeta\Omega \cos \omega t}{(1 - \Omega^2)^2 + (2\zeta\Omega)^2} \cdot \frac{P_0}{k}$$

where Ω is the *frequency ratio*, defined as:

$$\Omega = \frac{\omega}{\omega_n} \tag{13.20}$$

It can be shown by differentiation that the peak displacement, or amplitude, of the motion is:

$$U = \frac{1}{\sqrt{(1-\Omega^2)^2 + (2\zeta\Omega)^2}} \cdot \frac{P_0}{k}$$

Now P_0/k is simply the displacement that would result if the load P_0 were applied statically to the mass. So the coefficient between U and P_0/k provides a measure of the increase in peak displacement caused by the dynamics of the system. This ratio is known is the *dynamic amplification factor* (DAF):

$$\text{DAF} = \frac{kU}{P_0} = \frac{1}{\sqrt{(1-\Omega^2)^2 + (2\zeta\Omega)^2}} \tag{13.21}$$

Returning to the displacement expression, we can now write:

$$u_s = \frac{(1-\Omega^2)\sin\omega t - 2\zeta\Omega\cos\omega t}{\sqrt{(1-\Omega^2)^2 + (2\zeta\Omega)^2}} \cdot U$$

With a little trigonometric manipulation, this can be expressed as:

$$u_s = U\sin(\omega t - \phi) \tag{13.22}$$

where

$$\tan\phi = \frac{2\zeta\Omega}{1-\Omega^2} \tag{13.23}$$

Figure 13.9 shows the loading function and the steady-state displacement response (equation (13.22)) plotted against ωt. While both are sinusoidal, the peaks in the response lag behind those in the loading, with the magnitude of the lag defined by the angle ϕ, which is known as the *phase angle*. From equation (13.23), ϕ depends on both the damping and the frequency ratio.

Figure 13.10 shows (a) the dynamic amplification factor and (b) the phase angle plotted as functions of the frequency ratio, for various values of damping. In Figure 13.10(a) we can identify three quite distinct regimes of behaviour:

- If the system is very stiff or the loading very slow ($\Omega \ll 1$) then no dynamic effects occur. The DAF ≈ 1 and the peak displacement is therefore the same as if the load P_0 were applied statically.
- When the natural frequency of the system is similar to the loading frequency ($\Omega \approx 1$) resonance occurs; the DAF becomes large and the displacements are much greater than would occur under static loading. For the case when the two frequencies are exactly the same, the dynamic amplification factor is equal to $1/(2\zeta)$. The DAF at resonance is thus very sensitive to damping, and is infinite for the theoretical case of zero damping. As the damping becomes very large (for example, $\zeta = 0.5$) a slightly higher peak occurs at $\Omega < 1$, but this difference is not normally significant.
- If the system is very flexible or the loading very fast ($\Omega \gg 1$) then the DAF becomes less than unity and less displacement occurs than in the static case.

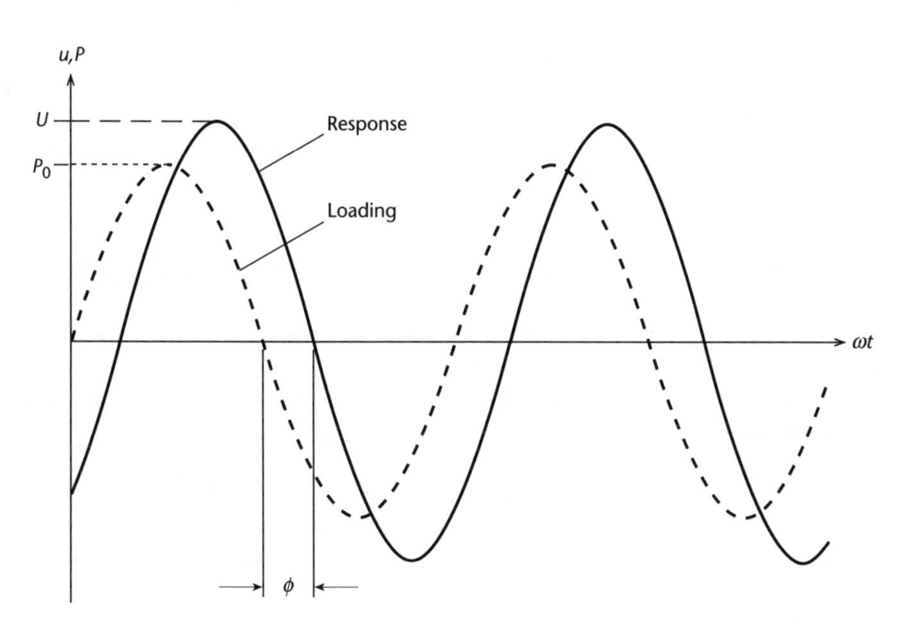

Figure 13.9
Steady-state response of a damped SDOF system to harmonic loading.

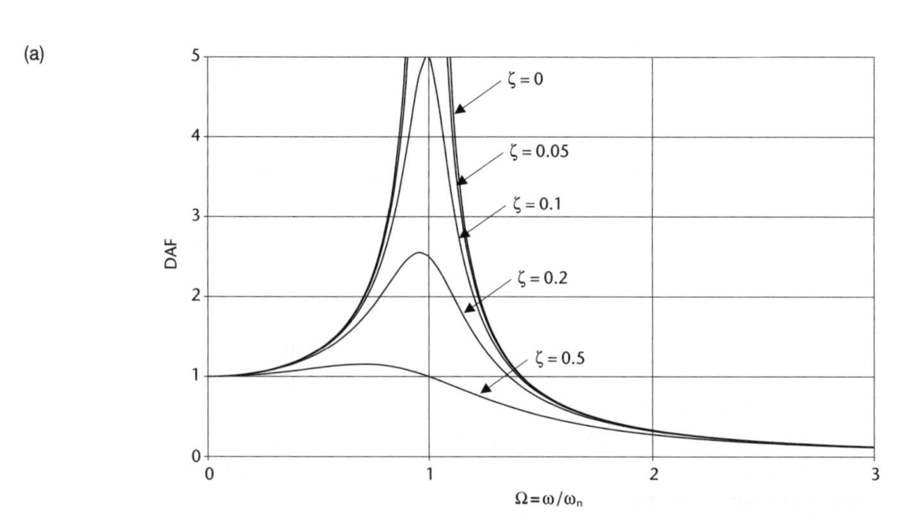

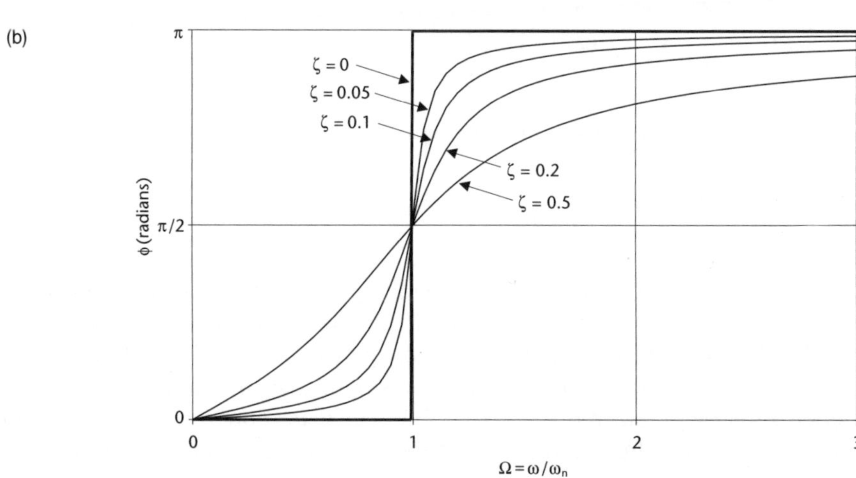

Figure 13.10
(a) Dynamic amplification factor and (b) phase angle of a harmonically loaded SDOF system.

Figure 13.10(b) shows that the damping ratio also has a large influence on the phase angle in the frequency region near resonance. For an undamped system the phase angle is zero for forcing frequencies below resonance, 90° at resonance and 180° above resonance. For more heavily damped systems the transition from $\phi = 0$ to $\phi = 180°$ is more gradual.

EXAMPLE 13.4

A machine weighing 2.0 Tonnes is positioned at the midspan of a simply supported floor of length 8 m and flexural stiffness $EI = 250 \times 10^6$ Nm^2, with $\zeta = 0.05$. Determine the maximum dynamic displacement of the floor under the following two conditions:

(a) during normal operation of the machine, when the rotating parts produce a sinusoidal out-of-balance force of magnitude 40 kN and frequency 20 Hz;

(b) during start-up, when the frequency increases steadily from zero and the magnitude of the out-of-balance force increases in proportion to the square of the frequency.

Figure 13.11

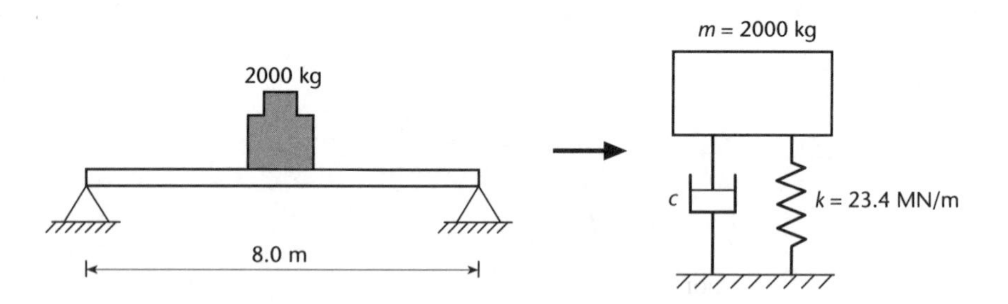

First we must idealise the machine and floor into a SDOF system, as shown in Figure 13.11. Assuming that the machine is rigid and heavy compared to the floor, then the machine provides all the mass and the floor the stiffness. The stiffness can be found using the midspan deflection formula for a centrally loaded, simply supported beam (see Table 5.2):

$$\delta = \frac{Pl^3}{48EI} \rightarrow k = \frac{P}{\delta} = \frac{48EI}{l^3} = \frac{48 \times 250 \times 10^6}{8^3} = 23.4 \times 10^6 \text{ N/m}$$

The natural frequency of the system is then:

$$f_n = \frac{1}{2\pi}\sqrt{\frac{23.4 \times 10^6}{2000}} = 17.2 \text{ Hz}$$

(a) When the machine is at its normal operating speed: the static displacement of the floor due to a load of 40 kN is simply $P_0/k = 40 \times 10^3/23.4 \times 10^6 = 0.00171$ m, or 1.71 mm, and the frequency ratio $\Omega = 20/17.2 = 1.16$. Therefore, using the $\zeta = 0.05$ curve in Figure 13.10(a), the dynamic amplification factor is approximately 2.7 (a more accurate DAF could be obtained using equation (13.21), but the greater accuracy is probably not justified, given the approximate nature of many other aspects of the analysis). The peak dynamic displacement at the operating speed is therefore:

$$u_{max} = \text{DAF} \times \frac{P_0}{k} = 2.7 \times 1.71 = 4.6 \text{ mm}$$

(b) During start-up: the machine starts from rest, so to reach its operating speed it must pass through every frequency between 0 and 20 Hz. The greatest dynamic amplification will occur at a frequency of 17.2 Hz ($\Omega = 1$), giving a DAF of $1/(2\zeta) = 10$. Now the amplitude of the force at this frequency is:

$$P_0 = 40 \text{ kN} \times \frac{17.2^2}{20^2} = 29.6 \text{ kN}$$

which gives a static displacement $P_0/k = 1.26$ mm. The peak dynamic displacement is then:

$$u_{max} = \text{DAF} \times \frac{P_0}{k} = 10 \times 1.26 = 12.6 \text{ mm}$$

13.1.4 General dynamic loading

While the loading functions considered above arise quite frequently in practice, many real loadings are more irregular in nature. The response to a time-varying load taking any form can be derived using *Duhamel's integral*, which is derived by treating the loading as a series of discrete impulses and then summing the response to each impulse.

Consider first the response of a SDOF system to single impulse. At time $t = 0$ we apply a force of magnitude P for a very short duration $\mathrm{d}t$. The impulse applied to the system is simply the area under the force–time curve (Figure 13.12(a)), that is $P \cdot \mathrm{d}t$. Using impulse = change in momentum:

$$P\mathrm{d}t = m\dot{u}(0) \rightarrow \dot{u}(0) = \frac{P\mathrm{d}t}{m}$$

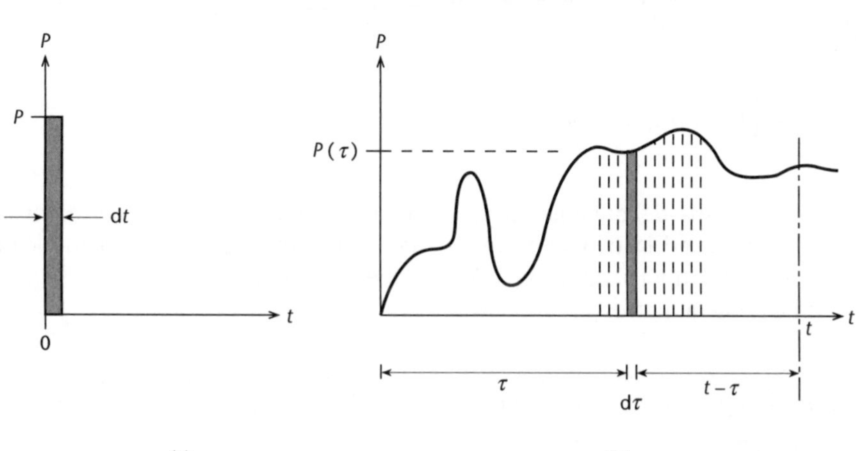

Figure 13.12
Representation of a loading function as a series of impulses.

After the impulse has been applied, no further loading occurs and the system undergoes free vibrations. The impulsive load case is therefore equivalent to the free vibration case with initial conditions $u(0) = 0$, $\dot{u}(0) = Pdt/m$. Substituting into the general solution for a damped SDOF system (equation (13.14)) gives:

$$u = \frac{Pdt}{m\omega_d} \sin \omega_d t \cdot e^{-\zeta\omega_n t} \tag{13.24}$$

We now move on to the case of an arbitrary, time-varying loading function (Figure 13.12(b)). Let the impulse occurring at some time τ be $P(\tau) \cdot d\tau$. The displacement at time $t(> \tau)$ caused by this impulse can be found from equation (13.24), except that the time elapsed after the impulse is now $(t - \tau)$ instead of just t:

$$du = \frac{P(\tau)d\tau}{m\omega_d} \sin \omega_d (t - \tau) e^{-\zeta\omega_n (t-\tau)}$$

The total displacement at time t is found by summing the displacements due to all the impulses occurring up to time t; this done by integrating with respect to τ, giving:

$$u = \frac{1}{m\omega_d} \int_0^t P(\tau) \sin \omega_d (t - \tau) e^{-\zeta\omega_n (t-\tau)} d\tau \tag{13.25}$$

Equation (13.25) is Duhamel's integral, a fundamental tool in much modern dynamic analysis. Integrals of this type, in which an excitation function is combined with a time-shifted impulse response function, are known as *convolution integrals*, and arise in many different areas of engineering analysis.

EXAMPLE 13.5

The pressure wave due to an explosive blast can be approximated as shown in Figure 13.13(a), in which a virtually instantaneous rise to a peak load is followed by a roughly linear decay. Calculate the displacement response of the SDOF system in Figure 13.13(b) to this loading. For simplicity, neglect damping.

With no damping $\zeta = 0$ and $\omega_d = \omega_n$. Noting that $1/(m\omega_n) = \omega_n/k$, equation (13.25) can be written:

$$u = \frac{\omega_n}{k} \int_0^t P(\tau) \sin \omega_n (t - \tau) d\tau$$

To cope with the discontinuity in the loading function at $t = 0.5$ s, the response must be evaluated in two separate regimes; first during application of the load ($t \leq 0.5$ s) and then after completion of the loading ($t > 0.5$ s). For $t \leq 0.5$ s Duhamel's integral gives:

$$u = \frac{\omega_n}{k} \int_0^t P_0(1 - 2\tau) \sin \omega_n (t - \tau) d\tau = \frac{P_0}{k}\left(1 - 2t - \cos \omega_n t + \frac{2}{\omega_n} \sin \omega_n t\right)$$

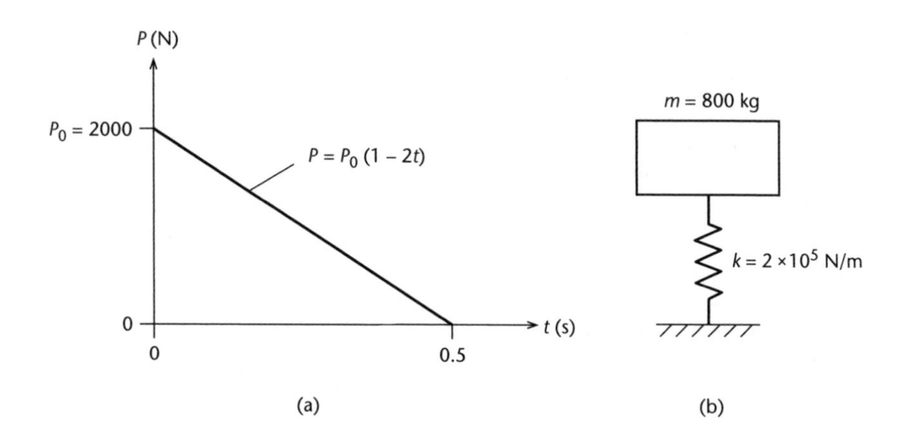

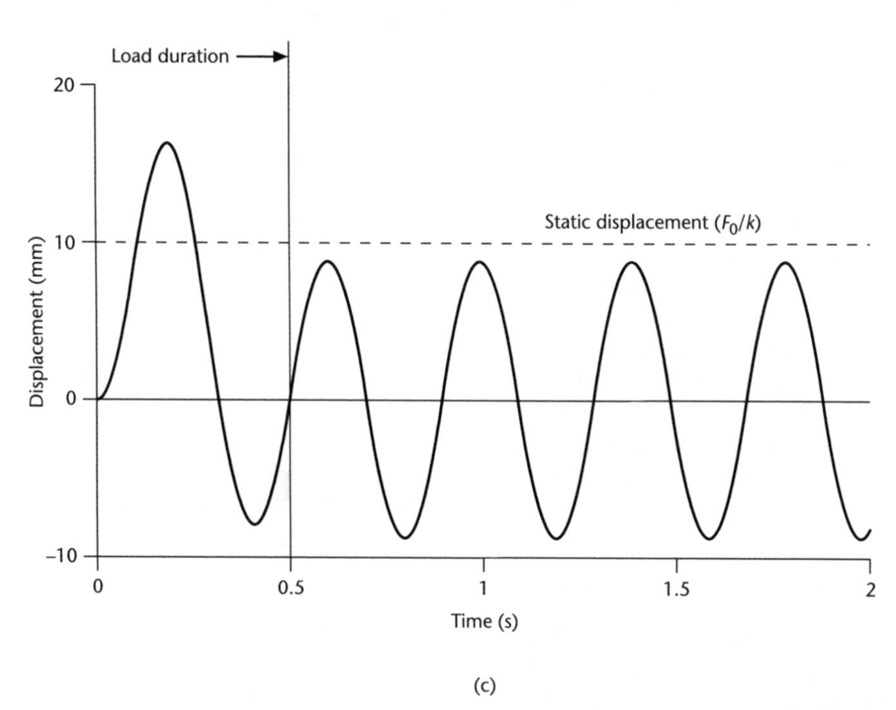

Figure 13.13

For the system and loading shown $\omega_n \quad k/m \quad 15.8$ rad/s and $P_0/k \quad 0.01$ m, and the displacement is then:

$$u \quad 0.01 \; 1 \quad 2t \quad \cos 15.8t \quad 0.127 \sin 15.8t$$

For $t > 0.5$:

$$u \quad \frac{\omega_n}{k} \int_0^{0.5} P_0 \; 1 \quad 2\tau \; \sin \omega_n \; t \quad \tau \; dt \quad \int_{0.5}^{t} 0dt$$

$$\frac{P_0}{k} \left(\quad \cos \omega_n t \quad \frac{2}{\omega_n} \sin \omega_n \; t \quad 0.5 \quad \frac{2}{\omega_n} \sin \omega_n t \right)$$

and substituting in numerical values gives:

$$u \quad 0.01 \quad \cos 15.8t \quad 0.127 \sin 15.8 \; t \quad 0.5 \quad 0.127 \sin 15.8t$$

The results are plotted in Figure 13.13(c). A peak displacement of approximately 16 mm is achieved during the load application, and after the loading is complete the structure undergoes steady-state oscillations of amplitude approximately 8 mm. In a more realistic, damped system, these oscillations would, of course, die away quite quickly.

Except for very simplified cases such as the one in Example 13.5, direct evaluation of Duhamel's integral is difficult. Often it is not possible to write P in analytical form, and even when this can be done the resulting integration may be complex. It is therefore normal to perform the calculations using a numerical integration procedure.

13.2 Multi-degree-of-freedom systems

Obviously not all structures can be realistically modelled as SDOF systems. Structures with distributed mass and stiffness may undergo significant deformations in several modes of vibration and therefore need to be analysed as multi-degree-of-freedom (MDOF) systems. In reality, all structures have an infinite number of degrees of freedom. However, it is usually possible to obtain reasonably accurate results by idealising them as assemblages of masses and springs, that is, as *lumped* rather than *distributed parameter systems*. This allows us to reduce the number of degrees of freedom to a manageable level and also renders the model amenable to computational solution.

There are many types of MDOF system, but for simplicity we shall concentrate here on the example of multi-storey buildings with relatively light, flexible columns (the springs in our mass–spring system) and heavy, rigid floors (the masses).

13.2.1 Equations of motion

Figure 13.14(a) shows a plane representation of a simple multi-storey frame building having n floors, subjected to a distributed horizontal load that varies with both position and time. By making some assumptions about the structural behaviour, it is possible to idealise this as a mass–spring system of the form shown in Figure 13.14(b); here the springs have been drawn vertically for diagrammatic convenience, but they should, of course, be regarded as providing resistance to horizontal movement.

The idealisation can be carried out as follows. If the floors are stiff compared to the columns, then the frame will deform as indicated by the dashed lines in Figure 13.14(a), with the floors remaining approximately straight and parallel to each other. With this mode of deformation, the bending stiffness provided by each column is $12EI/l^3$ and the total lateral stiffness of a particular storey is found by summing the column stiffnesses for that storey. The mass at each floor level consists of the floor masses plus a proportion of the masses of the attached columns. A simple way of distributing the masses of the columns is simply to lump half at the top end and half at the bottom. Lastly, the distributed load can be lumped at the floor levels by integrating the load acting over an area half a storey either side of the floor in question.

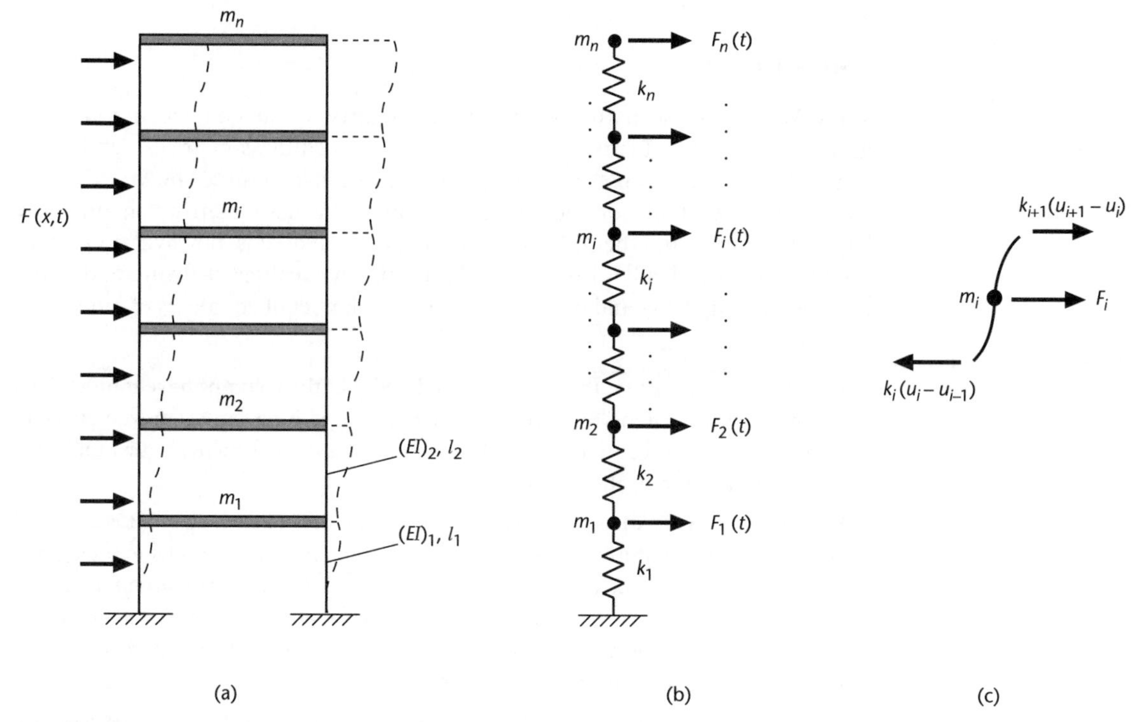

Figure 13.14
Idealisation of a multi-storey frame building.

We could also include a series of viscous dampers in our model, but in practice this is not usually done as the damping coefficients are difficult to define accurately. Instead, we normally incorporate damping in the form of a global damping ratio ζ for the whole structure. We shall see how this is done when we go on to look at the calculation of forced response.

Having reduced the structure to mass–spring form, we can now set up the equations of motion for the idealised system. Figure 13.14(c) shows the forces acting on a single mass m_i, with the directions of the spring forces drawn on the assumption that each floor has deflected further than the one below it. Resolving horizontally and applying Newton's second law:

$$m_i\ddot{u}_i = F_i + k_{i+1}(u_{i+1} - u_i) - k_i(u_i - u_{i-1})$$

Rearranging:

$$m_i\ddot{u}_i - k_i u_{i-1} + (k_i + k_{i+1})u_i - k_{i+1}u_{i+1} = F_i$$

Repeating for all the masses in the structure and writing in matrix form:

$$\begin{bmatrix} m_1 & & & & & \\ & m_2 & & & & \\ & & m_3 & & & \\ & & & \ddots & & \\ & & & & \ddots & \\ & & & & & m_n \end{bmatrix} \begin{bmatrix} \ddot{u}_1 \\ \ddot{u}_2 \\ \ddot{u}_3 \\ \cdot\cdot \\ \cdot\cdot \\ \ddot{u}_n \end{bmatrix} + \begin{bmatrix} (k_1+k_2) & -k_2 & & & & \\ -k_2 & (k_2+k_3) & -k_3 & & & \\ & -k_3 & (k_3+k_4) & -k_4 & & \\ & & \cdot\cdot & \cdot\cdot & \cdot\cdot & \\ & & & \cdot\cdot & \cdot\cdot & \cdot\cdot \\ & & & & -k_n & kn \end{bmatrix} \begin{bmatrix} u_1 \\ u_2 \\ u_3 \\ \cdot\cdot \\ \cdot\cdot \\ u_n \end{bmatrix} = \begin{bmatrix} F_1 \\ F_2 \\ F_3 \\ \cdot\cdot \\ \cdot\cdot \\ F_n \end{bmatrix}$$

or

$$\mathbf{M\ddot{U}} + \mathbf{KU} = \mathbf{F} \qquad (13.26)$$

where **M** is the mass matrix, **K** the stiffness matrix, **F** the load vector and **U** the displacement vector. Blank spaces in the mass and stiffness matrices signify zero terms while the dots represent additional terms similar to those shown.

It can be seen that, for the structure shown, the mass matrix contains only diagonal elements and the stiffness matrix is tri-diagonal. It is, however, important to remember that this has been derived for a highly idealised structure and that a more realistic model would result in a more complex set of equations of motion. In particular:

- It has been assumed that the floors are rigid and that the columns have infinite axial stiffness, so that all joints displace only horizontally, with no vertical movements and no rotations. Relaxing these assumptions would considerably increase the number of degrees of freedom in the problem.
- A *lumped* mass matrix has been used, in which the mass of each element is assumed to be concentrated at its two ends. This approximation always results in a diagonal mass matrix. If the structure is modelled using the finite element method, then it is possible to derive a more accurate, *consistent* mass matrix by a similar approach to that used for deriving stiffness matrices in Chapter 10. Use of this approach would introduce off-diagonal terms into the mass matrix.
- For a less simple geometry than the one shown, the bandwidth of the stiffness matrix may be higher.

It will be noted that, if either the mass or stiffness matrix has off-diagonal terms, the individual equations of motion which make up (13.26) are coupled together, giving n simultaneous differential equations for the displacements. The first step in the solution of these equations is to analyse the free vibrations of the structure, by putting the right-hand side of equation (13.26) equal to zero. We can then use the resulting natural frequencies and mode shapes of the system to solve the forced vibration case. This procedure will be outlined in subsequent sections.

EXAMPLE 13.6

Figure 13.15(a) shows a simple two-storey building in which the floors are assumed rigid and have mass 800 kg/m^2 and the columns are 305 305 158 kg/m Universal Column sections with flexural rigidity EI 81.35 10^6 Nm^2. The structure is subjected to a uniform lateral wind pressure of amplitude 300 N/m^2 which is assumed (rather artificially) to vary sinusoidally with time, with frequency 2 Hz 4π rad/s . Formulate the equations of motion for lateral sway vibrations.

First we must reduce the structure to mass–spring form. Using the lumped mass approach, the mass at each floor level is the actual floor mass plus half the mass of the attached columns:

m_1 800 5 8 6 158 2.5 6 158 2 36,266 kg

m_2 800 5 8 6 158 2 33,896 kg

Figure 13.15

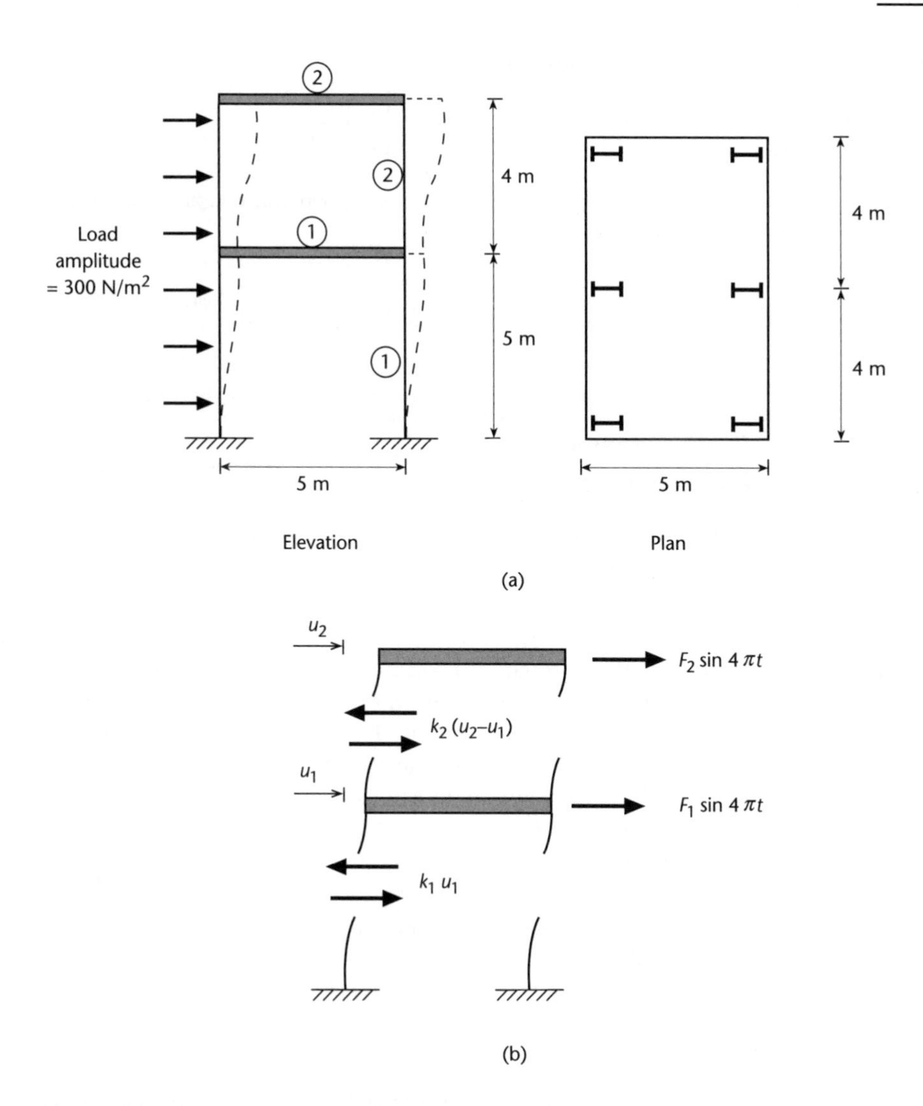

The lateral stiffness of a column with a point load at one end and zero rotation at each end is $12EI/l^3$, so the storey stiffnesses are:

$$k_1 \quad 6 \quad 12 \quad 81.35 \quad 10^6/5^3 \quad 46.9 \quad 10^6 \text{ N/m}$$

$$k_2 \quad 6 \quad 12 \quad 81.35 \quad 10^6/4^3 \quad 91.5 \quad 10^6 \text{ N/m}$$

The equivalent load amplitudes are found by integrating the distributed load over an area half a storey-height either side of the floor under consideration:

$$F_1 \quad 300 \quad 8 \quad 2.5 \quad 2 \quad 10{,}800 \text{ N}$$

$$F_2 \quad 300 \quad 8 \quad 2 \quad 4{,}800 \text{ N}$$

Now consider the forces acting on each floor level (Figure 13.15(b)). Resolving horizontally and applying Newton's second law:

$$F_1 \sin 4\pi t \quad k_2\, u_2 \quad u_1 \quad k_1 u_1 \quad m\ddot{u}_1$$

$$F_2 \sin 4\pi t \quad k_2\, u_2 \quad u_1 \quad m\ddot{u}_2$$

Rearranging and writing in matrix form:

$$\begin{bmatrix} m_1 & 0 \\ 0 & m_2 \end{bmatrix}\begin{bmatrix} \ddot{u}_1 \\ \ddot{u}_2 \end{bmatrix} \quad \begin{bmatrix} k_1 \quad k_2 & k_2 \\ k_2 & k_2 \end{bmatrix}\begin{bmatrix} u_1 \\ u_2 \end{bmatrix} \quad \begin{bmatrix} F_1 \\ F_2 \end{bmatrix} \sin 4\pi t$$

So, substituting in the numerical values calculated above and dividing through by 10^3, the equations of motion are:

$$\begin{bmatrix} 36.3 & 0 \\ 0 & 33.9 \end{bmatrix}\begin{bmatrix} \ddot{u}_1 \\ \ddot{u}_2 \end{bmatrix} \quad 10^3 \quad \begin{bmatrix} 138.4 & 91.5 \\ 91.5 & 91.5 \end{bmatrix}\begin{bmatrix} u_1 \\ u_2 \end{bmatrix} \quad \begin{bmatrix} 10.8 \\ 4.8 \end{bmatrix} \sin 4\pi t$$

13.2.2 Free vibrations of MDOF systems

Free vibrations of MDOF systems are governed by equations of motion of the form of (13.26), but with no external forcing, that is:

$$\mathbf{M}\ddot{\mathbf{U}} + \mathbf{K}\mathbf{U} = 0 \tag{13.27}$$

As stated previously, if the damping ratio is small (less than about 0.1) then damping will have little influence on the free vibration characteristics and can be neglected at this stage. Now the elements of $\mathbf{U}$ are functions of both position (storey level) and time. To solve, we assume that $\mathbf{U}$ can be expressed as the product of a vector $\boldsymbol{\psi}$ whose elements are functions solely of position, and a scalar which is a function only of time:

$$\mathbf{U} = \boldsymbol{\psi} \sin \omega t \tag{13.28}$$

$\boldsymbol{\psi}$ thus defines the overall shape of the vibrating structure at its maximum deflection, and the $\sin \omega t$ term describes how the vibration varies with time. Differentiating this expression and substituting into (13.27) gives:

$$(\mathbf{K} - \omega^2\mathbf{M})\boldsymbol{\psi} = 0 \tag{13.29}$$

It follows that either $\boldsymbol{\psi} = 0$, in which case the solution is trivial, or the matrix $(\mathbf{K} - \omega^2\mathbf{M})$ is singular, which requires that its determinant is zero:

$$|\mathbf{K} - \omega^2\mathbf{M}| = 0 \tag{13.30}$$

For an n degree of freedom problem there will be n roots ω_i of equation (13.30); these are the natural frequencies of the system. If each ω_i is separately substituted back into (13.29) then a corresponding vector $\boldsymbol{\psi}_i$ will be obtained; this is the deflected shape associated with a particular natural frequency and is known as the *mode shape*. Thus an n-DOF system is able to vibrate in n different modes, each having a distinct deformed shape and each occurring at a particular natural frequency. The modes of vibration are system properties, independent of the external loading.

Figure 13.16 shows the sway modes of vibration of a four-storey building, with the modes numbered in order of ascending natural frequency. In the first mode all floors are moving in the same direction at any time; in the second the upper and lower storeys move in opposite directions; in the third the middle part of the building moves in the opposite direction to the floors above and below, and so on.

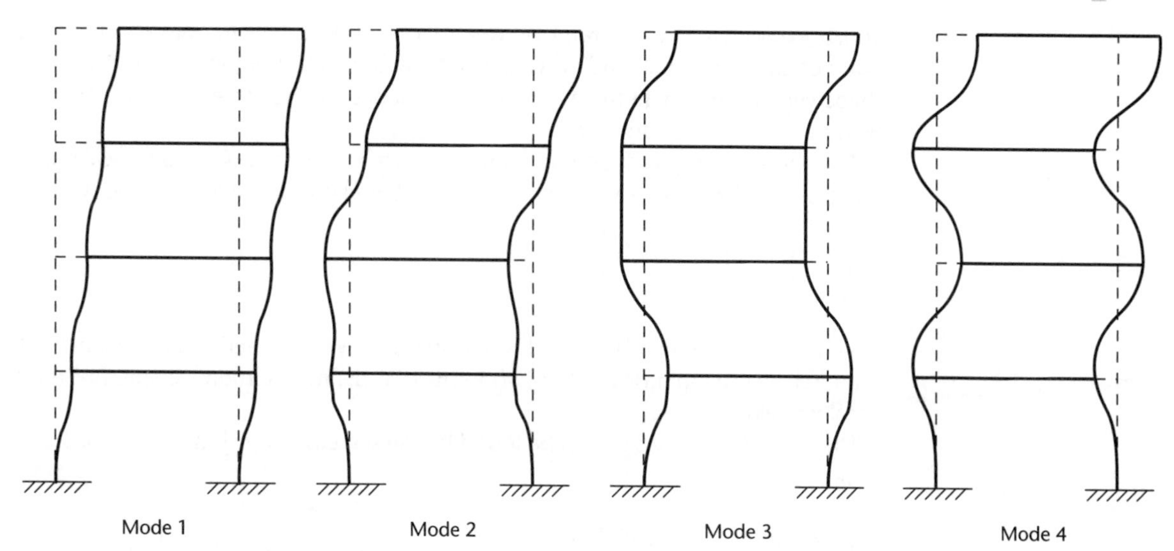

Figure 13.16
Mode shapes of a four-storey building.

It is always the case that modes with relatively simple deformed shapes have lower natural frequencies, while those involving many reversals of curvature have higher natural frequencies.

EXAMPLE 13.7

Find the natural frequencies and mode shapes for the structure in Example 13.6.

Using the mass and stiffness matrices derived in Example 13.6, we can write:

$$\mathbf{K} \quad \omega^2\mathbf{M} \quad \begin{bmatrix} 138{,}400 \quad 36.3\omega^2 & 91{,}500 \\ 91{,}500 & 91{,}500 \quad 33.9\omega^2 \end{bmatrix}$$

Putting the determinant equal to zero:

$$\left|\mathbf{K} \quad \omega^2\mathbf{M}\right| \quad 138{,}400 \quad 36.3\omega^2 \;\; 91{,}500 \quad 33.9\omega^2 \quad 91{,}500^2 \quad 0$$

This is a quadratic equation in ω^2. Solving gives:

$\omega_1 \quad 24.3$ rad/s $f_1 \quad 3.9$ Hz

$\omega_2 \quad 77.0$ rad/s $f_2 \quad 12.2$ Hz

We can now find the corresponding mode shapes. Let the mode shape vector be $\boldsymbol{\psi} \quad a\ b^{\mathrm{T}}$. Then, substituting numerical values for $\mathbf{K}$, $\mathbf{M}$ and ω_1 into equation (13.29) gives:

$$\begin{bmatrix} 117{,}036 & 91{,}500 \\ 91{,}500 & 71{,}548 \end{bmatrix} \begin{bmatrix} a \\ b \end{bmatrix} \quad \begin{bmatrix} 0 \\ 0 \end{bmatrix} \quad \frac{a}{b} \quad 0.78$$

Note that the two lines of this matrix equation both give exactly the same relationship between a and b. This means that we cannot determine the values of a

and b explicitly, only the ratio between them. This is, in fact, always the case. Without some external loading or initial conditions, the magnitudes of the mode shape vectors are indeterminate. The mode shapes are therefore just that – shapes, with no particular magnitude.

To express the mode shapes in numerical form we must impose some scaling. In this case we shall scale so that b 1, so that the first mode shape can be written:

$$\boldsymbol{\psi}_1 \quad \begin{bmatrix} 0.78 \\ 1 \end{bmatrix}$$

It is important to remember that this scaling is purely for numerical convenience – it is the relative magnitudes of the two terms that define the mode shape, not their absolute values.

The second mode shape can be found by the same process, but using ω_2 instead of ω_1:

$$\begin{bmatrix} 76{,}599 & 91{,}500 \\ 91{,}500 & 109{,}284 \end{bmatrix} \begin{bmatrix} a \\ b \end{bmatrix} \quad \begin{bmatrix} 0 \\ 0 \end{bmatrix} \quad \frac{a}{b} \quad 1.19 \quad \boldsymbol{\psi}_2 \quad \begin{bmatrix} 1.19 \\ 1 \end{bmatrix}$$

The two mode shapes are sketched in Figure 13.17.

Figure 13.17

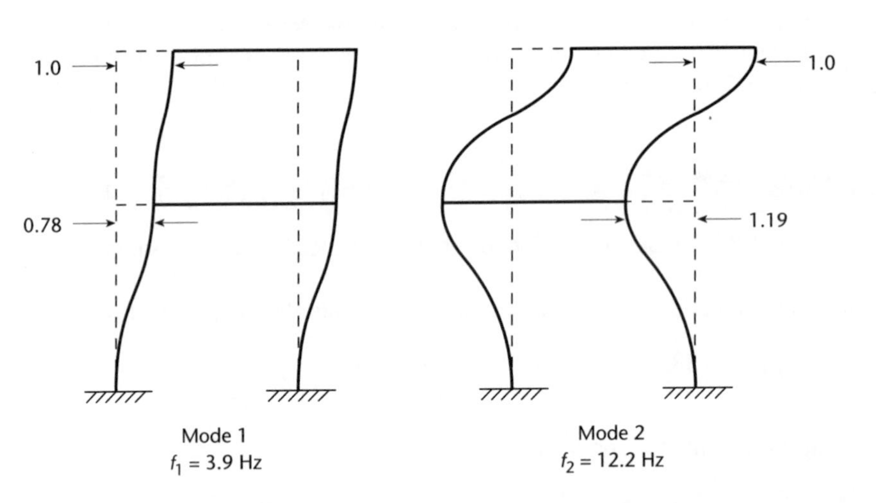

Readers familiar with matrix algebra will have recognised that equation (13.29) is an eigenproblem. This is more obvious if we pre-multiply by $\mathbf{M}^{-1}$ giving:

$$(\mathbf{M}^{-1}\mathbf{K} - \omega^2\mathbf{I})\boldsymbol{\psi} = 0$$

where $\mathbf{I}$ is the unit matrix. This now looks very similar to the classical eigenvalue equation for a matrix $\mathbf{A}$ having eigenvalues λ and eigenvectors $\boldsymbol{\psi}$:

$$(\mathbf{A} - \lambda\mathbf{I})\boldsymbol{\psi} = 0$$

Comparing these two equations, we see that the natural frequencies of our system are the square roots of the eigenvalues of $\mathbf{M}^{-1}\mathbf{K}$ and the mode shapes are the corresponding eigenvectors. This is important for two reasons:

- Determining the natural frequencies by expanding the determinant is very cumbersome for systems with more than a few degrees of freedom. However, there are many extremely efficient numerical methods for determining eigenvalues and eigenvectors for large matrices (these will not be described here – see the Further Reading in section 13.4 for details).
- Eigenvectors have certain orthogonality properties which will be used in the analysis of forced vibrations in the next section.

In preliminary calculations, it is common to make use of simple, 'rule of thumb' formulae to give a rough estimate of structural behaviour. A well-known rule of thumb for the first (or *fundamental*) natural frequency of multi-storey buildings is:

$$f_1 = \frac{10}{n} \tag{13.31}$$

where n is the number of storeys. For a two-storey building this formula gives a fundamental frequency of 5 Hz, which is not too bad an approximation to the value of 3.9 Hz calculated in Example 13.7.

13.2.3 Forced vibration response

Having determined the natural frequencies and mode shapes of our system, we can go on to analyse the response to some external forcing function. At this stage we need to include damping in the analysis, so that the equations of motion have the form:

$$\mathbf{M}\ddot{\mathbf{U}} + \mathbf{C}\dot{\mathbf{U}} + \mathbf{K}\mathbf{U} = \mathbf{F} \tag{13.32}$$

where $\mathbf{M}$, $\mathbf{C}$ and $\mathbf{K}$ are $n \times n$ matrices. As we shall see, it is not necessary to formulate the damping matrix $\mathbf{C}$ explicitly, but let us assume for the moment that it exists. Since the matrices contain non-zero off-diagonal terms, the equations of motion in (13.32) are coupled together, making them difficult to solve.

Now, it can be shown mathematically that any displaced shape of the structure can be expressed as a linear combination of its mode shapes. We can therefore write:

$$\mathbf{U} = z_1\boldsymbol{\psi}_1 + z_2\boldsymbol{\psi}_2 + z_3\boldsymbol{\psi}_3 + \ldots + z_n\boldsymbol{\psi}_n$$

Here the coefficients z_i, which are functions only of time, are known as the *generalised coordinates*, while the mode shapes are functions only of position. This equation can be written rather more compactly as:

$$\mathbf{U} = \boldsymbol{\Psi}\mathbf{Z} \tag{13.33}$$

where $\boldsymbol{\Psi}$ is a matrix containing the mode shapes written as column vectors (often called the *modal matrix*) and $\mathbf{Z}$ is a vector of generalised coordinates. Substituting into equation (13.32) and then pre-multiplying each term by $\boldsymbol{\Psi}^{\mathrm{T}}$ gives:

$$(\boldsymbol{\Psi}^{\mathrm{T}}\mathbf{M}\boldsymbol{\Psi})\ddot{\mathbf{Z}} + (\boldsymbol{\Psi}^{\mathrm{T}}\mathbf{C}\boldsymbol{\Psi})\dot{\mathbf{Z}} + (\boldsymbol{\Psi}^{\mathrm{T}}\mathbf{K}\boldsymbol{\Psi})\mathbf{Z} = \boldsymbol{\Psi}^{\mathrm{T}}\mathbf{F}$$

or

$$\mathbf{M}^*\ddot{\mathbf{Z}} + \mathbf{C}^*\dot{\mathbf{Z}} + \mathbf{K}^*\mathbf{Z} = \mathbf{F}^* \tag{13.34}$$

where $\mathbf{M}^*$, $\mathbf{C}^*$ and $\mathbf{K}^*$ are the generalised mass, damping and stiffness matrices and $\mathbf{F}^*$ is the generalised loading:

$$\mathbf{M}^* = \mathbf{\Psi}^{\mathrm{T}}\mathbf{M}\mathbf{\Psi}, \qquad \mathbf{C}^* = \mathbf{\Psi}^{\mathrm{T}}\mathbf{C}\mathbf{\Psi}, \qquad \mathbf{K}^* = \mathbf{\Psi}^{\mathrm{T}}\mathbf{K}\mathbf{\Psi}, \qquad \mathbf{F}^* = \mathbf{\Psi}^{\mathrm{T}}\mathbf{F} \tag{13.35}$$

Consider the generalised mass and stiffness matrices. Typical elements in row i, column j of the matrices are related to the ith and jth mode shapes by:

$$\mathbf{m}_{ij}^* = \mathbf{\psi}_i^{\mathrm{T}}\mathbf{M}\mathbf{\psi}_j, \qquad \mathbf{k}_{ij}^* = \mathbf{\psi}_i^{\mathrm{T}}\mathbf{K}\mathbf{\psi}_j$$

We now recall that the mode shapes are eigenvectors of $\mathbf{M}$ and $\mathbf{K}$. It follows from the orthogonality property of the eigenvectors that:

$$\mathbf{\psi}_i^{\mathrm{T}}\mathbf{M}\mathbf{\psi}_j = \mathbf{\psi}_i^{\mathrm{T}}\mathbf{K}\mathbf{\psi}_j = 0 \qquad i \neq j$$

So all the off-diagonal terms in $\mathbf{M}^*$ and $\mathbf{K}^*$ are zero. If we *assume* that a similar orthogonality condition exists for the damping matrix, then all three of the matrices in equation (13.34) contain diagonal elements only:

$$\begin{bmatrix} m_1^* & & & \\ & m_2^* & & \\ & & \ddots & \\ & & & m_n^* \end{bmatrix}\begin{bmatrix} \ddot{z}_1 \\ \ddot{z}_2 \\ \vdots \\ \ddot{z}_n \end{bmatrix} + \begin{bmatrix} c_1^* & & & \\ & c_2^* & & \\ & & \ddots & \\ & & & c_n^* \end{bmatrix}\begin{bmatrix} \dot{z}_1 \\ \dot{z}_2 \\ \vdots \\ \dot{z}_n \end{bmatrix} + \begin{bmatrix} k_1^* & & & \\ & k_2^* & & \\ & & \ddots & \\ & & & k_n^* \end{bmatrix}\begin{bmatrix} z_1 \\ z_2 \\ \vdots \\ z_n \end{bmatrix} = \begin{bmatrix} F_1^* \\ F_2^* \\ \vdots \\ F_n^* \end{bmatrix}$$

Writing the equations of motion in terms of the generalised coordinates has thus caused them to be *uncoupled*, that is, we now have n independent differential equations, each having the form:

$$m_i^*\ddot{z}_i + c_i^*\dot{z}_i + k_i^*z_i + = F_i^*$$

where $m_i^* = \mathbf{\psi}_i\mathbf{M}\mathbf{\psi}_i$ etc. This looks much like the equation for a damped SDOF system (13.18) and, by analogy with the SDOF case, can be written as:

$$\ddot{z}_i + 2\zeta_i\omega_i\dot{z}_i + \omega_i^2 z_i = \frac{F_i^*}{m_i^*} \tag{13.36}$$

This now gives us an easy way of incorporating damping into the analysis. Instead of calculating the damping matrix explicitly at the start, we simply specify a *modal damping ratio* for each mode of vibration, and this then allows us to formulate n equations of the form of (13.36). Each of these can then be solved using Duhamel's integral (equation (13.25)), if the loading is irregular, or the simpler methods given earlier in this chapter if the loading function is very simple.

The solution procedure for forced vibration response can be summarised as follows:

1. First formulate the mass and stiffness matrices and solve the undamped free vibration problem as described in section 13.2.2. This will result in n natural frequencies ω_i and n corresponding mode shape vectors $\mathbf{\psi}_i$.
2. Find the generalised mass matrix $\mathbf{M}^* = \mathbf{\Psi}^{\mathrm{T}}\mathbf{M}\mathbf{\Psi}$; this will contain leading diagonal terms m_i^* only.
3. Find the generalised load vector $\mathbf{F}^* = \mathbf{\Psi}^{\mathrm{T}}\mathbf{F}$.
4. Hence formulate n independent equations of motion in terms of the generalised coordinates z_i, of the form of equation (13.36). Solve these using Duhamel's integral (equation (13.25)), or a simpler method if the loading function is very simple.

5 Having found the generalised coordinates, multiply them by the corresponding mode shapes to give the actual displacement in each mode under the given loading, then add these modal displacements together to give the total displacement response of the structure – equation (13.33).

6 Other information such as member forces can now be found by back-substituting the displacements into the original equations of motion.

Since it involves summing the structural responses in the individual modes, this is often known as the *mode superposition* method.

EXAMPLE 13.8

Use the mode superposition approach to determine the displacement response of the structure introduced in Example 13.8. For simplicity, neglect damping.

From Example 13.7, the modal matrix for the structure is:

$$\boldsymbol{\Psi} \quad \begin{bmatrix} 0.78 & 1.19 \\ 1 & 1 \end{bmatrix}$$

The generalised mass, stiffness and loading can be found from equation (13.35) and hence the uncoupled equations of motion, in terms of the generalised coordinates, can be written:

$$\begin{bmatrix} 55.99 & 0 \\ 0 & 85.30 \end{bmatrix} \begin{bmatrix} \ddot{z}_1 \\ \ddot{z}_2 \end{bmatrix} \quad \begin{bmatrix} 32{,}963 & 0 \\ 0 & 505{,}258 \end{bmatrix} \begin{bmatrix} z_1 \\ z_2 \end{bmatrix} \quad \begin{bmatrix} 13.22 \\ 8.05 \end{bmatrix} \sin 4\pi t$$

The solution to each of these decoupled equations consists of a transient part of the form of equation (13.3) and a steady-state term of the form of equation (13.22). Assuming zero initial displacement and velocity, we therefore get:

$$\begin{bmatrix} z_1 \\ z_2 \end{bmatrix} \quad \begin{bmatrix} 5.47 \sin 4\pi t \quad 0.517 \sin 24.3t \\ 0.016 \sin 4\pi t \quad 0.163 \sin 77.0t \end{bmatrix}$$

In each case the first sine term is the steady-state response, in which the structure oscillates at the forcing frequency $4\pi \quad 12.6\ \mathrm{rad/s}$, and the second is the transient term, which occurs at the natural frequency (24.3 rad/s for mode 1, 77.0 rad/s for mode 2). In a damped analysis this second term would, of course, become negligible after a few cycles.

The two generalised coordinates are plotted in Figure 13.18(a) and (b). For z_1 the two components occur at similar frequencies, so that the interaction is quite complex. For z_2 the natural frequency is much higher than the forcing frequency, so that the transient part appears as a high-frequency noise superimposed on the steady-state response. Note that the magnitude of z_2 is *much* less than z_1.

Finally, we can find the displacements of the actual structure by multiplying the generalised coordinates by the corresponding mode shapes (equation (13.33)), giving:

$$\begin{bmatrix} u_1 \\ u_2 \end{bmatrix} \quad \begin{bmatrix} 0.78 \\ 1 \end{bmatrix} z_1 \quad \begin{bmatrix} 1.19 \\ 1 \end{bmatrix} z_2$$

Figure 13.18

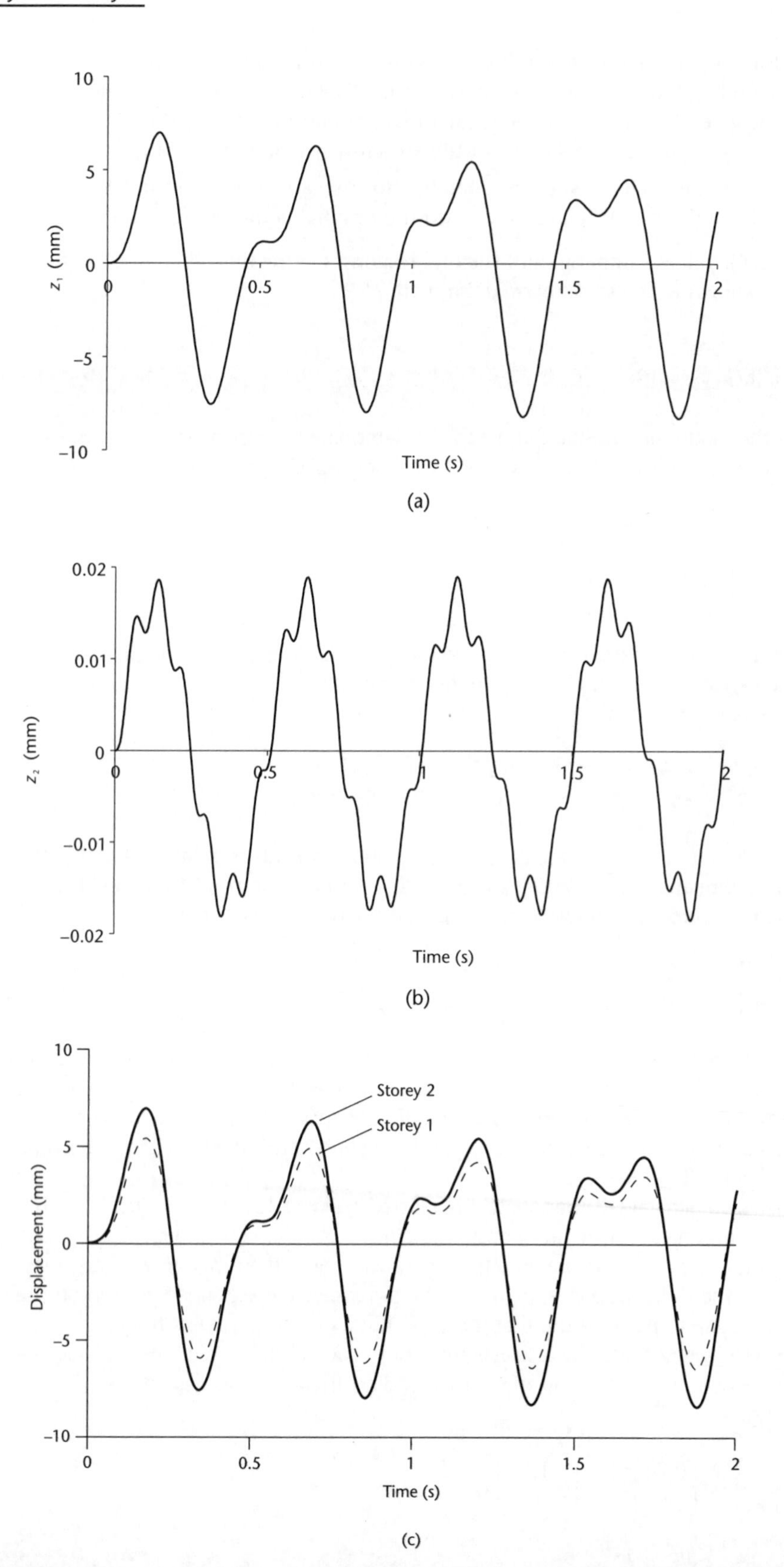
10
5
0
−5
−10
z_1 (mm)
0
0.5
1
1.5
2
Time (s)
(a)
0.02
0.01
0
−0.01
−0.02
z_2 (mm)
0
0.5
1
1.5
2
Time (s)
(b)
10
5
0
−5
−10
Displacement (mm)
Storey 2
Storey 1
0
0.5
1
1.5
2
Time (s)
(c)

The results are plotted in Figure 13.18(c). In this case, since z_2 is so small, the structural response is completely dominated by the first mode. Thus the displacement of the top storey is almost exactly equal to z_1 and the displacement of the lower storey is very nearly $0.78z_1$.

13.3 Earthquake loading

As an example of a particular dynamic loadcase of interest to civil engineers, we shall look very briefly at earthquake loading.

Earthquakes are sudden releases of energy caused by slip between sections of the earths crust, along fault lines. They are usually focused several kilometres below ground level and result in a violent shaking of the ground for typically ten to thirty seconds. In broad terms, most large earthquakes relate to one of two major fault systems. The first runs round the perimeter of the Pacific Ocean, affecting the west coast of America (both north and south), the east coast of Asia (most notably Japan) and New Zealand. The second runs roughly east–west from the Himalayas through the Middle East and Turkey to southern Europe. However, it is quite possible for small or moderate earthquakes to occur in many other regions, including northern Europe and the East Coast of North America.

Seismologists quantify earthquakes using the Richter scale, a logarithmic measure of source energy. However, this alone is not sufficient for engineers, who are interested in the characteristics of the earthquake motions experienced by buildings and centres of population some distance away from the source. The particular features of interest are the magnitude, frequency content and duration of the motion, and these are best determined by placing monitoring instruments on the ground surface in the region of interest. Besides the source energy, the motion at a point will depend on:

- the horizontal and vertical distance of the point under consideration from the earthquake source;
- the properties of the rock and soil through which the seismic waves are transmitted (the motion may be attenuated or amplified with distance, depending on the soil properties);
- the topography of the region, which can cause local amplifications.

A typical ground acceleration record is shown in Figure 13.19 – this is one of the earliest recordings of earthquake motion, from El Centro, California in 1940. As is usual, the motion has been recorded as three perpendicular components (north–south, east–west and vertical) of which only the north–south component is shown here. It can be seen that the shaking lasts approximately 30 seconds, has a peak acceleration of approximately 0.3 *g*, and has an irregular, oscillatory form. The record can be thought of as the superposition of a large number of acceleration histories, each occurring at a different frequency. In this instance the dominant frequencies are in the range 0.5 to 5 Hz, and so large dynamic responses would be expected from structures having natural frequencies within this band. Measurements of earthquake ground motions have been made for many years now, so that a large database of records exists.

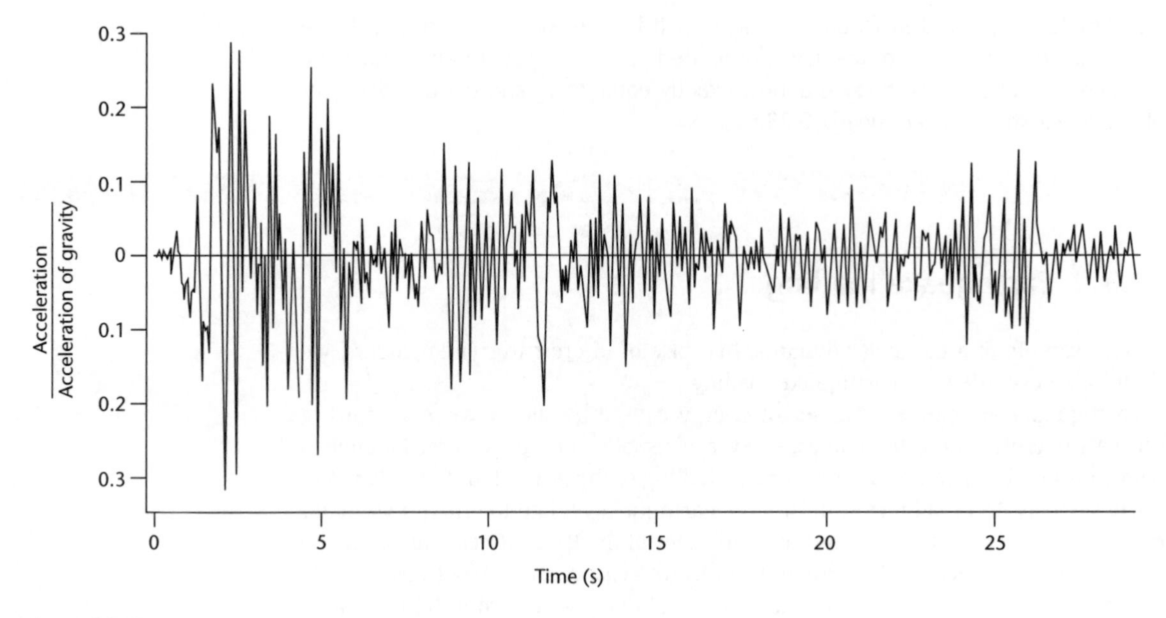

Figure 13.19
Accelerogram from El Centro earthquake, 18 May 1940 (NS component).

In choosing appropriate earthquake design parameters, it is necessary first to study the geology and the earthquake history of the region of interest, over as long a period as possible. This should enable an estimate to be made of the likely magnitude of earthquakes and their frequency of occurrence. The design ground motion can then be defined by selecting an accelerogram recorded in a previous earthquake of the appropriate type and magnitude.

Having defined the input ground motion, let us now consider the loading that this imposes on a structure. For simplicity we shall consider only a SDOF system and shall apply only a uni-directional, horizontal ground motion. All real earthquakes are, of course, three dimensional, but the horizontal components are usually the largest and also the ones to which structures tend to be most vulnerable.

Figure 13.20 shows an idealised, damped SDOF structure subjected to a horizontal ground displacement u_g. Now the forces generated in the columns and damper are proportional to the *relative* displacement and velocity across them whereas the inertia force of the mass is related to its *absolute* acceleration. Applying Newton's second law therefore gives:

$$-k(u - u_g) - c(\dot{u} - \dot{u}_g) = m\ddot{u}$$

Let the relative motion between the mass and the ground be $v = u - u_g$. Then, substituting and rearranging gives:

$$m\ddot{v} + c\dot{v} + kv = -m\ddot{u}_g$$

or

$$\ddot{v} + 2\zeta\omega_n\dot{v} + \omega_n^2 v = -\ddot{u}_g \tag{13.37}$$

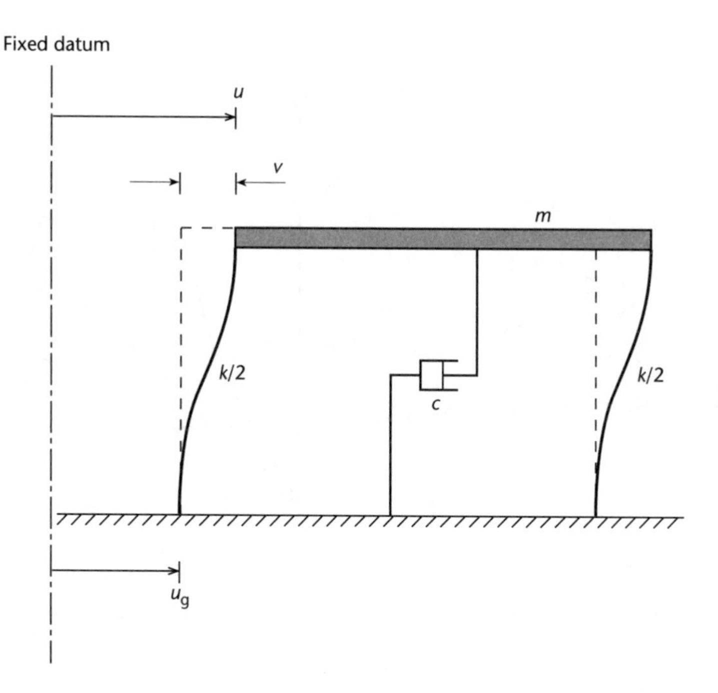

Figure 13.20
Damped SDOF structure subjected to a seismic base motion.

So the earthquake results in a conventional equation of motion, but in terms of *relative* rather than *absolute* motions, and with the forcing simply equal to the mass times the ground acceleration. This can be solved using Duhamel's integral (equation (13.25)), giving the relative displacement as a function of time, and this can be differentiated to give velocities and accelerations if required.

A convenient way of characterising an earthquake loading is by plotting a *response spectrum*. Suppose that, for a given earthquake record, we repeated the Duhamel's integral calculation with numerous SDOF structures having different combinations of natural frequency and damping ratio, and in each case determined the peak acceleration of the structure. If we now plot these peak accelerations against natural frequency, we obtain an acceleration response spectrum for the earthquake (Figure 13.21(a)). This shows us at a glance the peak acceleration achieved by a SDOF system having a certain natural frequency and damping ratio when subjected to a particular earthquake input.

Although it is a plot of structural response to the loading, the spectrum tells us something about the frequency content of the loading itself – the largest responses are for structures with natural frequencies coinciding with the dominant frequencies in the earthquake record. Earthquakes with different characteristics will therefore result in quite different spectra.

Of course, earthquakes and the resulting ground motions are highly unpredictable and it would be extremely risky to design an earthquake-resistant structure on the basis of a single accelerogram. This problem can be conveniently dealt with using the response spectrum approach. Figure 13.21(b) shows response spectra for several different earthquakes, for a single value of structural damping. If these are averaged and smoothed, we can come up with a single *design spectrum* which is representative of all the likely earthquake inputs (the bold line).

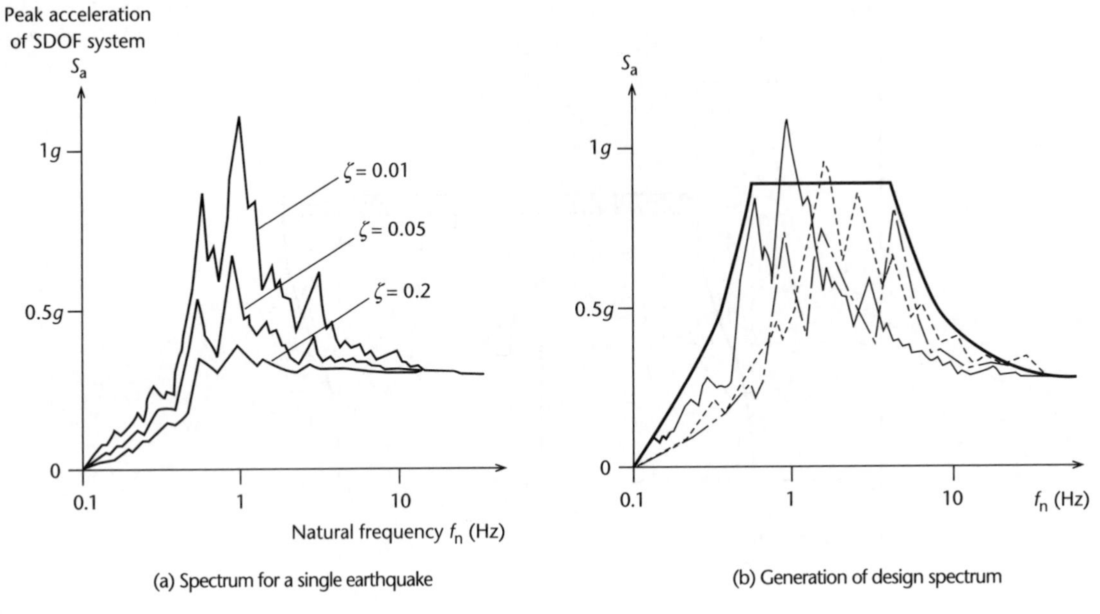

Figure 13.21
Typical earthquake acceleration response spectra.

13.4 Further reading

The following texts all provide an excellent account of the theory of structural dynamics:

- R. W. Clough and J. Penzien, *Dynamics of Structures*, 2nd edn, McGraw-Hill, 1993.
- R. R. Craig, *Structural Dynamics: An Introduction to Computer Methods*, Wiley, 1981.
- W. T. Thompson, *Theory of Vibration with Applications*, 4th edn, Chapman & Hall, 1993.

For complex structures, dynamic analysis is often performed using the finite element method (see Chapter 10). This combination of topics is very well covered by:

- M. Petyt, *Introduction to Finite Element Vibration Analysis*, Cambridge University Press, 1998.

More applied books relating specifically to earthquake-resistant design include:

- D. Dowrick, *Earthquake Resistant Design for Engineers and Architects*, 2nd edn, Wiley, 1987.
- D. Key, *Earthquake Design Practice for Buildings*, Thomas Telford, 1988.

13.5 Problems

Where appropriate, answers are given at the end of the book. The questions marked with an asterisk are a little more challenging.

13.1. A spring is hung vertically from a rigid support. When a mass of 1 kg is attached to the spring it extends by 5 mm. If the mass is given a small displacement from this equilibrium position and then released, calculate the natural frequency of the resulting oscillations.

13.2. Figure 13.22 shows a simple, undamped pendulum, comprising a mass m on a light, rigid bar of length l. Assuming that the angle θ is always small (so that $\sin\theta \approx \theta$ and $\cos\theta \approx 1$), write down the equation of motion in terms of θ. If the pendulum is to be used in a clock, with a period of oscillation of one second, choose a suitable value of l.

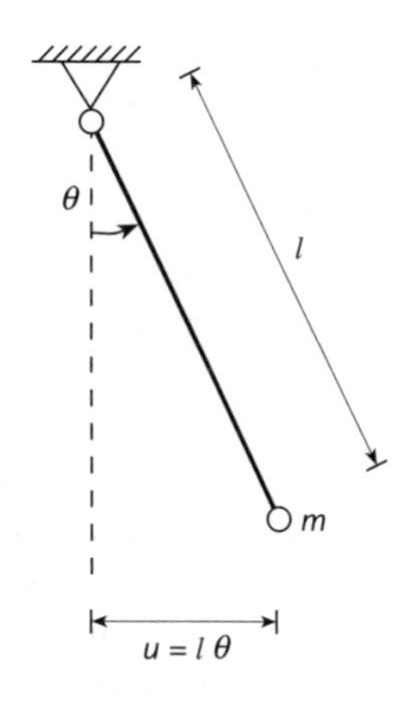

Figure 13.22

13.3. Figure 13.23 shows a concrete bridge pier 8 m high and having a 600 mm square cross-section. It is fixed at its base and supports a length of bridge deck of mass 10 Tonnes. Young's modulus E for concrete may be taken as 25 GPa. Neglecting the mass of the pier itself, determine the undamped natural frequency of sway vibrations if (a) the pier is free to rotate at the top and (b) the pier is constrained by the deck to have zero rotation at the top (see Figure 13.23).

If the structural damping ratio is 0.05, determine the peak displacement δ caused by a horizontal sinusoidal force at cap beam level of amplitude 50 kN and frequency 3.5 Hz.

*The real behaviour is likely to lie somewhere between the two extremes in Figure 13.23. Explain why it is not acceptable to treat the displacements calculated for these two cases as upper and lower bounds to the correct solution.

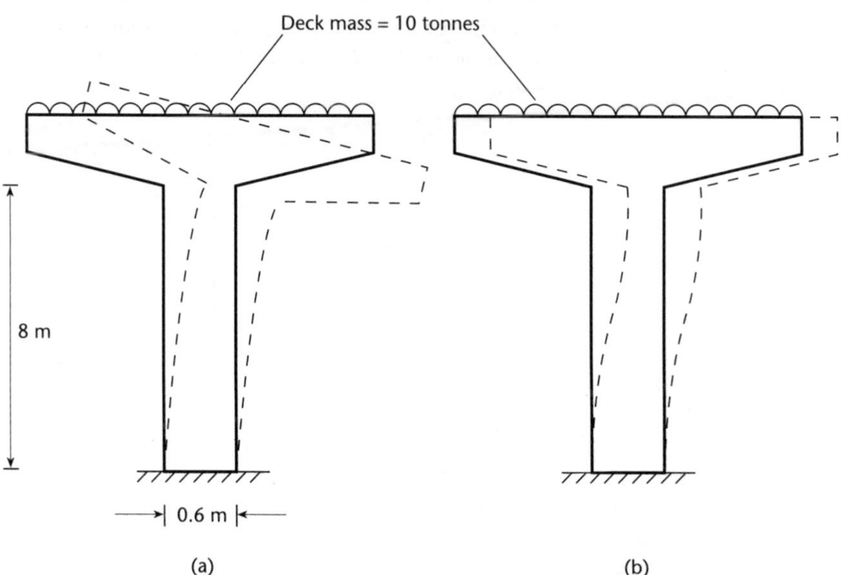

Figure 13.23

13.4. A lightly damped SDOF system has natural frequency ω_n and is subjected to a dynamic load $P(t)$. Without detailed calculation, sketch graphs showing how the displacement of the system varies with time for the following cases:

(a) $P(t)$ is a short impulse at time $t = 0$.
(b) At $t = 0$, $P(t)$ increases very rapidly to a value P_0 and is then held constant indefinitely.
(c) At $t = 0$, $P(t)$ increases rapidly to a value P_0, is held constant for a few seconds and then reduces rapidly to zero.
(d) $P(t)$ varies sinusoidally with amplitude P_0 and frequency $0.7\omega_n$.

For (a) and (b) sketch also the variation of displacement with time if the system were critically damped.

13.5.* A set of scales consists of a platform of mass 2 kg supported on a spring of stiffness 100 kN/m and a damper chosen so as to provide critical damping when there is no load on the scales. Masses are weighed by placing them on the platform and measuring the change in length of the spring. A mass of 5 kg is dropped onto the scales from a small height. It lands without bouncing, and the mass and platform then commence oscillations with initial velocity 1 m/s.

(a) Determine the damping coefficient.
(b) Write down the equation of motion for free vibration of the platform and mass.
(c) Solve this equation to find the position of the mass at time t.
(d) How long does it take for the load reading to settle down to within 1% of the correct value? [*Hint: for motion of the form* $u = A\sin\omega t + B\cos\omega t$, *the amplitude is* $(A^2 + B^2)^{1/2}$.]

13.6.* Figure 13.24 shows a simple, undamped SDOF structure subjected to a forcing function $P(t)$ caused by wind gusts. Using Duhamel's integral, determine the displacement $u(t)$ of the system both during and after the loading. Find also the largest and smallest displacements during each phase of the response. [*Hints: (i) It will be necessary to perform separate calculations for each of the three phases of response, that is,* $0 \le t < 1.5,\ 1.5 \le t < 5,\ t \ge 5$ s. *(ii) Maxima and minima can be found either by plotting the results or by using trigonometric manipulation to express the results in the form* $u = A\sin\omega_n t + B\cos\omega_n t$ *and then using the hint in Problem 13.5.*]

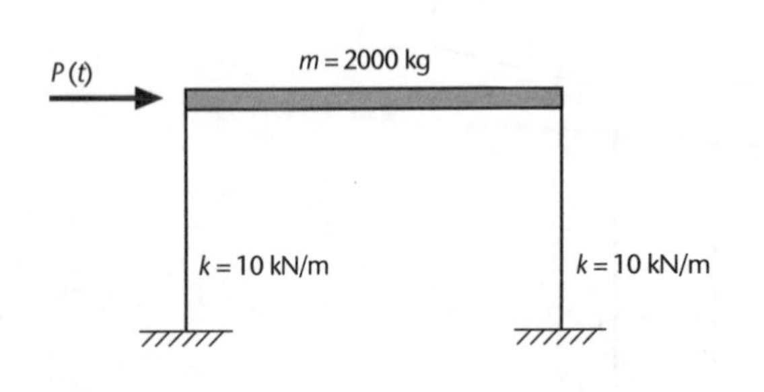

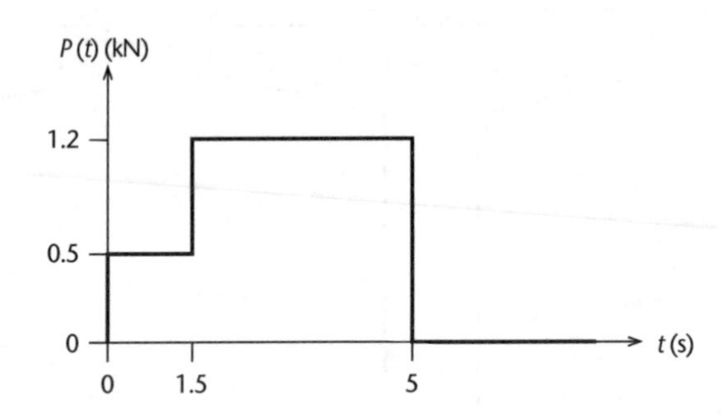

Figure 13.24

13.7. Write in matrix form the equations of motion for the three-storey frame whose storey masses and stiffnesses are shown in Figure 13.25. Find also the equations of motion for the same structure when subjected to a horizontal ground motion $u_g(t)$ instead of the forces $F_i(t)$.

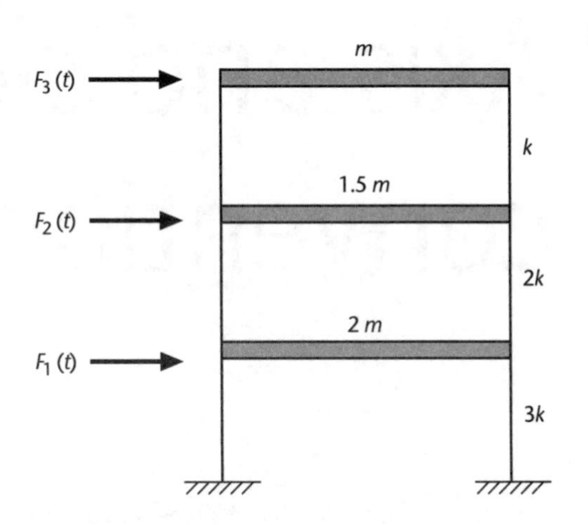

Figure 13.25

13.8. If the ratio k/m in Problem 13.7 is 400 s^{-2}, then the natural frequencies in rad/s are $\omega_1 = 11.83$, $\omega_2 = 25.36$ and $\omega_3 = 37.67$. Check these results by back substituting into $|\mathbf{K} - \omega^2\mathbf{M}| = 0$, and find and sketch the corresponding mode shapes.

13.9. The structure of Problems 13.7 and 13.8 is subjected to a horizontal load, which is constant over the building height but varies sinusoidally with time. Give a qualitative description of the structural response if the circular frequency of the loading is:

(a) 1 rad/s;
(b) 25 rad/s;
(c) 100 rad/s.

APPENDIX A

Axis and sign convention

A consistent sign convention has been used as far as possible in this book. The convention is stated and explained at numerous locations within the text, but is summarised here for ease of reference.

A.1 Three-dimensional axis system

A right-handed (x, y, z) coordinate system is used, with the y-axis oriented vertically upwards (Figure A.1). Positive moments and rotations obey the right-handed screw rule, which states that a positive rotation about an axis is in the direction of twist of a right-handed screw advancing in the positive axis direction. This means that to perform a positive rotation about the x-axis (that is, in the y–z plane) we turn from the positive y towards the positive z-axis, about the y-axis we

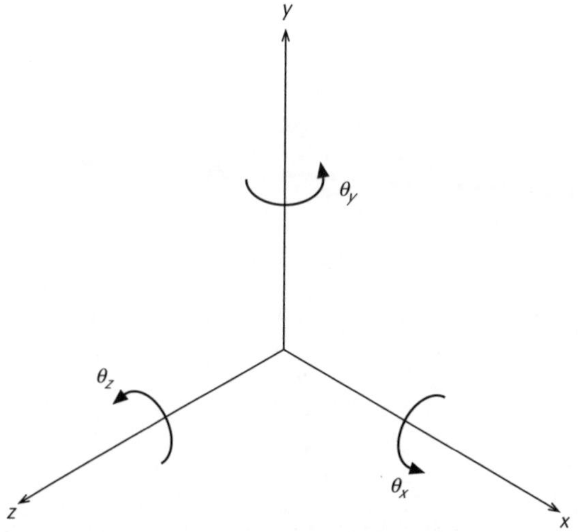

Figure A.1
Positive axis directions and rotations in three dimensions.

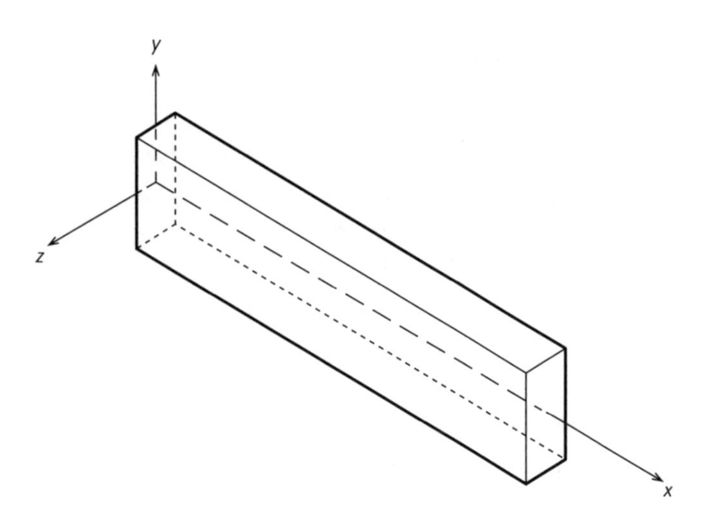

Figure A.2
Three-dimensional axis system for a single member.

turn from positive z towards positive x and about the z-axis we turn from the positive x towards positive y.

The way we choose to orient the x and y axes relative to the structure or element being analysed can be varied from case to case for convenience, as described below.

A.2 Bars and beams

For analysis of individual bars or beams, the x coordinate is measured along the member's longitudinal axis, the y-axis is vertically upwards and the z-axis is then a horizontal axis in the plane of the member cross-section (Figure A.2).

In general, a member may be subjected to forces and moments along or about any axis, but a very important special case is that of a beam subjected to vertical (y-direction) loads. In this case the distribution of internal forces and moments can be calculated by considering the elevation of the member in the x–y plane (Figure A.3). Note that the positive z-axis is coming out of the page, so that a positive rotation is anti-clockwise.

External loads on a member can be divided into vertical *gravity* loads due to the weight of the member and objects placed on it, and horizontal loads caused by wind etc. Since a gravity load acts downwards, its magnitude ought really to be prefaced by a minus sign, indicating that the load is in the negative y direction. In practice, however, this tends to cause confusion. Therefore, in hand calculation methods, external loads are represented simply by magnitudes, with arrows

Figure A.3
Elevation of a beam in the x–y plane.

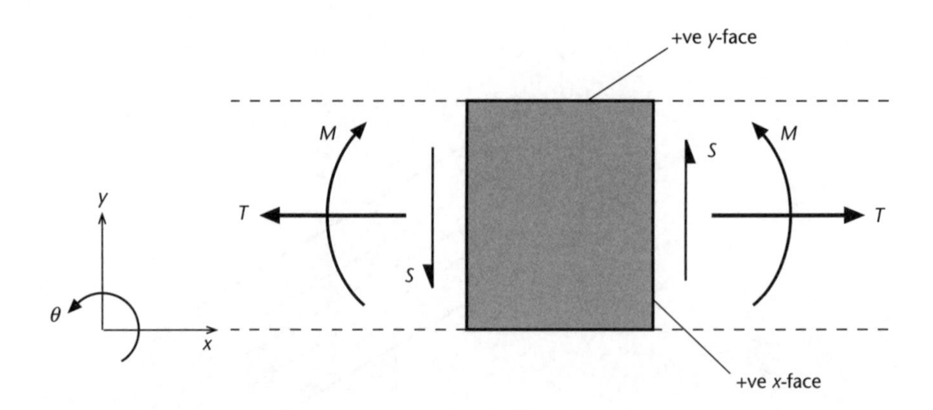

Figure A.4
Positive internal forces.

indicating their directions. For computer methods, of course, a rigorous adherence to the sign convention is essential – see section A.5.

If a small element is cut from a beam, the positive directions of the force resultants acting across the cuts (the axial force T, shear force S and bending moment M) are as shown in Figure A.4. All of these internal forces obey the rule that a positive force acts *either* in a positive direction on a positive face *or* in a negative direction on a negative face. A positive face is defined as one where a normal drawn outwards from the face is in a positive axis direction. Thus, for example, the right-hand face of the element in Figure A.4 is referred to as the positive x-face because its outward drawn normal is in the positive x direction. A positive shear force on this face therefore acts in the positive y direction. The reverse is true for the shear force on the left-hand face. Note that these sign conventions lead to the following results:

- for axial forces, tension is positive, compression negative;
- positive shear forces on either end of the element form an anti-clockwise couple;
- a sagging bending moment is positive.

When drawing the shear force diagram, we plot a positive value of S upwards. For the bending moment diagram, however, we draw a positive value of M downwards. This means that, if the bending moment diagram is superimposed on the outline of the beam, the bending moment always plots on the side of the beam that is in *tension* (Figure A.5). This convention becomes particularly useful when we come to plotting bending moment diagrams for frames.

If we now wish to go on and calculate the stresses and deflections caused by the bending moments in the beam, these will be governed by the second moment of area about a horizontal axis through the cross-section, that is, I_{zz}.

A.3 Two-dimensional frames

A 2D frame is assumed to lie in the x–y plane, as shown in Figure A.6. Note again that all external loads are simply indicated by a directional arrow and a magnitude. There are no negative signs attached to the vertical loads, even though they act in the negative y direction.

Other conventions are the same as for beams and bars. In particular, the convention of plotting bending moments on the tension side of a member now

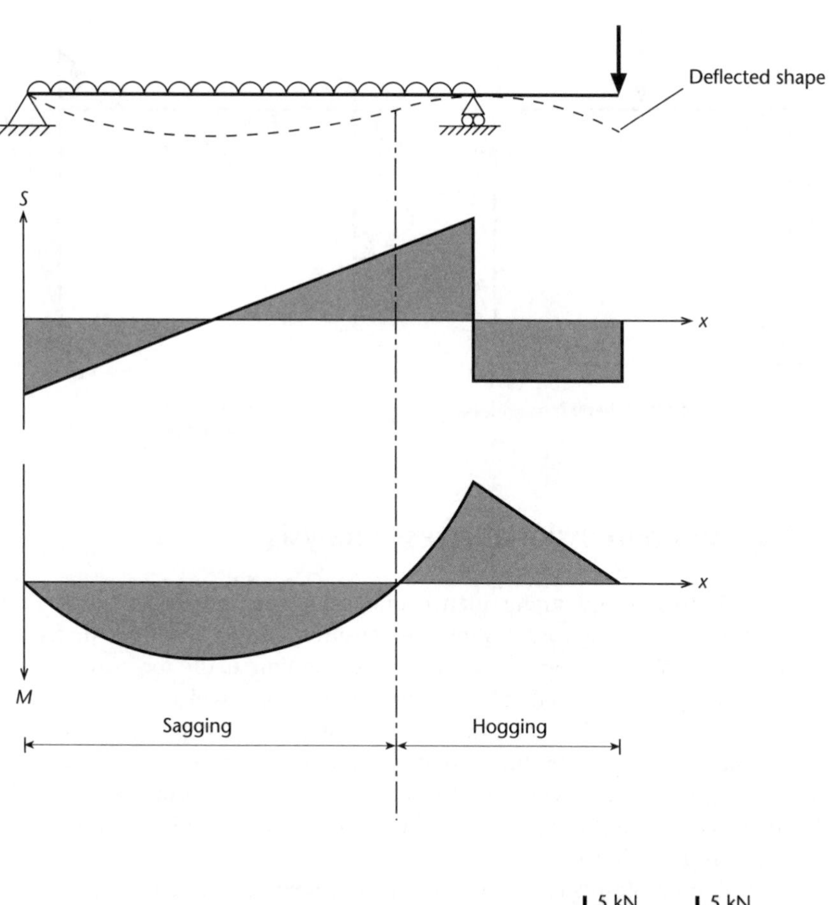

Figure A.5
Shear force and bending moment diagrams for a beam.

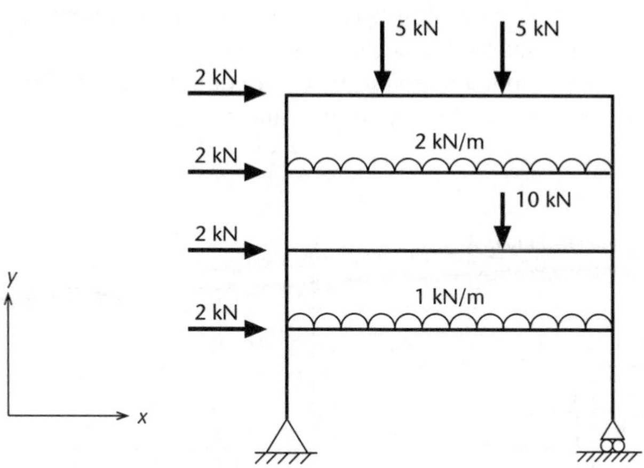

Figure A.6
Axis system for a 2D frame.

comes into its own. Consider the simple frame shown in Figure A.7. The concepts of sagging and hogging are of little use here since they depend on the viewpoint of the analyst. For example, the column AB appears to be sagging (positive moment) when viewed from the outside of the frame, but is hogging (negative moment) if viewed from the inside. However, plotting the bending moment diagram with lines on the tension sides of the members is unambiguous since, no matter where it is viewed from, the outside face of AB is in tension and the inside in compression.

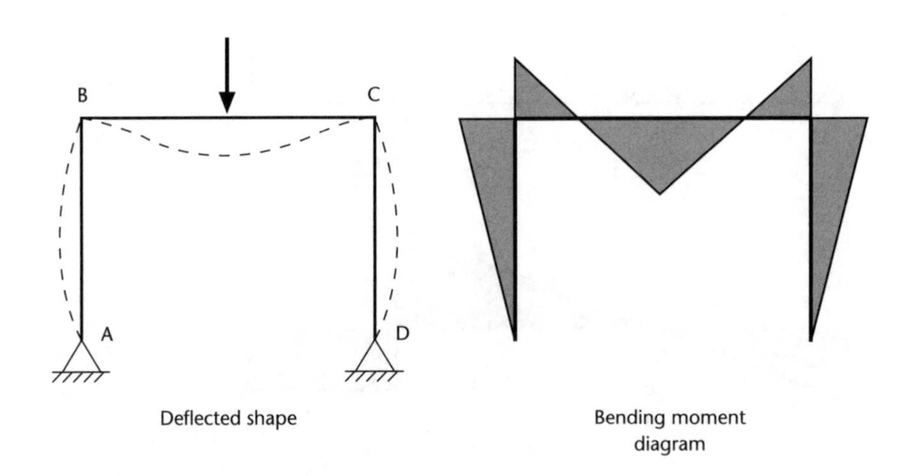

Figure A.7
Bending moment diagram for a 2D frame.

A.4 Two-dimensional stress analysis

Systems of stresses and strains in two dimensions are referred to conventional Cartesian (x, y) axes. Figure A.8 shows the positive stresses acting on the faces of a small element. These follow the same sign convention as the forces on the beam element in Figure A.4, that is, a positive force is one which acts in a positive direction on a positive face, and *vice-versa*. Of the two subscripts, the first refers to the face of the element and the second to the direction of the force. Thus σ_{xx} is a stress acting in the x direction on a face whose outward normal is also in the x direction, while τ_{yx} is a stress acting in the x direction but on a face whose outward normal is in the y direction.

When plotting Mohr's circle for a 2D stress system, a different sign convention must be used for the shear stresses. We take shear stresses as positive if they cause a *clockwise* moment about the centre of the element. Thus, for the shear stresses in Figure A.8, τ_{yx} plots as positive and τ_{xy} as negative. In all other respects, the sign

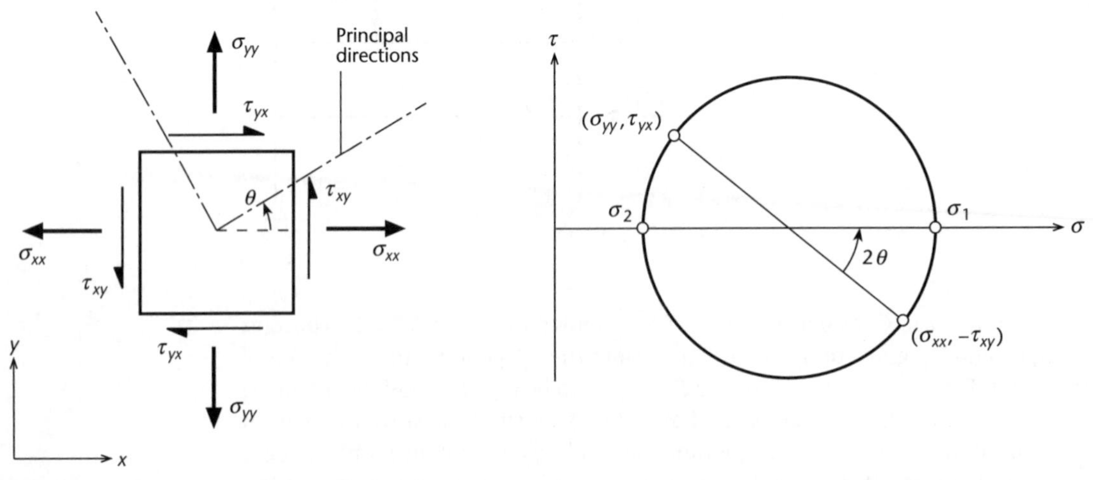

Figure A.8
Axis system and Mohr's circle for 2D stress analysis.

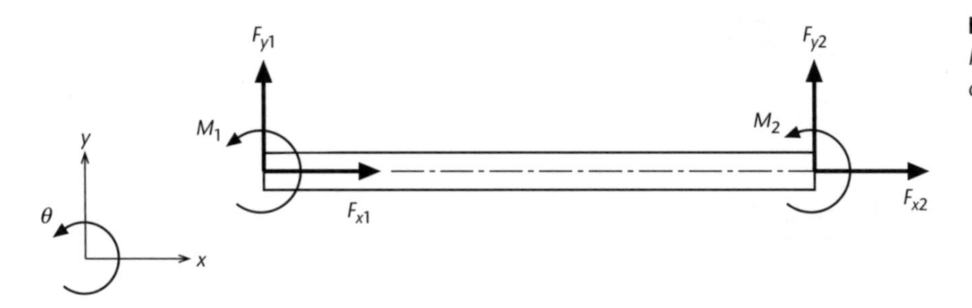

Figure A.9
Positive internal forces for computer applications.

convention used in the Mohr's circle is the same as used in the stress system: tensile direct stresses are positive, and angles turned in the Mohr's circle are in the same direction as on the physical plane.

A.5 Computer applications

For computer applications (stiffness matrices, finite elements), the axis systems used are in most respects the same as those described above. For a computer, however, complete consistency is essential, and this requires changes to one or two of the more 'intuitive' conventions used in hand methods.

First, the directionality of the external loads must be dealt with more rigorously; if a load acts in a negative axis direction then it *must* be assigned a negative value. (This was avoided in the chapters on hand calculation methods by using arrows to define loading directions.)

A further difference from the system outlined above is that we no longer make use of the concept of positive and negative faces of an element. The sign of a displacement, force or moment is governed solely by its direction in space, and does not depend on the face on which it acts. Thus the forces and moments on the 2D beam element in Figure A.9 are all positive.

APPENDIX B

Glossary

Words in italics are other glossary terms.

arch a one-dimensional curved member which carries loads by developing axial compressive forces along the line of the arch.

bar a one-dimensional element which carries only axial forces (tension or compression).

beam a (usually) horizontal, one-dimensional element which carries loads primarily by bending.

beam-column a one-dimensional, vertical element which sustains both axial compression and bending loads (sometimes simply called a *column*).

bending moment the internal moment required to balance the moment created by a set of external forces on a *beam* segment.

buckling lateral instability of an element under axial compression.

bulk modulus constant of proportionality between the mean *stress* on a three-dimensional element and its *volumetric strain*.

cable a slender, flexible element, capable of carrying only axial tension.

cable-stayed bridge a bridge in which the deck is supported by straight stay *cables* tied back to towers.

cantilever a beam that is held by a *fixed support* at one end and is free at the other.

column a one-dimensional, vertical element which carries loads primarily by axial compression, sometimes accompanied by significant *bending moments* (see also *beam-column*).

concentrated load an idealised form of loading in which a force is applied over an infinitesimally short distance – also known as a *point load.*

critical damping the minimum level of *damping* required to prevent any oscillatory motion of a dynamic system.

damping the dissipation of energy within a structural system which causes its vibrations to decay with time (in the absence of any external loading). Damping may arise from many sources, but is usually represented by a viscous dashpot, which provides a retarding force proportional to the velocity of the structure.

dead load the permanent load on a structure due to self-weight, cladding etc.

deflection translational (as opposed to rotational) movement.

direct strain change in length of a member divided by its original length.

direct stress direct (that is, tensile or compressive) force per unit area.

displacement any form of movement, that is, a *deflection* or a rotation.

displacement function function used in *finite element* analysis to describe how the displacements at points within an element are related to the nodal displacements.

dome a two-dimensional, curved structure which carries loads principally by developing compressive forces along radial lines within the structure.

effective length the length of an equivalent pin-ended strut having the same *buckling* load as a strut with some other end conditions.

elastic modulus constant of proportionality between the *bending moment* on a section and the largest resulting axial stress.

elasticity the analysis of the relationships between forces and deformations in linear, elastic materials.

equation of condition additional equilibrium equation which can be obtained from the geometry of a structure – for example, if the two halves of a structure are joined together by a single pin, then an equation of condition can be obtained by noting that the moment at the pin must be zero.

equilibrium matrix matrix of coefficients relating internal bar forces to applied loads.

finite element method numerical analysis approach in which a continuum structure is analysed by treating it as a set of discrete elements joined together at nodal points.

fixed-end moment the moment which would be developed at the end of a member if both ends were rigidly fixed.

fixed support a point at which a structure is held in such a way as to prevent any deflection or rotation.

flexural rigidity or **flexural stiffness** constant of proportionality between a *bending moment* at a point in a member and the corresponding curvature (equal to the product of the *Young's modulus* of the material and the *second moment of area* of the cross-section).

free body diagram a line diagram of all or part of a structure, with forces indicated by arrows.

grillage frame in which all members lie within a single (usually horizontal) plane and all loads are applied perpendicular to this plane.

hogging curvature of a *beam* in which the top face is in tension and the bottom in compression.

Hooke's law the linear relationship between *direct stress* and *strain* in an elastic material, which forms the fundamental basis of the theory of *elasticity*.

imposed load the variable part of the load on a structure, due to traffic, wind, building contents etc. – also known as *live load*.

influence line a plot of how a quantity such as shear force or bending moment at a fixed point varies as a unit load traverses the structure. Should not be confused with, say, a bending moment diagram, which shows the variation in bending moment over the structure under a fixed set of loads.

isoparametric element a *finite element* formulated using a single set of functions for both the *shape functions* and the *displacement functions*.

lateral-torsional buckling lateral and twisting instability of an element subjected to a bending moment.

live load the variable part of the load on a structure, due to traffic, wind, building contents etc. – also known as *imposed load*.

load factor the factor by which the applied loads on a structure would have to be increased to cause complete collapse.

major axis axis within the plane of a cross-section about which the *second moment of area* takes its maximum value.

mechanism a structure having too few members or support reactions to prevent collapse.

method of sections determination of internal forces by consideration of the equilibrium of a section of a structure.

minor axis axis within the plane of a cross-section about which the *second moment of area* takes its minimum value.

mode shape the deformed shape that a structure adopts when undergoing vibrations at one of its *natural frequencies.*

Mohr's circle simple graphical construction which allows systems of stresses or strains in two dimensions to be resolved in any direction; often used for the determination of *principal stresses* or *strains.*

moment distribution iterative technique for calculating bending moment distributions in redundant, rigid-jointed frames.

moment frame frame in which at least some of the connections are *rigid joints,* so that the members carry *bending moments* and *shear forces,* as well as axial forces.

multi-degree-of-freedom system structural system whose motion cannot be adequately described by a single displacement variable. A system having n degrees of freedom will have n possible modes of vibration, each occurring at a particular *natural frequency* and having a characteristic *mode shape.*

natural frequency the number of oscillations per second which a structure undergoes when set in motion and then left to vibrate free of any external loading – the inverse of the *natural period.*

natural period the time taken for a structure to undergo a single oscillation when set in motion and then left to vibrate free of any external loading – the inverse of the *natural frequency.*

neutral axis line or plane within a beam whose length remains unchanged during bending.

neutral equilibrium condition of a structure in which the second differential of its potential energy is zero. If a structure in neutral equilibrium is given a small disturbance, it neither returns to its equilibrium position nor undergoes large displacements, but stays in the disturbed position – see also *stable equilibrium and unstable equilibrium.*

pin joint a joint between two or more members which leaves them free to rotate relative to each other, and is incapable of transmitting a moment.

pin-jointed frame a frame made up of *bars,* connected by *pin joints,* so that all the members carry only axial forces (also known as a *truss*).

pinned support a point at which a structure is held so that it is prevented from deflecting in any direction, but is free to rotate.

plastic hinge cross-section which has yielded completely in bending, so that it can carry no additional moment and therefore behaves similarly to a *pin joint.*

plastic modulus constant of proportionality between the *bending moment* required to cause complete plasticity of a cross-section and the yield stress of the material from which it is made.

plasticity the analysis of the relationships between forces and deformations of systems in which yielding occurs.

plate a two-dimensional element which carries loads primarily by bending.

point load an idealised form of loading in which a force is applied over an infinitesimally short distance – also known as a *concentrated load.*

point of contraflexure point in a *beam* at which the *bending moment* (and hence also the curvature) is zero.

Poisson's ratio factor to account for the change in dimensions of a member in a direction perpendicular to the loading direction; if an element is loaded in direction A, resulting in a strain ε_A in that direction, then the strain in a perpendicular direction caused by the load in direction A is $-\nu\varepsilon_A$, where ν is Poisson's ratio.

polar second moment of area *second moment of area* about an axis normal to the plane of a cross-section, which governs its torsional behaviour.

principal stresses (or **strains**) the direct stresses (or strains) acting on planes on which the shear stresses (or strains) are zero. In a 2D stress system, the principal stresses are the largest and smallest stresses present.

propped cantilever a beam having a *fixed support* at one end and a *roller support* at the other.

radius of gyration the square root of the ratio of *second moment of area* to cross-sectional area.

reaction the force or moment exerted on a structure by a support.

redundant structure a structure with more than the minimum number of members and support *reactions* to prevent collapse, and which cannot be fully analysed by *statics* alone – also known as a statically indeterminate structure.

rigid joint a joint between two or more members, which does not permit any relative movement between the members.

roller support a point at which a structure is prevented from moving at right angles to the ground, but is free to move parallel to the ground, and to rotate.

safety factor the factor by which the applied loads would have to be increased to cause yielding to commence at some point in the structure.

sagging curvature of a beam in which the top face is in compression and the bottom in tension.

second moment of area geometric property of a plane cross-section which governs its behaviour under the action of a *bending moment* – see also *polar second moment of area*.

shape factor the ratio of the moment required to cause a section to become fully plastic to the moment required to cause initial yielding.

shape function function used to define the geometry of a *finite element* in terms of its nodal coordinates.

shear centre the point through which a beam must be loaded in order to prevent it from twisting; for a symmetrical section the shear centre always coincides with the centroid of the cross-section, but for an unsymmetrical section it is likely to be offset.

shear flow shear stress multiplied by thickness, mostly used in the analysis of torsion in thin-walled sections.

shear force the force normal to the axis required to balance the external loads acting on a *beam* segment.

shear modulus constant of proportionality between *shear stress* and *strain*.

shear strain change in angle between two initially perpendicular lines in an element.

shear stress *shear force* per unit area.

shear wall a two-dimensional, vertical element which resists horizontal loads by in-plane shearing action.

shell a two-dimensional element, usually having a curved profile, which carries loads mainly by in-plane tension or compression.

simply supported beam a *beam* having a *pinned support* at one end and a *roller support* at the other.

single-degree-of-freedom system structural system whose motion can be described entirely in terms of one displacement variable.

slenderness ratio ratio of *effective length* to *radius of gyration*, a non-dimensional measure of the susceptibility of a member to *buckling*.

stable equilibrium condition of a structure in which its potential energy takes a minimum value. If a structure in stable equilibrium is given a small disturbance, it returns to its equilibrium position – see also *neutral equilibrium* and *unstable equilibrium*.

statically determinate structure a structure having the minimum number of members and support reactions to prevent collapse, and for which all the internal forces can be found by *statics*.

statics the analysis of structures based on the principle of equilibrium of forces.

stiffness the force at a given point and in a certain direction required to create a unit deflection at the same point and in the same direction (or the moment required to create a unit rotation).

stiffness coefficient an element of a *stiffness matrix*; the stiffness coefficient in row i, column j is the force at point i required to produce a unit displacement at point j.

stiffness matrix a square matrix relating the forces at a set of locations in a structure to the corresponding displacements.

strain normalised change in dimension of a member (see also *direct strain*, *shear strain* and *volumetric strain*).

strain energy the potential energy stored in a loaded member as a result of its elastic deformation.

strength (for a structure or member) the force required to cause failure; (for a material) the *stress* required to cause failure – see also *yield strength* and *ultimate strength*.

stress force per unit area – see also *direct stress* and *shear stress*.

structure a body whose purpose is to carry a set of forces from a point in space either to the ground or to another structure.

strut a *bar* loaded in compression.

support a contact point between a structure and the ground – see also *fixed support*, *pinned support* and *roller support*.

suspension bridge bridge in which the deck is suspended from a large, approximately parabolic *cable* slung between two towers.

tension coefficient the axial force in a bar divided by its length, used primarily in the analysis of 3D pin-jointed structures.

tie a *bar* loaded in tension.

torque moment acting about the longitudinal axis of a member.

torsion a twisting deformation caused by a *torque*.

torsional rigidity or **torsional stiffness** constant of proportionality between a *torque* at a point in a member and the corresponding twist (equal to the product of the *shear modulus* of the material and the *polar second moment of area* of the cross-section).

truss a frame made up of *bars*, connected by *pin joints*, so that all the members carry only axial forces (also known as a *pin-jointed frame*).

ultimate strength (for a structure or member) the force required to cause complete collapse; (for a material) the *stress* required to cause complete failure.

Universal Beam a rolled steel *beam* having an I-shaped cross-section.

unstable equilibrium condition of a structure in which its potential energy takes a maximum value. If a structure in unstable equilibrium is given a small disturbance, then very large displacements may occur – see also *neutral equilibrium* and *stable equilibrium*.

virtual work principle which states that, when a set of compatible displacements is imposed on a body in equilibrium, the work done is zero. The term *virtual work* is used because there is no requirement for the forces and displacements to be related to each other.

volumetric strain change in volume divided by original volume.

warping longitudinal distortion of a member, so that an initially planar cross-section becomes curved.

yield strength (for a structure or member) the force above which the load–deformation relationship ceases to be linear, and permanent deformations occur; (for a material) the stress above which the stress–strain relationship ceases to be linear and permanent deformations occur (also known as the yield stress).

Young's modulus constant of proportionality between *direct stress* and *strain* for a linear elastic material.

Answers to problems

Chapter 2

2.1. (a) 1
(b) 1
(c) $H_A = 0$, $V_A = 3wl/2 \uparrow$, $M_A = wl^2$ anti-clockwise, $V_C = wl/2 \uparrow$
(d) $H_A = V_A = 0$, $V_C = 9wl/4 \uparrow$, $V_D = wl/2 \downarrow$, $V_E = wl/4 \uparrow$

2.2. (a) $H_A = 0$, $V_A = 3$ kN $\uparrow$
(b) $H_A = 5$kN $\leftarrow$, $V_A = 7.5$ kN $\downarrow$
(c) $H_A = 0$, $V_A = 7.85$ kN $\uparrow$, $M_A = 31.39$ kNm anti-clockwise

2.3. $V_A = 19.5$ kN $\uparrow$, $V_B = 25.5$ kN $\uparrow$, $S_{max} = 19.5$ kN at A, $M_{max} = 46.0$ kNm at midpoint of AB

2.4. $P = wl/4$

2.7. $S_{max} = \pm wl/4$ at ends, $M_{max} = wl^2/12$ at centre

2.8. $V_A = 9.0$ kN $\uparrow$, $V_C = 22.25$ kN $\uparrow$, $V_E = 5.75$ kN $\uparrow$, $V_H = 3.0$ kN $\uparrow$, $S_max = 15.0$ kN at C, $M_max = -24.0$ kNm at C

2.9. $P = wl/2$

2.10. $P = 0.586wl$

Chapter 3

3.1. (a) statically determinate
(b) redundant
(c) statically determinate
(d) part mechanism, part redundant

3.2. $H_D = 5$ kN $\leftarrow$, $V_D = 4.33$ kN $\downarrow$, $V_F = 4.33$ kN $\uparrow$, $T_{AB} = T_{BD} = 5$ kN, $T_{AC} = T_{CF} = -5$ kN, $T_{DE} = T_{EF} = 2.5$ kN, rest $= 0$

3.3. $T_{BC} = -8$ kN, $T_{CH} = -1.41$ kN, $T_{HI} = 9$ kN, $T_{CI} = 2$ kN

3.4. (a) statically determinate unless $\theta = 0°$ or $180°$
(b) is statically determinate

3.5. $T_{AE} = -5.1$ kN, $T_{BE} = -7.8$ kN, $T_{CE} = 0$, $T_{AF} = -6.0$ kN, $T_{CF} = -7.9$ kN, $T_{EF} = 6.7$ kN, $T_{AD} = 12.9$ kN, $T_{DE} = -7.2$ kN, $T_{DF} = -7.2$ kN, $T_{DG} = 21.3$ kN, $T_{EG} = -14.8$ kN, $T_{FG} = -14.8$ kN

3.6. (a) $H_A = 20$ kN $\leftarrow$, $V_A = 14$ kN $\uparrow$, $V_D = 10$ kN $\uparrow$, $V_E = 36$ kN $\uparrow$
(b) $T = 20$ kN in AB, -36 kN in BE, $S_{max} = -25$ kN in BC at B
(c) $M_{max} = 60$ kNm in BE at B

3.7. (b) $V_D = 7wl/6 \uparrow$, $V_E = wl/12 \uparrow$, $V_F = 7wl/4 \uparrow$, $H_F = 2wl \leftarrow$
(c) $M_{max} = 3wl^2/2$ at C

3.9. $V = 24$ kN upwards, $H = 19.2$ kN inwards

3.10. $V = wR$ upwards, $H = wR/2$ inwards , $S = (wR/2) \cdot (\cos\theta - \sin 2\theta)$,
$T = (-wR/2) \cdot (1 + \cos 2\theta + \sin\theta)$, $M = (wR^2/4) \cdot (1 - 2\sin\theta - \cos 2\theta)$

Chapter 4

4.1. $\tau_{xy} = 4$ MPa, $\delta = 0.6$ mm

4.2. $k_{outer} = 536$ kN/mm, $k_{inner} = 845$ kN/mm, 138.1 kN

4.3. $\sigma_{outer} = -7.7$ MPa, $\sigma_{inner} = 14.3$ MPa, extension = 0.025 mm

4.4. (a) P/σ
(b) $(P/\sigma)\exp(wy/\sigma)$
(c) $\sigma l/E$

4.5. (a) ± 14.14 MPa, 22.5°
(b) 15 MPa, 5 MPa, 45°
(c) 16.18 MPa, −6.18 MPa, 31.7°
(d) 12.22 MPa, 2.78 MPa, 119°

4.6. $\varepsilon_1 = 234.4\mu\varepsilon$ at 20.5° clockwise from A, $\varepsilon_2 = -47.8\mu\varepsilon$, $\gamma_m ax = 282.2\mu\varepsilon$

4.7. 32 mm

4.8. Volumetric strains: (a) $-2.88\mu\varepsilon$, (b) $86.4\mu\varepsilon$

4.9. (a) 185.1 kN, 0.134 mm
(b) 266.5 kN, 0.286 mm
(c) 0.093 mm
(d) 65.1 MPa

Chapter 5

5.2. $I = 4.505 \times 10^6$ mm^4, $Z_e = 59.51 \times 10^3$ mm^3

5.3. (a) 191.3 MPa tension, 422.2 MPa compression
(b) 252.1 MPa tension, 114.2 MPa compression

5.4. Timber, $M = 12.9$ kNm

5.6. $M_A = wl^2/8$, $V_A = 5wl/8$, $V_B = 3wl/8$

5.7. (b) $v_B = -3wl^4/4EI$, $v_C = 5wl^4/6EI$

5.8. (a) $P = wl/4$
(b) $5wl^4/192EI$
(c) $3wl^4/8(3EI + kl^3)$

5.10. $M_y = 126.1$ kNm, $M_p = 149.1$ kNm, $\alpha = 1.182$

Chapter 6

6.1. (a) $r = 63.1$ mm, (b) $r = 54.2$ mm, weight of (a) is 43% of that of (b)

6.2. (a) $r = 63.1$ mm, (b) $r = 56.3$ mm, weight of (a) is now 40% of that of (b)

6.3. 199.4 MPa at 39.3° to the shaft axis

6.4. (a) $T = 34.0$ kNm, $\phi = 0.0264$ rads
(b) $T = 39.8$ kNm, $\phi = 0.0396$ rads
(c) $T = 38.3$ kNm, $\phi = 0.0316$ rads

6.5. Ratio of torsional stiffnesses = 1.621, ratio of yield torques = 1.273 – the circular section has the higher value in each case

6.7. τ_{max}(flange) = 46.3 MPa, τ_{max}(web) = 152.7 MPa, $K = 1.39$

6.8. $\tau = S \sin\theta / \pi a t$

6.10. $e = 13d/6$

6.11. (a) 0.56 mm
(b) 0.24 mm
(c) 0.0255 rads

Chapter 7

7.1. (a) $11wl/4$
(b) wl
(c) $3M/l$

7.2. 6.29 mm down, 4.13 mm to the right

7.3. $11Pl/6AE$

7.4. (a) $\delta_{VB} = 4.22$ mm, $\delta_{HD} = 2.85$ mm
(b) $V_B = 37.1$ kN, $\delta_{HD} = 2.12$ mm

7.5. $T_{BD} = 0.435P + 0.653AE\varepsilon/l$, $T_{AD} = T_{CD} = 0.326P - 0.377AE\varepsilon/l$

7.6. $\delta_{VG} = 10.0$ mm

7.7. $V_A = -3wl/4$, $V_B = 7wl/4$, $M_B = wl^2/4$

7.8. $0.0708Wa^3/EI$

7.9. Normal force = $0.379P$, $k_{comp} = 4.6EI/a^3$

7.11. V_B(max) = 14.3 kN, M_B(max) = 17.0 kNm

Chapter 8

8.1. $M_A = 37.5$ kNm anti-clockwise, $V_A = 17.2$ kN, $V_B = 2.8$ kN, both upwards

8.2. $M_{AB} = 19.4$ kNm, $M_{BA} = -M_{BC} = -23.5$ kNm,
$M_{CB} = -M_{CD} = -11.1$ kNm, $M_{DC} = 5.5$ kNm

8.4. Move D 1.6 mm upwards

8.5. $M_{AB} = 7.0$ kNm, $M_{AC} = -7.0$ kNm, $M_{BA} = -14.5$ kNm, $M_{CA} = -8.5$ kNm,
$M_{CD} = 15.1$ kNm, $M_{CE} = -6.6$ kNm, $M_{DC} = -28.5$ kNm, $M_{EC} = 3.3$ kNm

8.6. $M_{AB} = 34.5$ kNm, $M_{BA} = -M_{BC} = -11.2$ kNm, $M_{CB} = -M_{CD} = 3.5$ kNm, $M_{DC} = -1.7$ kNm, $H_A = 22.9$ kN $\leftarrow$, $H_C = 16.4$ kN $\leftarrow$, $H_D = 0.7$ kN $\rightarrow$

8.7. $M_{AB} = -59.9$ kNm, $M_{BA} = -M_{BC} = -6.7$ kNm, $M_{CB} = -M_{CD} = 14.4$ kNm, $M_{DC} = -23.6$ kNm

Chapter 9

9.2. $\theta_A = -781.3/EI$, $\theta_B = 312.5/EI$, $\theta_C = -195.3/EI$, $M_{max} = 281.3$ kNm at B

9.3. $M_{AB} = -19.0$ kNm, $M_{BA} = -M_{BC} = -47.6$ kNm, $M_{CB} = -M_{CD} = -41.3$ kNm, $M_{DC} = 25.4$ kNm

9.4. $5Ml/72EI$

9.6. (a) $u_{XA} = -2.63$ mm, $u_{YA} = -2.25$ mm
(b) $u_{XA} = 21.02$ mm, $u_{YA} = 17.29$ mm

9.7. $F_{XB} = 100$ kN, $F_{YB} = 200$ kN, $M_{ZB} = 50$ kNm

9.8. (a) $[P \quad -wl/2 \quad -wl^2/12 \quad 0 \quad wl/2 \quad wl^2/12]^T$,
(b) $[0 \quad -Wb/(a+b) \quad -Wab^2/(a+b)^2 \quad 0 \quad Wa^2b/(a+b)^2]^T$

9.9. $u_{XB} = 4.98$ mm, $u_{YB} = 0$, $\theta_{ZB} = -1.294 \times 10^{-3}$ rads

9.10. $$\mathbf{K} = \frac{K}{l^3}\begin{bmatrix} 36 & 6l & -12l \\ 6l & 5l^2 & 0 \\ -12l & 0 & 9l^2 \end{bmatrix}\begin{matrix} u_{YA} \\ \theta_{XA} \\ \theta_{ZA} \end{matrix}$$

Chapter 10

10.1. $S_3 = (-1 + \alpha^2 + \beta^2 + \alpha\beta + \alpha^2\beta + \alpha\beta^2)/4$, $S_8 = (1 - \alpha - \beta^2 + \alpha\beta^2)/2$

10.2. Element 1: $\varepsilon_{XX} = \varepsilon_{YY} = 0$, $\gamma_{XY} = -0.001$. Element 2: $\varepsilon_{XX} = 0.00125$, $\varepsilon_{YY} = 0$, $\gamma_{XY} = -0.003$

10.3. (a) $wl/6, wl/3$
(b) $0, wl/3, wl/6$
(c) $wl/6, wl/3$

10.4. Loads in kN at nodes 1–8: 0, 7.5, 7.5, 22.5, 15.0, 37.5, 22.5, 52.5

Chapter 11

11.1. $\theta = 0$ – unstable if $P > kl$; $\theta = \pi$ – unstable if $P < -kl$; $\cos\theta = P/kl$ – unstable if $0 < \theta < \pi$

11.2. (a) Buckling at $P = 1117$ kN
(b) Yielding at $P = 1650$ kN

11.6. Wl/π

11.7. $v_{max} = 22.3$ mm, $\sigma_{max} = 6.2$ MPa (2.2 MPa if initially straight)

11.8. $P_R = 22.5EI_0/l^2$, $P_E(I = 2I_0) = 19.7EI_0/l^2$

11.9. $$P_R = \frac{\pi^2 EI}{4l^2} + \frac{8kl}{\pi^2}$$

11.10 (a) Extreme fibres yield at $w = 59.9$ kN/m
(b) Lateral-torsional buckling at $w = 15.6$ kN/m

Chapter 12

12.1. Number of possible mechanisms: (a) 1, (b) 3, (c) 3, (d) 7

12.2. Simply supported ends: $P = 120$ kN. Built-in ends: $P = 160$ kN

12.6. Length of end spans $= 0.315l$

12.7. $P = 144$ kN

12.8. $\lambda = 24/11$

12.9 $P = \left|\dfrac{12M_p}{l(k\sqrt{3} - 1)}\right|$

12.10. $\lambda = 2.02$

12.11. (a) 2 independent mechanisms, 1 combined
(b) 2 independent, 1 combined
(c) 3 independent, 4 combined

Chapter 13

13.1. 7.0 Hz

13.2. 0.248 m

13.3. (a) $f_n = 2.0$ Hz, $\delta = 15.3$ mm
(b) $f_n = 4.0$ Hz, $\delta = 31.6$ mm

13.5. (a) 894.4 Ns/m
(c) $u = e^{-63.9t}(9.59 \sin 101t - 0.49 \cos 101t)$ (measured in mm from the final at-rest position)
(d) 0.12 s

13.6. $0 \leq t < 1.5$: $u = 25(1 - \cos \omega_n t)$, $u_{max} = 50.0$ mm, $u_{min} = 0$,
$1.5 \leq t < 5$: $u = 60 - 35 \cos \omega_n(t - 1.5) - 25 \cos \omega_n t$, $u_{max} = 103.6$ mm,
$u_{min} = 16.4$ mm,
$t \geq 5$: $u = 60 \cos \omega_n(t - 5) - 35 \cos \omega_n(t - 1.5) - 25 \cos \omega_n t$,
$u_{max} = 90.5$ mm, $u_{min} = -90.5$ mm

13.8. $\boldsymbol{\psi}_1 = [0.30 \quad 0.65 \quad 1.0]^T$, $\boldsymbol{\psi}_2 = [-0.68 \quad -0.61 \quad 1.0]^T$,
$\boldsymbol{\psi}_3 = [2.43 \quad -2.55 \quad 1.0]^T$

Index